Lecture Notes in Computer Science 7034

Commenced Publication in 1973
Founding and Former Series Editors:
Gerhard Goos, Juris Hartmanis, and Jan van Leeuwen

Marc van Kreveld   Bettina Speckmann (Eds.)

Graph Drawing

19th International Symposium, GD 2011
Eindhoven, The Netherlands, September 21-23, 2011
Revised Selected Papers

 Springer

Volume Editors

Marc van Kreveld
Utrecht University, Department of Information and Computing Sciences
P.O. Box 80 089, 3508 TB Utrecht, The Netherlands
E-mail: m.j.vankreveld@uu.nl

Bettina Speckmann
TU Eindhoven, Department of Mathematics and Computer Science
P.O. Box 513, 5600 MB Eindhoven, The Netherlands
E-mail: speckman@win.tue.nl

ISSN 0302-9743   e-ISSN 1611-3349
ISBN 978-3-642-25877-0   ISBN 978-3-642-25878-7 (eBook)
DOI 10.1007/978-3-642-25878-7
Springer Heidelberg Dordrecht London New York

Library of Congress Control Number: 2011944514

CR Subject Classification (1998): E.1, G.2.2, G.1.6, I.5.3, H.2.8

LNCS Sublibrary: SL 1 – Theoretical Computer Science and General Issues

Typesetting: Camera-ready by author, data conversion by Scientific Publishing Services, Chennai, India

Printed on acid-free paper

Springer is part of Springer Science+Business Media (www.springer.com)

Preface

The 19th Symposium on Graph Drawing was held during September 21–23, 2011, in Eindhoven, The Netherlands, at the Technical University Eindhoven.

In response to the call for papers the Program Committee received 79 submissions of which 13 were short-paper submissions. The Program Committee reviewed these submissions carefully and selected 34 regular papers and 3 short papers. Furthermore, 6 poster abstracts were selected from 9 submissions. The revised versions of the accepted submissions can be found in the proceedings. Furthermore, the proceedings contain the abstracts of two invited talks, given by Jarke J. van Wijk and Günter Rote. To commemorate Kozo Sugiyama and his pioneering research in graph drawing, the proceedings include an obituary.

A unique and fun part of the symposium is the Graph Drawing Contest, which is part of the Graph Drawing Challenge. This year was the 18th edition. The Graph Drawing Contest Committee together with Kevin Verbeek made sure that it was a smoothly running and successful event. A report on the contest is included at the end of the proceedings.

We express our thanks to all contributors of papers to the proceedings. We also thank the Program Committee members for their substantial work, providing reviews of high quality. Furthermore, we thank the GD Contest Committee, Anne Driemel, Marcel Roeloffzen and Wouter Meulemans who provided local help, and especially Kevin Buchin and Dirk Gerrits, the other two members of the local Organizing Committee.

We acknowledge the generous support of NWO (Netherlands Organisation for Scientific Research), KNAW (Royal Netherlands Academy of Arts and Sciences), Tom Sawyer Software (Gold sponsor), and Microsoft (Silver sponsor). Furthermore, the TU Eindhoven provided lecture rooms and other spaces free of charge, and de TUimelaar offered child daycare free of charge. These contributions made it possible to keep the registration costs low, allowing more researchers to attend and exchange ideas on graph drawing. We hope that the graph drawing symposium can keep the tradition of low registration costs.

Next year the 20th International Symposium on Graph Drawing will take place during September 19–21, 2012, and will be hosted by Microsoft Research in Redmond, Washington, USA.

October 2011

Marc van Kreveld
Bettina Speckmann

Organization

Program Committee

Ulrik Brandes	University of Konstanz, Germany
Kevin Buchin	TU Eindhoven, The Netherlands
Sergio Cabello	University of Ljubljana, Slovenia
Walter Didimo	University of Perugia, Italy
Vida Dujmović	Carleton University, Canada
Fabrizio Frati	University of Rome III, Italy
Éric Fusy	LIX Ecole Polytechnique, France
Michael Goodrich	University of California, Irvine, USA
Carsten Gutwenger	TU Dortmund, Germany
Marc van Kreveld (Co-chair)	Utrecht University, The Netherlands
Takao Nishizeki	Kwansei Gakuin University, Japan
Martin Nöllenburg	Karlsruhe Institute of Technology, Germany
Bettina Speckmann (Co-chair)	TU Eindhoven, The Netherlands
Roberto Tamassia	Brown University, USA
Alexandru Telea	University of Groningen, The Netherlands
Csaba D. Tóth	University of Calgary, Canada
Dorothea Wagner	Karlsruhe Institute of Technology, Germany

Organizing Committee

Kevin Buchin	TU Eindhoven, The Netherlands
Dirk Gerrits	TU Eindhoven, The Netherlands
Marc van Kreveld (Co-chair)	Utrecht University, The Netherlands
Bettina Speckmann (Co-chair)	TU Eindhoven, The Netherlands

Graph Drawing Contest Committee

Christian Duncan	Louisiana Tech University, USA
Carsten Gutwenger (Chair)	TU Dortmund, Germany
Lev Nachmanson	Microsoft, Redmond, USA
Georg Sander	IBM, Bad Homburg, Germany

Additional Reviewers

Eyal Ackerman	Reinhard Bauer
Marie Albenque	Laurent Beaudou
Patrizio Angelini	Carla Binucci

Luca Castelli Aleardi
Markus Chimani
Emilio Di Giacomo
David Eppstein
Gasper Fijavz
Radoslav Fulek
Andreas Gemsa
Daniel Goncalves
Luca Grilli
Robert Görke
S. Mehdi Hahemi
Tanja Hartmann
Herman Haverkort
Petr Hlineny
Seok-Hee Hong
Clemens Huemer
Karsten Klein
Stephen Kobourov
Nils Kriege
Marcus Krug
Kazuyuki Miura
Bojan Mohar
Fabrizio Montecchiani
Pat Morin

Sonoko Moriyama
Petra Mutzel
Tomoki Nakamigawa
Yoshio Okamoto
Thomas Pajor
Maurizio Patrignani
Dominique Poulalhon
Md. Saidur Rahman
Vincenzo Roselli
Ignaz Rutter
Laura Sanità
André Schulz
Andrea Schumm
Rodrigo I. Silveira
Claudio Squarcella
Daniel Stefankovic
Andrew Suk
Yusuke Suzuki
Markus Völker
Hoi-Ming Wong
David Wood
Hsu-Chun Yen
Xiao Zhou

Financial Support and Sponsors

Netherlands Organisation for Scientific Research

K O N I N K L I J K E   N E D E R L A N D S E
A K A D E M I E   V A N   W E T E N S C H A P P E N

Gold Sponsor   Silver Sponsor

   **Microsoft**

Table of Contents

Obituary

Papers

Posters

Graph Drawing Contest

Kozo Sugiyama 1945 - 2011

Kozo Sugiyama was born in Gifu Prefecture Japan on September 17, 1945. He received his B.S., M.S., and Dr. Sci. at Nagoya University in 1969, 1971, 1974 respectively. For 23 years from 1974 he was a researcher at Fujitsu. During this time he spent a year at the International Institute for Applied Systems Analysis in Laxenburg in Austria. In the mid 1990s he served as the Director of the Information Processing Society of Japan. In 1997 he moved from Fujitsu to the newly-created Japan Advanced Institute of Science and Technology. His first position there was Professor of the School of Knowledge Science, but he soon became Director of the Center for Knowledge Science, and then Dean of the School of Knowledge Science. His last few years at JAIST were spent as a Vice President of the University.

Everyone in the Graph Drawing community knows the name Sugiyama for his famous paper *"Methods for visual understanding of hierarchical system structures"*. The paper defines the *layered* method for drawing directed graphs; it is often simply called "Sugiyama's method". The paper was published as a technical report from Fujitsu Laboratories in the late 1970s, and appeared in a journal in 1981. It describes a very general framework for drawing directed graphs. This is one of the most influential papers in Graph Drawing; it has spawned many algorithms, many theorems, and many implementations; it forms the backbone of many commercial systems.

In 1992, Kozo wrote the first book on Graph Drawing, in Japanese. It was translated into English in 2002.

Kozo is also known for a number of other ground-breaking papers in Graph Drawing: on compound graph layout (predating clustered graphs), the magnetic spring model, and on the mental map problem.

However, Kozo's interests went far beyond Graph Drawing. He was passionate about finding better ways to assist the human process of knowledge creation. Indeed, his motivation for doing graph drawing research was to support the *KJ Method*, which is a way of creating and organizing ideas, commonly used in business in Japan. He developed a number of thinking support systems. He published papers on creating puzzles, and on technology to assist corporations in exchanging tacit knowledge. He wrote two books on Knowledge Science, the more recent one in 2008. In all his work, Kozo showed the courage and ability to address big problems.

Older members of the Graph Drawing community remember Kozo personally. He attended the first Graph Drawing conference in Rome. Many of us visited his laboratory either in Fujitsu or in JAIST, and he has visited our laboratories. He was an inspiration for us.

Kozo passed away on June 10, 2011. We have lost a valuable colleague and a good friend.

Peter Eades, Seok-Hee Hong, and Kazuo Misue,
on behalf of the Graph Drawing community,
September 2011

M. van Kreveld and B. Speckmann (Eds.): GD 2011, LNCS 7034, p. 1, 2012.
© Springer-Verlag Berlin Heidelberg 2012

Confluent Hasse Diagrams

David Eppstein and Joseph A. Simons

Department of Computer Science, University of California, Irvine, USA

Abstract. We show that a transitively reduced digraph has a confluent upward drawing if and only if its reachability relation has order dimension at most two. In this case, we construct a confluent upward drawing with $O(n^2)$ features, in an $O(n) \times O(n)$ grid in $O(n^2)$ time. For the digraphs representing series-parallel partial orders we show how to construct a drawing with $O(n)$ features in an $O(n) \times O(n)$ grid in $O(n)$ time from a series-parallel decomposition of the partial order. Our drawings are optimal in the number of confluent junctions they use.

1 Introduction

One of the most important aspects of a graph drawing is that it should be readable: it should convey the structure of the graph in a clear and concise way. Ease of understanding is difficult to quantify, so various proxies for it have been proposed, including the number of crossings and the total amount of ink required by the drawing [1, 18]. Thus given two different ways to present information, we should choose the more succinct and crossing-free presentation.

Confluent drawing [7, 8, 9, 15, 16] is a style of graph drawing in which multiple edges are combined into shared tracks, and two vertices are considered to be adjacent if a smooth path connects them in these tracks (Figure 1). This style was introduced to re-duce crossings, and in many cases it will also improve the ink requirement by represent-ing dense subgraphs concisely. However, it

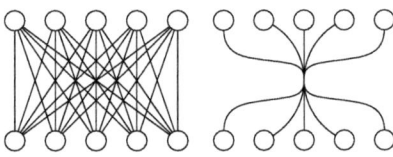

Fig. 1. Conventional and confluent drawings of $K_{5,5}$

has had a limited impact to date, as there are only a few specialized graph classes for which we can either guarantee the existence of a confluent drawing or test for confluence efficiently. A closely related graph drawing technique, edge bundling [10], differs from confluence in emphasizing the visualization of high level graph structure, but does not necessarily seek to reduce the number of edge crossings.

Hasse diagrams are a type of upward drawing of transitively reduced directed acyclic graphs (DAGs) that have been used since the late 19th century to visu-alize partially ordered sets. To maximize the readability of Hasse diagrams, as with other types of graph drawing, we would like to draw them without cross-ings. Thus upward planar graphs (DAGs that can be drawn so that all edges go upwards and no edges cross) have been an important thread of research in

M. van Kreveld and B. Speckmann (Eds.): GD 2011, LNCS 7034, pp. 2–13, 2012.

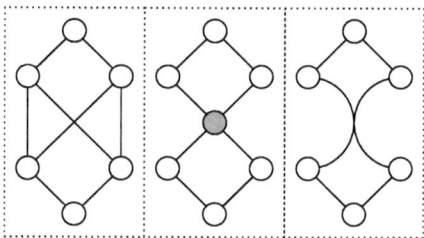

Fig. 2. A simple DAG P (left) that is not upward planar, although its underlying graph is planar. Its Dedekind–MacNeille completion (middle) is upward planar, with an added element (shaded). Replacing that element with a junction creates an upward confluent drawing of P (right).

graph drawing. A DAG is upward planar if and only if it is a subgraph of a planar st-graph, i.e. a planar DAG with one source and one sink, both on the outer face [6]. Testing upward planarity is NP-complete [12] but for DAGs with a single source or a single sink it may be tested efficiently [17,4]. However, many DAGs (even planar DAGs such as the one in Figure 2) are not upward planar.

In this paper, we bring these threads together by finding efficient algorithms for upward confluent drawing of transitively reduced DAGs. We show that a graph has an upward confluent drawing if and only if it represents a partial order P with order dimension at most two, and that these drawings correspond to two-dimensional lattices containing P. We construct the smallest lattice containing P (its Dedekind–MacNeille completion) in worst-case-optimal time, and draw it confluently in area $O(n^2)$, using as few confluent junctions as possible. For series-parallel partial orders, the time and number of junctions can be reduced to linear.

2 Preliminaries

2.1 Posets and Lattices

A partially ordered set (partial order, or poset) $P = (V, \leq)$ is a set V with a reflexive, antisymmetric, and transitive binary relation $\leq$. We adopt the convention that $n = |V|$ unless otherwise stated. We also use $a < b$ to denote that $a \leq b$ and $a \neq b$. We say that a *covers* b in P if $b < a$ and $\nexists x \in P$ such that $b < x < a$. Elements $a, b \in P$ are *comparable* if $a \leq b$ or $b \leq a$; otherwise, we write $a \| b$ to indicate that they are *incomparable*. A *total order* or *linear order* is a partial order in which every pair of elements in P is comparable. If R is a set of linear orders R_i, we can define a poset P as the intersection of R: that is, $a \leq b$ in P if and only if $a \leq b$ in every linear order R_i. If P can be defined from R in this way, then R is called a *realizer* of P. Every partial order P has a realizer; the *dimension* $\dim(P)$ is the smallest number of linear orders in a realizer of P.

If $X \subseteq P$ is any subset of P, then an element $a \in P$ is called a *lower bound* of X if it is less than or equal to every element of X. Similarly, an element b is called an *upper bound* of X if it is greater than or equal to every element of X.

If X has a lower bound a that belongs to X itself, then a is the (unique) *least element* in X, and similarly if X has an upper bound b that belongs to X then b is the (unique) *greatest element* in X. If the set A of lower bounds of X has a greatest element a, then a is the *greatest lower bound* or *infimum* of X, and similarly if the set B of upper bounds of X has a lowest element b then b is the *least upper bound* or *supremum* of X. If P itself has an infimum or a supremum, these elements are typically denoted by 0 and 1 respectively. If P contains both an infimum and a supremum, it is said to be *bounded*.

A poset L is a *lattice* if for every pair of elements x and y in L the set $\{x, y\}$ has both an infimum and a supremum. In this context, the supremum of $\{x, y\}$ is called the *meet* of x and y and denoted $x \wedge y$, and similarly the infimum is called the *join* and denoted $x \vee y$. A lattice L is *complete* if every subset of L has an infimum and supremum in L. Every finite lattice is complete and bounded.

2.2 Hasse Diagrams and Upward Planarity

Every poset $P = (V, \leq)$ can be represented by a directed acyclic graph G which has a vertex for each element in P and an edge uv for each pair (u, v) with $u \leq v$ in P. However, when we draw a poset it is more common to draw a different DAG, the *transitive reduction* G' of G, in which there is an edge from u to v in G' if and only if v covers u in P. A *Hasse diagram* of P is an upward drawing of G', meaning that the y coordinate of the head of each edge is greater than the y coordinate of the tail of each edge, so that the drawing "flows" upward from smaller elements to larger elements. In a Hasse diagram, we do not need to explicitly draw the edges as directed edges: the direction of an edge is represented implicitly by the relative position of its endpoints. There is an upward path from a to b in a Hasse diagram of P if and only if $a \leq b$. A poset is *planar* if it has a Hasse diagram that is upward planar, i.e. its transitive reduction has an upward drawing in which none of the edges intersect except at a shared vertex.

A finite lattice is planar if and only if its transitive reduction is a planar st-graph, a DAG which contains exactly one source s and one sink t both of which belong to the outer face of an upward planar drawing [28]. More generally, any DAG is upward planar if and only if it is a subgraph of a planar st-graph [6]. In the other direction, every planar finite bounded poset must be a lattice [3,5,19]. This implies that a two-dimensional bounded poset that is not a lattice (such as the one on the left of Figure 2) cannot have an upward planar drawing, and that planarity (a crossing-free drawing) and two-dimensionality (realization by a pair of linear orders) are distinct for non-lattice posets.

2.3 Lattice Completion of a Poset

The Dedekind–MacNeille completion of a poset P is the smallest complete lattice containing P [22]. For any subset X of P, let X^- and X^+ denote the set of lower bounds and upper bounds of X respectively. A *cut* of P is a pair $A, B \subseteq P$ such that $A^+ = B$ and $A = B^-$; the completion of P has these cuts as its elements. The completion is partially ordered by set containment: if (A, B) and (C, D) are cuts, then $(A, B) \leq (C, D)$ if and only if $A \subseteq C$ and $B \supseteq D$. The element of the completion corresponding to an element x of P is the cut $(\{x\}^-, \{x\}^+)$, and the new elements added to P to make it into a lattice come from cuts (A, B) for which $A \cap B = \emptyset$. The completion automatically has the same dimension as the partial order from which it was constructed [27].

Ganter and Kuznetsov [11] give a stepwise algorithm for constructing the completion of P. Given a poset P and its completion L they show how to complete a one-element extension of P in time $O(|L| \cdot |P| \cdot \omega(P))$, where $\omega(P)$ denotes the width of P. To compute the completion of a large poset, they begin with a single-element poset (whose completion is trivial) and use this subroutine to add elements one at a time; therefore, the total time is $O(|L| \cdot |P|^2 \cdot \omega(P))$. Nourine and Raynaud [26] give an algorithm with running time $O((|P| + |B|) \cdot |B| \cdot |L|)$ where B is a *basis* of P (a set of subsets of P which generate L). As part of our drawing algorithm, we improve these results in the case of two-dimensional posets: we show for such sets how to construct the completion in time $O(|P|^2)$, optimal in the worst case since (as we also show) there exist two-dimensional posets whose completion has a quadratic number of elements.

2.4   Confluent Drawing

Confluent drawing is a technique for drawing non-planar diagrams without crossings [7, 8, 9, 15, 16] by merging together groups of edges and drawing them as *tracks* that, like train tracks, meet smoothly at junction points but do not cross. A *confluent drawing* consists of a set of labeled points (*vertices* and *junctions*) and curves (*track segments*) in the Euclidean plane, such that the two endpoints of each track segment are vertices or junctions, such that no two track segments intersect except at a shared endpoint, and such that all track segments that meet at a junction share a common tangent line at that point. The graph represented by a confluent drawing has as its vertices the vertices of the drawing; two vertices u and v are adjacent if and only if there is a smooth curve in the plane from u to v that is a union of track segments and that does not pass through any other vertex. (Some papers on confluence require that this curve also be non-self-intersecting but that requirement is irrelevant for upward drawings since monotone curves cannot self-intersect.) An undirected graph G is *confluent* if and only if there exists a confluent drawing that represents it.

We define a *confluent diagram* of a poset to be a drawing of its transitive reduction in a way that is both confluent and upwards. In other words, if G is a directed acyclic graph representing a poset P, then we define a confluent diagram of P to be an upward confluent drawing of the transitive reduction of G in which all tracks are oriented upwards (monotonic in the y direction), and therefore all smooth curves passing through the tracks are similarly oriented. For each pair of elements $a, b \in P$, the drawing should have a smooth track from a upwards to b if and only if a is covered by b. For technical reasons we also require that for each source there exists an unbounded y-monotone curve downwards that does not cross the diagram – that is, that each source can be seen from below – and symmetrically that each sink can be seen from above.

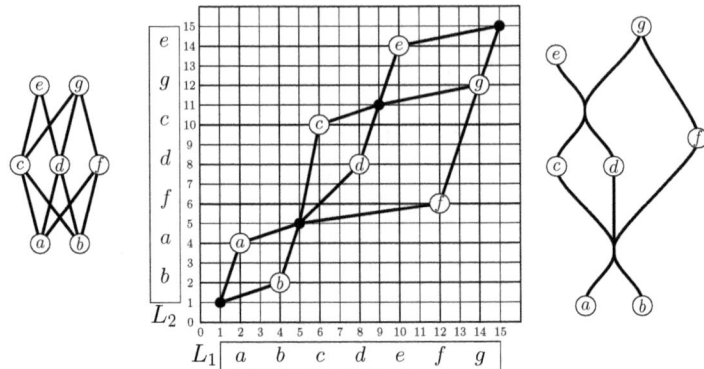

Fig. 3. Example of our algorithm. *Left*: Input poset P. *Middle*: Grid embedding with added points and dominance pairs. *Right*: Completion points replaced by confluent junctions and rotated $45°$.

In the application to visualization of partial orders, this is a natural restriction as it makes the minimal and maximal elements easy to find in the drawing.

3   The Algorithm

Let G be a poset with dimension at most two. We now describe an $O(n^2)$ algorithm to embed a confluent diagram of P in an $O(n) \times O(n)$ grid. That is, we will generate an upward confluent drawing of the transitive reduction of a DAG representing P such that each vertex in the drawing has integer coordinates.

Our algorithm has three phases. In the first phase, we embed the elements of P in a $(2n+1) \times (2n+1)$ grid. Recall that since P has dimension two, it is realized by two linear orders, which correspond to two different total orderings of the same n elements in P. Thus, the first steps of our algorithm are:

1. (a) Find two linear orders L_1 and L_2 that realize P. This can be done in $O(n^2)$ time from any graph whose transitive closure is P by Algorithm 1 of [21].
   (b) For each element p of P, having position p_1 in L_1 and p_2 in L_2 with $1 \le p_i, p_j \le n$, place a vertex representing p in the grid with coordinates $(2i, 2j)$.

After this step, the even rows and columns in the grid each contain exactly one element of P, and the dominance relationship of these points corresponds to the order of the elements in P. Recall that for two elements p and q in the plane, p *dominates* q if and only if $p_i \ge q_i$ for each coordinate i and $p \ne q$.

In the second phase, we insert additional points representing elements of the completion of P; these completion nodes correspond to confluent junctions in the confluent diagram of P. We defer to a later section the proof that the dominance order on the points generated in the first two phases gives the completion of P.

2. For each odd pair of indices (i, j), in $[3, 2n - 1]$ insert a junction in the grid with coordinates (i, j) if all of the following four conditions hold:
   - The poset point with x-coordinate $i - 1$ has y-coordinate less than $j - 1$.
   - The point with x-coordinate $i + 1$ has y-coordinate greater than $j + 1$.
   - The point with y-coordinate $j - 1$ has x-coordinate less than $i - 1$.
   - The point with y-coordinate $j + 1$ has x-coordinate greater than $i + 1$.

   In addition if P does not already have a least or a greatest element, then insert invisible points at $(1, 1)$ and $(2n + 1, 2n + 1)$ respectively.

In the third phase, we generate the segments of the confluent diagram. These segments correspond to direct dominance pairs of points from the first two phases. It is possible to find all dominance pairs in a set of N points in time $O(N \log N + k)$ [13] where k is the number of dominance pairs, but in our case this would only lead to an $O(n^2 \log n)$ time bound. Instead, we leverage the fact that the vertices are embedded in an $O(n) \times O(n)$ grid, and use the following $O(n^2 + k)$ time method to generate dominance pairs using a stack-based algorithm related to Graham scan within each row. We prove later that the diagram is planar and therefore that the number of dominance pairs $k = O(n^2)$.

3. Initialize for each column c a value t_c, the topmost element seen so far in column c.

   Then, for each row r from 1 to $2n + 1$:

   (a) Initialize an empty stack S.

   (b) For each column c from 1 to $2n + 1$:

      i. If there is a vertex or junction p at (r, c), add an edge from every element of S to p, add an edge from t_c to p (if t_c is non-empty), and set t_c to p.

      ii. If t_c is non-empty, pop all items from S whose row number is less than or equal to the row number of t_c, and push t_c onto S.

Thus we have computed the coordinates of all elements, confluent junctions, and edges in the confluent diagram. When we render the drawing, we rotate it 45° counterclockwise to make it upward confluent (Figure 3).

Examples of non-confluent and confluent drawings of the same 100-element set are shown in Figure 4. Our Python implementation renders the confluent track segments as cubic Bézier curves with control points at a small fixed distance directly above and below each confluent junction. Two such curves cannot cross each other: for pairs of edges that do not share an endpoint, this follows from the fact that the convex hulls of the control points are disjoint and that the curves lie within the convex hulls, while for pairs of curves that share an endpoint it follows from the fact that the two curves are images of each other under an affine transformation of the plane and that (for pairs of edges sharing an endpoint) the direction that any point on the curve is translated by this affine transformation is transverse to the tangent direction of the curve at that point.

If the input is provided as a realizer rather than as a graph, and its completion has few elements, then it is possible to construct the diagram more efficiently. To do so, construct for each odd-indexed row or column of the integer grid an

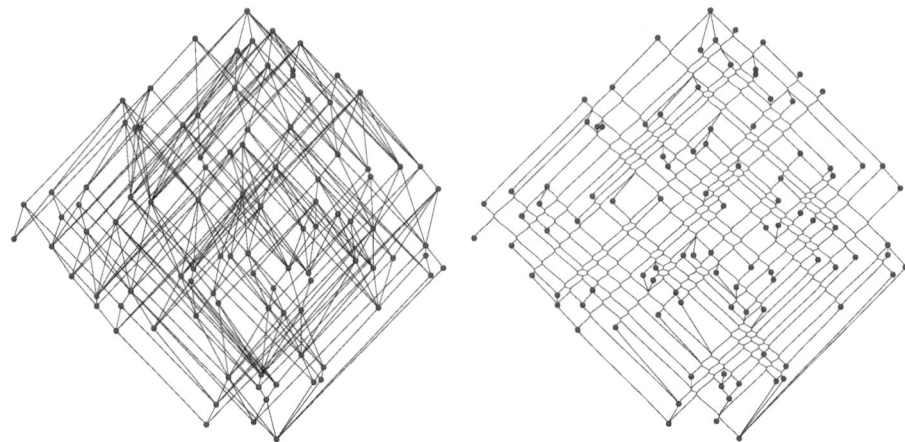

Fig. 4. A 100-element partially ordered set, the intersection of two random permutations, drawn as a conventional Hasse diagram with crossings (left), and as a confluent Hasse diagram (right)

axis-parallel line segment that passes through a grid point if and only if that point meets two of the four conditions for adding a junction in phase two of our algorithm. The junctions can be recovered as the intersections of these line segments, and we may compute the edges of the diagram using an output-sensitive algorithm for dominance pairs. By using integer searching data structures the total time for this algorithm may be reduced to $O((n + k) \log \log n)$, where k is the number of confluent junctions; we omit the details.

4   Algorithm Correctness and Minimality

In this section we prove that the algorithm of Section 3 is correct and has optimal running time. Our analysis also shows that a poset P has a confluent diagram if and only if it has dimension at most two.

Lemma 1 (Baker, Fishburn and Roberts [3]). *Let P be a bounded finite planar poset. Then P is a lattice and has dimension at most 2.*

Lemma 2. *Let P be a finite poset with a confluent Hasse diagram D. Then $\dim(P) \leq 2$, and there exists a two-dimensional lattice C containing P such that the elements of $C \setminus P$ (other than the top and bottom element, if they do not belong to P) correspond one-for-one with the confluent junctions of D.*

Proof: Replace the confluent junctions of D with vertices, and re-interpret the confluent segments as edges between these vertices. If there is more than one minimal vertex of P, add a vertex below all minimal vertices, connected to the minimal vertices by upward edges, and similarly if there is more than one maximal vertex of P, add a vertex above all maximal vertices connected to them by edges. The modified drawing is st-planar and hence by Lemma 1 represents a lattice, which clearly contains P.   □

Lemma 3. *Let P be a finite poset with order dimension at most two, let C be the completion of P, and let S be the set of elements of $C \setminus P$ (other than the top and bottom element, if P itself is not bounded). Then the elements of S coincide with the junction points added in phase 2 of our algorithm, and the dominance ordering on these points coincides with the lattice ordering in C.*

Proof: In one direction, let p be a junction point added in phase 2 of our algorithm, and p^- and p^+ be the sets of points from phase 1 that are dominated by p and that dominate p respectively. Then it follows from the four conditions according to which phase 2 adds a point that (p^-, p^+) forms a cut in P. The equivalence of the dominance and lattice orderings on pairs consisting of a junction point and a point from P follows immediately, and the same equivalence for pairs of junction points is also easy to verify.

In the other direction, we must show that we add a junction point for every element of S, that is, every cut (L, U) where L has more than one maximal element and U has more than one minimal element. Let i be one less than the minimum x-coordinate of a point in U, and let j be one less than the minimum y-coordinate; then (because the coordinates of points in P are their positions in the two orderings of a realizer) the set L of points dominated by every point in U equals the set of points below and to the left of (i, j). Two of the four conditions of phase 2 are automatically met at (i, j): the points with x-coordinate $i + 1$ and with y-coordinate $j + 1$ are both in U and are distinct because U has more than one minimal point. The other two conditions must also be met, for if they were not then the point violating the condition would dominate L, contradicting the fact that all points that dominate L belong to U. □

Theorem 1. *A given partial order P has a confluent diagram if and only if $\dim(P) \leq 2$. If P has a confluent diagram, the algorithm of Section 3 computes a valid confluent diagram of P, and embeds that diagram in a $O(n) \times O(n)$ grid in worst case optimal $O(n^2)$ time. The number of confluent junctions in the drawing is the minimum possible for any confluent diagram of P.*

Proof: If a poset P has dimension three or more, then so does any lattice containing it, and by Lemma 1 and Lemma 2 there can be no confluent diagram of P. Otherwise, we may assume that P has dimension at most two.

By Lemma 3, the dominance ordering on the points computed by our algorithm coincides (except possibly for the removal of the top and bottom elements) with the completion of P. In this set of points, there can be no crossing pairs of dominance relations, for if the edges (L_1, U_1)–(L_2, U_2) and (L_3, U_3)–(L_4, U_4) crossed (where (L_i, U_i) is a cut either added in the completion or corresponding to an original point of P) then $(L_1 \cup L_3, U_2 \cup U_4)$ would also be a cut whose point would lie between the other four points, contradicting the assumption that these edges represent minimal dominance pairs. Therefore, the diagram constructed by our algorithm is planar, and by Lemma 1 it must represent a lattice superset of P. The added elements belong to the completion, so the diagram must represent a subset of the completion, and since the completion has no proper

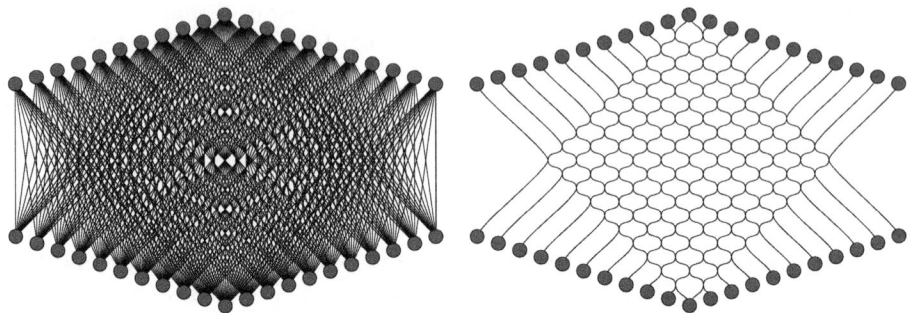

Fig. 5. A poset P with $O(n)$ elements and dimension 2 whose completion has size $\Omega(n^2)$. On the left is the normal Hasse diagram, and on the right is the confluent version as drawn by our algorithm. The two permutations L_1 and L_2 generating P are the identity and the permutation $(3n, 3n - 2, \ldots, n; 4n + 1, n - 1, 4n, n - 2, \ldots, 3n + 2, 0; 3n + 1, 3n - 1, \ldots, n + 1)$.

lattice subsets it must represent the completion itself. The completion gives the minimum number of added elements (and therefore, by Lemma 2, the minimum number of junctions) of any diagram for P.

Our algorithm spends $O(n^2)$ time in its first two phases as it iterates over $O(n^2)$ grid cells spending constant time per cell. In the third phase, it uses constant time per edge and by planarity there are $O(n^2)$ edges, so the time is again $O(n^2)$. This time bound is optimal since (as shown in Figure 5) there exist two-dimensional posets whose completion has $\Omega(n^2)$ elements.   □

Although our method produces drawings in a grid of linear dimensions, it may be possible in some cases to compact our drawings into a smaller grid. An algorithm of de la Higuera and Nourine [14] may be used to find the smallest grid into which a drawing produced by our algorithm can be compacted.

5   Confluent Drawings of Series-Parallel Posets

A *series-parallel partial order* is a poset that can be built up from single elements by two simple composition operations:

- The *series composition* $P; Q$ of posets P and Q is the order on the set $P \cup Q$ in which $p \leq q$ for every $p \in P$ and $q \in Q$.
- The *parallel composition* $P||Q$ is the order on $P \cup Q$ in which every pair of an element from P and an element from Q are incomparable.

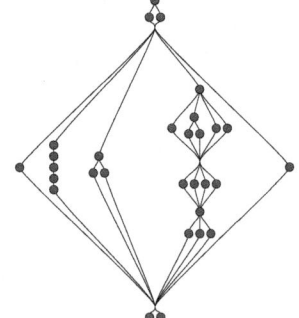

Fig. 6. A series-parallel poset

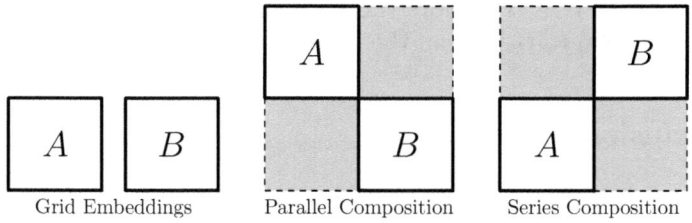

Grid Embeddings   Parallel Composition   Series Composition

Fig. 7. Series and parallel composition operations on two drawings A and B

Pairs of elements that are both from P or both from Q retain their ordering in the larger set.

Series-parallel partial orders are attractive because many important computational problems can be solved more easily in them than in more general posets, and because they have applications to a wide variety of problems including scheduling [25], concurrency [20], data mining [23], networking [2], and more (see [24]).

Series-parallel partial orders can be represented naturally by a binary tree, known as a decomposition tree of the order. The leaves of the tree correspond to single element sets and the internal nodes of the tree correspond to series or parallel composition operations. As the following theorem shows, given a decomposition tree T for a series-parallel partial order P, we can draw the confluent diagram of P in linear time by traversing T, performing the corresponding composition operations, and inserting confluent junctions when necessary.

Theorem 2. *Let P be a series-parallel partial order, given as its decomposition tree. Then a confluent diagram of P with a linear number of junctions can be drawn in an $O(n) \times O(n)$ grid in linear time.*

Proof: We traverse the decomposition tree in postorder, recursively finding embeddings for each subtree. For each tree node, we do the following:

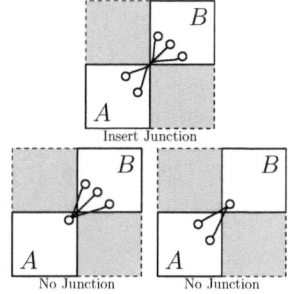

1. If the node is a leaf, then we embed the corresponding element in a single grid cell.
2. Otherwise, if the node is a series or parallel node, then we translate the grid embeddings of its two children so that their bounding boxes meet corner to corner (Figure 7).
3. For a series composition $A; B$ we also insert a confluent junction at the shared corner of A and B if and only if A has more than one maximal element and B has more than one minimal element (Figure 8).

Fig. 8. Series composition $A; B$ has a confluent junction if and only if A has no unique upper bound and B has no unique lower bound

By using a linked list of the maximal and minimal nodes for the current subtrees, we can perform these operations in time proportional to the number of leaves in the decomposition tree. Therefore the total time is linear. The size of the grid will be proportional to the size of the decomposition tree, i.e., $O(n) \times O(n)$ □

6   Conclusions

We have designed, analyzed, and implemented an algorithm for drawing confluent Hasse diagrams using a minimum number of confluent junctions. It would be of interest to test experimentally how many crossings it eliminates, and how much ink it saves. Also, upward planarity may be tested even for non-st-planar graphs that have only one source or one sink; can similar conditions be extended to the case of upward confluent drawings? Can we efficiently find upward planar drawings of graphs that are not transitively reduced? If a partially ordered set must be drawn with crossings, can we use confluence in a principled way to keep the number of crossings small? We leave these questions to future research.

Acknowledgements. This work was supported in part by NSF grant 0830403 and by the Office of Naval Research under grant N00014-08-1-1015.

References

1. Aeschlimann, A., Schmid, J.: Drawing orders using less ink. Order 9(1), 5–13 (1992)
2. Amer, P., Chassot, C., Connolly, T., Diaz, M., Conrad, P.: Partial-order transport service for multimedia and other applications. IEEE/ACM Transactions on Networking 2(5), 440–456 (1994)
3. Baker, K.A., Fishburn, P.C., Roberts, F.S.: Partial orders of dimension 2. Networks 2(1), 11–28 (1972)
4. Bertolazzi, P., Di Battista, G., Mannino, C., Tamassia, R.: Optimal upward planarity testing of single-source digraphs. SIAM J. Comput. 27(1), 132–169 (1998)
5. Birkhoff, G.: Lattice Theory. American Mathematical Society, Providence (1967)
6. Di Battista, G., Tamassia, R.: Algorithms for plane representations of acyclic digraphs. Theoret. Comput. Sci. 61(2-3), 175–198 (1988)
7. Dickerson, M.T., Eppstein, D., Goodrich, M.T., Meng, J.Y.: Confluent drawings: visualizing non-planar diagrams in a planar way. J. Graph Algorithms Appl. 9(1), 3–52 (2005), http://jgaa.info/accepted/2005/Dickerson+2005.9.1.pdf
8. Eppstein, D., Goodrich, M.T., Meng, J.Y.: Delta-Confluent Drawings. In: Healy, P., Nikolov, N.S. (eds.) GD 2005. LNCS, vol. 3843, pp. 165–176. Springer, Heidelberg (2006)
9. Eppstein, D., Goodrich, M.T., Meng, J.Y.: Confluent layered drawings. Algorithmica 47(4), 439–452 (2007)
10. Gansner, E., Hu, Y., North, S., Scheidegger, C.: Multilevel agglomerative edge bundling for visualizing large graphs. In: IEEE Pacific Visualization Symposium (PacificVis), pp. 187–194 (2011)
11. Ganter, B., Kuznetsov, S.O.: Stepwise Construction of the Dedekind-MacNeille Completion. In: Mugnier, M.-L., Chein, M. (eds.) ICCS 1998. LNCS (LNAI), vol. 1453, pp. 295–302. Springer, Heidelberg (1998)
12. Garg, A., Tamassia, R.: On the computational complexity of upward and rectilinear planarity testing. SIAM J. Comput. 31(2), 601–625 (2002)

13. Güting, R.H., Nurmi, O., Ottmann, T.: Fast algorithms for direct enclosures and direct dominances. J. Algorithms 10(2), 170–186 (1989)
14. de la Higuera, C., Nourine, L.: Drawing and encoding two-dimensional posets. Theoret. Comput. Sci. 175(2), 293–308 (1997)
15. Hirsch, M., Meijer, H., Rappaport, D.: Biclique Edge Cover Graphs and Confluent Drawings. In: Kaufmann, M., Wagner, D. (eds.) GD 2006. LNCS, vol. 4372, pp. 405–416. Springer, Heidelberg (2007)
16. Hui, P., Pelsmajer, M.J., Schaefer, M., Štefankovič, D.: Train tracks and confluent drawings. Algorithmica 47(4), 465–479 (2007)
17. Hutton, M.D., Lubiw, A.: Upward planar drawing of single source acyclic digraphs. SIAM J. Comput. 25(2), 291–311 (1996)
18. Jourdan, G.V., Rival, I., Zaguia, N.: Upward Drawing on the Plane Grid Using Less Ink. In: Tamassia, R., Tollis, I.G. (eds.) GD 1994. LNCS, vol. 894, pp. 318–327. Springer, Heidelberg (1995)
19. Kelly, D., Rival, I.: Planar lattices. Canad. J. Math. 27(3), 636–665 (1975)
20. Lodaya, K., Weil, P.: Series-Parallel Posets: Algebra, Automata and Languages. In: Meinel, C., Morvan, M. (eds.) STACS 1998. LNCS, vol. 1373, pp. 555–565. Springer, Heidelberg (1998)
21. Ma, T.H., Spinrad, J.: Transitive closure for restricted classes of partial orders. Order 8(2), 175–183 (1991)
22. MacNeille, H.M.: Partially ordered sets. Trans. Amer. Math. Soc. 42(3), 416–460 (1937)
23. Mannila, H., Meek, C.: Global partial orders from sequential data. In: Proceedings of the sixth ACM SIGKDD International Conference on Knowledge Discovery and Data Mining, KDD 2000, pp. 161–168. ACM, New York (2000)
24. Möhring, R.H.: Computationally tractable classes of ordered sets. In: Rival, I. (ed.) Algorithms and Order, pp. 105–193. Kluwer Academic Publishers (1989)
25. Möhring, R.H., Schäffter, M.W.: Scheduling series-parallel orders subject to 0/1-communication delays. Parallel Comput. 25(1), 23–40 (1999)
26. Nourine, L., Raynaud, O.: A fast algorithm for building lattices. Inform. Process. Lett. 71(5-6), 199–204 (1999)
27. Novák, V.: Über eine Eigenschaft der Dedekind-MacNeilleschen Hülle. Math. Ann. 179, 337–342 (1969)
28. Platt, C.R.: Planar lattices and planar graphs. J. Combinatorial Theory, Ser. B 21(1), 30–39 (1976)

Planar Open Rectangle-of-Influence Drawings with Non-aligned Frames

Soroush Alamdari and Therese Biedl

David R. Cheriton School of Computer Science, University of Waterloo
{s26hosse,biedl}@uwaterloo.ca

Abstract. A straight-line drawing of a graph is an *open weak rectangle-of-influence (RI) drawing*, if there is no vertex in the relative interior of the axis-parallel rectangle induced by the end points of each edge. No algorithm is known to test whether a graph has a planar open weak RI-drawing, not even for inner triangulated graphs.

In this paper, we study RI-drawings that must have a *non-aligned frame*, i.e., the graph obtained from removing the interior of every filled triangle is drawn such that no two vertices have the same coordinate. We give a polynomial algorithm to test whether an inner triangulated graph has a planar open weak RI-drawing with non-aligned frame.

1 Background

The rectangle-of-influence (RI for short) drawability problem was introduced by Liotta *et al.* [8]. In a strong RI drawing of a graph, there is an edge between two vertices of the graph if and only if there is no other vertex in the axis-parallel rectangle defined by the two ends of every edge. There are two variants of RI-drawings: In a *closed RI-drawing*, the rectangle required to be empty is closed, whereas in an *open RI-drawing*, only the relative interior of the rectangle is required to be empty.

Biedl *et al.* [3] introduced the concept of *weak RI drawings* in which graphs are drawn such that for any edge the rectangle is empty, but not for all empty rectangles the edge is necessarily present. They proved that a plane graph has a planar weak closed RI drawing if and only if it has no *filled triangle* (i.e., a triangle that has vertices in its interior.) Furthermore, they presented an algorithm to find such a drawing in an $(n-1) \times (n-1)$ grid in linear time. The grid size can be improved to $(n-3) \times (n-3)$ [12].

For open RI drawings, better bounds are known. Miura and Nishizeki [11] presented an algorithm to find a small weak open RI drawing of a given 4-connected graph. Their grid size is $W \times H$ where $W + H \leq n$. Zhang and Vaidya [15] also provided small weak open RI drawings for inner triangulated 4-connected graphs with quadrangular outer face. They do this by proving that the drawing presented by Fusy [4] is a weak open RI drawing.

However, as opposed to (weak planar) closed RI-drawings of planar graphs, no necessary and sufficient conditions or testing algorithms are known for the

M. van Kreveld and B. Speckmann (Eds.): GD 2011, LNCS 7034, pp. 14–25, 2012.

existence of (weak planar) open RI-drawings, even for inner triangulated graphs. This study was initiated by Miura, Matsuno and Nishizeki [10]. They first gave necessary and sufficient conditions for planar weak open RI-drawability of triangulated planar graphs. Here all faces including the outer-face are triangles, so the outer-face is a filled triangle, which severely restricts the placement of interior vertices and facilitates testing the existence of a weak open RI-drawing.

Miura *et al.* [10] also aimed to develop necessary and sufficient condition for all inner triangulated graphs, but did not succeed. It is clear that such a drawing imposes conditions on how filled triangles are drawn; a natural first step is hence to remove the interior of all filled triangles and try to draw the resulting *frame graph* while satisfying these conditions. Miura *et al.* then changed their model a bit and only considered what they called *oblique drawings* where no edges of the frame graph are drawn horizontally or vertically. They gave one set of conditions that are clearly necessary, and showed that adding one condition made them sufficient. (See later for more details.)

In this paper, we use a slight variant of oblique drawings that we call drawings with *non-aligned frame*, which means that no two vertices of the frame graph have the same x-coordinate or the same y-coordinate. We give necessary and sufficient conditions for a graph to have a planar weak open RI-drawing with non-aligned frames. Our proof is algorithmic and yields a test whether a graph has a planar weak open RI-drawing with non-aligned frame; it also constructs such a drawing if one exists. Also, the algorithm works via a detour into rectangular drawings and proves a correspondence between RI-drawings and rectangular drawings that may be of independent interest.

Due to space limitations, some details have been omitted; a full version can be found in [1].

2   Preliminaries

Let $G = (V, E)$ be a graph with n vertices V and m edges E. The graph G is called *simple* if it has no loops or multiple edge. It is called *planar* if it can be drawn in the plane without crossing. A planar drawing of G can be specified by giving for each vertex the cyclic order of edges around it. A planar drawing divides the plane into regions called *faces*. The unbounded region is called the *outer face*, all other faces are called *inner faces*. Any vertex not on the outer face is called an *inner vertex*. A *plane graph* is a planar graph with a planar embedding and the outer face specified. An *inner triangulated* graph is a plane graph in which every inner face is a triangle; it is called *triangulated* if the outer face is also a triangle. In this paper, all graphs are assumed to be simple, plane and inner triangulated, and we occasionally omit these quantifiers.

In a plane graph, a triangle is called *filled* if there is at least one vertex inside the triangle. Crucial for our study is the *frame graph*, which is the graph obtained by removing the inside of every filled triangle (see Fig. 2). Also crucial is the concept of angles of a plane graph. Each instance of a vertex appearing in a face is called an *angle*. The angles on the outer face are *outer angles* and the angles on the inner faces are called *inner angles*.

Given a plane graph, the *dual graph* is obtained by creating a vertex v_f for every face f, and adding an edge (v_f, v_g) whenever faces f and g share an edge. The angles in the dual graph are in natural 1-1 correspondence with the angles of the original graph: The angle at vertex v in face f corresponds in the dual graph to the angle at vertex v_f in the face formed where v used to be.

A *planar straight-line drawing* of a planar graph is a drawing without crossing where all edges are straight-line segments. Such a drawing is called a *planar weak open rectangle-of-influence* (RI for short) drawing if for every edge (v, w), the relative interior of the axis-parallel rectangle defined by the v and w contains no other vertex. The drawing in Fig. 1(a) is a planar weak open RI drawing. Since we do not consider any other type of RI-drawing, we omit the classifiers "planar", "weak" and "open" occasionally.

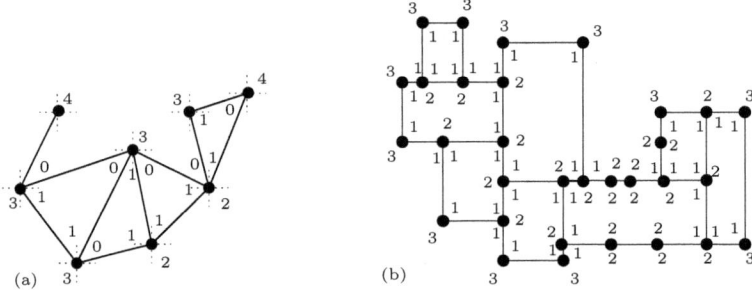

Fig. 1. (a) An oblique RI-drawing with RI-labels. (b) A rectangular drawing with corresponding RD-labels in the interior.

A straight-line drawing of a graph is *oblique* if no edge in the drawing is axis-parallel. It is *non-aligned* if no axis-parallel line intersects two or more vertices of the graph. Every non-aligned drawing is oblique, but not vice versa. An inner triangulated graph has a non-aligned RI-drawing if and only if it has no filled triangle, since a non-aligned RI-drawing has no vertices on the boundaries of rectangles and hence is a closed RI-drawing.

An oblique drawing of a graph G naturally induces a labeling of the angles with $\{0, 1, 2, 3, 4\}$ by assigning to each angle the number of coordinate axes contained in the angle. Since we use this concept only for RI-drawings, we call it an *RI-labeling*. The following is known.

Lemma 1. *[10] In an oblique RI-drawing of an inner triangulated graph, the RI-labels of any inner face consists of two 1s and one 0.*

An inner triangulated graph G is said to have a *inner rectangular dual drawing* if G can be represented as the touching graph of a set of interior-disjoint axis-aligned rectangles such that their union is simply connected (i.e., has no holes.) Fig. 1(b) shows an inner rectangular dual drawing of the graph in Fig. 1(a) (ignore the circles on the lines.) A *rectangular dual drawing* is an inner rectangular dual drawing where the union of the rectangles is also a rectangle.

A graph has an inner rectangular dual drawing if and only if it does not have a filled triangle [14,7,6]. Recall that a graph has a non-aligned RI-drawing if and only if it has no filled triangle, which suggests a relationship between these two types of drawings. We prove this formally in this paper, arguing via a third, closely related, type of drawing. A drawing of a plane graph is called an *inner rectangular drawing* if every edge is drawn as a horizontal or vertical line segment so that every inner face boundary is a rectangle. A *rectangular drawing* is an inner rectangular drawing in which the outer face is a rectangle too. See Fig. 1(b) and 5 for examples. Note that any (inner) rectangular dual drawing of a graph G is an (inner) rectangular drawing of a graph that is the dual graph of G except for some changes near the outer-face.

An inner rectangular drawing of a graph G induces a labeling of the (graph-theoretic) angles that we call an *RD-labeling*: If the angle is drawn with (geometric) angle $i\pi/2$, then assign it label $i \in \{1, 2, 3, 4\}$. Such a labeling can be used to characterize graphs that have a rectangular drawing. Call an RD-labeling *admissable* if (a) each inner angle is labeled 1 or 2, (b) each inner face has exactly 4 angles of label 1, (c) for each vertex, the labels of incident angles sum to 4, and (d) the sum of the labels on the outer-face is $2k + 4$, where k is the number of angles on the outer-face.

Lemma 2. *[9] A plane graph has an inner rectangular drawing if and only if it has an admissible RD-labeling.*

3   Results

Let G be an inner triangulated graph. Let F be the frame graph of G. In this section we give a constructive algorithm to decide whether G admits an open RI drawing such that F is non-aligned.

Overview: Like the result by Miura *et al.* [10], our algorithm is based on testing whether the frame-graph F of G has an RI-labeling that satisfies certain restrictions, and if so, compute an RI-drawing from it. We hence review their approach first and explain the changes with our algorithm.

Miura *et al.* first test for every filled triangle T whether the graph inside T has an RI-drawing. If this fails for any T then clearly G has no RI-drawing either. So in the following we always assume that all interiors of all filled triangles of T have an RI-drawing, at least under some restrictions on the drawing of T. Next, Miura *et al.* compute the restrictions made by a filled triangle T.

Lemma 3. *[10] If $T = \{a, b, c\}$ is a triangle of the frame graph that is a filled triangle in G, and if a is not adjacent to all vertices inside T, then in any open RI-drawing of G with oblique frame, the induced oblique RI-drawing of the frame has RI-label 1 at a.*

So there is a set A of inner angles of the frame graph F that must be labeled 1 in any non-aligned (hence oblique) RI-drawing of F induced by an RI-drawing of G. Moreover, if we can find a non-aligned RI-drawing of F that has these RI-labels, then it can be expanded into an open RI-drawing of graph G.

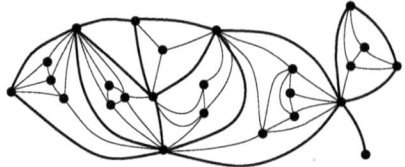 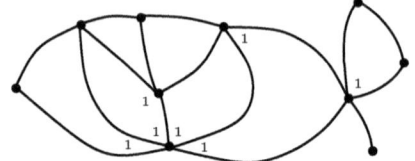

Fig. 2. Graph G (left) and its frame graph F with forced RI-labels (right)

Definition 1. *(based on [10]) A labeling of the angles of the frame graph F with $\{0, 1, 2, 3, 4\}$ is a decent RI-labeling if (a) the labels at every vertex sum to 4, and (b) every inner triangle has labels $\{0, 1, 1\}$, and every angle in A is labeled 1, where A is the set of restriction implied from the filled triangles.*

A decent RI-labeling is called good *if (c) the outer angles have labels $\{2, 3, 4\}$.*

A decent RI-labeling is called admissable *if (c') the sequence of labels on the outer angles does not contain $01*0$ as a subsequence.*

Miura *et al.* showed that if G has an open RI-drawing with oblique frame, then F has a decent RI-labeling. However, they also showed a graph where this is not sufficient. Hence they added condition (c) which forces the outer-face to consist of four chains that are monotone in x and y. This condition is not necessary, but they show that adding it gives sufficient conditions: any graph that has a good RI-labeling has an oblique RI-drawing.

We show here that using the restriction (c') gives conditions that are both necessary and sufficient, at least for the closely related concept of drawings with non-aligned frame.

Theorem 1. *An inner triangulated graph G has a planar weak open RI-drawing with non-aligned frame if and only if the frame graph F has an admissible RI-labeling.*

To prove this theorem, first consider necessity. Miura *et al.* already showed that conditions (a) and (b) of an admissible RI-labeling are necessary. We only sketch the proof of the necessity of condition (c'). Assume in a (planar, oblique) RI-drawing the sequence of RI-labels on the outer angles contain 00 as a subsequence, say at vertices v_1 and v_2 and edges e_0, e_1, e_2. Edge e_1 is not axis-aligned so the axis-aligned rectangle defined by its endpoints is non-trivial and must not contain the other endpoints of e_0 and e_2. But then the RI-labels of 0 force e_0 and e_2 to cross each other. Hence no (planar, oblique) RI-drawing can exist. The proof for 01^+0 as a subsequence is similar but more intricate; it is vital for this proof that the RI-drawing is non-aligned. See [1] for details.

We do not prove sufficiency directly; instead we give an algorithm that tests whether an inner triangulated G has a planar weak open RI-drawing with non-aligned frame, and the steps of the algorithm imply sufficiency of an admissible RI-labeling. We outline here our algorithm:

(i) Compute the frame graph F (see Fig. 2).

(ii) For every triangle T of F that was filled in G, compute whether the interior of T is realizable in an open RI-drawing [10]. If this fails for any triangle, then G has no open RI-drawing. Else, let A be the set of inner angles of F that must have RI-label 1 (Lemma 3.) See Fig. 2.

(iii) Construct D (see Fig. 3), which is roughly the dual graph of F after adding one vertex in the outer-face.

(iv) Find an admissible RD-labeling of D that respects A in some sense. See Fig. 5. If there is none, stop: F does not have a non-aligned RI-drawing (as we will show in Lemma 4.) . Otherwise, convert the RD-labeling to an inner rectangular drawing by Lemma 2.

(v) Expand the inner rectangular drawing Γ_D into a rectangular drawing $\Gamma_{D'}$ of a super-graph D', by adding more rectangular faces in the outside. $\Gamma_{D'}$ also respect A (see Fig. 5).

(vi) Construct the dual graph of D' and then remove the outer face vertex. The resulting graph F' is a super-graph of the frame-graph F (see Fig. 6).

(vii) From the RD-labeling of D', extract an RI-labeling of F'. This RI-labeling is decent, but in fact, it is good. See Fig. 6.

(viii) Using this good RI-labeling, create a non-aligned RI-drawing of F' using a variant of the algorithm presented in [10]. See Fig. 7.

(ix) Then insert the filled triangles (which is possible by choice of A) to obtain an open RI-drawing with non-aligned frame of a super-graph G' of G.

(x) Remove the vertices of $V_{G'} \setminus V_G$ from the drawing (see Fig. 7).

Steps (i), (ii), (ix) and (x) are either taken from [10] or are straightforward. We give definitions and details for the other steps below.

Definition of D: We first clarify how graph D is defined. Let F be the frame-graph, i.e., F is an inner triangulated graph without any filled triangle. Let F^+ be the graph obtained from F by adding one vertex v_o in its outer-face. For every outer angle α at a vertex v, we add three edges from v to v_o in F^+ at the place (in the cyclic order around v) where α was. Thus, a vertex that appears on the outer-face of F twice would have 6 edges to v_o, though not all of them would be consecutive. Now let D be the dual graph of F^+. See Fig. 3.

Recall that there is a 1-1-correspondence between angles in a planar graph and its dual. So for every inner angle α of F there is a corresponding inner angle β of D. For every outer angle α_i of F, there are four corresponding inner angles $\beta_i^1, \beta_i^2, \beta_i^3, \beta_i^4$ of D at the duals of the three edges from the vertex at α_i to the added vertex v_o in F^+. See Fig. 3.

From Admissible RI-Labelings to Rectangular Drawings: Recall that we assume the existence of a set A of inner labels of F that must be labeled 1 in any decent RI-labeling. We use the same set A to restrict rectangular drawings of D. More precisely, we say that a rectangular drawing Γ_{RD} of D *respects A* if for every angle $\alpha \in A$ (which is an inner angle of F), the corresponding angle in Γ_{RD} has RD-label 1.

In this part, we aim to show that step (iv) is correct: If D does not have a rectangular drawing that respects A, then F does not have a non-aligned open

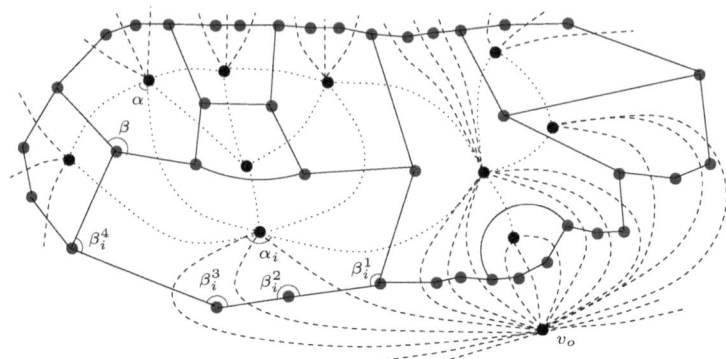

Fig. 3. The graph F from Fig. 2 (dotted), the added vertex v_o (dashed) and the graph D (solid)

RI-drawing. We prove this by showing that any non-aligned open RI-drawing of F can be converted to a rectangular drawing of D with corresponding angles.

Definition 2. *We say that an RI-labeling ℓ_{RI} of F and an RD-labeling ℓ_{RD} of D have the same inner structure if for any two corresponding inner angles α and β of F and D, $\ell_{RI}(\alpha) = 1$ if and only if angle $\ell_{RD}(\beta) = 1$.*

Lemma 4. *For any admissible RI-labeling ℓ_{RI} of F, there exists an admissible RD-labeling ℓ_{RD} of D that has the same inner structure.*

Proof. Given $\ell_{RI}(.)$ we define $\ell_{RD}(.)$ as follows: If α is an inner angle of F with corresponding inner angle β of D, then set $\ell_{RD}(\beta) = 2 - \ell_{RI}(\alpha)$. Since α has label 0 or 1, hence β has label 1 or 2, and it has label 1 if and only if α has label 1, so the two sets of labels have the same inner structure.

If α is an outer angle of F, then assigning labels to its corresponding 4 angles of D is more complicated (and in particular, not always a local operation.) Let $\alpha_0, \ldots, \alpha_{k-1}$ be the outer angles of F in clockwise order; addition in the following is modulo k. For each α_i, let $\beta_i^1, \ldots, \beta_i^4$ be the four corresponding inner angles of D, in clockwise order around the face. Now for each i (see also Fig. 4):

- If $\ell_{RI}(\alpha_i) = 0$, then assign labels $2, 2, 2, 2$ to $\beta_i^1, \beta_i^2, \beta_i^3, \beta_i^4$.
- If $\ell_{RI}(\alpha_i) = 2$, then assign labels $1, 2, 2, 1$ to $\beta_i^1, \beta_i^2, \beta_i^3, \beta_i^4$.
- If $\ell_{RI}(\alpha_i) = 3$, then assign labels $1, 1, 2, 1$ to $\beta_i^1, \beta_i^2, \beta_i^3, \beta_i^4$.
- If $\ell_{RI}(\alpha_i) = 4$, then assign labels $1, 1, 1, 1$ to $\beta_i^1, \beta_i^2, \beta_i^3, \beta_i^4$.
- If $\ell_{RI}(\alpha_i) = 1$, then we assign $1, 2, 2, 2$ or $2, 2, 2, 1$ to $\beta_i^1, \beta_i^2, \beta_i^3, \beta_i^4$, but the choice between these depends on the neighborhood.
  Explore from angle α_i both clockwise and counter-clockwise along the outer-face until we obtain a maximal subsequence where all RI-labels are 1. Say this sequence is $\alpha_j, \ldots, \alpha_l$. Since the RI-labeling is admissible, by condition (c') the sequence $\alpha_{j-1}, \alpha_j, \ldots, \alpha_l, \alpha_{l+1}$ does *not* have the form 01^+0, so

one of α_{j-1} and α_{l+1} has label ≥ 2. If $\ell_{RI}(\alpha_{j-1}) \geq 2$, then assign labels $1, 2, 2, 2$ to $\beta_i^1, \beta_i^2, \beta_i^3, \beta_i^4$ (and also to all other corresponding angles in that subsequence), else assign labels $2, 2, 2, 1$ to $\beta_i^1, \beta_i^2, \beta_i^3, \beta_i^4$.

Finally, for all outer angles of D, we set the RD-label such that the sum of labels around the vertex is 4. We verify that the labeling is admissible:

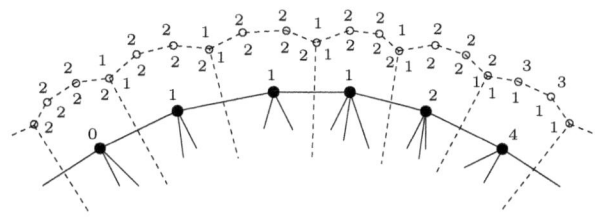

Fig. 4. Conversion of RI-labels of F to RD-labels of D

- Each inner angle of D has RD-label 1 or 2 since RI-label of inner angles of F are 0 or 1.
- Every inner face f of D has exactly 4 angles that have RD-label 1. For f corresponds to some vertex v_f of F, and the RI-labels at v_f sum to 4. By construction, an RI-label i at vertex v_f gives rise to i angles with RD-label 1 at f (this holds even if v_f is an outer vertex of F.)
- The RD-labels at every vertex v of D sum to 4. For if v is an inner vertex, then it corresponds to a triangle T of F which had RI-labels $\{0, 1, 1\}$, which correspond to RD-labels $\{2, 1, 1\}$. If v is an outer vertex, then by construction of the RD-labels at outer angles of D the total is 4.
- We claim that every outer angle α of D has RD-label $\{1, 2, 3\}$. Recall that $\ell_{RD}(\alpha)$ is defined as 4 minus the sum of other labels at the vertex v that supports α. Since there is at least one other label at v, and it is 1 or 2, hence $\ell_{RD}(\alpha) \leq 3$.
  Assume for contradiction that $\ell_{RD}(\alpha) \leq 0$. Since there are at most two inner angles at v, hence there must be exactly two (say β_i^4 and β_{i+1}^1) and they must both have RD-label 2. From the construction, this can happen only if $\ell_{RI}(\alpha_i) = 0 = \ell_{RI}(\alpha_{i+1})$. But an admissable RI-labeling does not have consecutive labels 0 on the outer-face by (c'), so this cannot happen.
- Finally we must show that the number of labels on the outer face sum to $2k + 4$, where k is the number of angles on the outer-face of D. This is a simple (but lengthy) counting-argument, which we omit for brevity's sake.

Hence the RD-labeling is admissible as desired.   □

Remark 1. Note that Lemma 2 implies a correspondence between inner rectangular drawings and RI-drawings: Any non-aligned RI-drawing defines an admissible RI-labeling, which implies an admissible RD-labeling, which implies an inner rectangular drawing, and they all have the same inner structure. The other direction also holds, and is proved implicitly with our algorithm.

The contrapositive of Lemma 4 proves correctness of step (iv). If D does not have an admissable RD-labeling that respects A, then F cannot have a non-aligned RI-drawing with all angles in A having RI-label 1. Fig. 5 shows an admissable RD-labeling for the restrictions of Fig. 3, and the corresponding inner rectangular drawing.

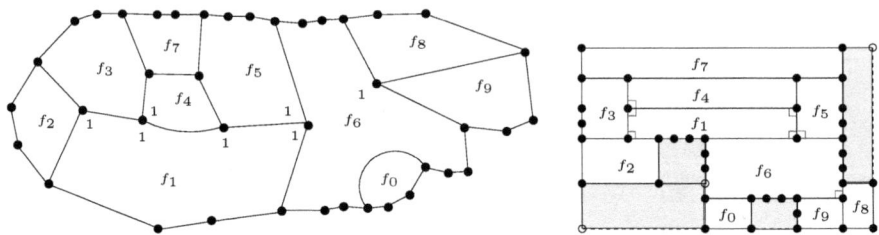

Fig. 5. Graph D with the restrictions on RD-labels (left), and an inner rectangular drawing expanded to a rectangular drawing by adding shaded rectangles (right)

From Inner Rectangular Drawing to Non-aligned RI-Drawing

Lemma 5. *Any inner rectangular drawing Γ_D of D that respects A can be expanded into a rectangular drawing $\Gamma_{D'}$ of a graph D' of size $O(|D|)$ such that inner angles of D are inner angles of D', and $\Gamma_{D'}$ respects A.*

Proof. As part of his orthogonal-shape approach to orthogonal graph drawing, Tamassia ([13], see also [2]) provided an algorithm to add a linear number of vertices and edges to an orthogonal drawing to turn it into a rectangular drawing without changing directions of edges. The algorithm does not create any vertex of degree 4. Applying this algorithm to the inner rectangular drawing Γ_D gives a rectangular drawing $\Gamma_{D'}$ of a graph D' and only adds vertices and edges in the outer-face, since all inner faces are rectangles already. Hence all inner angles (and their RD-labels) are preserved.   □

Lemma 6. *If D' has a rectangular drawing $\Gamma_{D'}$ that respects A, then there is a super graph F' of F that has a good RI-labeling.*

Proof. We prove this by converting the RD-labeling of $\Gamma_{D'}$ into an RI-labeling of F', hence more or less the reverse of the proof of Lemma 4. Let F' be the dual of D' minus the outer face vertex. For every angle α of F', let i be the number of angles in D' that correspond to α and that have RD-label 1 (i.e., their geometric angle is $\pi/2$.) Set $\ell_{RI}(\alpha) = i$. See Fig. 6.

Since every inner vertex of D' has RD-labels $\{1, 1, 2\}$ at its angles, every inner triangle of F' receives RI-labels $\{1, 1, 0\}$. Since every face of the RD-drawing is a rectangle, the RI-labels at any vertex of F' sum to 4. Also, any angle in A obtains RI-label 1 since its corresponding label had RD-label 1, so the resulting RI-labeling is decent. But in fact it is good: in a rectangular drawing (where the outer-face is a rectangle), any rectangle adjacent to the outer-face has at least

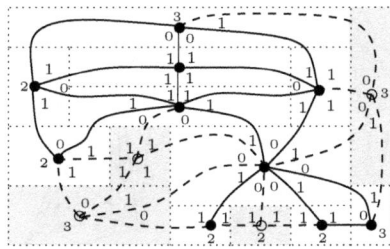

Fig. 6. The drawing $\Gamma_{D'}$ (dotted edges) and the graph F' (left) and the corresponding good RI-labeling of F' (right)

two angles of value $\pi/2$ on the outer-face, and so any outer angle of F' receives RI-label $2, 3$ or 4.   □

Lemma 7. *If F' has a good RI-labeling, then F' has a non-aligned RI-drawing with this RI-labeling.*

Proof. We can apply Miura *et al.*'s algorithm to construct an RI-drawing. However, their algorithm only promises an oblique drawing; it need not be non-aligned. But we can modify their algorithm to make the drawing non-aligned. Briefly, they can show that valid coordinates can be found by solving a system of constraints. All constraints have the form of an acyclic digraph where edge-weights express lower bounds on the differences of x-coordinates. Since there are no upper bounds on relative x-coordinates, we can find a solution to this system of constraints where all x-coordinates are distinct (e.g. by adding edges to turn the digraph into a total order (a complete acyclic digraph) and enforcing a minimum weight of 1 on all edges.) Similarly we can compute distinct y-coordinates. Hence we obtained a non-aligned RI-drawing with the same RI-labels.   □

Putting It All Together: If a graph G has an open RI-drawing with non-aligned frame F, then it has an admissable RI-labeling, hence D has an admissable RD-labeling (Lemma 4), hence D has an inner rectangular drawing (Lemma 2). and it respects A. Expand the inner rectangular drawing to a rectangular drawing (Lemma 5), extract a good RI-labeling from it (Lemma 6), and create a non-aligned RI-drawing from it (Lemma 7). See also Fig. 7. Insert the filled triangles and delete the added vertices and edges then results in the desired open RI-drawing with non-aligned frame of G. This proves correctness of the algorithm.

Our proof was constructive and gives rise to an algorithm to test whether G has an open RI-drawing with non-aligned frame. It remains to analyze the run-time of this algorithm. Most steps are clearly doable in linear time. The bottleneck is the time to test whether D has an RD-labeling that respects A.

We do this with a flow-approach inspired by Tamassia [13]. We only sketch the details here. Tamassia created a flow network of a plane graph that encodes

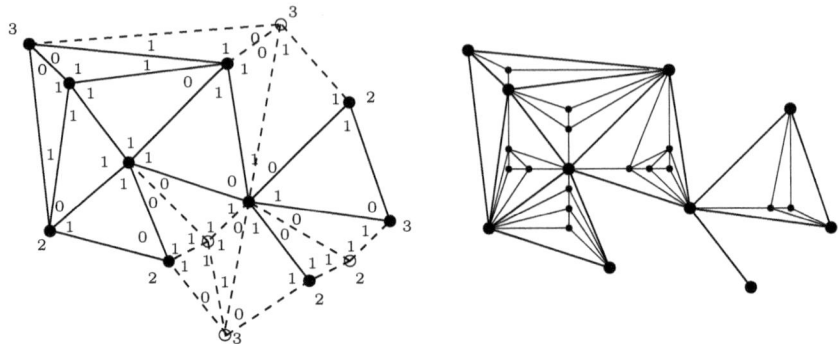

Fig. 7. The RI-drawing of G obtained by the algorithm

the shapes (i.e., abstract descriptions via bends and angles) of all possible plane orthogonal drawings. It is easy to add upper and lower bounds that forbid bends on the edges, forbids reflex angles at interior faces, and forces angles in A to have geometric angle $\pi/2$. The feasible integral flows in this network then correspond to the desired RD-labels. Tamassia's result required finding a minimum-cost flow, but since we forbid bends on edges, we only need to find a feasible flow, which can be done in $O(n^{1.5} \log n)$ time [5].

Theorem 2. *Let G be a plane inner triangulated graph. In $O(n^{1.5} \log n)$ time, we can test whether G has a planar weak open RI-drawing with non-aligned frame, and if so, construct it.*

We briefly return to the sufficiency for Theorem 1. If F has an admissable RI-labeling, then as mentioned after Lemma 4, D has a rectangular drawing that respects A. Steps (iv-x) of the algorithm then construct a planar weak open RI-drawing of G with non-aligned frame, proving Theorem 1.

4   Conclusion

We presented an algorithm to find an open RI drawing with non-aligned frame of a given inner triangulated graph G, if there exists such a drawing. We also characterized existence of such drawings in terms of properties of RI-labelings.

Our results also imply a correspondence between non-aligned RI-drawings and inner rectangular drawings. Lemma 4 shows that any non-aligned RI-drawing can be converted to an inner rectangular drawing with the same inner structure. Steps (iv)-(x) of our algorithm show that any inner rectangular drawing can be converted to a non-aligned RI-drawing, that preserves the inner structure. So apart from modifications near the outer-face (rectangles can "slide outward"), there is a 1-1-correspondence between non-aligned RI-drawings and inner rectangular drawings.

The most pressing open problem is what happens when we want to drop "with non-aligned frame". Can we efficiently test whether a given inner triangulated graph has a weak open RI-drawing? We note here that the concept of RI-labeling can be generalized quite easily to the case when the drawing is not necessarily non-aligned, if we add labels in $\{0, 1\}$ to each edge with an edge labeled 1 if it is parallel to a coordinate axis. It is quite easy to find necessary conditions for such a labeling, but are they sufficient? And if they are sufficient, how easy is it to test whether a graph has a labeling that satisfies these conditions? Neither of these questions appears straight-forward to answer.

Secondly, what is the situation for planar graph that are not inner triangulated? How quickly can we test whether they have a weak open RI-drawing (perhaps under some restrictions on the frame graph)?

References

1. Alamdari, S., Biedl, T.: Planar Open Rectangle-of-Influence Drawings with Non-Aligned Frames. Technical Report CS-2011-17, David R. Cheriton School of Computer Science, University of Waterloo (2011)
2. Battista, G.D., Eades, P., Tamassia, R., Tollis, I.G.: Graph Drawing: Algorithms for the Visualization of Graphs. Prentice-Hall (1998)
3. Biedl, T.C., Bretscher, A., Meijer, H.: Rectangle of Influence Drawings of Graphs without Filled 3-Cycles. In: Kratochvíl, J. (ed.) GD 1999. LNCS, vol. 1731, pp. 359–368. Springer, Heidelberg (1999)
4. Fusy, E.: Transversal structures on triangulations: A combinatorial study and straight-line drawings. Discrete Mathematics 309(7), 1870–1894 (2009)
5. Goldberg, A.V., Rao, S.: Beyond the flow decomposition barrier. J. ACM 45, 783–797 (1998)
6. Kozminski, K., Kinnen, E.: Rectangular dual of planar graphs. Networks 5, 145–157 (1985)
7. Leinwand, S.M., Lai, Y.-T.: An algorithm for building rectangular floor-plans. In: 21st Design Automation Conference, pp. 663–664. IEEE Press (1984)
8. Liotta, G., Lubiw, A., Meijer, H., Whitesides, S.H.: The rectangle of influence drawability problem. Computational Geometry 10(1), 1–22 (1998)
9. Miura, K., Haga, H., Nishizeki, T.: Inner rectangular drawings of plane graphs. Int. J. Comput. Geometry Appl. 16(2-3), 249–270 (2006)
10. Miura, K., Matsuno, T., Nishizeki, T.: Open rectangle-of-influence drawings of inner triangulated plane graphs. Discrete & Computational Geometry 41(4), 643–670 (2009)
11. Miura, K., Nishizeki, T.: Rectangle-of-influence drawings of four-connected plane graphs. In: Asia-Pacific Symposium on Information Visualization (APVIS). CR-PIT, vol. 45, pp. 75–80 (2005)
12. Sadasivam, S., Zhang, H.: Closed rectangle-of-influence drawings for irreducible triangulations. Comput. Geom. Theory Appl. 44, 9–19 (2011)
13. Tamassia, R.: On embedding a graph in the grid with the minimum number of bends. SIAM J. Comput. 16, 421–444 (1987)
14. Ungar, P.: On diagrams representing maps. J. London Mathematical Society 28(3), 336–342 (1953)
15. Zhang, H., Vaidya, M.: On open rectangle-of-influence and rectangular dual drawings of plane graphs. Discrete Mathematics, Algorithms and Applications 1, 319–333 (2009)

Proportional Contact Representations of Planar Graphs

Muhammad Jawaherul Alam[1,*], Therese Biedl[2,**], Stefan Felsner[3],
Michael Kaufmann[4], and Stephen G. Kobourov[1,*]

[1] Department of Computer Science, University of Arizona, Tucson, AZ, USA
[2] David R. Cheriton School of Computer Science, University of Waterloo, Waterloo, Canada
[3] Institut für Mathematik, Technische Universität Berlin, Berlin, Germany
[4] Wilhelm-Schickhard-Institut für Informatik, Universität Tübingen, Tübingen, Germany

Abstract. We study contact representations for planar graphs, with vertices represented by simple polygons and adjacencies represented by point-contacts or side-contacts between the corresponding polygons. Specifically, we consider proportional contact representations, where pre-specified vertex weights must be represented by the areas of the corresponding polygons. Several natural optimization goals for such representations include minimizing the complexity of the polygons, the cartographic error, and the unused area. We describe constructive algorithms for proportional contact representations with optimal complexity for general planar graphs and planar 2-segment graphs, which include maximal outerplanar graphs and partial 2-trees.

1 Introduction

For both theoretical and practical reasons, there is a large body of work about representing planar graphs as *contact graphs*, where vertices are represented by geometrical objects with edges corresponding to two objects touching in some fashion. Typical classes of objects might be curves, line segments, or polygons. An early result is Koebe's theorem [15] that all planar graphs can be represented by touching disks.

In this paper we consider contact graphs, with vertices represented by simple polygons with disjoint interiors, and adjacencies represented by point-contacts or side-contacts between corresponding polygons; see Fig. 1. In the weighted version of the problem, the input is not only a planar graph but also a weight function $w : V(G) \rightarrow R^+$ that assigns a weight to each vertex of $G = (V, E)$. A graph G admits a *proportional contact representation* with the weight function w if there exists a contact representation of G where the area of the polygon for each vertex v of G is proportional to $w(v)$. Such representations have practical applications in cartography, VLSI Layout, and floor-planning.

Using adjacency of regions to represent edges in a graph can lead to a more compelling visualization than drawing a line segment between two points [4]. In such representations of planar graphs it is desirable, for aesthetic, practical and cognitive reasons, to limit how complicated the polygons are. In practical areas like VLSI layout, it is also desirable to minimize the unused area in the representation. With these considerations in mind, we study the problem of constructing proportional point-contact and

* Research funded in part by NSF grants CCF-0545743 and CCF-1115971 and supported by NSERC.
** Research partially supported by EUROGIGA project GraDR and DFG Fe 340/7-2.

M. van Kreveld and B. Speckmann (Eds.): GD 2011, LNCS 7034, pp. 26–38, 2012.
© Springer-Verlag Berlin Heidelberg 2012

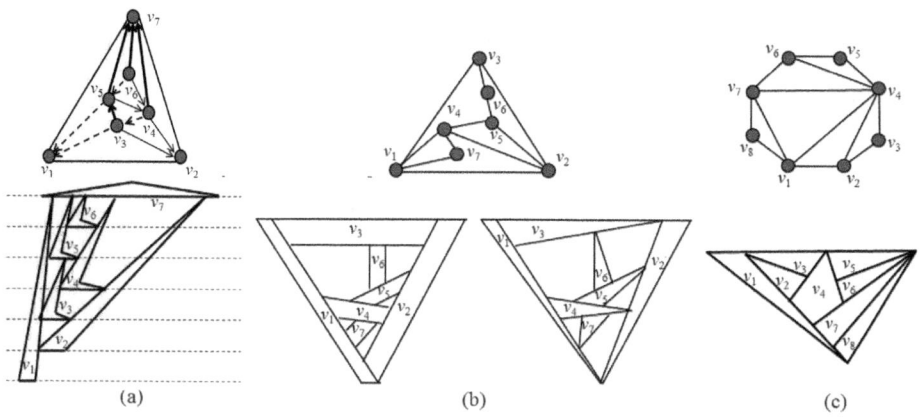

Fig. 1. (a) A planar graph and its proportional point-contact representation with 4-sided non-convex polygons; (b) A 2-tree and its proportional side-contact representation with trapezoids and proportional point-contact representation with triangles; (c) A maximal outerplanar graph and its hole-free proportional side-contact representation with 4-sided convex polygons

side-contact representations of planar graphs w.r.t. the following parameters, partially taken from the cartography-oriented literature, e.g. [13,20] :

- *complexity*: maximum number of sides in a polygon representing a vertex;
- *cartographic error*: $\max_{v \in V} |A(v) - w(v)|$, where $A(v)$ is v's area, $w(v)$ its weight;
- *holes*: total unused area of the representation that is in the interior.

1.1   Related Work

Koebe's theorem [15] is an early example of point-contact representation and shows that a planar graph can be represented by touching circles. Any planar graph also has a contact representation where all the vertices are represented by triangles [5] and with cubes in 3D [8]. Badent *et al.* [3] show that partial planar 3-trees and some series-parallel graphs also have contact representations with homothetic triangles. Recently, Gonçalves *et al.* [11] proved that any 3-connected planar graph and its dual can be simultaneously represented by touching triangles.

While the above results deal with point-contacts, the problem of constructing side-contact representations is less studied. Gansner *et al.* [9] show that any planar graph G has a side-contact representation with convex hexagons. Moreover, they show that 6 sides are necessary if convexity is required. For maximal planar graphs, the representation obtained by the algorithm in [9] is hole-free. Buchsbaum *et al.* [4] give an overview on the state of the art concerning rectangle contact graphs. The characterization of graphs admitting a hole-free side-contact representation with rectangles was obtained by Kozmiński and Kinnen [16] or in the dual setting by Ungar [19]. There is a also a simple linear time algorithm for constructing triangle side-contact representations for outerplanar graphs [10].

Note that in all the contact representation results mentioned above, the areas of the circles or polygons are not considered. That is, these results deal with the unweighted version of the problem. Furthermore, previous works on side-contact representations rarely focused on the presence or absence of holes, or the actual area taken by such holes. In our work we take both the area of regions and the presence of holes into account. For example, we show that representations by triangles or any convex shapes are not possible for certain planar graphs with pre-specified weights.

Motivated by the application in VLSI layouts, contact representations of planar graphs with rectilinear polygons and no holes have also been studied and it is known that 8 sides are sometimes necessary and always sufficient [21]. Very recently, we showed that 8 sides are also sufficient for the weighted case [1].

1.2   Our Results

In this paper we study the problem of proportional contact representation of planar graphs, with the goal to minimize the complexity of the polygons, the cartographic error, and the unused area. The four main results in our paper are optimal (with respect to complexity) algorithms for proportional contact representations for general planar graphs, outerplanar graphs, and partial 2-trees. We say k-sided polygons are sometimes necessary and always sufficient for representations of a particular class of planar graphs when there is an algorithm to construct a representation for any graph of this class with k-sided polygons and there is at least one example of a graph in this class that requires a (non-degenerate) k-sided polygons for any representation. Specifically, we show that: (a) 4-sided polygons are sometimes necessary and always sufficient for a point-contact proportional representation for any planar graph; (b) triangles are necessary and sufficient for point-contact proportional representation of partial 2-trees; (c) trapezoids are sometimes necessary and always sufficient for side-contact proportional representation of partial 2-trees; (d) quadrilaterals (convex 4-sided polygons) are sometimes necessary and always sufficient for hole-free side-contact proportional representation for maximal outerplanar graphs. In Table 1, we summarize the main results.

Table 1. The entries in this table correspond to results that are proven this paper, except one marked ($*$), which is trivial to see since any polygon with area > 0 requires at least three sides, and another marked ($**$), which follows from [10]. All the upper bound results are obtained by algorithm for representations that have no cartographic error. Note that some related results not in this table do have cartographic error.

Class of Graphs	Convexity	Complexity Lower Bound	Complexity Upper Bound	Hole-Free	Type of Contact
Planar	×	4	4	×	point
Partial 2-Trees	√	3*	3	×	point
Partial 2-Trees	√	4**	4	×	side
Maximal outerplanar	√	4	4	√	side

2 Preliminaries

In a *point-contact representation* of a planar graph $G = (V, E)$, we construct a set P of closed simple interior-disjoint polygons with an isomorphism $\mathcal{P} : V \to P$ where for any two vertices $u, v \in V$, the boundaries of $\mathcal{P}(u)$ and $\mathcal{P}(v)$ touch at a *contact point* if and only if (u, v) is an edge. A *side-contact representation* of a planar graph is defined analogously, where instead of a contact point, we have a *contact side* between $\mathcal{P}(u)$ and $\mathcal{P}(v)$, which is a non-degenerate line segment in the boundary of both. Let Γ be a contact (point-contact or side-contact) representation of G. Then each interior face of G corresponds to a bounded hole (possibly empty) in Γ and the exterior face of G corresponds to the unbounded hole in Γ.

In the weighted version of the problem, the input also includes a weight function $w : V(G) \to R^+$ that assigns a positive weight to each vertex of G. We say that G admits a *proportional contact representation* with the weight function w if there is a contact representation of G where the area of the polygon for each vertex v of G is proportional to its weight $w(v)$. We define the *complexity of a polygonal region* as the number of sides it has. In this paper, we also consider a polygon with less than k sides to be a (degenerate) k-sided polygon for convenience.

A *plane graph* is a planar graph with a fixed embedding. A plane graph is *fully triangulated* or *maximally planar* if all its faces including the outerface are triangles. Both the concept of "canonical order" [6] and "Schnyder realizer" [18] are defined for fully triangulated plane graphs in the context of straight-line drawings of planar graphs on an integer grid. We briefly review the two concepts below:

Let $G = (V, E)$ be a fully triangulated plane graph with outerface u, v, w in clockwise order. Then G has a *canonical order* of the vertices $v_1 = u$, $v_2 = v$, v_3, ..., $v_n = w$, $|V| = n$, which satisfies for every $4 \leq i \leq n$:

- The subgraph $G_{i-1} \subseteq G$ induced by v_1, v_2, ..., v_{i-1} is biconnected, and the boundary of its outer face is a cycle C_{i-1} containing the edge (u, v).
- The vertex v_i is in the exterior face of G_{i-1}, and its neighbors in G_{i-1} form an (at least 2-element) subinterval of the path $C_{i-1} - (u, v)$.

A *Schnyder realizer* of a fully triangulated graph G is a partition of the interior edges of G into three sets T_1, T_2 and T_3 of directed edges such that for each interior vertex v, the following conditions hold:

- v has out-degree exactly one in each of T_1, T_2 and T_3,
- the counterclockwise order of the edges incident to v is: entering T_1, leaving T_2, entering T_3, leaving T_1, entering T_2, leaving T_3.

The first condition implies that each T_i, $i = 1, 2, 3$ defines a tree rooted at exactly one exterior vertex and containing all the interior vertices such that the edges are directed towards the root. The following well-known lemma (for example, see [5]) shows a profound connection between canonical orders and Schnyder realizers.

Lemma 1. *Let G be a fully triangulated plane graph. Then a canonical order of the vertices of G defines a Schnyder realizer of G, where the outgoing edges of a vertex v are to its first and last predecessor (where "first" is w.r.t. the clockwise order around v), and to its highest-numbered successor.*

3   Proportional Point-Contact Representations of Planar Graphs

In this section we show that 4-sided non-convex polygons are sometimes necessary and always sufficient for a proportional contact representation of a planar graph. We first describe an algorithm to obtain proportional point-contact representations of planar graphs using 4-sided non-convex polygons. We then show that there exists a planar graph with a given weight function that does not admit a proportional point-contact representation with convex polygons, thus making our 4-sided construction optimal.

Theorem 1. *Let $G = (V, E)$ be a planar graph and let $w : V \to R^+$ be a weight function. Then G admits a proportional point-contact representation with respect to w in which each vertex of V is represented by a 4-sided polygon.*

Proof. We prove this claim constructively, showing how to generate a proportional contact representation of G with respect to w. We first take a planar embedding of G and assume that it is fully triangulated, for if it is not, we can add dummy vertices to make it so, and later remove those dummy vertices from the obtained proportional contact representation.

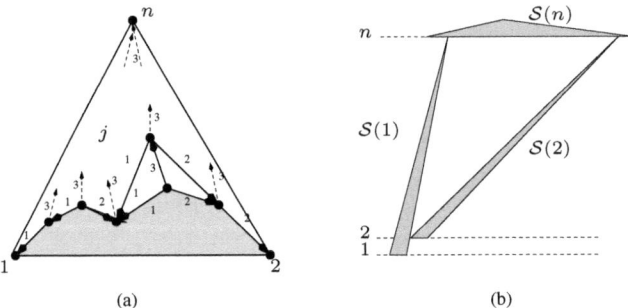

(a)   (b)

Fig. 2. (a) The canonical order and T_i (marked by labels); (b) the placement of 1,2,n

Assume after possible scaling that $w(v) \leq 1/n^2$ for all $v \in V$ and fix an arbitrary outer-face. We construct the drawing incrementally, following a canonical ordering $v_1, \ldots, v_n$. We prescribe what the polygon assigned to j looks like before even placing it (here and in the rest of the paper we use j as a shorthand for v_j). So let T_1, T_2, T_3 be the Schnyder realizer defined by the canonical ordering, where T_1 is rooted at 1, T_2 is rooted at 2 and T_3 is rooted at n; see Fig. 2(a). Let $\Phi_i(j)$ be the parent of j in tree T_i.

It is easy to show that $T_2^{-1} \cup T_1$ is an acyclic graph on the vertex set $V - \{n\}$, where T_2^{-1} is the tree T_2 with the direction of all its edges reversed. For every vertex $j \neq n$, let $\pi(j)$ be the index of j in a topological order of this graph. Then $n \geq \pi(\Phi_1(j)) > \pi(j) > \pi(\Phi_2(j)) \geq 1$. Now for every vertex $j \neq 1, 2, n$, we define the *spike* $\mathcal{S}(j)$ to be a 4-sided polygon with one reflex vertex. One segment (the *base*) is horizontal with y-coordinate j. Its length will be determined later, but it will always be at least $2/n^2 \geq 2w(j)$. From the left endpoint of the base, the spike continues with the *upward segment*, which has slope $\pi(j)$ and up to its *tip* which has y-coordinate $y = \Phi_3(j)$. Next comes the *downward segment* until the reflex vertex, and from there to the right endpoint of the base; see Fig. 3(a). The placement of the reflex vertex is arbitrary, as long as the resulting shape has area $w(j)$ and the down-segment has positive slope. Note that since the base has length $\geq 2w(j)$ and y-coordinate j, the reflex vertex will have y-coordinate at most $j + 1$. We first place $1, 2, n$, and then add $3, \ldots, n - 1$ (in this order):

- Vertex 1 is represented by a triangle $\mathcal{S}(1)$ whose base has length $2w(1)/(n - 1)$, placed arbitrarily with y-coordinate 1. The tip of $\mathcal{S}(1)$ has y-coordinate n.
- Vertex 2 is represented by a triangle $\mathcal{S}(2)$ whose base has length $2w(2)/(n - 2)$, placed at y-coordinate 2 and with its left endpoint abutting $\mathcal{S}(1)$. The tip of $\mathcal{S}(2)$ has y-coordinate n.
- Vertex n is represented by a triangle whose base is at y-coordinate n and long enough to cover the tips of $\mathcal{S}(1)$ and $\mathcal{S}(2)$. We choose the height of $\mathcal{S}(n)$ such that the area is correct.

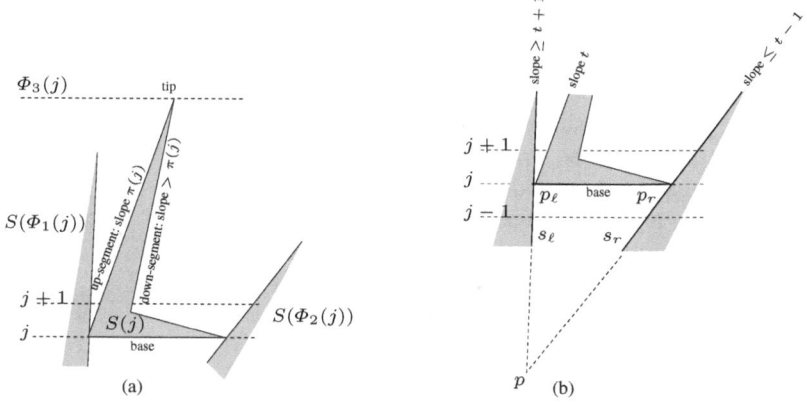

Fig. 3. (a) Adding j; (b) computing the width of the base

We maintain the following invariant: For $j \geq 2$, after vertex j has been placed, the horizontal line with y-coordinate $j + 1$ intersects only the spikes of the vertices on the outer-face of G_j, and in the order in which they occur on the outer-face.

To place $j \geq 3$, we place the base of $\mathcal{S}(j)$ with y-coordinate j, and extend it from the down-segment of $\Phi_1(j)$ to the up-segment of $\Phi_2(j)$. Recall that $\Phi_2(j)$ and $\Phi_1(j)$ are exactly the first and last predecessor of j, and $j = \Phi_3(i)$ for all other predecessors $i \neq j$. Hence $\mathcal{S}(j)$ touches $\mathcal{S}(\Phi_1(j))$ and $\mathcal{S}(\Phi_2(j))$ at the ends of the base, and all other predecessors i of j have their tips at the base. So this creates a contact between j and all its predecessors. The rest of $\mathcal{S}(j)$ is then as described above. It is easy to verify the invariant, and therefore $\mathcal{S}(j)$ does not intersect any other spikes. To see that the base of $\mathcal{S}(j)$ is long enough, let p_ℓ and p_r be its left and right endpoints, and s_ℓ and s_r be the other segments containing them. Imagine that we extend s_ℓ and s_r until they meet in a point p. Since s_r contains a point with y-coordinate $\leq j - 1$ (at the base of $\mathcal{S}(\Phi_2(j))$), triangle $\Delta\{p, p_\ell, p_r\}$ has height $h \geq 1$; see Fig. 3.

Let $t = \pi(v_j)$ be the slope of the up-segment of $\mathcal{S}(v_j)$. Since $\pi(\Phi_2(v_j)) < \pi(v_j) = t$, we have that s_r has slope at most $t - 1$ and $x(p_r) \geq x(p) + \frac{h}{t-1}$. On the other hand, the slope of s_ℓ is positive by construction, and must exceed the slope of the up-segment of $\Phi_1(v_j)$, which has slope $\pi(\Phi_1(v_j)) > \pi(v_j) = t$. So s_ℓ has slope $\geq t + 1$ and $x(p_\ell) \leq x(p) + \frac{h}{t+1}$. Therefore,

$$x(p_r) - x(p_\ell) \geq \frac{h}{t-1} - \frac{h}{t+1} = \frac{h(t+1-(t-1))}{t^2-1} \geq \frac{2h}{t^2} \geq \frac{2}{n^2} \geq 2w(v_j)$$

where the last inequality holds since weights are small enough. Therefore the base of $S(j)$ is wide enough, which ends the proof of the theorem.   □

Our construction used non-convex shapes. This is sometimes required.

Lemma 2. *There exists a planar graph and a weight function such that the graph does not admit a proportional point-contact representation with respect to the weight function with convex shapes for all vertices.*

Sketch of proof: We aim to show that the graph in Figure 4 has no proportional representation with convex polygons if the small vertices have weight δ and the larger vertices have weight $D > 3\delta$. Assume for contradiction that we had such a representation; by symmetry we may assume that d is in the outer-face.

For $i = 0, 1, 2$, let p_i be a point of contact between $P(a_i)$ and $P(a_{i+1})$ (where addition is modulo 3.) Further, let q_i be a point of contact between $P(a_i)$ and $P(b)$. Define T_0 to be the triangle $\Delta\{p_0, p_1, p_2\}$ and T_2 to be the triangle $\Delta\{q_0, q_1, q_2\}$. We will only consider the case where T_2 is circumscribed by T_0 (i.e., q_0, q_1, q_2 lie on three different sides of T_0); the other case is more intricate and requires defining and analyzing a third triangle T_1 (details can be found in [2].)

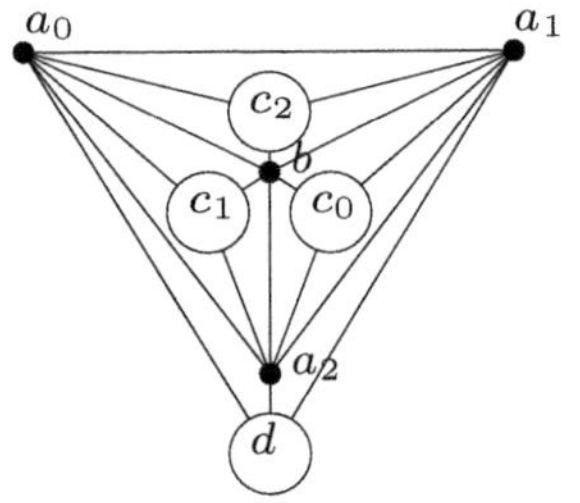

By convexity, the three sides of T_0 lie inside $P(a_0), P(a_1)$ and $P(a_2)$, respectively. In particular, all of $P(c_0), P(c_1), P(c_2)$ are inside T_0. On the other

Fig. 4. Graph without proportional convex contact representation

hand, by convexity all side of T_2 lies inside $P(b)$, so all of $P(c_0), P(c_1), P(c_2)$ are outside T_2. But the region between T_0 and T_2 consists of three triangles, and to maintain the planar embedding each of $P(c_0), P(c_1), P(c_2)$ must be in one of these triangles. So now we have a triangle T_2 of area at most δ that is circumscribed by a triangle T_0 such that the three triangles of $T_0 - T_2$ each have area at least $\Delta > \delta$. This is impossible by a very old result from geometry; see e.g. [7].   □

Lemma 2 implies that 3-sided polygons are not always sufficient for proportional contact representations of planar graphs. On the other hand, Theorem 1 implies that any planar graph has a proportional contact representation with any given weight function on the vertices so that each of the vertices is represented by a non-convex 4-sided polygon. Summarizing these two results we have the following theorem.

Theorem 2. *4-sided non-convex polygons are always sufficient and sometimes necessary for proportional point-contact representation of a planar graph with a given weight function on the vertices.*

4   Subclasses of Planar Graphs with Convex-Shape Representations

In this section we address the problem of proportional contact representations for sub-classes of planar graphs. The lower bound in Lemma 2 shows that for planar triangula-tions, the complexity in any proportional contact representation must be at least 4 and the polygons must be non-convex. We hence focus on planar graphs with fewer edges. In the next subsection we deal with proportional contact representations using triangles (or convex quadrilaterals for side-contacts.) Then we describe an algorithm for hole-free representation of maximal outerplanar graphs.

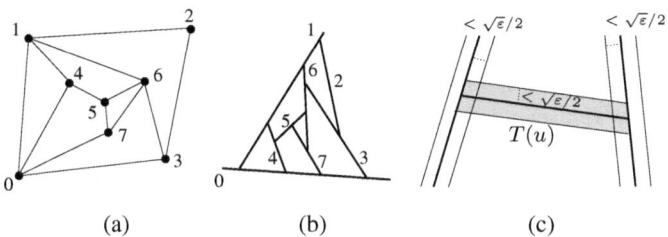

Fig. 5. (a&b) A 2-segment graph and representation; (c) converting to trapezoids

4.1   2-Segment Graphs and Partial 2-Trees

Call a planar graph a *2-segment graph* if it can be represented by assigning interior-disjoint line segments to vertices such that line segments share a point if and only if the corresponding vertices are adjacent, and no 3 line segments share a point. See Fig. 5. 2-segment graphs include 2-trees, maximal outerplanar graphs, partial 2-trees, and series-parallel graphs. We show that 4-sided convex polygons are always sufficient and sometimes necessary for side-contact representations of these graphs. For point-contact representations we show that 3 sides are sufficient (and, of course, necessary) for proportional contact representations of 2-segment graphs.

Theorem 3. *Let $G = (V, E)$ be a planar 2-segment graph. Then for any weight func-tion $w : V \to R^+$ and any $\varepsilon > 0$, G has a proportional side-contact representation where each vertex v is represented by a trapezoid with area between $w(v) - \varepsilon$ and $w(v)$.*

Proof. Let $\ell(v)$ be the line segment that represents v. We assume that ε is small enough such that "off-setting" any $\ell(v)$ by distance $\sqrt{\varepsilon}/2$ preserves adjacencies and does not create intersections. Here, *off-setting* $\ell(v)$ means moving it in parallel while shorten-ing/lengthening it so that it still touches the segments at its ends. We also assume (after possible scaling) that $||\ell(v)|| \geq 2w(v)/\sqrt{\varepsilon} + \sqrt{\varepsilon}$ for all vertices v.

For any vertex v, create two copies of $\ell(v)$ that are off-set in parallel in both direc-tions so that the trapezoid $T(v)$ between the two off-set lines has area $w(v)$. By the assumption on $||\ell(v)||$, this will require an off-set of les than $\sqrt{\varepsilon}/2$, hence adjacencies are preserved. This yields a proportional side-contact representation, except that $T(u)$ and $T(v)$ intersect for any edge (u, v).

To remove these unwanted intersections, let (u, v) be an edge, and assume that in the 2-segment representation, $\ell(u)$ ended at an interior point of $\ell(v)$. We then "retract"

$T(u)$, i.e., we replace it by $T(u) - T(v)$. It remains to show that this does not disturb the area too much. Note that $T(u) \cap T(v)$ is a parallelogram, defined by $\ell(v)$ and one off-set line of $\ell(v)$, as well as the two off-set lines of $\ell(u)$, where the pairs of parallel lines have distance less than $\sqrt{\varepsilon}/2$ and $\sqrt{\varepsilon}$, respectively. Therefore, the area of $T(u) \cap T(v)$ is less than $\varepsilon/2$, and we remove such an area at each end of $T(u)$. Thus, the area of the retracted trapezoid is more than $w(u) - \varepsilon$, as desired. ☐

It is natural to ask for a characterization of 2-segment graphs. Thomassen gave one (Theorem 4) at Graph Drawing 1993 but never published his proof.

Theorem 4. *A planar graph $G = (V, E)$ is a 2-segment graph if and only if $|E[W]| \leq 2|W| - 3$ for every $W \subseteq V$, where $E[W]$ is the set of edges with both ends in W.*

We provide a new proof of Theorem 4 based on rigidity theory in [2]. The condition stated in the theorem can efficiently be checked, for example Lee and Streinu [17] provide a simple algorithm. (In contrast, Hliněný [14] showed that the recognition of general contact graphs of segments is NP-complete.) So 2-segment graphs can be easily recognized.

However, the representations we gave for 2-segment graphs have a small carto-graphic order, which seems unavoidable if the incidences endpoints of segments to the other segments are circular, as for example for vertices $\{5, 6, 7\}$ in Fig. 5. This error can be avoided if G is *2-shellable*, which means that it is planar has a vertex order $v_1, \ldots, v_n$ such that for $i \geq 3$ vertex v_i has at most two neighbors in $v_1, \ldots, v_{i-1}$. Such graphs have at most $2n - 3$ edges, hence by Theorem 4 a 2-shellable graph is a 2-segment graph. Moreover, it is easy to see that we may assume that the endpoints of segment $\ell(v)$ are adjacent to the predecessors of v for all vertices v. We can then create a proportional side-contact representation as above but without cartographic error by creating trapezoids in this vertex order. For each vertex v_i, first shorten $\ell(v_i)$ so that it ends at the off-set lines of v_i's predecessors. Then off-set $\ell(v_i)$ so that the resulting trapezoid has area $w(v_i)$. It is easy to verify that all off-sets are still at most $\sqrt{\varepsilon}/2$, and thus the adjacencies are preserved.

Theorem 5. *Let $G = (V, E)$ be a 2-shellable graph and $w : V \to R^+$ be a weight function. Then G admits a proportional side-contact representation where each vertex of G is represented by a trapezoid with area $w(v)$.*

We derive two corollaries from Theorem 3 and 5. First, it is known that planar bipartite graphs are 2-segment graphs (we can even restrict the segments to be horizontal or vertical) [12]. Hence they have proportional side-contact representations with arbitrarily small cartographic error with trapezoids (in fact, rectangles.)

Second, a *2-tree* is either an edge or a graph G with a vertex v of degree two in G such that $G - v$ is a 2-tree and the neighbors of v are adjacent. A *partial 2-tree* is a subgraph of a 2-tree; partial 2-trees are the same as series-parallel graphs. Every partial 2-tree is planar. Directly from the definition we see that 2-trees (and hence partial 2-trees) are also 2-shellable. Therefore they have a proportional side-contact representation with trapezoids. We also show that 4 sides are sometimes required.

Theorem 6. *Four-sided convex polygons are always sufficient and sometimes neces-*
sary for a proportional side-contact representation of a 2-shellable graph, in particular
of a partial 2-tree, with a given weight function.

Proof. Sufficiency follows from Theorem 5, since partial 2-trees are 2-shellable. To establish necessity, consider the 2-tree obtained from $K_{2,4}$ by adding an edge between the
vertices of the partition of size two. These two vertices have four common neighbors,
but as was proved in [10], in any side-contact representation with triangles, any pair of
adjacent vertices has at most three common neighbors. Hence this graph has no side-
contact representation with triangles, let alone one that respects the weights.   □

Note that if we switch from side-contact representations to point-contact representations, we can reduce the complexity of the regions from four to three. Specifically, we
can replace line-segments by triangles so that only one endpoint of $\ell(v)$ is moved (in
both directions). Using a similar approach as that in Theorem 3 we can prove:

Theorem 7. *Let $G = (V, E)$ be a 2-segment graph and $w : V \to R^+$ be a weight*
function. Then for any $\varepsilon > 0$, G admits a proportional point-contact representation
where each vertex of G is represented by a triangle with area between $w(v) - \varepsilon$ and
$w(v)$. If G is a 2-shellable graph, then the area of the triangle of v is exactly $w(v)$.

4.2   Maximal Outerplanar Graphs

In this section, we study maximal outerplanar graphs, i.e., planar graphs whose outer-
face is a cycle and all interior faces are triangles. These are 2-trees, so the results from
the previous subsection apply, but (using a different construction) we can construct a
side-contact representation using triangles that has no holes.

Let G be a maximal outerplanar graph. For any two vertices u, v denote by $G(u, v)$
the graph induced by the vertices that are between u to v (ends excluded) while walking
along the outer-face in counterclockwise order, and let $w(G(u, v))$ be the sum of the
weights of all these vertices.

Define an *aligned triangle* to be one with horizontal base and tip below the base.
This naturally defines a *left* and *right side* of the triangle. We will use the observation
that an outerplanar graph can be represented inside *any* aligned triangle of suitable area.

Lemma 3. *Let $G = (V, E)$ be a maximal outerplanar graph and (u, v) an edge on*
the outer-face of G, with u before v in counterclockwise order. Let $w : V \to R^+$ be
a weight-function. Then for any aligned triangle T of area $w(G(v, u))$, there exists a
hole-free proportional side-contact representation of $G(v, u)$ inside T such that the left
[right] side of T contains segments of the neighbors of u [v] and of no other vertices.

Proof. We proceed by induction on the number of vertices in G. In the base case, G is
a 3-cycle $\{u, v, x\}$. Use T itself to represent x; this satisfies all conditions.

In the inductive step, let x be the unique common neighbor of u and v. Divide T
with a segment s from the tip to the base such that the region T_ℓ left of s has area
$w(G(x, u)) + \frac{1}{2}w(x)$, and the region T_r right of ℓ has area $w(G(v, x)) + \frac{1}{2}w(x)$. Cut
off triangles of area $\frac{1}{2}w(x)$ each from the tips of T_ℓ and T_r; the combination of these
two triangles forms a convex quadrilateral of area $w(x)$ which we use for x; see Fig. 6.

Recursively place $G(x, u)$ and $G(v, x)$ (if non-empty) in the remaining triangles of T; it is easy to verify that these have the correct area, which yields the desired side-contact representation.   □

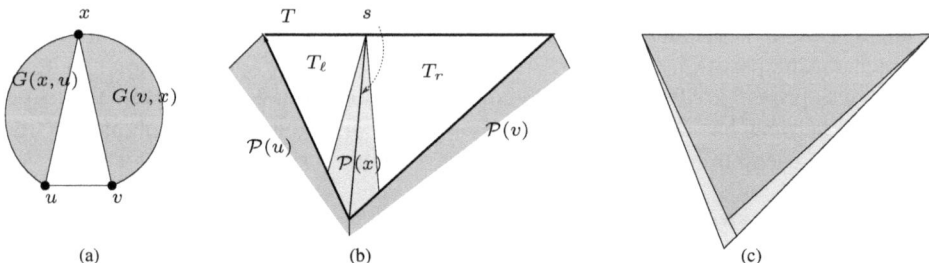

Fig. 6. The construction for maximal outerplanar graphs: (a) the graph; (b) splitting triangle T suitably; (c) adding u and v in the outer-most recursion

Apply this lemma for an arbitrary edge (u, v) on the outer-face and an arbitrary triangle T with area $w(G(v, u))$. We can then add triangles for u and v to it to complete the drawing into a contact representation of G; see Fig. 6(c). So we obtain:

Corollary 1. *Let $G = (V, E)$ be a maximal outerplanar graph and let $w : V \to R^+$ be a weight function. Then G admits a hole-free proportional side-contact representation where vertices are represented by triangles or convex quadrilaterals.*

We now show that the representation obtained by this algorithm is also optimal for a maximal outerplanar graph with respect to complexity. To do this we use the *snowflake graph* S, which is the general name given to an infinite family of outerplanar graphs obtained from a triangle by repeatedly walking around the outer-face and adding a vertex of degree 2 at each edge; each complete walk around the boundary gives a new snowflake graph; see Fig. 7(a).

Lemma 4. *A snowflake graph S has no hole-free side-contact representation with triangles that all have the same area.*

Sketch of Proof (a detailed proof is given in [2].) Assume for contradiction that there is such a representation Γ.

Let the *i-th level vertices* be those added when we walk around the outer-face for the i-th time. One can observe that all the angles in the outer-boundary of Γ_i are concave but for at most four convex corners. Then between any two consecutive convex corners, the triangles corresponding to the $(i + 1)$-th level vertices are inserted in concave corners.

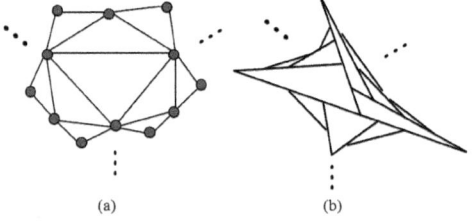

Fig. 7. (a) The snowflake graph S; (b) illustration for the proof of Lemma 4

Since the number of vertices doubles on each level, for sufficiently large i there must be a triangle T on the i-th level and its adjacent triangles T' and T'' on $(i+1)$-th level such that the *base* of T (the side that was exposed after adding T) has length greater than both the bases of T' and T''. Since all triangles have equal area, a simple calculation involving adjacent angles shows that this is a contradiction; see Fig. 7(b).   □

By Corollary 1 and Lemma 4, we have the following theorem.

Theorem 8. *Convex quadrilaterals are always sufficient and sometimes necessary for hole-free proportional side-contact representations of maximal outerplanar graphs.*

5   Conclusion and Open Problems

We described several constructive algorithm for proportional point-contact and side-contact representations of planar graphs, outerplanar graphs, and 2-trees. We focused on the complexity of the polygons representing vertices, and provided bounds on this complexity that are tight, for a variety of graph classes and drawing models.

However, many problems still remain open. What is the complexity of side-contact proportional representations of maximal planar graphs? We can achieve 7-sided polygons easily (essentially by cutting the convex corners of the 4-sided spikes), but can we do better? Likewise, what is the complexity for hole-free proportional representations of maximal planar graphs? Here, a bound of 8 is known (and the polygons are orthogonal) [1], but can we do better if polygons need not be orthogonal?

Acknowledgment. This work was initiated at the Dagstuhl Seminar 10461 on Schematization. We thank Marcus Krug, Ignaz Rutter, Henk Meijer, Emilio Di Giacomo, and Andreas Gerasch and several anonymous referees for useful discussions and remarks.

References

1. Alam, M.J., Biedl, T., Felsner, S., Gerasch, A., Kaufmann, M., Kobourov, S.G., Ueckert, T.: Computing cartograms with optimal complexity (submitted, 2011)
2. Alam, M.J., Biedl, T., Felsner, S., Kaufmann, M., Kobourov, S.G.: Proportional contact representations of planar graphs. Technical Report CS-2011-11. University of Waterloo (2011)
3. Badent, M., Binucci, C., Giacomo, E.D., Didimo, W., Felsner, S., Giordano, F., Kratochvíl, J., Palladino, P., Patrignani, M., Trotta, F.: Homothetic triangle contact representations of planar graphs. In: CCCG 2007, pp. 233–236 (2007)
4. Buchsbaum, A.L., Gansner, E.R., Procopiuc, C.M., Venkatasubramanian, S.: Rectangular layouts and contact graphs. ACM Transactions on Algorithms 4(1) (2008)
5. de Fraysseix, H., de Mendez, P.O., Rosenstiehl, P.: On triangle contact graphs. Combinatorics, Probability and Computing 3, 233–246 (1994)
6. de Fraysseix, H., Pach, J., Pollack, R.: How to draw a planar graph on a grid. Combinatorica 10(1), 41–51 (1990)
7. Debrunner, H.: Aufgabe 260. Elemente der Mathematik 12 (1957)
8. Felsner, S., Francis, M.C.: Contact representations of planar graphs with cubes. In: Proc. ACM Symposium on Computational Geometry (2011)
9. Gansner, E.R., Hu, Y.F., Kaufmann, M., Kobourov, S.G.: Optimal Polygonal Representation of Planar Graphs. In: López-Ortiz, A. (ed.) LATIN 2010. LNCS, vol. 6034, pp. 417–432. Springer, Heidelberg (2010)

10. Gansner, E.R., Hu, Y., Kobourov, S.G.: On Touching Triangle Graphs. In: Brandes, U., Cornelsen, S. (eds.) GD 2010. LNCS, vol. 6502, pp. 250–261. Springer, Heidelberg (2011)
11. Gonçalves, D., Lévêque, B., Pinlou, A.: Triangle Contact Representations and Duality. In: Brandes, U., Cornelsen, S. (eds.) GD 2010. LNCS, vol. 6502, pp. 262–273. Springer, Heidelberg (2011)
12. Hartman, I., Newman, I., Ziv, R.: On grid intersection graphs. Discrete Mathematics 97, 41–52 (1991)
13. Heilmann, R., Keim, D.A., Panse, C., Sips, M.: Recmap: Rectangular map approximations. In: 10th IEEE Symp. on Information Visualization (InfoVis 2004), pp. 33–40 (2004)
14. Hliněný, P.: Contact graphs of line segments are NP-complete. Discr. Math. 235, 95–106 (2001)
15. Koebe, P.: Kontaktprobleme der konformen Abbildung. Berichte über die Verhandlungen der Sächsischen Akademie der Wissenschaften zu Leipzig. Math.-Phys. Kl. 88, 141–164 (1936)
16. Koźmiński, K., Kinnen, E.: Rectangular duals of planar graphs. Networks 15, 145–157 (1985)
17. Lee, A., Streinu, I.: Pebble game algorithms and sparse graphs. Discrete Mathematics 308(8), 1425–1437 (2008)
18. Schnyder, W.: Embedding planar graphs on the grid. In: SODA, pp. 138–148 (1990)
19. Ungar, P.: On diagrams representing graphs. J. London Math. Soc. 28, 336–342 (1953)
20. van Kreveld, M.J., Speckmann, B.: On rectangular cartograms. Computational Geometry 37(3), 175–187 (2007)
21. Yeap, K.-H., Sarrafzadeh, M.: Floor-planning by graph dualization: 2-concave rectilinear modules. SIAM Journal on Computing 22, 500–526 (1993)

Embedding Plane 3-Trees in $\mathbb{R}^2$ and $\mathbb{R}^3$

Stephane Durocher[1,*], Debajyoti Mondal[1], Rahnuma Islam Nishat[2],
Md. Saidur Rahman[3], and Sue Whitesides[2,**]

[1] Department of Computer Science, University of Manitoba
[2] Department of Computer Science, University of Victoria
[3] Graph Drawing and Information Visualization Laboratory,
Department of Computer Science and Engineering,
Bangladesh University of Engineering and Technology
{durocher,jyoti}@cs.umanitoba.ca, {rnishat,sue}@cs.uvic.ca,
saidurrahman@buet.ac.bd

Abstract. A point-set embedding of a planar graph G with n vertices on a set P of n points in $\mathbb{R}^d$, $d \geq 1$, is a straight-line drawing of G, where the vertices of G are mapped to distinct points of P. The problem of computing a point-set embedding of G on P is NP-complete in $\mathbb{R}^2$, even when G is 2-outerplanar and the points are in general position. On the other hand, if the points of P are in general position in $\mathbb{R}^3$, then any bijective mapping of the vertices of G to the points of P determines a point-set embedding of G on P. In this paper, we give an $O(n^{4/3+\epsilon})$-expected time algorithm to decide whether a plane 3-tree with n vertices admits a point-set embedding on a given set of n points in general position in $\mathbb{R}^2$ and compute such an embedding if it exists, for any fixed $\epsilon > 0$. We extend our algorithm to embed a subclass of 4-trees on a point set in $\mathbb{R}^3$ in the form of nested tetrahedra. We also prove that given a plane 3-tree G with n vertices, a set P of n points in $\mathbb{R}^3$ that are not necessarily in general position and a mapping of the three outer vertices of G to three different points of P, it is NP-complete to decide if G admits a point-set embedding on P respecting the given mapping.

1 Introduction

A *plane graph* is a planar graph with a fixed planar embedding. A *straight-line drawing* of a plane graph G in $\mathbb{R}^d$, $d \geq 1$, is a planar drawing of G, where the vertices of G are drawn as points in $\mathbb{R}^d$ and edges of G are drawn as noncrossing straight line segments. Although two straight line segments meet at their common endpoints if their corresponding edges are adjacent, we do not consider such a meeting point to be a crossing point. Given a plane graph G with n vertices and a set P of n points in $\mathbb{R}^d$, a *point-set embedding* of G on P is a straight-line drawing of G, where each vertex of G is mapped to a distinct point of P. See Figure 1 for an illustration of point-set embeddings in $\mathbb{R}^2$ and $\mathbb{R}^3$.

[*] Work of the author is supported in part by the Natural Sciences and Engineering Research Council of Canada (NSERC).

[**] Work of the author is supported by the Natural Sciences and Engineering Research Council of Canada (NSERC) and the University of Victoria.

M. van Kreveld and B. Speckmann (Eds.): GD 2011, LNCS 7034, pp. 39–51, 2012.

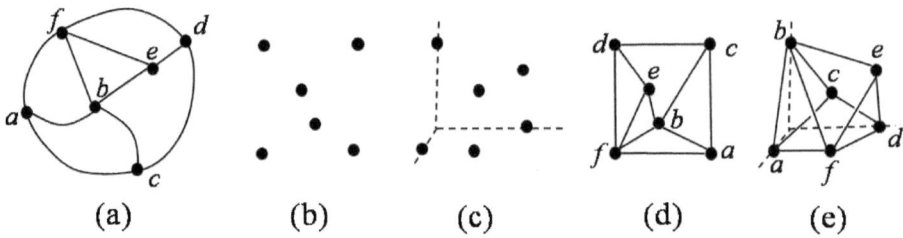

Fig. 1. (a) A plane graph G, (b) a set P of points in $\mathbb{R}^2$, (c) a set P' of points in $\mathbb{R}^3$, (d) a point-set embedding of G on P, and (e) a point-set embedding of G on P'

The problem of embedding planar graphs on fixed vertex locations has been studied for many years [1,4,8,9,12]. Every outerplanar graph with n vertices admits a point-set embedding on any set of n points in $\mathbb{R}^2$, where the points are in general position, i.e, no three points are collinear [4]. Bose et al. gave efficient algorithms to compute point-set embeddings of trees and outerplanar graphs in $O(n \log n)$-time [2] and $O(n \log^3 n)$-time [1], respectively. Recently, Nishat et al. [11] gave an $O(n^2 \log n)$-time algorithm that can decide if a plane 3-tree admits a point-set embedding on a given set of points in $\mathbb{R}^2$, even when the points are not in general position, and computes such an embedding if it exists. Although the point-set embeddability problem in $\mathbb{R}^2$ is polynomial-time solvable for outerplanar graphs and plane 3-trees, Cabello [3] proved that this problem is NP-complete for 2-outerplanar graphs, even when the given points are in general position. On the other hand, given a graph G with n vertices and a set P of n points in $\mathbb{R}^3$, where the points are in general position, i.e., no four points are coplanar, G always admits a point-set embedding on P.

In this paper, we give an $O(n^{4/3+\epsilon})$-expected time algorithm to compute a point-set embedding of a plane 3-tree with n vertices on a set of n points in $\mathbb{R}^2$ if such an embedding exists, for any fixed $\epsilon > 0$. We extend the algorithm to embed a subclass of 4-trees on a point set in $\mathbb{R}^3$ in the form of nested tetrahedra. We also prove that given a plane 3-tree G with n vertices, a set P of n points in $\mathbb{R}^3$ not necessarily in general position and a mapping of the three outer vertices of G to three points of P, it is NP-complete to decide whether G admits a point-set embedding on P for the given mapping of the outer vertices. This negative result is interesting since the problem is solvable in polynomial time in $\mathbb{R}^2$ [11]. Cabello [3] also asked: What is the complexity of the point-set embeddability problem for 3-connected plane graphs in $\mathbb{R}^2$? Since a plane 3-tree is 3-connected, our hardness result answers the analogous question for $\mathbb{R}^3$.

2   Preliminaries

In this section we give some definitions that will be used throughout the paper.

A plane graph divides the plane into connected regions called *faces*. The unbounded region is the *outer face* and all other faces are *inner faces*. The vertices on the outer face are *outer vertices* and all other vertices are *inner vertices*. A *triangular face* contains only three vertices on its boundary. If all the faces of a plane graph G are triangular, then G is a *triangulated plane graph*.

For a cycle C in G, $G(C)$ denotes the subgraph of G induced by the vertices inside and on the boundary of C. If a cycle contains only three vertices a, b, c on its boundary then we denote the cycle by C_{abc}. A graph G with $n \geq 3$ vertices is a *plane 3-tree* if it satisfies the following properties. (a) G is a triangulated plane graph. (b) If $n > 3$, then G has a vertex of degree three whose removal gives a plane 3-tree with $n - 1$ vertices.

Any plane 3-tree G has exactly one inner vertex p, which is the common neighbor of the outer vertices of G. We call p the *representative vertex* of G. Plane 3-trees are also known as Apollonian networks and stacked polytopes [6].

Let P be a set of points. We denote by $|P|$ the number of points in P. Let a, b and c be three points that do not necessarily belong to P. By $P(abc)$ we denote the points of P, which are on the boundary and inside of triangle abc.

3   Point-Set Embeddings of Plane 3-Trees in $\mathbb{R}^2$

In this section we give an $O(n^{4/3+\epsilon})$-expected time algorithm to embed a plane 3-tree with n vertices on a set of n points in general position in $\mathbb{R}^2$, where $\epsilon > 0$ is fixed.

Nishat et al. [11] gave an $O(n^2)$-time algorithm for computing a point-set embedding of a plane 3-tree with n vertices on a set of n points in general position. Recently, Moosa et al. [10] tried to give a faster algorithm for computing point-set embeddings of plane 3-trees using a range search data structure of Chazelle et al. [5]. Their algorithm takes $O(n^{4/3+\epsilon} \log n + n^{4/3+\epsilon} \log(l/s))$ time, where $\epsilon > 0$, l is the largest distance between any two points in the point-set and s is the distance between the closest pair of points. Consequently, finding an algorithm for computing point-set embeddings of plane 3-trees with improved running time, where the time complexity is only a function of n, was open.

Like Moosa et al. we also use the range search data structure of Chazelle et al. [5]. Using randomization, however, the expected running time of our algorithm is bounded by a function of n alone for any set of n points in general position in $\mathbb{R}^2$ and independent of the corresponding parameters l and s. Before describing our algorithm we need the following lemma.

Lemma 1. *Let abc be a triangle with a set P of $n > 0$ points in its proper interior, where the points are in general position and preprocessed to answer any triangular range counting query in $f(n)$ time. Let $k \leq n$ be a positive integer. Then in $O(f(n) \log n)$ expected time we can find a point q on bc such that $|P(abq)| = k$.*

Proof. We first set $x = b$ and $y = c$. We then execute the following steps.

Step 1. Randomly choose a point[1] $w \in P(axy)$. Let z denote the intersection point of xy and the line passing through a and w.

Step 2. If $|P(abz)| < k$, then set $x = z$ and go to Step 1. If $|P(abz)| > k$, then set $y = z$ and go to Step 1. Otherwise, set $q = z$.

It is straightforward to observe that Steps 1–2 correctly find the required point q. We now analyze the running time. Consider some iteration i of Steps 1–2. Let P_i be the points of $P(axy)$ at the beginning of the i-th iteration. Let X_j be the indicator random variable such that $X_j = 1$ if point $p_j \in P_i$ remains inside triangle axy after the i-th iteration, and $X_j = 0$ otherwise. Since any point p_j is removed from further consideration with probability $1/2$, therefore $E[X_j] = 1/2$. Consequently, the expected number of points that remains in axy after the i-th iteration is $E[X] = \sum_{\forall p_j \in P_i} E[X_j] = |P_i|/2$. Since at each iteration the number of points to consider is reduced by a factor of $1/2$, the expected number of iterations is $O(\log n)$. At each iteration, Steps 1–2 take $O(f(n))$-time. Therefore, the total expected running time is $O(f(n) \log n)$. □

Theorem 1. *Let G be a plane 3-tree with n vertices and let P be a set of n points in general position in $\mathbb{R}^2$. We can decide in $O(n^{4/3+\epsilon})$ expected time, for any fixed $\epsilon > 0$, whether G admits a point-set embedding on P and compute such an embedding if it exists.*

Proof. Let a, b and c be the three outer vertices of G and let p be the representative vertex of G. We use the following steps of Nishat et al. [11] to test and compute point-set embedding of G on P.

Step 1. Let C be the convex hull of P. If the number of points on the boundary of C is not exactly three, then G does not admit a point-set embedding on P.

Step 2. For the possible six different mappings of vertices a, b, c to the three points x, y, z on C, execute Step 3.

Step 3. Let n_1, n_2 and n_3 be the number of vertices of $G(C_{abp})$, $G(C_{bcp})$ and $G(C_{cap})$, respectively. Without loss of generality assume that the current mapping of a, b and c is to x, y and z, respectively. Find the unique mapping of the representative vertex p of G to a point $w \in P$ such that the triangles xyw, yzw and zxw properly contain exactly n_1, n_2 and n_3 points, respectively. If no such mapping of p exists, then G does not admit a point-set embedding on P for the

---

[1] A simplex range searching data structure based on partition trees and cutting trees (such as that of Chazelle et al. [5]) can be augmented to return a range selection query in $f(n)$ time without any asymptotic increase in space or preprocessing time. That is, each of the t distinct range selection queries on triangle pqr, where $t = |P(pqr)|$, returns a distinct element of $P(pqr)$. The ordering of elements is determined by the trees' internal structures; the specific order is unimportant, so long as there is a bijection between selection queries and elements returned for a given query triangle. By choosing a value uniformly at random in $\{1, 2, \ldots, t\}$ and retrieving the corresponding element using a range selection query, we can select a point $w \in P(pqr)$ at random.

current mapping of a, b, c to x, y, z; hence go to Step 2 for the next mapping. Otherwise, recursively compute point-set embeddings of $G(C_{abp}), G(C_{bcp})$ and $G(C_{cap})$ on $P(xyw), P(yzw)$ and $P(zxw)$, respectively.

The time complexity is dominated by the cost of Step 3 and the bottleneck is the recursive computation of the mappings of the representative vertices. It is straightforward to observe that the recurrence relation for the time taken in Step 3 is $T(n) = T(n_1) + T(n_2) + T(n_3) + \mathcal{T}$, where $\mathcal{T}$ denotes the time required to find the mapping of the representative vertex.

We speed up the mapping of the representative vertex as follows: We use a data structure to preprocess the points of P in $O(g(n))$ time to answer any triangular range reporting query in $O(f(n)+k)$ time and triangular range counting query in $O(f(n))$ time, where k is the number of points reported. Let the outer vertices a, b, c be mapped to points x, y, z, respectively, and let n_1, n_2 and n_3 be the number of vertices of $G(C_{abp}), G(C_{bcp})$ and $G(C_{cap})$, respectively. We need to find a mapping of p to a point $w \in P$ such that triangles xyw, yzw and zxw properly contain exactly n_1, n_2 and n_3 points, respectively. Without loss of generality assume that $n_2 \leq \min\{n_1, n_3\}$.

By Lemma 1, we find two points u and v on yz such that $P(xyu)=n_1 + 3$ and $P(xzv)=n_3 + 3$ in $O(f(n) \log n)$ time. It is straightforward to show that if $dist(z, v) > dist(z, u)$, then p does not have the required mapping. Otherwise, if p has the required mapping to a point $w \in P$, then $w \in P(xuv)$. Since $|P(xuv)|=O(n_2)$, we can enumerate all the points of $P(xuv)$ in $O(f(n) + n_2)$ time. For each point $q \in P(xuv)$, we check if $|P(xyq)|=n_1 + 3, |P(yzq)|=n_2 + 3$ and $|P(zxq)|=n_3+3$ in $O(f(n))$ time. Hence, $\mathcal{T} = O(f(n) \log n)+O(f(n)+n_2)+ O(n_2 \cdot f(n))$ and $T(n) = T(n_1) + T(n_2) + T(n_3) + O(\min\{n_1, n_2, n_3\}f(n) \log n)$. This recurrence solves to $T(n) = O(nf(n) \log^2 n)$.

For n points in $\mathbb{R}^d$, the data structure of Chazelle et al. [5] takes $g(n) = O(m^{1+\epsilon})$ preprocessing time and $f(n) = O(n^{1+\epsilon}/m^{1/d})$ time for range counting queries, where $n < m < n^d$ and $\epsilon > 0$. Here $d = 2$ and for the best bound, we choose $m = n^{4/3}$. We thus get $T(n) = (n^{4/3+\epsilon} \log^2 n)$ and $g(n) = O(n^{4/3+4\epsilon/3})$. Therefore, we need $O(n^{4/3+\epsilon'} \log^2 n)$ time in total, where $\epsilon' = 4\epsilon/3 > 0$.

Observe that for any $\epsilon' > 0$, $n^{4/3+\epsilon'} \log^2 n = O(n^{4/3+\epsilon''})$ for any $\epsilon'' > \epsilon'$.   □

4   Tetrahedral Embeddings of Tetrahedral 4-Trees

In this section we introduce tetrahedral 4-trees and extend Theorem 1 to $\mathbb{R}^3$.

Let a, b, c and d be four points in general position in $\mathbb{R}^3$. By $T(abcd)$ we denote the tetrahedron defined by points a, b, c and d. A *vertex insertion* operation on $T(abcd)$ places a vertex p interior to $T(abcd)$ and adds edges from p to a, b, c, d, such that $T(abcp), T(abdp), T(bcdp)$ and $T(cadp)$ define four new tetrahedra. By a *tetrahedral embedding* we denote a straight-line embedding formed by starting with a tetrahedron and then applying vertex insertion operations recursively on zero or more newly generated tetrahedra. A graph G with $n \geq 4$ vertices is a *tetrahedral 4-tree* if it admits a tetrahedral embedding. A *tetrahedral point-set embedding* of G on a set P of n points is a tetrahedral embedding of G, where the vertices of G are mapped to distinct points of P.

Let G be a tetrahedral 4-tree with n vertices. Then by definition, G satisfies the following properties.

(a) G is a 4-tree.

(b) Let Γ be a tetrahedral embedding of G. Then the convex hull of the points of Γ is a tetrahedron $T(s_1 s_2 s_3 s_4)$, where s_1, s_2, s_3, s_4 are the four points on the convex hull. By the *surface vertices* of G we denote the vertices u_1, u_2, u_3, u_4 of G that correspond respectively to the points s_1, s_2, s_3, s_4.

(c) If $n > 4$, then there exists a point p in Γ which is adjacent to the points s_1, s_2, s_3, s_4. By the *core vertex* of G we denote the vertex v that corresponds to p.

(d) Removal of v, u_1, u_2, u_3, u_4 splits G into four (possibly empty) components C_1, C_2, C_3 and C_4, respectively. Then the vertices of C_i along with $\{v, u_1, u_2, u_3, u_4\} \setminus \{u_i\}$ induce a tetrahedral 4-tree, which is placed inside $T(pabc)$ in Γ, where $\{a, b, c\} \subseteq \{\{s_1, s_2, s_3, s_4\} \setminus \{s_i\}\}$.

If G admits a tetrahedral point-set embedding on a given set of points in $\mathbb{R}^3$, then we can prove that the mapping of the core vertex is unique. Using the range search data structure of Chazelle et al. [5] we can preprocess the points in $O(n^{(1+\epsilon)9/4})$ time, where any triangular range counting query takes $O(n^{1/4+\epsilon})$ time, $\epsilon > 0$. Therefore, we can find the mapping of the core vertex in $O(n \cdot n^{1/4+\epsilon}) = O(n^{5/4+\epsilon})$ time. Since we need to find $O(n)$ such mappings in a recursive fashion, the total time required is $O(n^{9/4+\epsilon})$. We thus have the following theorem.

Theorem 2. *Let G be a tetrahedral 4-tree with n vertices and let P be a set of n points in general position in $\mathbb{R}^3$. We can decide in $O(n^{9/4+\epsilon})$ time, for any fixed $\epsilon > 0$, whether G admits a tetrahedral point-set embedding on P and compute such an embedding if it exists.*

5   Point-Set Embeddings of Plane 3-Trees in $\mathbb{R}^3$

Given a plane 3-tree G with n vertices, a set P of n points (not necessarily in general position) in $\mathbb{R}^2$ and a mapping for the outer vertices of G to three points in P, Nishat et al. [11] gave an $O(n^2 \log n)$-time algorithm for testing whether G admits a point-set embedding on P for the given mapping of the outer vertices. In this section we prove that the corresponding decision problem is NP-complete when the points are in $\mathbb{R}^3$. A formal definition of the problem is as follows:

Problem: Three Dimensional Point-Set Embedding (3DPSE)

Instance: A plane 3-tree G with n vertices, a set P of n points (not necessarily in general position) in $\mathbb{R}^3$ and a mapping of the three outer vertices of G to three different points in P.

Question: Does G admit a point-set embedding on P that respects the given mapping of the outer vertices?

We prove NP-hardness of 3DPSE by reduction from a strongly NP-complete problem 3-Partition [7], which is defined as follows.

Instance: A set of $3m$ nonzero positive integers $S=\{a_1, a_2, \ldots, a_{3m}\}$ and an integer $B > 0$, where $a_1+a_2+\ldots+a_{3m} = mB$ and $B/4 < a_i < B/2, 1 \leq i \leq 3m$.

Question: Can S be partitioned into m subsets $S_1, S_2, \ldots, S_m$ such that $|S_1| = |S_2| = \ldots = |S_m| = 3$ and the sum of the integers in each subset is equal to B?

Here is an outline of our proof for NP-hardness. For a given instance $\mathcal{I} = \{S, m, B\}$ of 3-PARTITION, we construct a point set $\mathcal{P}$, a plane 3-tree $\mathcal{G}$ and a mapping of the three outer vertices of $\mathcal{G}$ to the three points of $\mathcal{P}$. We prove that $\mathcal{G}$ admits a point-set embedding on $\mathcal{P}$ respecting the mapping of the outer vertices if and only if $\mathcal{I}$ has an affirmative answer.

We first assume that $\mathcal{I}$ has an affirmative answer, and then show a construction of a point-set embedding of $\mathcal{G}$ on $\mathcal{P}$ respecting the mapping of the outer vertices. The other direction of the claim is: if $\mathcal{G}$ admits the required embedding on $\mathcal{P}$, then $\mathcal{I}$ has an affirmative answer. We prove the contrapositive. We assume that $\mathcal{I}$ has a negative answer, and then prove that $\mathcal{G}$ does not admit a point-set embedding on $\mathcal{P}$ respecting the mapping of the outer vertices. To prove this, we show that the mapping of the outer vertices of $\mathcal{G}$ restricts some vertices of G to map onto some special points of $\mathcal{P}$. This mapping leaves m groups of B points unmapped, where the remaining vertices of $\mathcal{G}$ are to be mapped. These remaining vertices of $\mathcal{G}$ correspond to the integers in S. If $\mathcal{G}$ admits the required embedding on $\mathcal{P}$, then those remaining vertices admit a mapping to the unmapped groups of points. Each group corresponds to a subset of the solution of $\mathcal{I}$. Since we assumed that $\mathcal{I}$ has a negative answer, this gives a contradiction.

We now describe the formal reduction. Let m and B be two nonzero positive integers. We first define a set $\mathcal{P}_{m,B}$ of $2mB + 10m - 4$ points as follows:

(a) Two points p and r at $(0, 5, 4m)$ and $(mB+2(m-1), 0, 5m)$, respectively.
(b) The set P_z of $4m$ collinear points on line $x = y = 0$, where $P_z = \{(0,0,i)|0 \leq i \leq 4m - 1\}$. By q we denote the point at $(0, 0, 0)$.
(c) The set P_y of $mB + 2(m - 1)$ points on line $y - 1 = z = 0$, where $P_y = \{(i, 1, 0)|1 \leq i \leq mB + 2(m - 1)\}$.
(d) Points $P_u = \{u_1, u_2, \ldots, u_{m-1}\}$, where point $u_i, 1 \leq i \leq m-1$, is the intersection point of the plane $z=1$ with the line joining p and the midpoint of the line segment between $(i(B+2)-1, 1, 0)$ and $(i(B+2), 1, 0)$. See Figure 2(a).
(e) Points $P_v = \{v_1, v_2, \ldots, v_{m-1}\}$, where point $v_i, 1 \leq i \leq m-1$, is the intersection point of the plane $z=4m+1$ with the line joining r and point $u_i \in P_u$.
(f) Points $P_w = \{w_1, w_2, \ldots, w_{mB+2(m-1)}\}$, where point $w_i, 1 \leq i \leq mB + 2(m - 1)$, is the intersection point of the plane $z = 4m$ with the line joining r with point $p_i \in P_y$. See Figure 2(b).

Observe that $|P_z|=4m$, $|P_y|=mB+2(m-1)$, $|P_u|=m-1$, $|P_v|=m-1$ and $|P_w| = mB+2(m-1)$. Thus the number of points in $\mathcal{P}_{m,B}$ along with p, r is $2mB+10m-4$. We now have the following lemma.

Lemma 2. Let l_1 be a line segment joining points a and b, where $a \in P_y$ and $b \in P_z$. Let l_2 be another line segment joining points a' and b', where $a' \in P_y, b' \in P_z$ and $\{a', b'\} \neq \{a, b\}$. Then l_1 and l_2 do not cross.

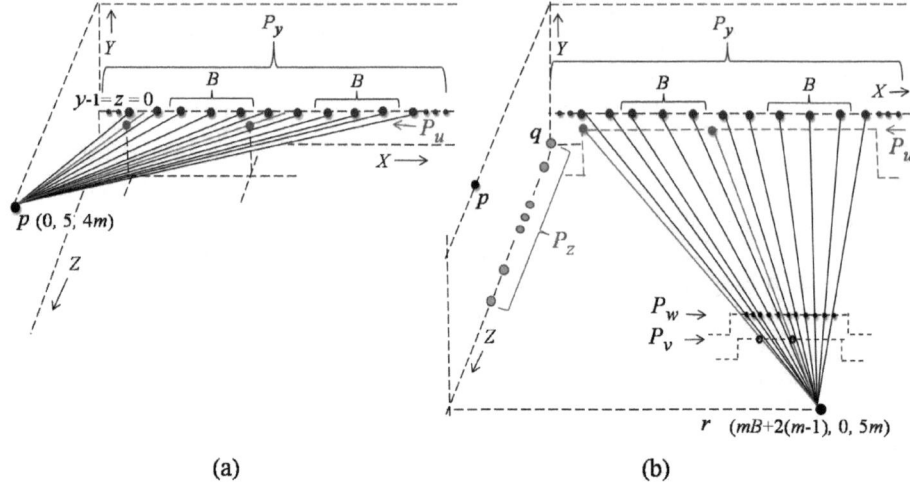

(a)   (b)

Fig. 2. $\mathcal{P}_{m,B}$, where sets P_u, P_z and P_y are shown in red, green and blue, respectively

Let $x_0, x_1, \cdots, x_n$ be a path of $n + 1$ vertices. We add two vertices l, r to the path by adding the edges $(l, x_i), (r, x_i)$, where $0 \leq i \leq n$. We call the resulting graph a *butterfly* and denote it by W_{n+1}. We call l, r the *wings of* W_{n+1} and path $x_0, x_1, \ldots, x_n$ the *spine of* W_{n+1}. We call x_0 and x_n the two *ends* of the spine. Figure 3(a) depicts a butterfly W_4. Let m and B be two nonzero positive integers and let $S = \{a_1, a_2, \ldots, a_{3m}\}$ be a set of $3m$ nonzero positive integers. We now construct a plane graph $\mathcal{G}_{m,B,S}$ with $2mB + 10m - 4$ vertices as follows:

1. Construct a butterfly W_{4m}. Let a and c be its wings. Add an edge between a and c. Any plane embedding Γ of W_{4m} keeping a and c on the outer face will have one end of the spine on the outer face, which we denote by b. Without loss of generality assume $b = x_0$. See Figure 3(b).

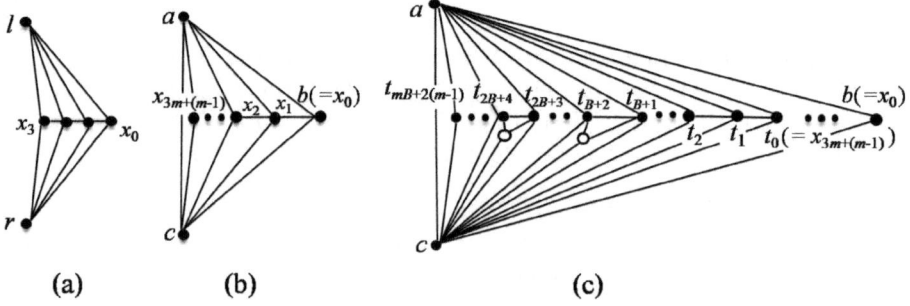

(a)   (b)   (c)

Fig. 3. (a) W_4, (b) W_{4m}, together with an edge (a, c) between the wings a and c, and (c) illustration for G', where the buds are shown by empty circles

2. Let the spine vertices of W_{4m} starting from b in Γ be $b(=x_0), x_1, \ldots, x_{3m+(m-1)}$. We now add a second butterfly $W_{mB+2m-1}$ in the triangular face $acx_{3m+(m-1)}$, where a, c are the wings of $W_{mB+2m-1}$ and $t_0(=x_{3m+(m-1)}), t_1, \ldots, t_{mB+2(m-1)}$ is the spine. Insert a vertex in each triangular face $ct_{i(B+2)-1}t_{i(B+2)}$, $1 \leq i \leq m-1$, and add three edges to connect the inserted point with $c, t_{i(B+2)-1}$ and $t_{i(B+2)}$. Let G' be the subgraph of $\mathcal{G}_{m,B,S}$ bounded by the triangular face $acx_{3m+(m-1)}$. We call each of these inserted vertices a *bud*. See Figure 3(c).
3. For each triangular face $ax_{3m+i}x_{3m+i-1}$, $1 \leq i \leq m-1$, in Γ, insert three vertices l_i, m_i, n_i inside that face and add edges $(l_i, m_i), (m_i, n_i), (n_i, l_i), (a, l_i)$, $(a, m_i), (a, n_i), (x_{3m+i}, l_i), (x_{3m+i}, n_i), (x_{3m+i-1}, n_i)$ avoiding crossing. See Figure 4(a). We call each of these inserted triples of vertices a *trigon*.
4. For each triangular face ax_ix_{i-1}, $1 \leq i \leq 3m$, in Γ, create a butterfly W_{a_i} inside that face with wings a and x_i. Then add an edge between x_{i-1} and one end of the spine of W_{a_i} avoiding crossing. See Figure 4(b). We denote all $W_{a_i}, 1 \leq i \leq 3m$, by *butterflies of* $\mathcal{G}_{m,B,S}$. The graph defined by the resulting embedding is $\mathcal{G}_{m,B,S}$. See Figure 4(c).

Note that $\mathcal{G}_{m,B,S}$ is an embedded plane graph (not necessarily a straight-line embedding). We used Γ only to define the plane embedding of $\mathcal{G}_{m,B,S}$.

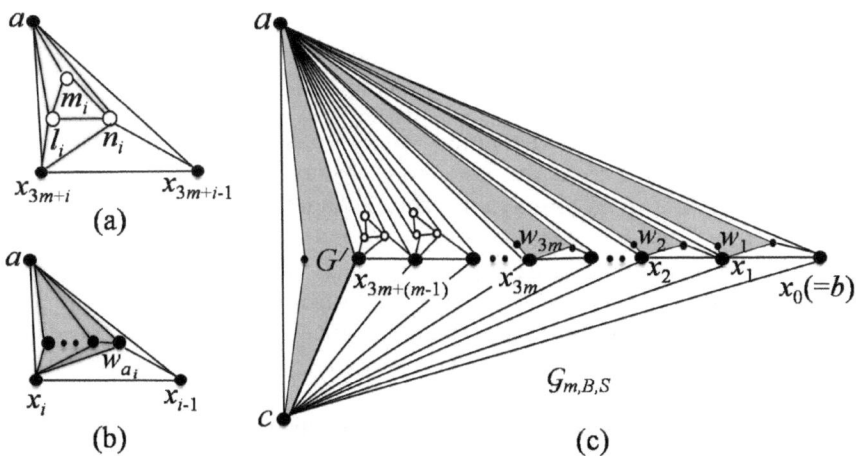

Fig. 4. (a) Insertion of a trigon, (b) illustration for W_{a_i}, and (c) $\mathcal{G}_{m,B,S}$, where vertices of the trigons are shown by empty circles

Observe that $x_0, x_1, \ldots, x_{3m+(m-1)}$ is a sequence of $|P_z|$ vertices of $\mathcal{G}_{m,B,S}$ and $t_1, t_2, \ldots, t_{mB+2(m-1)}$ is a sequence of $|P_w|$ vertices of $\mathcal{G}_{m,B,S}$. The number of buds in $\mathcal{G}_{m,B,S}$ is $|P_v|$ and the number of vertices in the trigons and spines of butterflies in $\mathcal{G}_{m,B,S}$ is $|P_y| + |P_u|$. Therefore, the number of vertices in $\mathcal{G}_{m,B,S}$ along with a, c is equal to the number of points in $\mathcal{P}_{m,B}$, i.e., $2mB + 10m - 4$. We now have the following lemma.

Lemma 3. $\mathcal{G}_{m,B,S}$ *is a plane 3-tree.*

We now use $\mathcal{P}_{m,B}$ and $\mathcal{G}_{m,B,S}$ to prove the following theorem.

Theorem 3. *3DPSE is NP-complete.*

Proof. Given a mapping of the vertices of a plane 3-tree G to the points of P, it is straightforward to check if the drawing determined by this mapping is a straight-line drawing of G in polynomial time. Therefore, the problem is in NP.

We now create an instance of 3DPSE from an instance $B, S=\{a_1, a_2, \ldots, a_{3m}\}$, of 3-PARTITION. We construct a point-set $\mathcal{P}_{m,B}$ and a plane 3-tree $\mathcal{G}_{m,B,S}$. For convenience we denote $\mathcal{P}_{m,B}$ and $\mathcal{G}_{m,B,S}$ by $\mathcal{P}$ and $\mathcal{G}$, respectively. Since 3-PARTITION is strongly NP-complete, i.e., it remains NP-complete even when B is bounded by a polynomial in m. Therefore, $\mathcal{G}$ has a polynomial number of vertices and $\mathcal{P}$ has a polynomial number of points. Furthermore, the coordinates of p are bounded by polynomials. Consequently, we can construct $\mathcal{P}$ and $\mathcal{G}$ in polynomial time. Recall the points p, q, r of $\mathcal{P}$ and vertices a, b, c of $\mathcal{G}$. We now ask whether $\mathcal{G}$ admits a point-set embedding on $\mathcal{P}$, where the vertices a, b and c are mapped respectively to the points p, q and r. In the following we prove that such a point-set embedding is possible if and only if the given instance of 3-PARTITION has an affirmative answer.

Case 1: The given instance of 3-PARTITION has an affirmative answer.

We construct a point-set embedding of $\mathcal{G}$ on $\mathcal{P}$, where the vertices a, b, c are mapped respectively to the points p, q, r, as follows:

1. Map the buds of G' to the points of P_v consecutively. Map the internal vertices of G' other than the buds of G' to the points of P_w consecutively. Since the points of P_v are visible from p and the points of P_v and P_w are visible from r, no two internal edges of G' cross. See Figure 5(a).

2. Map the vertices $b(= x_0), x_1, x_2, \ldots, x_{3m+(m-1)}$ to the points of P_z starting from $(0, 0, 0)$. The points of P_z are visible from points p and r since these visibilities are not occluded by the edges of G'. Therefore, we can draw the edges joining a and c to $b(= x_0), x_1, x_2, \ldots, x_{3m+(m-1)}$ without creating any crossing.

3. Map each trigon l_i, m_i, n_i of $\mathcal{G}$ respectively to the points $(i(B+2)-1, 1, 0), u_i, (i(B+2), 1, 0)$, where $u_i \in P_u$ and $1 \leq i \leq m-1$. Observe that the points of P_y and P_u are still visible from p. See Figure 5(b). Moreover, by Lemma 2, the edges joining points from P_z and P_y do not create any crossing. Therefore, we can draw the edges joining vertices $x_{3m}, x_{3m+1}, \ldots, x_{3m+(m-1)}$ and vertex a to the trigons without creating any crossing.

4. Observe that there are m groups of consecutive B points on P_y. Denote these groups by $B_1, B_2, \ldots, B_m$. Let $S_1, S_2, \ldots, S_m$ be the solution of the given instance of 3-PARTITION. Since each S_i, $1 \leq i \leq m$, contains three integers a_j, a_k and a_l that sum to B, we can map the spines of the corresponding three butterflies W_{a_j}, W_{a_k} and W_{a_l} to B_i. Observe that the points of B_i are visible to p. See Figure 5(b). Moreover, by Lemma 2, the edges joining

points from P_z and P_y do not create any crossing. Therefore, we can draw the edges joining vertices $x_0, x_1, \ldots, x_{3m}$ and a to the spine vertices of the butterflies without creating any edge crossing.

Case 2: The given instance of 3-PARTITION has a negative answer and hence the set S cannot be partitioned into m subsets, where each subset contains exactly three integers and the sum of the integers in each subset is equal to B.

In the following we prove that in this case $\mathcal{G}$ does not admit a point-set embedding on $\mathcal{P}$, where vertices a, b, c are mapped respectively to points p, q, r.

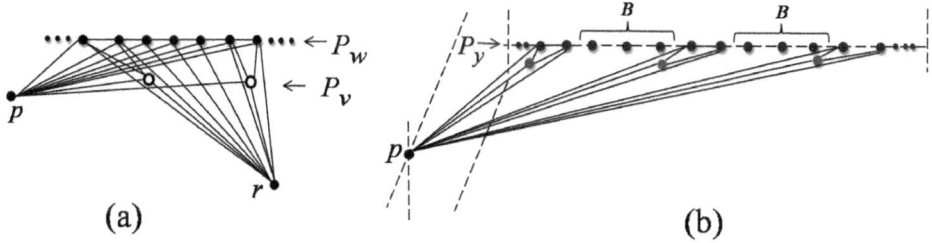

(a)   (b)

Fig. 5. Illustration for the proof of Theorem 3

1. Suppose for a contradiction that $\mathcal{G}$ admits a point-set embedding Γ on $\mathcal{P}$, where the vertices a, b and c are mapped respectively to the points p, q and r. We then claim that the trigons and spine vertices of the butterflies of $\mathcal{G}$ are mapped to the points of P_y and P_u in Γ. To justify the claim observe that c is not adjacent to the trigons and butterflies and the degree of c is $7m + mB - 2$. On the other hand, all the points of $\mathcal{P}$ other than the points of P_y and P_u are visible to r and there are $7m + mB - 2$ such points. Since c is mapped to r, the trigons and the spine vertices of the butterflies $W_{a_i}, 1 \leq i \leq 3m$, of $\mathcal{G}$ must be mapped to the points of P_y and P_u in Γ.
2. Recall that vertex a is mapped to point p, and the vertices of all trigons and butterflies are adjacent to a. There are $m-1$ trigons in $\mathcal{G}$ and $m-1$ points in P_u. Denote these trigons by $T_1, T_2, \ldots, T_{m-1}$. Since the points of P_y are collinear, one vertex of each trigon $T_j, 1 \leq j \leq m - 1$, will be mapped to a point of P_u. This mapping associates each trigon with a distinct point of P_u.
3. Let T_j be a trigon with three vertices l_j, m_j, n_j and without loss of generality assume that m_j is mapped to $u_i \in P_u$. We then claim that l_j and n_j must be mapped to the points $\{(i(B+2) - 1, 1, 0), (i(B+2), 1, 0)\}$ in Γ. Otherwise, assume that l_j or n_j is mapped to a point x, where $x \in P_y$ and $x \notin \{(i(B+2) - 1, 1, 0), (i(B+2), 1, 0)\}$. Then the edge xu_i must cross either the edge determined by $p, (i(B+2) - 1, 1, 0)$, or the edge determined by $p, (i(B+2), 1, 0)$. Therefore, the trigons must divide the points of $P_y \setminus \bigcup_{i=1}^{m-1}\{(i(B+2) - 1, 1, 0), (i(B+2), 1, 0)\}$ into m groups each containing consecutive B points. See Figure 5(b). Let these groups be $B_1, B_2, \ldots, B_m$. Consequently, the spine vertices of the butterflies must be mapped to these m groups in Γ.

4. Observe that the number of spine vertices in each butterfly is greater than $B/4$ and less than $B/2$. Therefore, four or more butterflies contain more than B spine vertices cumulatively, and hence cannot be mapped to a single B_i. Similarly, less than three butterflies contain less than B spine vertices cumulatively, and hence cannot cover the points of a single B_i. Therefore, each B_i must contain the spine vertices of exactly three butterflies in Γ and the corresponding three integers must sum to B. Consequently, if we form subsets $S_i, 1 \le i \le m$, where each S_i consists of three integers that correspond to B_i, then we can find m subsets $S_1, S_2, \ldots, S_m$, where the sum of the integers in each subset is equal to B.

Observe that subsets $S_1, S_2, \ldots, S_m$ correspond to a solution to the given instance of 3-PARTITION, which contradicts the assumption that the given instance has a negative answer.   □

6   Conclusion

In this paper we have given an $O(n^{4/3+\epsilon})$-expected time algorithm for computing point-set embeddings of plane 3-trees in $\mathbb{R}^2$. Since a planar 3-tree G has only a linear number of plane embeddings, we can check point-set embeddability for all the embeddings of G and determine whether G has a plane embedding on the given set of points in polynomial time. On the other hand, we have proved that this embeddability problem is NP-complete in $\mathbb{R}^3$, when a mapping for the outer vertices of the input graph is given and the given points are not necessarily in general position. The best known lower bound on time for computing point-set embeddings of plane 3-trees on the points in $\mathbb{R}^2$ is $\Omega(n \log n)$ [11]. Therefore, it would be interesting to find a faster algorithm as well as to improve the lower bound on the time required to find point-set embeddings of plane 3-trees in $\mathbb{R}^2$.

References

1. Bose, P.: On embedding an outer-planar graph in a point set. Computational Geometry: Theory and Applications 23(3), 303–312 (2002)
2. Bose, P., McAllister, M., Snoeyink, J.: Optimal algorithms to embed trees in a point set. Journal of Graph Algorithms and Applications 1(2), 1–15 (1997)
3. Cabello, S.: Planar embeddability of the vertices of a graph using a fixed point set is NP-hard. Journal of Graph Algorithms and Applications 10(2), 353–363 (2006)
4. Castañeda, N., Urrutia, J.: Straight line embeddings of planar graphs on point sets. In: Proc. of CCCG, pp. 312–318 (1996)
5. Chazelle, B., Sharir, M., Welzl, E.: Quasi-optimal upper bounds for simplex range searching and new zone theorems. Algorithmica 8(5&6), 407–429 (1992)
6. Demaine, E.D., Schulz, A.: Embedding stacked polytopes on a polynomial-size grid. In: Proc. of ACM-SIAM SODA, pp. 77–80 (2011)
7. Garey, M.R., Johnson, D.S.: Computers and intractability. Freeman, San Francisco (1979)

8. Giacomo, E.D., Didimo, W., Liotta, G., Meijer, H., Wismath, S.K.: Constrained point-set embeddability of planar graphs. International Journal of Computational Geometry and Applications 20(5), 577–600 (2010)
9. Kaufmann, M., Wiese, R.: Embedding vertices at points: Few bends suffice for planar graphs. Journal of Graph Algorithms and Applications 6(1), 115–129 (2002)
10. Moosa, T.M., Rahman, M.S.: Improved algorithms for the point-set embeddability problem for plane 3-trees. CoRR abs/1012.0230 (2010), http://arxiv.org/abs/1012.0230
11. Nishat, R.I., Mondal, D., Rahman, M. S.: Point-set embeddings of plane 3-trees. In: Brandes, U., Cornelsen, S. (eds.) GD 2010. LNCS, vol. 6502, pp. 317–328. Springer, Heidelberg (2011)
12. Pach, J., Wenger, R.: Embedding planar graphs at fixed vertex locations. Graphs and Combinatorics 17(4), 717–728 (2001)

Orthogeodesic Point-Set Embedding of Trees[*]

Emilio Di Giacomo[1], Fabrizio Frati[2,3,**], Radoslav Fulek[2,**],
Luca Grilli[1], and Marcus Krug[4]

[1] Dip. di Ingegneria Elettronica e dell'Informazione, Universitá degli Studi di Perugia, Italy
{digiacomo,grilli}@diei.unipg.it
[2] School of Basic Sciences, École Polytechnique Fédérale de Lausanne, Switzerland
{fabrizio.frati,radoslav.fulek}@epfl.ch
[3] School of Information Technologies, University of Sydney
[4] Institute of Theoretical Informatics, Karlsruhe Institute of Technology, Germany
marcus.krug@kit.edu

Abstract. Let S be a set of N grid points in the plane, and let G a graph with n vertices ($n \leq N$). An *orthogeodesic point-set embedding* of G on S is a drawing of G such that each vertex is drawn as a point of S and each edge is an orthogonal chain with bends on grid points whose length is equal to the Manhattan distance. We study the following problem. Given a family of trees $\mathcal{F}$ what is the minimum value $f(n)$ such that every n-vertex tree in $\mathcal{F}$ admits an orthogeodesic point-set embedding on every grid-point set of size $f(n)$? We provide polynomial upper bounds on $f(n)$ for both planar and non-planar orthogeodesic point-set embeddings as well as for the case when edges are required to be L-shaped chains.

1   Introduction

Let S be a set of N points in the plane, and let G be an n-vertex graph such that $n \leq N$. A *point-set embedding* of G on S is a drawing of G such that each vertex is drawn as a point of S. Point-set embeddings are a classical subject of investigation in graph drawing from both an algorithmic and a combinatorial point of view. Different types of point-set embeddings have been defined depending on the desired type of drawing.

Several algorithmic results are known on point-set embeddings in which edges are required to be straight-line segments. Deciding whether a planar graph admits a straight-line planar point-set embedding on a given point set is an NP-complete problem [5], while straight-line planar point-set embeddings of trees [3] and outerplanar graphs [2] can be computed efficiently. From the combinatorial perspective, Gritzmann et al. [12] prove that an n-vertex planar graph admits a straight-line planar point-set embedding on every set of n points in general position if and only if it is outerplanar. Kaufmann and Wiese show that every n-vertex planar graph admits a polyline planar point-set embedding on every set of n points with at most 2 bends per edge [14]. Colored versions of planar polyline point-set embeddings have also been investigated [1,7]. Special research efforts have been devoted to study *universal point sets* for planar graphs. A point set S is *universal* for a family $\mathcal{F}$ of graphs and for a type $\mathcal{D}$ of drawing if every

---

[*] Initiated during the "Bertinoro Workshop on Graph Drawing", Bertinoro, Italy, March 2011.
[**] Supported by the Swiss National Science Foundation Grant No.200021-125287/1.

M. van Kreveld and B. Speckmann (Eds.): GD 2011, LNCS 7034, pp. 52–63, 2012.

Table 1. Upper bounds on the value $f(n)$ obtained in this paper

	Planar L-Shaped	Non-Planar L-Shaped	Planar	Planar 2-spaced
Caterpillars $\Delta = 3$	n [Th. 8]	n [Th. 8]	n [Th. 8]	n [Th. 1]
Trees $\Delta = 3$	$n^2 - 2n + 2$ [Th. 6]	n [Th. 10][a]	n [Th. 3]	n [Th. 1]
Caterpillars $\Delta = 4$	$3n - 2$ [Th. 7]	$n + 1$ [Th. 11]	$\lfloor 1.5n \rfloor$ [Th. 4]	n [Th. 1]
Trees $\Delta = 4$	$n^2 - 2n + 2$ [Th. 6]	$4n - 3$ [Th. 9]	$4n$ [Th. 2]	n [Th. 1]

[a] Fink et al. [10] have independently obtained this result.

graph in $\mathcal{F}$ admits a point-set embedding of type $\mathcal{D}$ on S. Every universal point set for straight-line planar drawings of planar graphs has size at least $1.235 \cdot n$ [15] while there exist universal point sets of size $\frac{8}{9} n^2$ [4]. Universal point sets of size n exist for polyline drawings of planar graphs [9].

In this paper we study *orthogeodesic point-set embeddings* on the grid. Orthogeodesic point-set embeddings were introduced by Katz et al. [13] and require edges to be represented by *orthogeodesic chains*, i.e. by orthogonal chains whose total length is equal to the Manhattan distance between the endpoints. Since orthogeodesic chains correspond to shortest orthogonal connections in the L_1 metric, they can be considered as the counter part of straight lines in the L_2 metric. Katz et al. [13] considered orthogeodesic point-set embeddings from the algorithmic side and proved that it is NP-complete to decide whether an n-vertex planar graph with maximum degree 4 admits an orthogeodesic point-set embedding on n points, while the problem can be solved efficiently for cycles. Katz et al. [13] also show that, if the mapping between vertices and points is given and the bends are required to be at grid points, then the problem is NP-complete even for matchings, while the problem is polynomial-time solvable if bends need not be at grid points. Bi-colored planar orthogeodesic point-set embeddings have been studied by Di Giacomo et al. [6].

We consider orthogeodesic point-set embeddings *on the grid* from the combinatorial point of view. Let P be a set of grid points in the plane, i.e., $p = (i, j)$ with $i, j \in \mathbb{Z}$ for all $p \in P$. We write $x(p) := i$ and and $y(p) := j$. A set P of grid points with $x(p) \neq x(q)$ and $y(p) \neq y(q)$ for all $p, q \in P$ with $p \neq q$ is called *general*. For different classes of trees $\mathcal{F}$ and different drawing styles $\mathcal{D}$ we study the value $f(n)$ such that *every general pointset is universal for orthogeodesic point-set embeddings of all trees in $\mathcal{F}$ using $\mathcal{D}$*. The restriction to general point sets is necessary since there are arbitrarily large point sets that are not universal for orthogeodesic point-set embeddings of trees, e.g., a set of collinear points. We consider both planar and non-planar orthogeodesic point-set embeddings as well as the case when edges can be arbitrary orthogeodesic chains or when are edges required to be L-shaped chains. An *L-shaped chain* is an orthogonal chain with only one bend, thus, it is an orthogeodesic chain with the minimum number of bends for general point sets. Table 1 summarizes our results.

The rest of the paper is organized as follows. In Sects. 2, 3, and 4 we study planar, planar L-shaped, and non-planar L-shaped orthogeodesic point-set embeddings, respectively. Sect. 5 concludes and lists some open problems.

2   Planar Orthogeodesic Pointset Embeddings

In this section we consider planar orthogeodesic point-set embeddings of trees. First, we show that every tree with maximum degree 4 can be embedded on every general point set with n points with at most two bends per edge, if we require that the horizontal and vertical distance of any two points is at least two. We call point sets with this property *2-spaced*. This implies that we can embed every tree with n vertices on every general point set P with n points whose points are not horizontally or vertically aligned, if neither vertices nor bends are required to be grid points.

Theorem 1. *Every tree with n vertices and with maximum degree 4 admits a planar orthogeodesic point-set embedding on every general point set P with n points such that $\min\{|x(p) - x(q)|, |y(p) - y(q)|\} \geq 2$ for all $p, q \in P$ with $p \neq q$.*

Proof: Let T be any tree with n vertices and maximum degree 4. Root T at any node r of degree at most 3. We prove that T admits a planar orthogeodesic point-set embedding on every general point set P with n points in which: (i) each edge has two bends and (ii) no edge intersects a half-line arbitrarily chosen among the two horizontal and two vertical half-lines starting at r.

The statement is trivially true for $n = 1$. We inductively prove that T admits the required embedding for the case that no edge may intersect the horizontal half-line starting at r and directed rightward (the other constructions are analogous). Let $n_1 \geq 0$, $n_2 \geq 0$, and $n_3 \geq 0$ denote the number of vertices in the subtrees T_1, T_2, and T_3 rooted at children r_1, r_2, and r_3 of the root r of T, respectively. Refer to Fig. 1. Let P_1 denote the set of the n_1 bottommost points of P. Let P_2 denote the set of the n_2 leftmost points of $P \setminus P_1$. Let p be the bottommost point of $P \setminus (P_1 \cup P_2)$. Let $P_3 = P \setminus (P_1 \cup P_2 \cup \{p\})$. Embed r on p. Inductively embed T_i on P_i ($i = 1, 2, 3$) with no edge intersecting the vertical half-line starting at r_1 directed upward. Connect r with r_1 by an orthogeodesic edge vertically attached to r and to r_1 and having an intermediate segment s on the horizontal line one unit above the top side of the bounding box of P_1. Connect r with r_2 and r_3 analogously (see Fig. 1). It is easy to see that the constructed embedding is planar and that no edge intersects the horizontal half-line starting at r and directed rightward. Since $\min\{|x(p) - x(q)|, |y(p) - y(q)|\} \geq 2$ for all $p, q \in P$ and since the intermediate segment s occupies a grid line one unit above a point, this grid line does not contain any point from P.   □

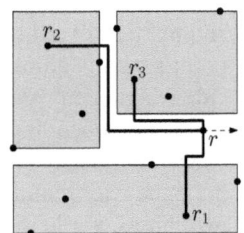

Fig. 1. Planar orthogeodesic point-set embedding of a tree on a general 2-spaced point set

As a consequence of Theorem 1 we obtain the following theorem for general point sets without the restriction on the horizontal and vertical distance of the points.

Theorem 2. *Every tree with n vertices and with maximum degree 4 admits a planar orthogeodesic point-set embedding on every general point set with $4n$ points.*

Proof: We prove that any set P of $4n$ points contains a subset of n points such that no two points have a horizontal or vertical distance of less than two. The theorem then

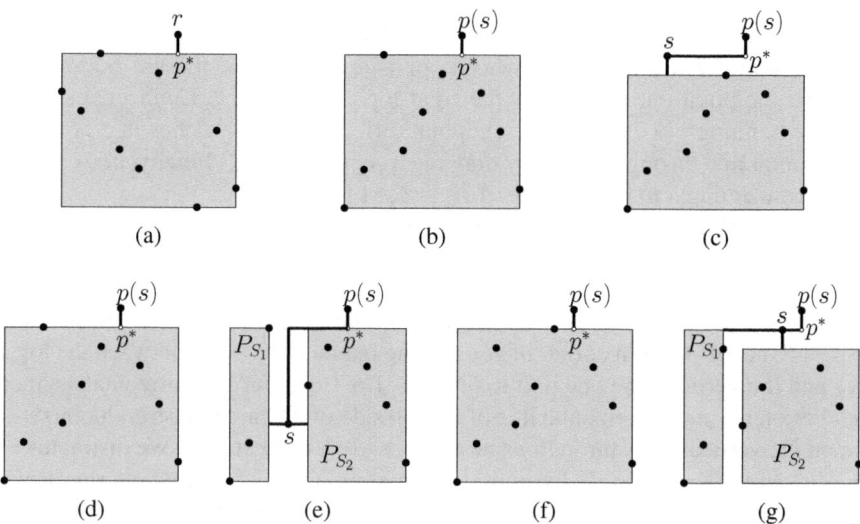

Fig. 2. Embedding a tree with maximum degree 3 on a set of n points. (a) Embedding r. (b)–(c) Embedding s with exactly one child. (d)–(g) Embedding s with two children.

directly follows from Theorem 1. Let the points in P be $p_1, \ldots, p_{4n}$ sorted from left to right. Let P_2 consist of the points p_{2i} ($1 \le i \le 2n$) and let $P_1 = P \setminus P_2$. Clearly, the points in P_1 and P_2 have the desired horizontal spacing and one of the sets, say P_1 must contain at least $2n$ points. Repeating the argument for P_1 vertically yields the claim. $\square$

For trees with maximum degree 3, we can improve this result.

Theorem 3. *Every tree with n vertices and with maximum degree 3 admits a planar orthogeodesic point-set embedding on every general point set with n points.*

Proof: Let T be any tree of degree 3 and let P be any general point set with n points. Root T at any leaf r. Construct a point-set embedding of T on P as follows. First, embed r on the topmost point p_t of P and assign to the subtree $T' = T \setminus \{r\}$ point set $P \setminus \{p_t\}$ and an axis-parallel rectangle $R_{T'}$ whose opposite corners are the left-bottom corner of the bounding-box of P and the point one unit below the right-top corner of the bounding-box of P. Connect r with the top border of $R_{T'}$ by drawing a vertical segment from p_t to the point p^* one unit below p_t (see Fig. 2a).

Second, we traverse T top-down. At each step we suppose that a point set P_S and an axis-parallel rectangle R_S have been assigned to a subtree S of T with root s so that the following invariants are satisfied: (i) $|S| = |P_S|$; (ii) P_S lies inside R_S; (iii) the parent $p(s)$ of s lies outside R_S and a horizontal or vertical segment $\overline{p(s), p^*}$ has been drawn connecting $p(s)$ to a point p^* on the border of R_S; (iv) let S and S' be two subtrees of T; if S is contained in S', then R_S is contained inside $R_{S'}$, if S' is contained in S, then $R_{S'}$ is contained inside R_S, and if neither S is contained in S' nor S' is contained in S, then $R_S \cap R_{S'} = \emptyset$. Suppose that p^* is on the top side of R_S; the cases in which p^* is on the bottom, left, or right side of R_S can be discussed analogously.

If s has exactly one child s_1, then denote by S_1 the subtree of T rooted at s_1. Refer to Figs. 2b–2c. Embed s on the topmost point p_t of P_S, assign to S_1 the point set $P_S \setminus \{p_t\}$ and the rectangle R_{S_1} whose opposite corners are the left-bottom corner of R_S and the point one unit below the right-top corner of R_S. Connect $p(s)$ to s by possibly extending $\overline{p(s), p^*}$ until its endpoint different from $p(s)$ lies in a point p_h on the horizontal line through p_t and by drawing a segment $\overline{p_t, p_h}$. Finally, draw a vertical segment connecting s to the top side of R_{S_1}. See Fig. 2c.

If s has two children s_1 and s_2 that are roots of subtrees S_1 and S_2, respectively, then denote by P_{S_1} (resp. by P_{S_2}) the point set composed of the leftmost $|S_1|$ points of P_S (resp. the rightmost $|S_2|$ points of P_S). Denote by p the only point of P_S that is neither in P_{S_1} nor in P_{S_2}. Assign to S_1 the point set P_{S_1} and the rectangle R_{S_1} whose opposite corners are the left-bottom corner of R_S and the intersection point between the top side of R_S and the vertical line one unit to the left of p. Consider the horizontal segment h which lies on the same horizontal line of the top side of P_S and whose endpoints p_1 and p_2 lie on the vertical lines through p and through $p(s)$, respectively. We distinguish two cases. *(1)* In the first case, h does not contain any point of P_S in its interior. Refer to Figs. 2d–2e. Embed s on p; assign to S_2 the point set P_{S_2} and the rectangle R_{S_2} whose opposite corners are the right-bottom corner of R_S and the intersection point between the top side of R_S and the vertical line one unit to the right of p. Connect $p(s)$ to s with an edge composed of $\overline{p(s), p^*}$, of a segment between p^* and the intersection point p' between the top side of R_S and the vertical line through p, and of segment $\overline{p', p}$. Finally, draw a horizontal segment connecting s with the right side of R_{S_1} and draw a horizontal segment connecting s with the left side of R_{S_2}. *(2)* In the second case, h contains a point $t(P_S)$ in its interior. Refer to Figs. 2f–2g. Embed s on $t(P_S)$; assign to S_2 the point set $P_{S_2} \setminus \{t(P_S)\} \cup \{p\}$ and the rectangle R_{S_2} whose opposite corners are the right-bottom corner of R_S and the intersection point between the horizontal line one unit below the top side of R_S and the vertical line through p. Connect $p(s)$ to s with an edge composed of $\overline{p(s), p^*}$ and of segment $\overline{p^*, t(P_S)}$. Finally, draw a horizontal segment connecting s with the right side of R_{S_1} and draw a vertical segment connecting s with the top side of R_{S_2}.

The only drawn edge $(p(s), s)$ is an orthogeodesic edge. Hence, the resulting drawing is an orthogeodesic point-set embedding of T on P. Moreover, it is easy to see that the invariants are maintained at every subtree of T and that such invariants imply the planarity of the point-set embedding. The statement of the theorem follows. $\square$

A *caterpillar* is a tree such that by removing all leaves we are left with a path, called *spine*. In Theorem 2 we show that every tree with maximum degree 4 has a planar orthogeodesic point-set embedding on every general point set with $4n$ points. For caterpillars with maximum degree 4, however, this result is not tight.

Theorem 4. *Every caterpillar with n vertices and with maximum degree 4 admits a planar orthogeodesic point-set embedding on every general point set with $\lfloor 1.5n \rfloor$ points.*

Proof: Let C be any caterpillar with n vertices and degree 4 and let n_i denote the number of vertices of C with degree $i = 1, \ldots, 4$. Let P^* be any point set with $\lfloor 1.5n \rfloor$ points. From P^* we arbitrarily choose a point set P of size $N = n + n_3 + n_4$ points on which we embed C. First, we show that $N \le 1.5n$, which implies $N \le \lfloor 1.5n \rfloor$ since N is a natural number. Suppose for contradiction that $n_3 + n_4 > n/2$. Since each vertex with degree at least 3 is incident to a leaf this yields $n_1 \ge n_3 + n_4$. Summing up we have $n \ge n_1 + n_3 + n_4 \ge 2(n_3 + n_4) > n$, a contradiction.

Next, we show how to embed C on P. Each vertex $v \in V$ is mapped to a point $\pi(v) \in P$. Let $S = (u_1, \ldots, u_k)$ be the spine of C. Remove from C all the leaves, except for one leaf u_0 incident to u_1 and one leaf u_{k+1} incident to u_k. Denote by S^+ the path $(u_0, u_1, \ldots, u_k, u_{k+1})$. For $i = 1, \ldots, k$, consider node u_i. If u_i has two adjacent leaves not in S^+, label one of them by "top" and one of them by "bottom"; if u_i has one adjacent leaf not in S^+, arbitrarily label it by "top" or by "bottom". Let B and T be the sets of leaves of C that have been labeled by bottom and by top, respectively.

Let P_T be the subset of the highest $|T|$ points of P and let P_B be the subset of the lowest $|B|$ points. Further, let $Q = P \setminus (P_T \cup P_B)$ be the remaining points. By construction Q contains $t = n_2 + 2(n_3 + n_4) + 2$ points. We embed C on P as follows: *(S1)* The leaves in T will be embedded on P_T, the leaves in B will be embedded on P_B and the vertices in S^+ will be embedded on a subset $P_{S^+} \subseteq Q$. *(S2)* The spine will be embedded as an x-monotone chain such that u_i is left of u_{i+1} for all $0 \le i \le k$. *(S3)* Edge $\{u_i, u_{i+1}\}$ occupies the horizontal segment incident to u_i on the right for all $0 \le i \le k$. If, additionally, the degree of u_i is at least 3, then edge $\{u_{i-1}, u_i\}$ occupies the horizontal segment incident to u_i on the left for all $1 \le i \le k$.

Let $q_1, \ldots, q_t$ be the points in Q sorted from left to right. First, we map u_0 to the leftmost point q_1 in Q. Suppose, we have mapped $u_0, \ldots, u_i$ for some $i < k + 1$ and let $q_j = \pi(u_i)$. If u_{i+1} has degree 2, then we map u_{i+1} to q_{j+1} and we connect u_i and u_{i+1} by an L-shaped orthogeodesic chain composed of a horizontal segment incident to u_i and a vertical segment incident to u_{i+1}. See Figs. 3a and 3b. If u_{i+1} has degree at least 3, then we map u_{i+1} to q_{j+2} skipping the point q_{j+1} in Q and we connect u_i by an orthogeodesic chain consisting of two horizontal segments incident to u_i and u_{i+1}, respectively, and a vertical segment in the column to the left of q_{j+2}. See Figs. 3c and 3d. By construction, u_{k+1} is mapped to a point q_j such that $j \le n_2 + 2(n_3 + n_4) + 2$ since we only skipped points for vertices with degree at least 3.

Now we describe how to embed the leaves in T on P_T. The leaves in B are embedded on P_B analogously. Let $w_1, \ldots, w_{|T|}$ be the vertices in T sorted such that their corresponding vertices on the spine are sorted from left to right and let T_i be the set of vertices in T that are incident to vertices u_j for $j < i$. For each i let P_i^- be the set of points in P_T to the left of $\pi(u_i)$ and let P_i^+ be the set of points in P_T to the right of $\pi(u_i)$, respectively. Each leaf w_i is mapped to a point $\pi(w_i)$ and is attached to the spine by an L-shaped orthogeodesic chain. We maintain the following invariant: *(L1)* If w_i is incident to u_j and $|P_i^-| > |T_j|$, then w_i is mapped to the lowest point $p \in P_i^- \setminus \bigcup_{l=1}^{i-1} \{\pi(w_l)\}$ by an L-shaped orthogeodesic chain consisting of the vertical segment incident to $\pi(u_j)$ and the horizontal segment incident to p. Otherwise, w_i is mapped to the highest unused point in $P_i^+ \setminus \bigcup_{l=1}^{i-1} \{\pi(w_l)\}$. See Figure 3e. The resulting point-set embedding is orthogeodesic by construction. Planarity follows from the invariants as follows.

Due to invariants (S1) and (S2) the spine is mapped to an x-monotone chain such that the angle at vertices with degree at least 3 is 180 degrees. This implies that the spine

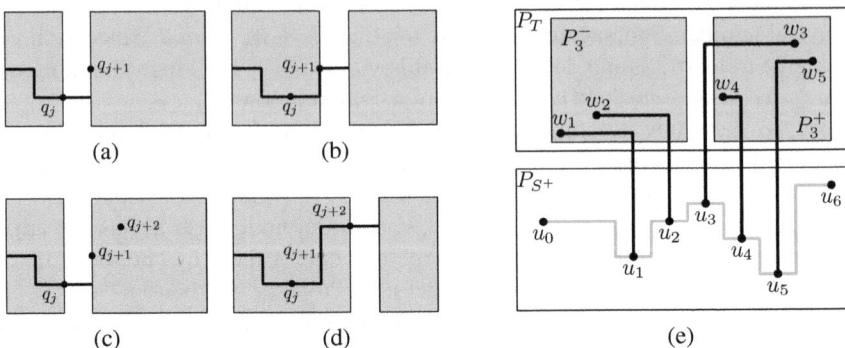

Fig. 3. Embedding a caterpillar on a set of $\lfloor 1.5n \rfloor$ points. (a)–(d) Embedding the spine S^+. (e) Embedding the leaves in T.

does not cross itself and that the vertical segments incident to the vertices with degree at least 3 are unoccupied by the spine. Since by invariant (S1) we attached the leaves in T above the spine and the leaves in B below the spine, there cannot be a crossing between two edges incident to a leaf in T and a leaf in B, respectively. Suppose for contradiction that there is a crossing between two edges e_i and e_j incident to two leaves w_i and w_j in T, respectively. Without loss of generality we assume $i < j$. If $\pi(w_i) \in P_i^-$ and $\pi(w_j) \in P_j^+$ there cannot be a crossing by construction. If $\pi(w_i) \in P_i^- \subseteq P_j^-$ and $\pi(w_j) \in P_j^-$, then a crossing can only occur if $\pi(w_j) \in P_i^-$ and $\pi(w_j)$ is below $\pi(w_i)$, which contradicts invariant (L1). Analogously, if $\pi(w_i) \in P_i^+$ and $\pi(w_j) \in P_j^+ \subseteq P_i^+$, then a crossing can only occur if $\pi(w_i) \in P_i^+$ and $\pi(w_i)$ is below $\pi(w_j)$, which contradicts invariant (L1). Finally, if $\pi(w_j) \in P_i^-$ and $\pi(w_i) \in P_j^+ \subseteq P_i^+$, then this contradicts invariant (L1), since w_i is only mapped to a point in P_i^+ if there is no unused point in P_i^-. Therefore, the embedding is crossing-free, which concludes the proof. $\square$

3   Planar L-Shaped Orthogeodesic Pointset Embeddings

Next, we consider planar L-shaped orthogeodesic point-set embeddings of trees. First, we prove that every tree with n vertices and with maximum degree 4 admits a planar L-shaped point-set embedding on every general point set with $n^2 - 2n + 2$ points. Every point set of this size contains a *diagonal* point set, which is universal for planar L-shaped point-set embeddings of trees with maximum degree 4. Let P be a point set and let $p_1, \ldots, p_n$ be the points in P ordered by increasing x-coordinates. Then P is diagonal if $y(p_{i+1}) > y(p_i)$ for every $i = 1, \ldots, n-1$ (then P is a *positive-diagonal point set*), or if $y(p_{i+1}) < y(p_i)$ for every $i = 1, \ldots, n-1$ (then P is a *negative-diagonal point set*). We have the following:

Theorem 5. *Every tree with n vertices and with maximum degree 4 admits a planar L-shaped point-set embedding on every diagonal point set with n points.*

Proof: We prove by induction a stronger statement. Let T be any tree with n vertices and with maximum degree 4. Root T in a vertex r of degree at most 3. We prove that T admits a planar L-shaped point-set embedding on every diagonal point set with n points with the further property that there is no edge overlapping or crossing a half-line arbitrarily chosen among the two horizontal half-lines and the two vertical half-lines starting at r.

In the base case $n = 1$ and the statement is trivially true. In the inductive case, we prove that T admits a planar L-shaped point-set embedding on every diagonal point set P with n points with the further property that no edge overlaps or crosses the vertical half-line starting at r and directed upward (the other constructions are analogous). We also suppose that P is a positive-diagonal point set, the case in which it is a negative-diagonal point set is analogous. Let $n_1 \geq 0$, $n_2 \geq 0$, and $n_3 \geq 0$ denote the number of vertices in the subtrees T_1, T_2, and T_3 rooted at the children r_1, r_2, and r_3 of r, respectively. Let P_1, P_2, and P_3 be the point sets consisting of the bottommost n_1 points of P, of the bottommost n_2 points of $P \setminus P_1$, and of the topmost n_3 points of P, respectively. Let p be the only point of P not in P_1, not in P_2, and not in P_3. Embed r on p. Inductively construct a non-planar L-shaped point-set embedding of T_1 on P_1 (resp. of T_2 on P_2, resp. of T_3 on P_3) such that no edge overlaps or crosses the vertical line through r_1 directed upward (resp. the horizontal line through r_2 directed rightward, resp. the vertical line through r_3 directed downward). Connect r with r_1 (resp. with r_2, resp. with r_3) by an L-shaped edge horizontally attached to r and vertically attached to r_1 (resp. vertically attached to r and horizontally attached to r_2, resp. horizontally attached to r and vertically attached to r_3). Since the embeddings of T_1 on P_1, of T_2 on P_2, and of T_3 on P_3 are L-shaped, all the edges of T_1, all the edges of T_2, and all the edges of T_3 lie inside the bounding boxes of P_1, of P_2, and of P_3, respectively. Hence, the vertical half-line through r directed upward has no overlapping or crossing edge, completing the induction. □

According to the Erdős-Szekeres theorem [8], every general point set with $n^2 - 2n + 2$ points contains either a positive-diagonal point set with n points or a negative-diagonal point set with n points. Hence, from Theorem 5 we have the following theorem.

Theorem 6. *Every tree with n vertices and with maximum degree 4 admits a planar L-shaped point-set embedding on every general point set with $n^2 - 2n + 2$ points.*

For caterpillars with maximum degree 4 we can improve the bound of Theorem 6 as follows:

Theorem 7. *Every caterpillar with n vertices and with maximum degree 4 admits a planar L-shaped point-set embedding on every general point set with $3n - 2$ points.*

Proof: We prove by induction a stronger statement. Let C be any caterpillar with n vertices and with maximum degree 4. Let $(u_2, \ldots, u_{k-1})$ be the spine of C. Let u_1 and u_k be two leaves of C adjacent to u_2 and to u_{k-1}, respectively. For any $i = 1, \ldots, k-1$ denote by C_i the subtree of C induced by nodes $u_1, \ldots, u_i$ and by their adjacent leaves in $C - u_k$ and denote $C_k := C$. Observe that C_i is a caterpillar, for $i = 1, \ldots, k$. We will prove that, for $i = 1, \ldots, k$, C_i admits a planar L-shaped point-set embedding on every general point set with $3|C_i| - 2$ points, so that the following invariant is satisfied:

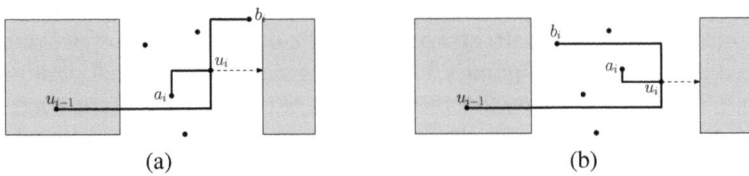

Fig. 4. Planar L-shaped point-set embedding of caterpillars on general point sets. (a) $y(p_1) < y(p_2) < y(p_3)$. (b) $y(p_1) > y(p_2) > y(p_3)$.

The horizontal half-line starting at u_i directed rightward does not intersect any edge of the constructed drawing of C_i. We use the fact that $(\star)$ every general point set with 5 points contains a diagonal point set with 3 points, which can easily be verified.

We now prove the statement. In the base case, $i = 1$. Then $|C_1| = 1$ and the statement is trivially true. Suppose the statement is true for C_{i-1}. Consider any point set P_i with $3|C_i| - 2$ points. Denote by P_{i-1} the point set consisting of the leftmost $3|C_{i-1}| - 2$ points of P_i. Construct, by induction, a planar L-shaped point-set embedding of C_{i-1} on P_{i-1} so that the horizontal half-line starting at u_{i-1} directed rightward does not intersect any edge of the constructed drawing of C_{i-1}. We distinguish three cases.

In the first case u_i has no adjacent leaf. Then, embed u_i on the rightmost point of P_i (such a point exists since $|P_i \setminus P_{i-1}| = 3$). Connect u_i with u_{i-1} by an L-shaped edge horizontally attached to u_{i-1} and vertically attached to u_i.

In the second case u_i has one adjacent leaf a_i. Then, consider the three leftmost points of $P_i \setminus P_{i-1}$ (such points exist since $|P_i \setminus P_{i-1}| = 6$). Then, either two of such three points are above the horizontal line $h(u_{i-1})$ through u_{i-1} or two are below. Suppose two points p_1 and p_2 are above $h(u_{i-1})$, the other case being analogous. Then, embed u_i on the rightmost of p_1 and p_2 and embed a_i on the leftmost of p_1 and p_2. Connect u_i with u_{i-1} by an L-shaped edge horizontally attached to u_{i-1} and vertically attached to u_i and connect u_i with a_i by an L-shaped edge horizontally attached to u_i and vertically attached to a_i.

In the third case u_i has two adjacent leaves a_i and b_i. Then, consider the nine leftmost points of $P_i \setminus P_{i-1}$ (such points exist since $|P_i \setminus P_{i-1}| = 9$). Then, either five of such nine points are above the horizontal line $h(u_{i-1})$ through u_{i-1} or five are below. Suppose five points $p_1, \ldots, p_5$ are above $h(u_{i-1})$, the other case being analogous. Then, by $(\star)$, three points, say without loss of generality p_1, p_2, and p_3, form a diagonal point set. Suppose, without loss of generality, that $x(p_1) < x(p_2) < x(p_3)$. Then, if $y(p_1) < y(p_2) < y(p_3)$ (see Fig. 4a) embed u_i on p_2, embed a_i on p_1, and embed b_i on p_3; otherwise, that is $y(p_1) > y(p_2) > y(p_3)$ (see Fig. 4b), embed u_i on p_3, embed a_i on p_2, and embed b_i on p_1. In both cases, connect u_i with u_{i-1} by an L-shaped edge horizontally attached to u_{i-1} and vertically attached to u_i, connect u_i with a_i by an L-shaped edge horizontally attached to u_i and vertically attached to a_i, and connect u_i with b_i by an L-shaped edge vertically attached to u_i and horizontally attached to b_i.

Since the planar L-shaped point-set embedding of C_{i-1} on P_{i-1} satisfies the invariant, the resulting L-shaped point-set embedding of C_i on P_i is planar. Moreover, such a point-set embedding clearly satisfies the invariant, thus completing the induction. □

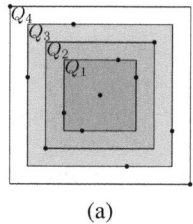 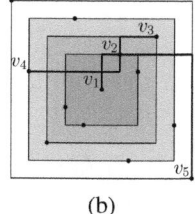

(a)   (b)

Fig. 5. Non-planar L-shaped point-set embedding of a tree with maximum degree 4

For caterpillars with maximum degree 3 we can improve this bound even further by showing that every such a caterpillar can be embedded on every general point set with n points using L-shaped edges. The proof of the following Theorem 8 is based on an inductive argument showing that we can embed a sub-caterpillar C_i of C, where C_i is defined according to the proof of Theorem 7, on the leftmost $|C_i|$ points of P such that, the horizontal half-line as well as one vertical half-line starting at the rightmost vertex of the spine of C_i do not intersect the drawing of C_i. A proof can be found in [11].

Theorem 8. *Every caterpillar with n vertices and with maximum degree 3 admits a planar L-shaped point-set embedding on every general point set with n points.*

4   Non-Planar L-Shaped Orthogeodesic Point-Set Embeddings

In this section we consider non-planar L-shaped orthogeodesic point-set embeddings. We start by showing that every tree with n vertices as a non-planar L-shaped orthogeodesic point-set embedding on every general point set with $4n - 3$ points.

Theorem 9. *Every tree with n vertices and with maximum degree 4 admits a non-planar L-shaped point-set embedding on every general point set with $4n - 3$ points.*

Proof: Let $T = (V, E)$ be a tree with n vertices and let P be a point set with $4n - 3$ points. Let T be rooted in a leaf $r \in V$ and let the vertices of T be labeled $r = v_1, \ldots, v_n$ according to a depth-first search in T. Let $Q_n = P$. For $n \geq i \geq 1$, let P_i consist of the points on the bounding box of Q_i, and for $n \geq i \geq 2$ let $Q_{i-1} = Q_i \setminus P_i$. By construction each P_i contains at least two and at most four vertices, except for P_1, which contains at least one vertex. See Figure 5a.

We embed T using L-shaped orthogeodesic chains such that vertex v_i is mapped to a point in P_i for all $1 \leq i \leq n$. We start by mapping the root v_1 to an arbitrary point $p^* \in P_1$. Suppose we have embedded all vertices $v_1, \ldots v_i$ for some $i \geq 1$ and we would like to embed v_{i+1}. Since the vertices are ordered according to a depth-first search, we have already embedded the parent v_j of v_{i+1}. Without loss of generality we may assume that the vertical segment above v_j is unoccupied (otherwise we can rotate the instance accordingly). By construction the points in P_i are on the bounding box of Q_{i+1}, which contains Q_i in its interior. Hence, P_i contains a point p_t above $f(v_j)$. We map v_{i+1} to p_t and connect it to v_j by a vertical segment incident to v_j and the horizontal segment incident to p_t. See Figure 5b.   □

We show that a point set of size n suffices for all trees with n vertices and maximum degree 3.

Theorem 10. *Every tree with n vertices and with maximum degree 3 admits a non-planar L-shaped point-set embedding on every general point set with n points.*

Proof: We prove by induction a stronger statement. Let T be any tree with n vertices and with maximum degree 3. Let r be the root of T. We prove that T admits a non-planar L-shaped point-set embedding on every general point set with n points with the further property that there is no edge overlapping a line arbitrarily chosen among the horizontal line and the vertical line through r.

In the base case $n = 1$ and the statement is trivially true. In the inductive case, we prove that T admits a non-planar L-shaped point-set embedding on every general point set P with n points with the further property that no edge overlaps the vertical line through r (the construction providing that no edge overlaps the horizontal line through r is analogous). Refer to Fig 6. Let $n_1 \geq 0$ and $n_2 \geq 0$ denote the number of vertices in the subtrees T_1 and T_2 rooted at the children r_1 and r_2 of r, respectively. Let P_1 and P_2 be the point sets consisting of the leftmost n_1 and the rightmost n_2 points of P, respectively. Let p be the only point of P not in P_1 and not in P_2. Embed r on p. Inductively construct a non-planar L-shaped point-set embedding of T_1 on P_1 (resp. of T_2 on P_2) such that no edge overlaps the vertical line through r_1 (resp. through r_2). Connect r with r_1 (resp. with r_2) by an L-shaped edge horizontally attached to r and vertically attached to r_1 (resp. to r_2). Since the embedding of T_1 (resp. of T_2) on P_1 (resp. on P_2) is L-shaped, all the edges of T_1 (resp. of T_2) lie inside the bounding box of P_1 (resp. of P_2). Hence, the vertical line through r has no overlapping edge, thus completing the induction.   □

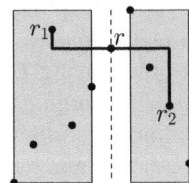

Fig. 6. Non-planar L-shaped point-set embedding of a tree

For caterpillars with maximum degree 4 we can improve this by showing that $n + 1$ points suffice for non-planar L-shaped orthogeodesic point-set embeddings. The proof is based on an inductive argument showing that every caterpillar has a non-planar L-shaped point-set embedding on every general point set such that the spine is embedded as an x-monotone chain with the property that each vertex of the spine has a spine edge incident to the right and such that all except possibly one point are used either by the spine or one of its adjacent leaves. The case analysis for the induction can be found in [11].

Theorem 11. *Every caterpillar with n vertices and with maximum degree 4 admits a non-planar L-shaped orthogeodesic point-set embedding on every general point set with $n + 1$ points.*

5   Conclusions

In this paper we studied orthogeodesic point-set embeddings of trees on the grid. For various types of drawings $\mathcal{D}$ and various families of trees $\mathcal{F}$ we proved upper bounds on the minimum value $f(n)$ such that every n-vertex tree in $\mathcal{F}$ admits a point-set embedding of type $\mathcal{D}$ on every point set of size $f(n)$. Since n is a trivial lower bound for $f(n)$ in all considered variants of the problem and since the upper bounds we provided are larger than n for some of the considered variants, it is an interesting topic for future research to close the gap between n and $f(n)$. The gap is especially large for planar L-shaped point-set embeddings of trees for which we only proved a quadratic upper bound. Hence it would be interesting to come up with a sub-quadratic upper bound or a non-trivial lower bound. Further, we restricted our attention to trees, but we may consider the same problem for different classes of graphs.

References

1. Badent, M., Di Giacomo, E., Liotta, G.: Drawing colored graphs on colored points. Theoretical Computer Science 408(2-3), 129–142 (2008)
2. Bose, P.: On embedding an outer-planar graph on a point set. Computational Geometry: Theory and Applications 23, 303–312 (2002)
3. Bose, P., McAllister, M., Snoeyink, J.: Optimal algorithms to embed trees in a point set. Journal of Graph Algorithms and Applications 2(1), 1–15 (1997)
4. Brandenburg, F.J.: Drawing planar graphs on $\frac{8}{9}n^2$ area. Electronic Notes in Discrete Mathematics 31, 37–40 (2008)
5. Cabello, S.: Planar embeddability of the vertices of a graph using a fixed point set is NP-hard. Journal of Graph Algorithms and Applications 10(2), 353–366 (2006)
6. Di Giacomo, E., Grilli, L., Krug, M., Liotta, G., Rutter, I.: Hamiltonian Orthogeodesic Alternating Paths. In: Iliopoulos, C.S. (ed.) IWOCA 2011. LNCS, vol. 7056, pp. 170–181. Springer, Heidelberg (2011)
7. Di Giacomo, E., Liotta, G., Trotta, F.: Drawing colored graphs with constrained vertex positions and few bends per edge. Algorithmica 57, 796–818 (2010)
8. Erdős, P., Szekeres, G.: A combinatorial problem in geometry. Compositio Mathematica 2, 463–470 (1935)
9. Everett, H., Lazard, S., Liotta, G., Wismath, S.: Universal sets of n points for one-bend drawings of planar graphs with n vertices. Discrete and Computational Geometry 43, 272–288 (2010)
10. Fink, M., Haunert, J.-H., Mchedlidze, T., Spoerhase, J., Wolff, A.: Drawing graphs with vertices at specified positions and crossings at large angles. pre-print, arXiv:1107.4970v1 (July 2011)
11. Di Giacomo, E., Frati, F., Fulek, R., Grilli, L., Krug, M.: Orthogeodesic point-set embedding of trees. Technical Report 2011-24, Kalrsruhe Institute of Technology, KIT (2011)
12. Gritzmann, P., Mohar, B., Pach, J., Pollack, R.: Embedding a planar triangulation with vertices at specified points. Amer. Math. Monthly 98(2), 165–166 (1991)
13. Katz, B., Krug, M., Rutter, I., Wolff, A.: Manhattan-Geodesic Embedding of Planar Graphs. In: Eppstein, D., Gansner, E.R. (eds.) GD 2009. LNCS, vol. 5849, pp. 207–218. Springer, Heidelberg (2010)
14. Kaufmann, M., Wiese, R.: Embedding vertices at points: Few bends suffice for planar graphs. Journal of Graph Algorithms and Applications 6(1), 115–129 (2002)
15. Kurowski, M.: A 1.235 lower bound on the number of points needed to draw all n-vertex planar graphs. Information Processing Letters 92(2), 95–98 (2004)

On Point-Sets That Support Planar Graphs[*]

Vida Dujmovic[1], William Evans[2], Sylvain Lazard[3], William Lenhart[4],
Giuseppe Liotta[5], David Rappaport[6], and Stephen Wismath[7]

[1] Carleton University, Canada
[2] University of British Columbia, Canada
[3] INRIA Nancy, LORIA, France
[4] Williams University, U.S.A.
[5] Universitá degli Studi di Perugua, Italy
[6] Queen's University, Canada
[7] University of Lethbridge, Canada

Abstract. A universal point-set supports a crossing-free drawing of any
planar graph. For a planar graph with n vertices, if bends on edges of
the drawing are permitted, universal point-sets of size n are known, but
only if the bend-points are in arbitrary positions. If the locations of the
bend-points must also be specified as part of the point-set, we prove that
any planar graph with n vertices can be drawn on a universal set $\mathcal{S}$ of
$O(n^2/\log n)$ points with at most one bend per edge and with the vertices
and the bend points in $\mathcal{S}$. If two bends per edge are allowed, we show that
$O(n \log n)$ points are sufficient, and if three bends per edge are allowed,
$\Theta(n)$ points are sufficient. When no bends on edges are permitted, no
universal point-set of size $o(n^2)$ is known for the class of planar graphs.
We show that a set of n points in balanced biconvex position supports
the class of maximum degree 3 series-parallel lattices.

1 Introduction

A set of points *supports* the drawing of a graph G if there is a one-to-one mapping
f of the vertices of G to the points so that for all pairs of edges $(a, b), (c, d)$ in G
(where a, b, c, d are distinct), segments $\overline{f(a)f(b)}$ and $\overline{f(c)f(d)}$ do not intersect.
A set of points that supports the drawing of all n-vertex graphs in some class
is called *universal* for that class, or simply *universal* if the class is all planar
graphs. The size of any universal point-set for planar graphs requires at least
$1.235n$ points as shown by Kurowski [10] (see also Chrobak and Karloff [3]). Early
graph drawing results, such as the canonical ordering technique of de Frasseix,
Pach, Pollack [4] and Schnyder's embedding [11] demonstrate that an $n \times n$ grid
of points is a universal point-set. However, no universal point-set of size $o(n^2)$ is
known.

[*] This paper was initiated at the 2011 McGill/INRIA/UVictoria Bellairs workshop.
Discussion with other participants is gratefully acknowledged. Research supported
by NSERC, and by MIUR of Italy under project AlgoDEEP prot. 2008TFBWL4.

M. van Kreveld and B. Speckmann (Eds.): GD 2011, LNCS 7034, pp. 64–74, 2012.
© Springer-Verlag Berlin Heidelberg 2012

Smaller universal point-sets for sub-classes of planar graphs are known. For example, any outerplanar graph can be drawn on *any* set of n points in general position [8]. Indeed, if the point-set is in convex position, then it supports exactly the family of outerplanar graphs. Determining other families of planar graphs for which universal point-sets of size n exist is an interesting problem. We examine a particular type of point-set, of size n, in which points are arranged in biconvex position, and show that it supports the drawing of all maximum degree 3 series-parallel lattices, a class of graphs that contains members that are not outerplanar. These notions are precisely defined in Section 2.

The main contributions in this paper are stated in Theorems 1 and 2, and pertain to universal point-sets for straight-line drawings, and drawings with bends respectively.

Theorem 1. *For all n, there exist universal point-sets of cardinality n that support the family of maximum degree 3 series-parallel lattices with n vertices.*

Suppose we relax the definition of *support* to allow edges of the graph to map to polylines composed of (at most) $k + 1$ line segments. In other words, we allow edges that "bend" at most k times. In this case, universal point-sets of size n exist for two bends [9] and even one bend [7]. However, these results assume that the bend-points can be placed in arbitrary locations and these bend-points are not included as part of the universal point-set. It is natural to ask if there exists a point-set that supports all planar graphs where each vertex *and each bend-point* occurs at a point in the set. As before, we require all pairs of edges (a, b) and (c, d) (where a, b, c, d are distinct) to map to non-intersecting polylines. Previous to this paper, no such point-set of cardinality $o(n^2)$ was known for any value of k. Extending the results of [7] and [9] in a straightforward manner imply point-sets of size $O(n^3)$. For $k = 3, 2, 1$, we present such universal point-sets of cardinality $O(n)$, $O(n \log n)$, and $O(n^2 / \log n)$ respectively.

Theorem 2. *For all n, there exist universal point-sets of cardinality $O(n)$, $O(n \log n)$, and $O(n^2 / \log n)$ that support the drawing of all n-vertex planar graphs with at most 3, 2, or 1 bend per edge, respectively.*

Table 1. Summary of results – cardinality of universal point-sets for classes of graphs. The first and last results are well-known. All other results are new.

Graphs	Number of Points	Number of Bends	Reference
outerplanar	n	0	[8]
3SP lattice	n	0	Thm 1
planar	$O(n)$	3	Thm 2
planar	$O(n \log n)$	2	Thm 2
planar	$O(n^2 / \log n)$	1	Thm 2
sub-Hamiltonian	$O(n)$	2	Lemma 3.1
sub-Hamiltonian	$O(n \log n)$	1	Lemma 3.2
planar	$O(n^2)$	0	[4],[11]

Table 1 summarizes our results in terms of which sets of planar graphs can be supported on point-sets of a given cardinality with a specified number of bends.

2   Preliminaries

We adopt standard notation from the graph drawing literature and we henceforth assume all graphs have n vertices. Many proofs are omitted for space reasons.

Two of our results rely on point-sets that have a specific form; see, for example, Fig. 2. Two non-intersecting non-linear arcs of curves λ_1 and λ_2 are defined to be *biconvex* if: each of these arcs is simple and convex, the convex hull of their 4 endpoints completely contains the two arcs, and the line segment joining any point a of λ_1 to any point b of λ_2 does not intersect either arc of curve except at a and b; for simplicity, we refer to such arcs of curves as curves. Without loss of generality, we assume the existence of a horizontal line separating the two curves with λ_1 below λ_2. A point-set all of whose points lie on two curves that are biconvex is in *biconvex position*. We note that point-sets in such a configuration have been used in other contexts under different names.

3   Universal Point-Sets for Drawing Planar Graphs with Bends

In this section we establish Theorem 2 by constructing universal point-sets for each of the three cases: 3, 2 or 1 bend per edge allowed. A fundamental tool in our constructions for universal point-sets with bends is the following result proving the existence of a book embedding of planar graphs in which the edges are permitted to cross the spine [6]. A *monotone topological book embedding* of a planar graph G is a planar drawing such that all vertices of G are represented as distinct points on a spine (i.e. the x-axis), and each edge is either represented as an arc in the bottom page (below the x-axis), or as an arc in the top page (above the x-axis), or as the concatenation of two arcs: the first (leftmost) in the bottom page and the second in the top page with their common crossing point between spine points. See Fig. 1.

Theorem 3 ([6]). *Every planar graph has a proper monotone topological book embedding which can be computed in linear time.*

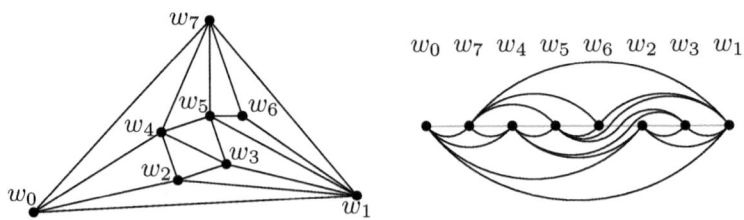

Fig. 1. A graph and a monotone topological book embedding of it

3.1 A Set of $\Theta(n)$ Points for Drawing Planar Graphs with Three Bends Per Edge

Lemma 1. *There exists a universal set of $10n - 18$ points that supports the drawing of planar graphs with 3 bends per edge.*

Proof. Before introducing the (fixed) universal point-set, we first outline how the graph will be processed. Consider a proper monotone topological book embedding of the input graph (see Fig. 1). For each edge that intersects the spine, introduce a dummy vertex creating an *augmented two page book embedding* with the vertices of the spine drawn on a horizontal line. There are at most $n + m \leqslant 4n - 6$ vertices on the spine. Imagine a horizontal line slightly above the spine that intersects all arcs in the top page – call these points of intersection from left to right $b_1, \ldots, b_{2a}$ where a is the number of arcs. Note that $a \leqslant 3n - 6$.

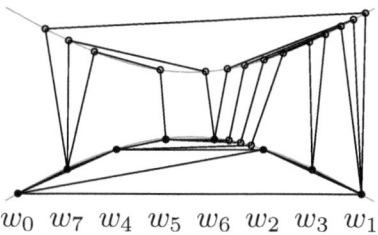

$$w_0 \quad w_7 \quad w_4 \quad w_5 \quad w_6 \quad w_2 \quad w_3 \quad w_1$$

Fig. 2. Example drawing of the graph of Fig. 1 on a biconvex point-set using three bends per edge. The curvature is exaggerated and only the first 14 of the 36 upper curve points and 11 of the 26 lower curve points in the universal point-set are shown.

Consider a point-set that lies on two curves in biconvex position and consists of $6n - 12$ points on the top curve and $4n - 6$ points on the bottom curve. Any such point-set is universal and supports the drawing of all planar graphs with at most three bends per edge. Refer to Fig. 2 for an example of the construction. For any specific graph, its augmented two page book embedding defines the drawing and requires at most $10n - 18$ points. The at most $4n - 6$ vertices on the spine (including dummy vertices) are assigned, in order, to the first points on the bottom curve. The bend-points $b_1, \ldots, b_{2a}$ are assigned to the first $2a$ points of the upper curve in left to right order and then each arc in the top page is drawn using the associated bend-points. These polylines do not intersect since the upper curve is convex and any segment joining the two curves does not properly intersect these curves. The arcs in the bottom page can be drawn with no bends – they are cords of the bottom curve. Each arc in the top page uses two bend-points. Substituting a bend-point for each of the dummy vertices results in a drawing with at most three bends per edge. □

Note that a sub-Hamiltonian planar graph corresponds exactly to a graph that has a two page (unaugmented) book embedding [6]. Since such graphs do not require dummy vertices, they can be drawn with at most two bends per edge.

3.2   A Set of $O(n \log n)$ Points for Drawing Planar Graphs with 2 Bends Per Edge

The geometric idea underlying our construction is as follows. Similar to Section 3.1, we draw the spine vertices of an augmented two page book embedding on a set of points that lie on a slightly concave curve close to the x-axis. This implies that all the arcs in the bottom page of the book embedding can be drawn as straight line segments. For arcs in the top page, if the arc is from the ith to the $(i+j)$th spine vertex, it is drawn to bend at a point at level j. We place approximately n/j bend-points approximately equally spaced in the x-dimension at level j, since only n/j top arcs can have "length" j. The bend-point that lies between the ith and $(i+j)$th spine vertices is used by this arc. Each level is at a y-coordinate that is large enough that the drawing of an arc that uses a bend-point at a lower level "nests" inside any drawing of an arc from the same vertex using a higher level bend-point. Of course, for each $j > n/2$, there can be only one arc of "length" j and it uses a single bend-point at level j. The total number of bend-points placed is $O(n \log n)$.

Lemma 2. *There exists a universal set of $O(n \log n)$ points that supports the drawing of planar graphs with 2 bends per edge.*

3.3   A Set of $O(n^2/\log n)$ Points for Drawing Planar Graphs with 1 Bend Per Edge

Lemma 3. *There exists a universal set of $O(n^2/\log n)$ points that supports the drawing of planar graphs with 1 bend per edge.*

Proof. The construction is similar to that in [7]. We recall briefly this construction, referring to Figures 1 and 3.

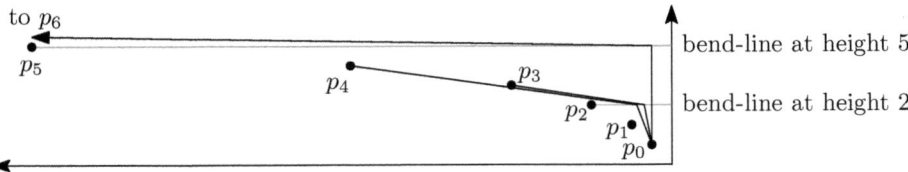

Fig. 3. The one-bend drawing of the three top-page edges adjacent to $v_0 = w_1$ in the graph of Fig. 1 following the construction of [7]. The points p_6 and p_7 are not shown since the figure is to scale.

Given a planar graph G with n vertices, we embed the graph on vertices $p_i = (-2^i, i)$ for $i = 0, \ldots, n-1$ with at most one bend per edge, as follows. We first compute a proper monotone topological book embedding, Γ of G. We relabel the vertices of that book embedding from right to left, as $v_0, \ldots, v_{n-1}$. We then map these vertices to $p_0, \ldots, p_{n-1}$, respectively. All the edges below the spine are drawn as straight-line segments. The others are drawn with a bend point as follows. Consider an edge whose rightmost vertex is v_i and that intersects the spine on the interval $(v_u, v_{u+1}]$ (inclusive v_{u+1} for the case where the leftmost endpoint of the edge is v_{u+1}). Such an edge is drawn with a bend point at the same height as v_u, and in the vertical strip delimited by v_i and v_{i+1}. A universal set of points for the bend location can easily be determined in this construction. However, this construction requires a set of size $\Theta(n^3)$ for the bend points since there need to be n bend points on each of n bend lines, and in each of $n-1$ vertical strips (delimited by v_i and v_{i+1}).

This construction can be modified to contain only a subquadratic universal set of points for the bends, while preserving a linear size universal set of points for the vertices. We consider as before a proper monotone topological book embedding, Γ of our input graph G. We then add on the spine extra isolated dummy vertices so that there is at most one edge crossing the spine between any two spine vertices. Since the number of edges of a planar graph is at most $3n - 6$, we add at most that number of isolated vertices and the total number of vertices is less than $4n$. Let G' be the resulting graph. Note that, if we use the construction of [7] with this augmented graph G', there is at most one bend point on each bend line; this yields that, for every bend line, we need only consider one candidate location for the bend points in each vertical strip, leading to a quadratic universal set of points for the bends.

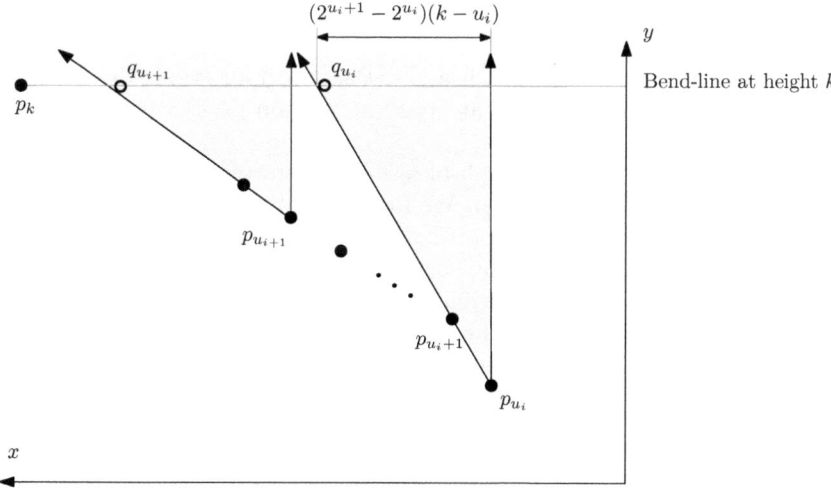

Fig. 4. Placement of bend points on the bend-line at height k. The figure is not to scale.

To obtain a subquadratic size we first construct a set of $4n$ points $p_0, \ldots, p_{4n-1}$ as in [7] that will support the vertices of any augmented graph G'. Now on the bend line at height k, we place $k/\log k$ candidate vertices for the bend points of the edges that intersect the spine through the window $p_k p_{k+1}$. The points on the bend-line at height k are defined as follows. See Fig. 4. The first (rightmost) point q_1 lies infinitesimally to the right of the line $p_0 p_1$. Let p_{u_1} be the rightmost point of $p_0, \ldots, p_{4n-1}$ that is to the left of q_1. The second point q_2 is infinitesimally to the right of the line $p_{u_1} p_{u_1+1}$ (at height k), and so on.

The addition of dummy points on the spine ensured that there is at most one edge that intersects the spine between two vertices on the spine. Hence, for any given graph, there is at most one bend point used on any bend line. This essentially implies that the construction of [7] works here.

How many bend points are there on a bend-line? It can be shown that there are $\Theta(k/\log k)$ candidate bend points on the bend line at height k, and $\Theta(n^2/\log n)$ candidate bend points in total. □

4 Biconvex Point-Sets and Series-Parallel Graphs

A *two terminal series-parallel digraph* (also called *TTSP-digraph*) is a planar digraph recursively defined as follows [5,12]: A directed edge joining two vertices forms a TTSP-digraph. Let G' and G'' be two TTSP-digraphs; the digraph obtained by identifying the sink of G' with the source of G'' (*Series Composition*) is also a TTSP-digraph. Let G' and G'' be two TTSP-digraphs; the digraph obtained by identifying the source of G' with the source of G'' and the sink of G' with the sink of G'' (*Parallel Composition*) is also a TTSP-digraph.

A TTSP-digraph has one source and one sink which are called its *poles*. Also, a TTSP-digraph is always acyclic and admits a planar embedding with the poles on the same face. A TTSP-digraph is a *TTSP lattice* if for every edge (u, v), there is no directed path from u to v that does not contain (u, v). Note that a TTSP lattice cannot have multiple edges.

The undirected underlying graph of a TTSP-digraph (resp. lattice) is called a *TTSP-graph* (resp. *TTSP-lattice*). We further shorten these terms and refer to them as series-parallel (SP).

Any point-set in general position supports the class of outerplanar graphs [8]. Indeed a point-set in convex position supports exactly the class of outerplanar graphs, and no other planar graphs. Motivated by this insight we now consider the class of planar graphs that are supported by a point-set in which $n/2$ points are on one convex curve and the remainder are on another convex curve – in biconvex position. Clearly outerplanar graphs can be supported by this point-set and efficient algorithms such as that developed by Bose [1] exist. We show that any $(n/2, n/2)$ biconvex point-set is universal for a subclass of the series-parallel graphs. Since our purpose is to exhibit universal point-sets for classes of planar graphs, the balancing condition is critical and since the number of vertices could be odd, the balancing must allow for one vertex to be placed arbitrarily.

A planar graph G is *biconvex* if there exists a crossing-free straight-line draw-ing Γ of G with all vertices located on the curves λ_1 and λ_2. A planar graph G is *balanced biconvex* if it is biconvex with a drawing Γ in which the numbers of vertices on the two curves differ by at most one; more formally if:
for n even, $n/2$ vertices are on λ_1 and $n/2$ vertices on λ_2 (called *uniform* and denoted as $\Gamma^=$), and for n odd, either:

- $\frac{n-1}{2}$ vertices are on λ_1 and $\frac{n+1}{2}$ vertices are on λ_2 (called *top-heavy* and denoted as Γ^+) or
- $\frac{n+1}{2}$ vertices are on λ_1 and $\frac{n-1}{2}$ vertices are on λ_2 (called *bottom-heavy* and denoted as Γ_+)

Our construction is recursive and attempts to contain the drawing of the SP lattice in a box spanning the biconvex curves with a balanced number of vertices on each curve and with s and t forming a diagonal of the box. Unfortunately, such a strong invariant cannot be maintained and slightly weaker conditions must be carefully considered.

A series-parallel digraph with poles s and t is *bottom-cornered* if it is balanced biconvex with a drawing Γ (n even) or Γ_+ (n odd) such that:

1. there exists a box (i.e. a convex quadrilateral) $B(s,t)$ with s on λ_1 and t on λ_2, $\overline{st}$ forms one diagonal of B, and the other diagonal has one corner on λ_1 and one on λ_2, and
2. the entire drawing lies inside B.

Similarly, a series-parallel digraph with poles s and t is *top-cornered* if it is balanced biconvex with a drawing Γ (n even), or Γ^+ (n odd) such that conditions 1 and 2 hold. If a series-parallel graph is both top-cornered and bottom-cornered, it is called *double-cornered* – i.e. if n is odd, there exist two drawings Γ^+ and Γ_+ both of which satisfy conditions 1 and 2.

In some situations, only weaker conditions on the drawings can be maintained, in which one of t or s is contained strictly inside a box rather than on the diagonal forming the box:

1'. there exists a box $B(s,x)$ with s on λ_1 and x on λ_2, $\overline{sx}$ forms one diagonal of $B(s,x)$, and the other diagonal has one corner on λ_1 and one on λ_2 and t is on λ_2 inside $B(s,x)$.
1''. there exists a box $B(x,t)$ with x on λ_1 and t on λ_2, $\overline{xt}$ forms one diagonal of $B(x,t)$, and the other diagonal has one corner on λ_1 and one on λ_2 and s is on λ_1 inside $B(x,t)$.

A series-parallel graph with source s and sink t is *bottom half-cornered* if it is balanced biconvex and conditions 1' and 2 hold; similarly, if conditions 1" and 2 hold, then the graph is *top half-cornered*.

In the lemmas that follow, we demonstrate only the existence of biconvex drawings for the graphs that we consider. The following lemma indicates that this is sufficient to claim a *universal* biconvex point-set of suitable size.

Lemma 4. *If a graph G on n vertices has a balanced biconvex drawing, then every balanced biconvex point-set of size n supports G.*

A series-parallel graph in which every vertex is of maximum degree 3 is denoted as 3SP. We distinguish between two critical cases. If both the source and sink of a 3SP lattice have degree $\leqslant 2$ then the graph is called *thin* and otherwise (i.e. if either pole has degree 3) it is called *thick*. It is the class of 3SP lattices that we show to be balanced biconvex. There are several cases to consider depending on whether the graph is biconnected or not, and whether the graph is thin or thick. Our proof is recursive in nature – interior components are replaced by appropriate balanced boxes. Lemmas 5 – 14 distinguish and organize these cases and Fig. 5 provides a simple example of each case, the type of drawing obtained, and the prerequisite lemmas used in the proof.

Lemma 5. *A simple path consisting of $n \geqslant 2$ vertices from s to t is double-cornered.*

First we present the lemmas used for subcases that are thin: biconnected and then not biconnected.

Lemma 6. *Let G be a biconnected thin 3SP lattice. Then G is double-cornered.*

Lemma 7. *Let G be a thin 3SP lattice with source s and sink t. If either $deg(s) = deg(t) = 1$ or $deg(s) = deg(t) = 2$ then G is double-cornered.*

Lemma 8. *Let G be a thin 3SP lattice with source s and sink t. If $deg(s) = 1$ and $deg(t) = 2$ then G is bottom-cornered.*

Lemma 9. *Let G be a thin 3SP lattice with source s and sink t. If $deg(s) = 2$ and $deg(t) = 1$ then G is top-cornered.*

The next sequence of lemmas pertains to the cases when the global poles have degree 3. The drawings obtained rely on the previous lemmas and are balanced biconvex, but may not be double-cornered.

Lemma 10. *Let G be a biconnected 3SP lattice with source s and sink t. If $deg(s) = deg(t) = 3$ and G consists of 3 series components combined in parallel, then G is balanced biconvex.*

Lemma 11. *Let G be a biconnected thick 3SP lattice with source s and sink t. If $deg(s) = deg(t) = 3$ then G is balanced biconvex.*

Lemma 12. *Let G be a biconnected thick 3SP lattice with source s and sink t. If $deg(s) = 2$ and $deg(t) = 3$ then G is bottom half-cornered.*

Lemma 13. *Let G be a biconnected thick 3SP lattice with source s and sink t. If $deg(s) = 3$ and $deg(t) = 2$ then G is top half-cornered.*

In the final case, G is not biconnected and at least one of the global poles has degree 3.

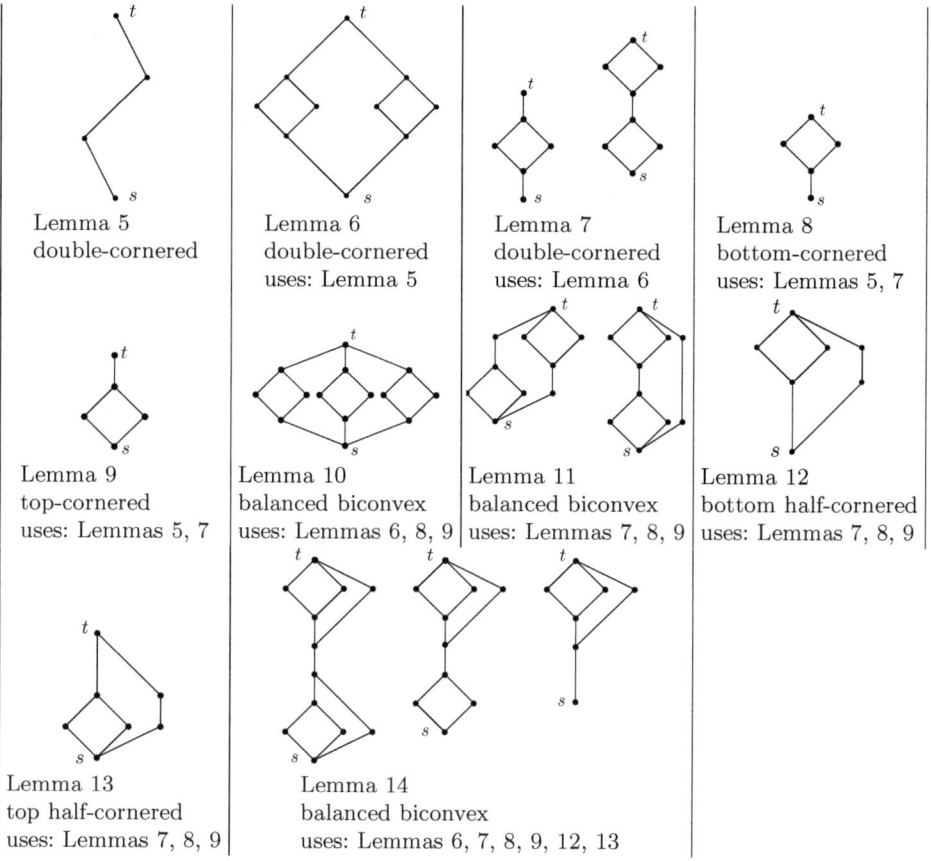

Fig. 5. Roadmap of the various cases

Lemma 14. *Let G be a thick 3SP lattice with source s and sink t. If deg(s) = 3 or deg(t) = 3 then G is balanced biconvex.*

Biconvex point-sets are the only known point-sets of size n that universally support some class of planar graphs other than the outerplanar graphs.

5   Conclusions and Open Problems

Our main contributions in this paper are stated in Theorems 1 and 2: that any balanced biconvex point-set supports the straight-line drawing of any 3SP lattice, and the specification of universal point-sets for drawings of any planar graph with a small number of bends per edge.

   Closing the gap between the upper and lower bounds of the cardinality of a universal point-set for planar graphs with no bends allowed remains an open problem. When k bends per edge are permitted, universal point-sets of smaller asymptotic cardinality may be determined for $k = 1, 2$.

References

1. Bose, P.: On embedding an outer-planar graph on a point set. Computational Geometry: Theory and Applications 23, 303–312 (2002)
2. Braß, P., Cenek, E., Duncan, C.A., Efrat, A., Erten, C., Ismailescu, D., Kobourov, S.G., Lubiw, A., Mitchell, J.S.B.: On simultaneous planar graph embeddings. Comput. Geom. 36(2), 117–130 (2007)
3. Chrobak, M., Karloff, H.: A lower bound on the size of universal sets for planar graphs. SIGACT News 20(4), 83–86 (1989)
4. de Fraysseix, H., Pach, J., Pollack, R.: How to draw a planar graph on a grid. Combinatorica 10, 41–51 (1990)
5. Di Battista, G., Eades, P., Tamassia, R., Tollis, I.G.: Graph Drawing. Prentice-Hall, NJ (1999)
6. Di Giacomo, E., Didimo, W., Liotta, G., Wismath, S.K.: Curve-constrained drawings of planar graphs. Computational Geometry 30, 1–23 (2005)
7. Everett, H., Lazard, S., Liotta, G., Wismath, S.: Universal sets of n points for one-bend drawings of planar graphs with n vertices. Discrete and Computational Geometry 43(2), 272–288 (2010)
8. Gritzmann, P., Mohar, B., Pach, J., Pollack, R.: Embedding a planar triangulation with vertices at specified points. Amer. Math. Monthly 98(2), 165–166 (1991)
9. Kaufmann, M., Wiese, R.: Embedding vertices at points: Few bends suffice for planar graphs. Journal of Graph Algorithms and Applications 6(1), 115–129 (2002)
10. Kurowski, M.: A 1.235 lower bound on the number of points needed to draw all n-vertex planar graphs. Inf. Process. Lett. 92(2), 95–98 (2004)
11. Schnyder, W.: Embedding planar graphs on the grid. In: Proc. 1st ACM-SIAM Sympos. Discrete Algorithms (SODA 1990), pp. 138–148 (1990)
12. Valdes, J., Tarjan, R.E., Lawler, E.L.: The recognition of series-parallel digraphs. SIAM J. Comput. 11(2), 298–313 (1982)

Small Point Sets for Simply-Nested Planar Graphs[*]

Patrizio Angelini[1], Giuseppe Di Battista[1], Michael Kaufmann[2],
Tamara Mchedlidze[3], Vincenzo Roselli[1], and Claudio Squarcella[1]

[1] Dip. di Informatica e Automazione, Roma Tre University, Italy
[2] Wilhelm-Schickard-Institut für Informatik, Universität Tübingen, Germany
[3] Dept. of Math. , National Technical University of Athens, Greece

Abstract. A point set $P \subseteq \mathbb{R}^2$ is universal for a class $\mathcal{G}$ if every graph of $\mathcal{G}$ has a planar straight-line embedding into P. We prove that there exists a $O(n(\frac{\log n}{\log \log n})^2)$ size universal point set for the class of simply-nested n-vertex planar graphs. This is a step towards a full answer for the well-known open problem on the size of the smallest universal point sets for planar graphs [1,5,9].

1 Introduction

A *planar straight-line embedding* of a graph G *into a point set* P is a mapping of each vertex of G to a distinct point of P and of each edge of G to the straight-line segment between the corresponding endpoints so that no two edges cross. Let $\mathcal{G}$ be a class of n-vertex planar graphs and P be a point set of size m, with $m \geq n$. Point set P is *universal* for the class $\mathcal{G}$ if for every $G \in \mathcal{G}$, G has a planar straight-line embedding into P.

Asymptotically, the smallest universal point set for general planar graphs is known to have size at least $1.235n$ [6,12], while the best known upper bound is $O(n^2)$ [7,10,13]. Characterizing the asymptotic size of the smallest universal point set is a well-known open problem also referred in [1,5,9].

A subclass of planar graphs for which a "small" universal point set is known is the class of outerplanar graphs, that is, the graphs that admit a straight-line planar embedding with all vertices incident to the outer face. Gritzmann et al. [11] and Bose [4] proved that any point set of size n is universal for outerplanar graphs. In [11] it is noticed that outerplanar graphs are the largest class of graphs for which any arbitrary point set is universal.

A generalization of outerplanar graphs are k-outerplanar graphs, $k \geq 2$. A planar embedding of a graph is k-*outerplanar* if removing the vertices of the outer face yields a $(k-1)$-outerplanar embedding, where 1-outerplanar is an outerplanar embedding. Vertices removed at the i-th step are at level i. A graph is k-*outerplanar* if it admits a

[*] Research partially supported by the MIUR project AlgoDEEP prot. 2008TFBWL4, by the ESF project 10-EuroGIGA-OP-003 GraDR "Graph Drawings and Representations", and by the European Union (European Social Fund - ESF) and Greek national funds through the Operational Program "Education and Lifelong Learning" of the National Strategic Reference Framework (NSRF) - Research Funding Program: Heraclitus II. Investing in knowledge society through the European Social Fund.

M. van Kreveld and B. Speckmann (Eds.): GD 2011, LNCS 7034, pp. 75–85, 2012.

k-outerplanar embedding. Note that no (arbitrarily large) convex point set is universal for k-outerplanar graphs, $k \geq 2$.

The decision question of whether a given planar graph admits a planar straight-line embedding into a given point set of the same size was proved to be $\mathcal{NP}$-hard, even for 2-outerplanar graphs and 3-level point sets [5].

A k-outerplanar graph is *simply-nested* [8] if levels 1 to $k - 1$ are chordless cycles and level k is either a cycle or a tree. A planar graph is *simply-nested* if it is k-outerplanar simply-nested for some $k \leq n$. Simply-nested graphs turned out to be useful to derive some properties of planar graphs. Cimikowski [8] proved hamiltonicity of simply-nested planar triangulations. Baker [3] used these graphs to derive approximation algorithms for various NP-complete problems on planar graphs. A variant of nested triangulations was explored by Yannakakis in his celebrated result on book embeddings of planar graphs [14].

In this paper we show a $O(n(\frac{\log n}{\log \log n})^2)$-size universal point set for simply-nested n-vertex graphs (Sect. 3). Such result is based on the construction of a $8n + 8$-size universal point set for simply-nested n-vertex graphs for which the number of vertices on each of level is known in advance (Sect. 2).

Our results find applications to another class of graphs, quite popular in Graph Drawing. In [2] Bachmaier *et al.* defined a graph to be (*proper*) k-*radial planar* if given a partition of its vertices into k concentric circles, its edges can be drawn as monotonic curves between (consecutive) circles without crossings and showed that radial planarity is decidable in linear time. Our results give a small universal point set for proper k-radial planar graphs, since they can be easily proved to be a subclass of simply-nested planar graphs.

2   A Universal Point Set for Simply-Nested Planar Graphs with n_i Vertices on Level i

In this section we describe a universal point set P of size $8(\sum_{i=1}^{k} n_i + k) = O(n)$ for simply-nested planar graphs in which the number n_i of vertices at each level i is known in advance. Note that, when this strong assumption is not possible, the same construction yields a point set with a quadratic number of points, namely $8(\sum_{i=1}^{k} n + k) = O(n^2)$, as $k = O(n)$. However, constructing the point set under this assumption is the basis of a construction, described in Sect. 3, that leads to subquadratic size in the general case.

We aim at placing the vertices of level i on a circle with a number of available points proportional to n_i. Then, we would like to place the vertices of level $i + 1$ greedily on a circle internal to the previous one. This is difficult for the following reason. If a vertex of level $i + 1$ is connected to many vertices of level i, the angle spanned by its connections gets close to 2π, and an arbitrary number of points of the internal circle become "unusable". See Fig. 1(a). Hence, we use a technique that places the vertices of each level on two concentric circles.

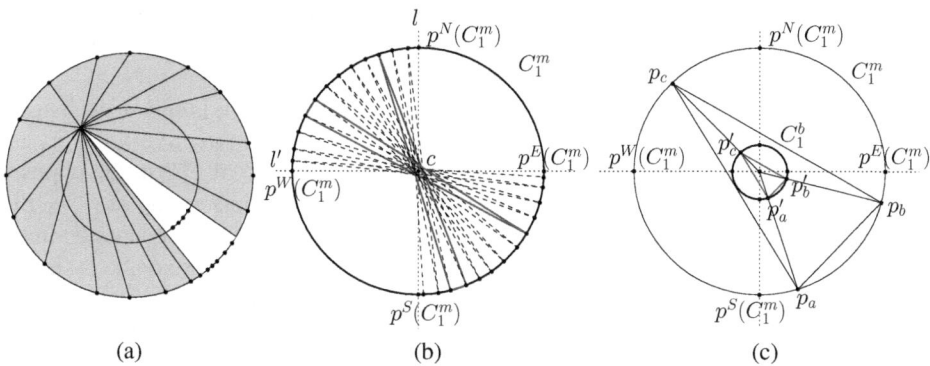

Fig. 1. (a) Problems in using one circle per level. (b) Construction of circle C_1^m. (c) Construction of circle C_1^b.

2.1 Construction of the Point Set

The points of P are on $2k$ concentric circles $C_1^m, C_1^b, \ldots, C_k^m, C_k^b$. For each level $i = 1, \ldots, k$, circles C_i^m and C_i^b are the *main circle* and the *back-up circle* of level i, respectively. Both have $4n_i + 4$ points. In the following we describe how to choose the radius and the distribution of the points for each circle.

Let l and l' be two orthogonal lines crossing at a point c, that is the center of circles C_i^m and C_i^b ($i = 1, \ldots, k$). The parts of the plane delimited by l and l' are the *quadrants*. For each circle C, denote by $p^N(C)$ and $p^S(C)$ the intersections between C and l, and by $p^W(C)$ and $p^E(C)$ the intersections between C and l'. Points $p^N(C)$, $p^S(C)$, $p^E(C)$, and $p^W(C)$ are the *cardinal* points of C.

Let C_1^m be a circle centered at c with any radius r_1^m. Place a point of P on each of $p^N(C_1^m)$, $p^S(C_1^m)$, $p^W(C_1^m)$, and $p^E(C_1^m)$. Then, place n_1 points of P in each arc of C_1^m determined by lines l and l', in such a way that for any two consecutive points p_a and p_b that are internal to a quadrant there exists a point p_c in the opposite quadrant, that is, its unique non-adjacent quadrant, such that triangle (p_a, p_b, p_c) contains c. Such a placement of points is always realizable. Namely, consider two opposite quadrants Q and Q'. Place a point p_a on C_1^m in Q and a point p'_a on C_1^m in Q' such that the center c is to the left of the oriented segment $\overrightarrow{p_a p'_a}$. Then place a point p_b on C_1^m in Q such that c is to the left of the oriented segment $\overrightarrow{p'_a p_b}$. Keeping on placing points in this way yields a point set with the desired property. See Fig. 1(b).

Let C_1^b be a circle centered at c with a radius $r_1^b < r_1^m$ such that, for every triangle (p_a, p_b, p_c) composed of three points of C_1^m, if (p_a, p_b, p_c) contains c, then it also contains C_1^b. Then, place $4n_1 + 4$ points on C_1^b in such a way that, for each point $p \in C_1^m$ there exists a point p' on the intersection between C_1^b and the radius of C_1^m to p. Note that, this implies that for any two consecutive points p'_a and p'_b of C_1^b that are internal to a quadrant there exists a point p'_c of C_1^b in its opposite quadrant such that (p'_a, p'_b, p'_c) contains c. See Fig. 1(c).

Then, for each level i, with $i = 2, \ldots, k$, construct the main circle C_i^m and the back-up circle C_i^b as follows.

Circle C_i^m is centered at c, has radius $r_i^m < r_{i-1}^b$, and for any triangle composed of two consecutive points p_a' and p_b' of C_{i-1}^b and a point p_c' in the opposite quadrant of C_{i-1}^b, if (p_a', p_b', p_c') contains c, then it also contains C_i^m.

Place a point of P on each cardinal point of C_i^m. Then, place n_i points in each arc of C_i^m determined by l and l' in such a way that: (a) for any two consecutive points p_a and p_b of C_i^m that are internal to a quadrant there exists a point p_c of C_i^m in the opposite quadrant such that (p_a, p_b, p_c) contains c; (b) for any two points p_1, p_2 of C_{i-1}^m that are in opposite quadrants, consider the quadrant Q that is completely contained in the wedge delimited by the half-lines from c to p_1 and from c to p_2 whose angle is smaller than π. Then, there exists a point p_3 of C_i^m in Q such that triangle (p_1, p_2, p_3) contains no point of C_i^m (see Fig. 2(a)); (c) the quadrilateral composed of points $p^N(C_{i-1}^m)$, $p^S(C_{i-1}^m)$, $p^W(C_i^m)$, and $p^E(C_i^m)$ contains all the points of C_i^m (see Fig. 2(b)); (d) the quadrilateral composed of points $p^E(C_{i-1}^m)$, $p^W(C_{i-1}^m)$, $p^N(C_i^m)$, and $p^S(C_i^m)$ contains all the points of C_i^m (see Fig. 2(b)). Note that a point set with these properties can always be constructed. Namely, a point set satisfying property (a) can be constructed analogously as for C_1^m (see Fig. 1(b)), while properties (b)–(d) can be easily satisfied by making the radius of C_i^m small enough.

Circle C_i^b is centered at c, has radius $r_i^b < r_i^m$, and is such that for every triangle (p_a, p_b, p_c) composed of three points placed on C_i^m, if (p_a, p_b, p_c) contains c, then it also contains C_i^b. Then, place $4n_i + 4$ points of P on C_i^b in such a way that, for each point $p \in C_i^m$ there exists a point p' on the intersection between C_i^b and the radius of C_i^m to p. Note that, this implies that for any two consecutive points p_a' and p_b' of C_i^b that are internal to a quadrant there exists a point p_c' of C_i^b in its opposite quadrant such that triangle (p_a', p_b', p_c') contains c.

2.2   Embedding a Simply-Nested Planar Graph on Point Set P

Let G be any simply-nested planar graph. We assume that G has only triangular faces; if it is not the case, we add dummy edges.

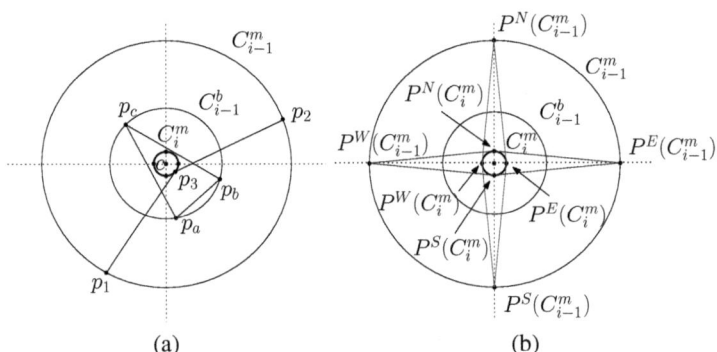

(a)   (b)

Fig. 2. Construction of circle C_i^m: (a) Triangle (p_1, p_2, p_3) contains no point of C_i^m; (b) quadrilaterals $(P^N(C_{i-1}^m), P^E(C_i^m), P^S(C_{i-1}^m), P^W(C_i^m))$ and $(P^N(C_i^m), P^E(C_{i-1}^m), P^S(C_i^m), P^W(C_{i-1}^m))$ contain all the points of C_i^m.

The drawing of G on P is constructed iteratively, starting by placing the vertices of level 1 on any n_1 points of circle C_1^m in such a way that the polygon representing the cycle composed of such vertices contains the center c. Note that, as any triangle composed of three points of C_1^m and containing c also contains C_1^b, the constructed polygon contains C_1^b, as well.

In order to describe how to embed the vertices of level $i = 2, \ldots, k$, we first give a further definition. We say that the drawing of the vertices of level i is 2-*radial* if it satisfies the following properties: (a) all the vertices of level i are on circle C_i^m, except for at most two vertices v_*' and v_*'', that are possibly drawn on two points of circle C_{i-1}^b. (b) Given the two lines tangent to C_i^b through v_*' (through v_*''), the triangle composed of their tangent points to C_i^b and v_*' (v_*'') does not contain any vertex of level i placed on a point of C_i^m.

Then, for each level $i = 2, \ldots, k$, we assume that a 2-radial drawing of level $i - 1$ is given, and we greedily construct a 2-radial drawing of level i, as follows.

Consider the vertices $v_1, \ldots, v_h$ of level i that have more than one neighbor in level $i - 1$. Observe that, the set of vertices that is the union of the neighbors of $v_1, \ldots, v_h$ coincides with the set of vertices of level $i - 1$. As the vertices of level $i - 1$ are already drawn, it is possible to determine, for each vertex v_j ($j = 1, \ldots, h$) of level i, the angle α_j of the smallest wedge W_j centered at c and containing all the neighbors $u_j^1, \ldots, u_j^{m(j)}$ of v_j. The wedge W_j of a vertex v_j is depicted as a shaded region in Fig. 3(a). Note that, $\sum_j \alpha_j = 2\pi$, and hence at most one angle α_j, with $1 \leq j \leq h$, can be greater than or equal to π.

First, we study the case (Case 1) when there exists one angle $\alpha_j \geq \pi$. Note that, there exists at least one quadrant Q such that Q is not completely contained into W_j, while the opposite quadrant of Q is. Refer to Fig. 3(a). Then, by construction, there exist two consecutive points p_a' and p_b' of C_{i-1}^b in Q that are not in W_j (they might be on the two delimiting half-lines of W_j) and a point p_c' of C_{i-1}^b in the opposite quadrant of Q such that triangle (p_a', p_b', p_c') contains circle C_i^m. This implies that triangle (p_a, p_b, p_c) contains C_i^m, as well, where p_a and p_b are the points of C_{i-1}^m on the same radius as p_a' and p_b', respectively.

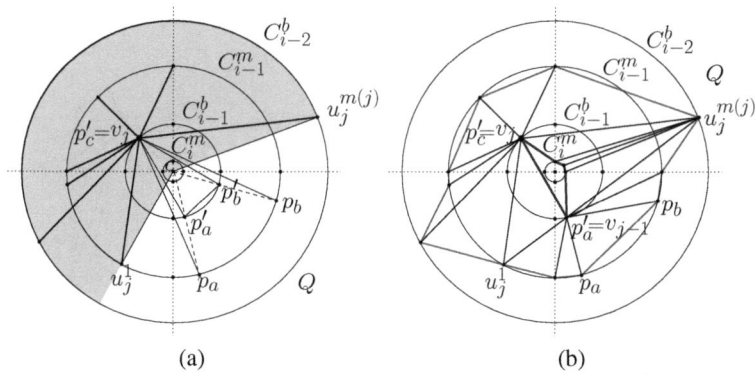

(a)   (b)

Fig. 3. (a) Case 1. Placement of a vertex v_j such that $\alpha_j \geq \pi$. (b) Case 1.1.1. There exists one angle $\alpha_j \geq \pi$, $v_{j-1} = v_{j+1}$, and $v' = v_{j-1}$.

Place vertex v_j on point p'_c and draw the edges between v_j and its neighbors $u^1_j, \ldots,$ $u^{m(j)}_j$. As (p_a, p_b, p'_c) contains C^m_i, none of such edges crosses C^m_i.

Note that, vertex u^1_j (vertex $u^{m(j)}_j$) has at least one neighbor v' (one neighbor v'') of level i different from v_j, possibly $v' = v_{j-1}$ (possibly $v'' = v_{j+1}$).

First (Case 1.1), suppose that $v_{j-1} = v_{j+1}$. We distinguish three cases, based on whether $v' = v_{j-1}$ (Case 1.1.1), $v'' = v_{j+1}$ (Case 1.1.2), or none of the two cases holds (Case 1.1.3). Cases 1.1.1 and 1.1.2 are mutually exclusive.

If $v' = v_{j-1}$ (Case 1.1.1), place v_{j-1} on p'_a. By construction, triangle (p'_a, p'_b, p'_c) contains C^m_i, which implies that edges (v_j, v_{j-1}), $(u^{m(j)}_j, v_{j-1})$, and (u^1_j, v_{j-1}) do not cross C^m_i. Also, all the vertices of level i that remain to be drawn are adjacent to $u^{m(j)}_j$. As such vertex, which lies in a quadrant Q on circle either C^m_{i-1} or C^b_{i-1}, has complete visibility to all the n_i points of circle C^m_i in the same quadrant Q, it is possible to draw all its neighbors on such points so that the polygon composed of vertices of level i contains C^m_i. See Fig. 3(b).

If $v'' = v_{j+1}$ (Case 1.1.2), then place v_{j+1} on p'_b and place the other vertices analogously to the previous case.

If none of the two cases holds (Case 1.1.3), we further distinguish three cases, based on whether u^1_j and $u^{m(j)}_j$ lie in opposite quadrants, in adjacent quadrants, or in the same quadrant. In the first case (see Fig. 4(a)), place v_{j+1} on either p'_a or p'_b and apply the same drawing algorithm as in the previous cases. If they lie in adjacent quadrants Q and Q' (see Fig. 4(b)), place v_{j-1} on the cardinal point, say $p^E(C^m_i)$, that is between Q and Q'. Note that, the wedge W centered at $p^E(C^m_i)$, delimited by the half-lines from $p^E(C^m_i)$ to u^1_j and from $p^E(C^m_i)$ to $u^{m(j)}_j$, and whose angle is smaller than π is external to quadrilateral $(p^N(C^m_{i-1}), p^E(C^m_i), p^S(C^m_{i-1}), p^W(C^m_i))$. As, by construction, such a quadrilateral contains all the points of C^m_i, W does not contain any of these points. Hence, both u^1_j and $u^{m(j)}_j$ have complete visibility to all the n_i points of quadrants Q and Q' of circle C^m_i, respectively, and it is possible to draw all their neighbors on such points. Finally, if u^1_j and $u^{m(j)}_j$ lie in the same quadrant (see Fig. 4(c)), they both have

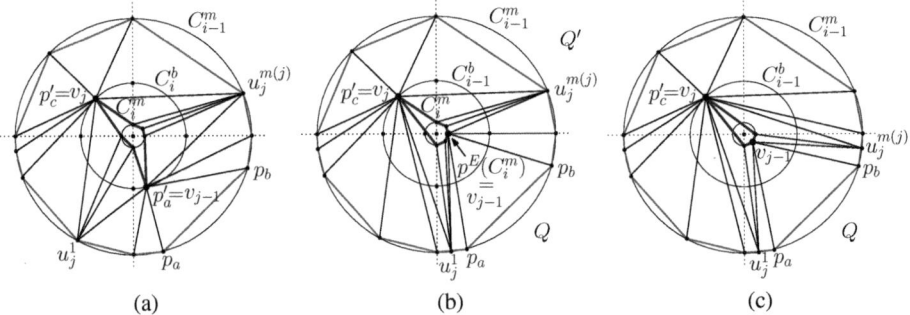

(a)   (b)   (c)

Fig. 4. Case 1.1.3. There exists one angle $\alpha_j \geq \pi$, $v_{j-1} = v_{j+1}$, $v' \neq v_{j-1}$, and $v'' \neq v_{j+1}$. Illustrations of the cases in which u^1_j and $u^{m(j)}_j$ lie (a) in opposite quadrants, (b) in adjacent quadrants, and (c) in the same quadrant.

visibility to all the points of C_i^m in such quadrant, and all their neighbors, including v_{j+1}, can be drawn on such points.

In each of the cases, all the vertices of level i are on the main circle C_i^m of level i, except for vertex v_j and, in one case, for vertex v_{j-1}, which are on the back-up circle C_{i-1}^b of level $i-1$. Also, no vertex is drawn on C_i^m in the same quadrant as the vertex (v_j or v_{j-1}) that is on C_{i-1}^b. Hence, given the two lines through v_j (through v_{j-1}) tangent to C_i^b, the triangle composed of v_j (of v_{j-1}) and of the two tangent points does not contain any vertex of level i placed on a point of C_i^m. It follows that the constructed drawings are 2-radial drawings.

Suppose (Case 1.2) that $v_{j-1} \neq v_{j+1}$. Let $u_{j-1}^1, \ldots, u_{j-1}^{m(j-1)}$ be the neighbors of v_{j-1} of level $i-1$. Note that $u_{j-1}^{m(j-1)} = u_j^1$. If u_{j-1}^1 is in the same quadrant as u_j^1 (Fig. 5(a)), place the first neighbor v_j^1 of u_j^1 on the first cardinal point of C_i^m encountered when rotating clockwise the radius to u_j^1. If it is in the adjacent quadrant (Fig. 5(b)), place v_{j-1} on the cardinal point of C_i^m between such two quadrants. Finally, if it is in the opposite quadrant (Fig. 5(c)), place v_{j-1} on a point p^* of C_i^m in its adjacent quadrant such that triangle (u_j^1, u_{j-1}^1, p^*) does not contain any point of C_i^m, which exists by construction. Then, place the first neighbor v_{j-1}^1 of u_{j-1}^1 different from v_{j-1} on the first cardinal point encountered when rotating clockwise the radius to u_{j-1}^1.

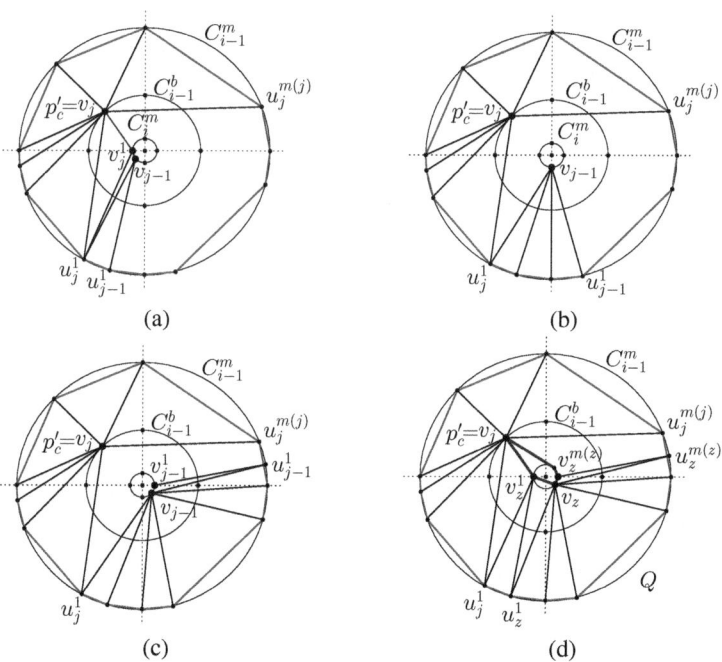

Fig. 5. Case 1.2. There exists an angle $\alpha_j \geq \pi$, $v_{j-1} \neq v_{j+1}$ and $v' \neq v_{j-1}$. Illustrations of the cases in which u_j^1 and u_{j-1}^1 lie (a) in the same quadrant, (b) in adjacent quadrants, and (c) in opposite quadrants. In (a), the placement of v_{j-1} is depicted, but it is not decided at this step. (d) Placement of vertices v_z such that u_z^1 and $u_z^{m(z)}$ are in opposite quadrants. Note that the first neighbor v_z^1 of u_z^1 coincides with v_{j-1}, while $v_z^{m(z)}$ does not coincide with v_{j+1}.

Then, consider each vertex v_z such that u_z^1 and $u_z^{m(z)}$ are in different quadrants. If such two quadrants are adjacent, place v_z on the cardinal point of C_i^m between them. If such two quadrants are opposite, then place v_z on a point p^* of C_i^m in the quadrant Q between them such that triangle $(u_z^1, u_z^{m(z)}, p^*)$ does not contain any point of C_i^m, and place the first neighbor v_z^1 of u_z^1 and the first neighbor $v_z^{m(z)}$ of $u_z^{m(z)}$ on the extremal points of Q, if such two vertices do not coincide with v_{j-1} and v_{j+1}, respectively. Note that, if they coincide with either v_{j-1} or v_{j+1}, the point where they had been placed in the previous step of the algorithm still allows for a planar drawing (see Fig. 5(d)).

Observe that, in each of the described cases all the vertices of level $i - 1$ whose neighbors of level i still remain to be placed have complete visibility to all the n_i points of a quadrant of circle C_i^m, and hence it is possible to draw all their neighbors on such points. Further, no vertex is drawn on C_i^m in the same quadrant as v_j. Hence, given the two lines through v_j (through v_{j-1}) tangent to C_i^b, the triangle composed of v_j (of v_{j-1}) and of the two tangent points does not contain any vertex of level i placed on a point of C_i^m. It follows that the constructed drawings are 2-radial drawings.

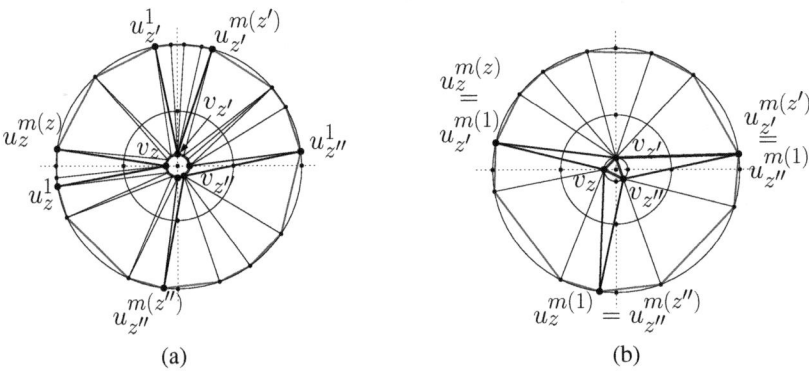

(a)   (b)

Fig. 6. Case 2. There exists no angle $\alpha_j \geq \pi$. Illustration for the cases when (a) $v_{z''}^1 \neq v_{z'}$ and $v_{z''}^{m(z'')} \neq v_z$, and (b) $v_{z''}^1 = v_{z'}$ and $v_{z''}^{m(z'')} = v_z$.

Suppose (Case 2) that there exists no angle $\alpha_j \geq \pi$. For each vertex v_z such that u_z^1 and $u_z^{m(z)}$ are in adjacent quadrants, place v_z on the cardinal point between them (see Fig. 6(a)). Then, for each vertex v_j such that u_j^1 and $u_j^{m(j)}$ are in opposite quadrants, place v_j on a point p^* of the quadrant Q between them such that triangle $(u_j^1, u_j^{m(j)}, p^*)$ does not contain any point of C_i^m, and place the first neighbors of u_j^1 and of $u_j^{m(j)}$ on the extremal points of Q, if such two vertices have not been already placed. Again, if this is the case, the point where they had been placed still allows for a planar drawing (see Fig. 6(a) and (b)).

Observe that, in each of the described cases all the vertices of level $i - 1$ whose neighbors of level i still remain to be placed have complete visibility to all the n_i points of a quadrant of circle C_i^m, and hence it is possible to draw all their neighbors on such points. The above discussion leads to the following.

Theorem 1. *Let $\mathcal{G}$ be the class of simply-nested planar graphs with k levels and such that each level i has n_i vertices. There exists a universal point set for $\mathcal{G}$ of size $8(\sum_{i=1}^{k} n_i + k)$.*

3 A Universal Point Set for Simply-Nested Planar Graphs

Let G be a simply-nested n-vertex planar graph. In Sect. 2 we described a universal point set of linear size provided that the number of levels of G and the number of vertices in each level is known. In this section we show how to limit the size even if such information is not known in advance.

3.1 A Simple Point Set of Size $O(n^{3/2})$

We group the levels of the graph into *dense levels* and *sparse levels*, depending on whether the level contains at least $\sqrt{n}$ vertices or not. Clearly, G contains at most $\sqrt{n}$ dense levels and at most n sparse levels.

Point set P is composed of $\sqrt{n}$ *dense levels*, each containing $8n + 8$ points, and n *sparse levels*, each containing $8\sqrt{n} + 8$ points. As in the point set of Sect. 2, levels of P are composed of a main and a backup circle. We start placing $\sqrt{n}$ outermost sparse levels. Then we place inside them a single dense level. Then again $\sqrt{n}$ sparse levels, followed by a dense level, and so on, until the total number of sparse levels reaches n and the number of dense levels reaches $\sqrt{n}$. This gives a point set of $n + \sqrt{n}$ levels and a total size of $O(n^{3/2})$ points.

Levels of G are assigned to levels of P as follows. Consider the levels of G starting from level 1 and the levels of P starting from the outermost one, proceeding inwards. Let i be the current level of G. If i is sparse, then assign it to the next available sparse level of P. Otherwise (i is dense), assign it to the next available dense level of P. Clearly, a dense level is skipped only if all the $\sqrt{n}$ sparse levels before it were already used. Hence, these previous sparse levels can account for the missing dense level. Summarizing, after scanning all n sparse and $\sqrt{n}$ dense levels of the graph, all its levels are assigned to the levels of the point set according to their size. We conclude with the following:

Lemma 1. *There is a universal point set of size $O(n^{3/2})$ for the class of simply-nested n-vertex planar graphs.*

3.2 Further Refinement

We refine now the classes of dense and sparse levels both of G and of P into m different classes $\mathcal{K}_i, 1 \le i \le m$. We say that level j of G, with n_j vertices, belongs to class $\mathcal{K}_i$, with $1 \le i \le m$, if $n^{(i-1)/m} \le n_j < n^{i/m}$. Hence the number of levels in class $\mathcal{K}_i$ is at most $n^{(m-i+1)/m}$, as G has n vertices. As discussed in Sect. 2, if the j-th level of the graph belongs to class $\mathcal{K}_i$, we can accommodate it in a level of P of size $8n^{i/m} + 8$. Hence, in what follows, a level of P containing $8n^{i/m} + 8$ points is called a level of the class $\mathcal{K}_i$.

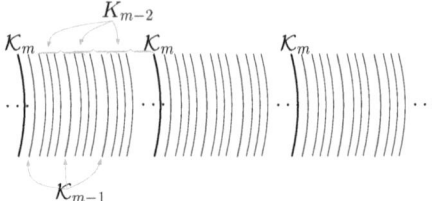

Fig. 7. Constructing the order of the levels

Now we discuss the number of levels and the size of P. The levels of P are first ordered and then placed on the plane one into the other according to the computed order. In order to construct such an order, we first place contiguously the $n^{1/m}$ levels of class $\mathcal{K}_m$ (each having $8n + 8$ points). Then, to the right of each level of class $\mathcal{K}_m$, we insert $n^{1/m}$ levels of class $\mathcal{K}_{m-1}$ (each having $8n^{m-1/m} + 8$ points), in total $n^{2/m}$ levels. We iterate this construction with increasing $i \leq m - 1$: to the right of each of the $n^{(i+1)/m}$ levels of class $\mathcal{K}_{m-i}$, we insert $n^{1/m}$ levels of class $\mathcal{K}_{m-i-1}$ (each having $8n^{m-i-1/m} + 8$ points), which gives in total $n^{(i+1)/m}$ levels. See Fig. 7. Finally, we scan the constructed order from right to left and construct the circles as in Sect. 2.

Summarizing, the total number of points for class $\mathcal{K}_i$ is $\Theta(n^{(m+1)/m})$. Thus, the overall number of points in P is $\Theta(mn^{(m+1)/m}) = \Theta(nmn^{1/m})$. Choosing m such that $mn^{1/m}$ is minimal, we get $m = \Theta(\frac{\log n}{\log \log n})$. Thus the total size of the constructed point set is $O(n(\frac{\log n}{\log \log n})^2)$.

Next we assign the levels of G of class $\mathcal{K}_i$ to the levels of P of class $\mathcal{K}_i$, $i = 1, \ldots, m$, by proceeding from the outside to the center. Intuitively we assign the next graph level of class $\mathcal{K}_i$ to the next unused point set level of class $\mathcal{K}_i$. To show the correctness we give a more formal description.

Let R_m be the minimal sequence of consecutive levels of G, starting from the outer level, that contains in total at least $n^{(m-1)/m}$ and at most n vertices. Note that sequence R_m ends latest at the outermost level of class $\mathcal{K}_m$. For the point set P, we similarly define a block of levels B_m to be the sequence of outer levels of P ending and including the outermost level of the class $\mathcal{K}_m$. We will describe below how to map the graph levels of R_m to the point set levels of B_m. Then, we shrink G by $G \setminus R_m$ and P by $P \setminus B_m$ and iterate. Note that by the structure of the graph and the point set we do this at most $n^{1/m}$ times.

If R_m contains a level of class $\mathcal{K}_m$, we map it to the single level of B_m of class $\mathcal{K}_m$, which is also the last level of B_m, by construction. The other levels of R_m have at most $n^{m-1/m}$ vertices. We repeat the above procedure: we identify a minimal initial sequence R_{m-1} of R_m that contains at least $n^{(m-2)/m}$ and at most $n^{(m-1)/m}$ vertices in total. Note that if $R_m = R_m \setminus R_{m-1}$ then this can be done at most $n^{1/m}$ times, as otherwise R_m would not be minimal. Concerning the point set, we set B_{m-1} to be the minimal sequence of outer levels of B_m that contains a single level of the class $\mathcal{K}_{m-1}$. Putting $B_m = B_m \setminus B_{m-1}$ this procedure can be applied exactly $n^{1/m}$ times, because of the structure of the point set. Finally, the graph levels B_{m-1} are mapped to the point-set levels R_{m-1} recursively. Summarizing the above we have the following theorem.

Theorem 2. *There is a universal point set of size $O(n(\frac{\log n}{\log \log n})^2)$ for the class of simply-nested n-vertex planar graphs.*

4   Concluding Remarks

In this paper we described a $O(n(\frac{\log n}{\log \log n})^2)$-size universal point set for simply-nested n-vertex planar graphs, doing a step towards answering the well-known open problem on the size of the smallest universal point set for planar graphs.

Several problems remain open in this field: (a) We use points with real coordinates. Is it possible to find a small point set for simply-nested planar graphs with points at integer coordinates and with an overall polynomial area? (b) Simply-nested planar graphs do not have chords between vertices of the same level. Is it possible to find a small point set if such chords are allowed? (c) Is there a small point set for k-outerplanar graphs if k is equal to 2 or 3?

References

1. Open problem garden, http://garden.irmacs.sfu.ca
2. Bachmaier, C., Brandenburg, F.J., Forster, M.: Radial level planarity testing and embedding in linear time. Journal of Graph Algorithms and Applications 9 (2005)
3. Baker, B.S.: Approximation algorithms for np-complete problems on planar graphs. J. ACM 41, 153–180 (1994)
4. Bose, P.: On embedding an outer-planar graph in a point set. Computat. Geom. Th. Appl. 23(3), 303–312 (2002)
5. Cabello, S.: Planar embeddability of the vertices of a graph using a fixed point set is NP-hard. J. Graph Alg. Appl. 10(2), 353–366 (2006)
6. Chrobak, M., Karloff, H.: A lower bound on the size of universal sets for planar graphs 20, 83–86 (1989)
7. Chrobak, M., Nakano, S.: Minimum-width grid drawings of plane graphs. Computational Geometry 11(1), 29–54 (1998)
8. Cimikowski, R.J.: Finding hamiltonian cycles in certain planar graphs. Information Processing Letters 35(5), 249–254 (1990)
9. Demaine, E.D., Mitchell, J.S.B., O'Rourke, J.: The open problems project, http://maven.smith.edu/~orourke/TOPP/
10. Fraysseix, H., Pach, J., Pollack, R.: How to draw a planar graph on a grid. Combinatorica 10, 41–51 (1990)
11. Gritzmann, P., Pach, B.M.J., Pollack, R.: Embedding a planar triangulation with vertices at specified positions. Amer. Math. Mont. 98, 165–166 (1991)
12. Kurowski, M.: A 1.235 lower bound on the number of points needed to draw all n-vertex planar graphs. Information Processing Letters 92(2), 95–98 (2004)
13. Schnyder, W.: Embedding planar graphs on the grid. In: Proceedings of the First Annual ACM-SIAM Symposium on Discrete Algorithms, SODA 1990, pp. 138–148 (1990)
14. Yannakakis, M.: Embedding planar graphs in four pages. J. Comput. Syst. Sci. 38, 36–67 (1989)

Graph Visualization

Jarke J. van Wijk

Eindhoven University of Technology, Eindhoven, The Netherlands
vanwijk@win.tue.nl
http://www.win.tue.nl/~vanwijk

Black and white node link diagrams are the classic method to depict graphs, but these often fall short to give insight in large graphs or when attributes of nodes and edges play an important role. Graph visualization aims obtaining insight in such graphs using interactive graphical representations. A variety of ingredients, including color, shape, 3D, shading, and interaction can be used to this end. In this invited talk an overview is given of work on graph visualization of the visualization group of Eindhoven University of Technology, The Netherlands. A wide variety of examples is shown and discussed using demos and animations.

One focus of the group has been software visualization, aiming towards the development of technology that makes it easier to understand the structure of large software artifacts. An early example were cushion treemaps, developed to visualize hierarchical data, in particular file systems (SequoiaView), another more frivolous example are our botanically inspired tree visualizations. State space models lead to very large, but also often symmetrical graphs, which can be exploited to obtain clear and compact visualizations. Combinations of hierarchical data and networks occur often in practice, a typical case is the visualization of call-graphs of software systems. Such data can be shown using an interactive incidence-matrix or using hierarchical edge bundles.

Besides the development of new methods and techniques, evaluation is an important and often difficult aspect. Details of visualizations can be evaluated using controlled user experiments, examples are the assessment of proper scales for icons and different ways to show edge direction.

Finally, some other work related to graphs is shown. Myriahedral projections are a new method to generate cartographic maps almost without distortion that lean heavily on graphs. Platonic solids, such as the cube and dodecahedron, are examples of so-called regular maps: highly symmetric graphs, embedded on a surface. Examples of such regular maps for surfaces of genus 2 and higher are presented.

M. van Kreveld and B. Speckmann (Eds.): GD 2011, LNCS 7034, p. 86, 2012.

Advances in the Planarization Method: Effective Multiple Edge Insertions

Markus Chimani[1,*] and Carsten Gutwenger[2]

[1] Inst. of Computer Science, FSU Jena
markus.chimani@uni-jena.de
[2] Dep. of Computer Science, TU Dortmund
carsten.gutwenger@tu-dortmund.de

Abstract. The planarization method is the strongest known method to heuristically find good solutions to the general crossing number problem in graphs: starting from a planar subgraph, one iteratively inserts edges, representing crossings via dummy nodes. In the recent years, several improvements both from the practical and the theoretical point of view have been made. We review these advances and conduct an extensive study of the algorithms' practical implications. Thereby, we present the first implementation of an approximation algorithm for the crossing number problem of general graphs, and compare the obtained results with known exact crossing number solutions.

1 Introduction

Given a graph $G = (V, E)$, the *crossing number* problem asks how to draw G into the plane with the fewest possible number of edge-crossings. The *planarization method* is the probably best known and most successful heuristic to tackle the crossing number problem in practice. In its simplest form it runs in two phases: first, a (large) planar subgraph $G' = (V, E') \subseteq G$ is computed. Then, the temporarily removed edges $F := E \setminus E'$ are re-inserted one after another, each time solving a *single edge insertion* problem. This problem can be stated as follows: Let H be a planar graph, and e an edge not yet in H. We search for a smallest planar graph H^+ which represents a drawing of $H + e$ where edge crossings are replaced by dummy nodes of degree 4, and all these crossings occur on the edge e. Hence, when removing the image of e from H^+, we obtain a planar embedded H. Using this method, each edge of F is inserted in a planar graph until we obtain a *planarization* G^+, representing G in a planar way by using dummy nodes for crossings.

In the first proposal [1] of this heuristic, the insertion problem was considered w.r.t. a *fixed* embedding (cyclic order of the edges around their incident nodes) of the planar graph H. (I.e., after obtaining the planar subgraph G', one embedding of G' is fixed and retained throughout the whole insertion phase.) A simple linear-time BFS-algorithm in the dual graph of H suffices to find an optimal solution.

---

* Markus Chimani was funded by a Carl-Zeiss-Foundation juniorprofessorship.

M. van Kreveld and B. Speckmann (Eds.): GD 2011, LNCS 7034, pp. 87–98, 2012.

Later, and rather surprisingly, it was shown in [14] that there exists a linear-time algorithm, using the SPQR-tree datastructure, which finds the optimal insertion path for e over all possible planar embeddings of H. In [13] it was shown that this approach is in practice vastly superior to the former in terms of the overall obtained number of crossings.

In recent years it was furthermore shown that there exists a (rather complex) insertion algorithm to optimally insert a vertex with all its incident edges into a planar graph [4] (*vertex insertion*), while it is NP-hard to insert an arbitrary set of edges simultaneously [18] (*multiple edge insertion*).

Most interestingly, the single edge insertion problem (over all possible embeddings of H) is known to approximate the crossing number of $H+e$ within a factor of $\Delta/2$ (where Δ is the graph's maximum degree) [15, 2], and also the vertex insertion problem approximates the crossing number of the resulting graph [6]. In particular, the proof of the latter can be generalized to show that an optimal multiple edge insertion solution—w.r.t. an edge set F—would approximate the crossing number of $G' + F$ within a factor only dependent on Δ and $|F|$ [6].

Hence, the question arose whether this multiple edge insertion problem can be efficiently approximated. After a rather complicated approach in [8], a simpler and at the same time approximation-wise stronger algorithm was presented only recently [5]. The algorithm reuses concepts of the SPQR-tree based single edge insertion and seems simple enough to be implemented and used in practice. The latter paper also shows that the traditional iterative single edge insertion algorithm cannot be an approximation strategy for the crossing number of G.

Contribution. In this paper we present recent advances of the planarization approach from a practical point of view. On the one hand, we show how to improve on the traditional approach of iteratively inserting single edges, via the use of strong postprocessing routines. On the other hand, we give the first practical implementation of a simultaneous multiple edge insertion algorithm—hence, this is also the first practical study of any crossing number approximation algorithm for arbitrary graphs. By considering graph classes of known crossing numbers (either from theory or from the application of the currently strongest branch-and-cut based exact crossing minimization algorithm [7]) we can deduce a practically very good performance of these heuristics, as they usually find optimum, or at least very-close-to-optimum, solutions.

2   Planarization Approach

In order to present our algorithmic choices and modifications, we first have to briefly introduce two central decomposition structures, used in all algorithms dealing with the insertion problem over all possible embeddings of H. In the above sketched planarization scheme, we can assume that the original graph G is connected—otherwise the crossing number problem decomposes into multiple independent problems. Furthermore the initial planar subgraph G' can be assumed to be maximal and hence also connected. For the single edge insertion algorithms, we will usually consider any intermediate graph H; for the multiple edge insertion algorithm we set $H := G'$.

First, we use the well known *BC-tree* $\mathcal{B} = \mathcal{B}(H)$ of H which is a tree with two different node types B and C: For each cut vertex (maximal two-connected subgraph or bridge, summarized under the term *block*) in H, $\mathcal{B}$ contains a unique corresponding C-node (B-node, respectively). Two nodes in $\mathcal{B}$ are adjacent if and only if they correspond to a block and a cut vertex, where the former contains the latter. We can construct such a linear-sized BC-tree $\mathcal{B}$ in linear time.

Based thereon, we can further decompose non-trivial blocks (i.e., non-bridges) via SPQR-trees [10]: While they are more complicated than BC-trees, they also only require linear size and can be constructed in linear time [16,12]. This datastructure is particularly interesting, as it directly encodes all (exponentially many) planar embeddings of its underlying block. We use the definition from [3,5] which does not use Q-nodes, and therefore call the decomposition tree $\mathcal{T} = \mathcal{T}(H')$ of a non-trivial block H' *SPR-tree* for conciseness. Chiefly summarizing, each tree node corresponds to a *skeleton*, which is a "sketch" of H' where certain subgraphs are replaced by virtual edges. By repeatedly merging the skeletons of adjacent nodes (at their virtual edges representing each other), we can obtain the original graph, and each virtual edge hence represents a 2-cut (*split pair*) in H'. Most importantly, a skeleton can only be one of three types: The skeleton of an S-node ("serial") is a simple cycle; the skeleton of a P-node ("parallel") consists of two vertices and multiple edges between them; the skeleton of an R-node is a simple triconnected graph. Note that a planar triconnected graph has a unique embedding (up to mirroring).

In the algorithmic description of the multiple edge insertion approximation algorithm [5], an amalgamated version of these trees, the so-called *con-tree*, is considered: a BC-tree, directly storing SPR-trees at the non-trivial B-nodes.

Single Edge Insertion. We will briefly recapitulate the central ingredients of the exact linear-time algorithm by Gutwenger et al. [14] to solve the single edge insertion problem over all possible embeddings of H. Let v_1, v_2 be the vertices we want to connect in H via a new edge. First consider a fixed embedding of H and let H_D be its dual. We define an *insertion path* to be a path in H_D connecting a face incident to v_1 with a face incident to v_2. The length of this path is then the number of edge crossings necessary to insert the edge $\{v_1, v_2\}$ into embedded H along this path; each dual edge in the insertion path corresponds to an edge in H that is to be crossed. We can directly compute the shortest insertion path via standard breadth-first search (BFS).

Now consider H with variable embedding. Let L be the unique shortest path in $\mathcal{B}(H)$ from a B-node containing v_1 to a B-node containing v_2. The optimal insertion path for $\{v_1, v_2\}$ in G can be obtained by concatenating the optimal insertion paths within the (non-trivial) blocks on this path L; we can always nest blocks at a common cut vertex into each other such that there arise no additional crossings. For a block H' represented by a B-node on L, let $v_i^{H'}$, $i = 1, 2$, denote v_i if $v_i \in V(H')$, or the cut vertex in H' closest to v_i otherwise. It remains to, for each non-trivial block H', find optimal insertion paths from any face incident to $v_1^{H'}$ to any face incident to $v_2^{H'}$.

Therefore, let $Q_{H'}$ be the unique shortest path in $\mathcal{T}(H')$ from a skeleton containing $v_1^{H'}$ to a skeleton containing $v_2^{H'}$. It was shown in [14] that only the embeddings of the skeletons along $Q_{H'}$ matter. In a nutshell, the algorithm walks along these skeletons and fixes suitable embeddings for the skeletons, one after another. Finally, an optimal embedding is found and fixed, and one can use the simple BFS algorithm on the dual graph to insert the edge $\{v_1^{H'}, v_2^{H'}\}$ optimally.

In the following, we can consider a *con-chain* Q of the edge $\{v_1, v_2\}$ as an extended version of L, where the "subpaths" $Q_{H'}$ are stored at each non-trivial block H' along L.

Multiple Edge Insertion. Let us briefly review the approximation algorithm for the multiple edge insertion problem by Chimani and Hliněný [5]. Let $H := G'$ be the initial planar subgraph of G into which to insert the edges $F = \{e_i\}_{1 \leq i \leq |F|}$. Assume we could independently insert each edge $e_i \in F$ into H. Using the above algorithm for single edge insertions, we would obtain a con-chain Q_i for each edge e_i, and therefore a so-called *embedding preference* for each node on Q_i w.r.t. e_i. Coarsely speaking, we obtain a common embedding of H via a voting scheme on the (possibly conflicting) embedding preferences per con-tree node, ensuring that at any node at least one preference is satisfied. After realizing the so-chosen embedding, we can once again use the simple BFS algorithm in the dual graph to insert the edges into this fixed embedding.

The prove-wise crucial part in the algorithm is that any two con-chains Q_i, Q_j are either disjoint or they intersect in one sub-chain. Hence, two con-chains (think of simple paths) "deviate" at at most two nodes in the con-tree (think of a regular tree): once when the two paths come together and once when they part. Roughly speaking, it is shown in [5] that the embedding preferences for the tree nodes can differ only at these two "places" (called *passes*), whereby the exact definition of *pass* is quite involved and might in fact span over up to three con-tree nodes. Yet, overall we can bound the number of nodes where some con-chains disagree on the embedding preference, as well as the additionally necessary number of crossings to route an edge through a skeleton that is differently embedded than desired. This gives an approximation factor for the optimal multiple edge insertion w.r.t. G' and F, and, subsequently, for the crossing number of G.

It remains to clarify what an *embedding preference* actually is: Observe that S-nodes do not allow different embeddings of their skeletons. For an R-node (a triconnected planar graph), we have only a unique planar embedding and its mirror. For a P-node, each inserted edge may want two particular edges of the skeleton to be cyclicly adjacent (in, say, clockwise direction). Finally, for a C-node each inserted edge may want a particular incident face in an adjacent block to be identified with a particular incident face in another adjacent block.

3   Engineering

Iterative Single Edge Insertion and Postprocessing. In the traditional planarization heuristic, we will "simply" insert the temporarily removed edges

F one after another into the planar subgraph. After each insertion, we replace the arising crossings by dummy nodes, and hence proceed with a planar graph. There are various ways to fine-tune the obtained result via postprocessing, as already discussed in [13]. The simplest—and in fact quite effective—variant is to start the insertion process multiple times, each time with a different, randomized order of F. Additionally, each such insertion run can be improved: After having inserted all edges, we can again remove some original edge e from the planarization (i.e., we remove all the subedges and dummynodes that represent e), and re-insert it, possibly requiring fewer crossings. For this operation, we can consider either the inserted edges F (*ins*), all edges (*all*), or the $x\%$ of the edges with the most crossings (*most*, for some constant x). In [13] it was shown, that these approaches lead to greatly improved results.

Herein, we propose a further improvement on these methods. The *incremental* (*inc*) strategy basically applies the *all* strategy after each single insertion step. I.e., after the insertion of an edge $e \in F$, we try to remove and reinsert every other edge already in the graph, in order to obtain a better crossing number, before proceeding with the next edge from F. We will see, that this approach again dominates the previously best strategy *all*, though at the cost of a vastly increased running time.

Note that all these strategies—when applied in a fixed embedding setting—are also applicable to the multi-edge insertion problem, after fixing an embedding into which all edges F need to be inserted. Formally, the *inc* setting has to restrict itself to only try to reinsert the edges F, in order to retain the approximation guarantee. Interestingly, after having obtained a postprocessed solution in the fixed embedding, we can run the *all* postprocessing where the graph's embedding may change, i.e., using the optimal edge insertion over all possible embeddings! As the solution value never decreases, the algorithm retains its approximation guarantee and improves the number of crossings in practice.

From the approximation point of view, we can observe that the first part of the algorithm (fixing a suitable overall embedding) tries to minimize the number of crossings between F and G', while the postprocessing routines most importantly try to reduce the number of crossings between edges of F—their quantity can only be estimated as $\binom{|F|}{2}$ in the formal quality guarantee.

Implementing Multiple Edge Insertion. In [5], certain aspects of the multiple edge insertion algorithm are described to be suitable for a comparatively smooth approximation proof. When implementing the algorithm, we take some different, though completely equivalent, routes. A main point of deviation is the consideration of *dirty passes*, i.e., con-tree node tuples where multiple insertion paths disagree on their preferred embedding. We highlight the two main divergent choices here. Overall, our viewpoint allows a quite simpler implementation than would be easily deduced from the theoretical proofs of [5] alone.

Con-Tree. Originally, an amalgamated version of BC- and SPR-trees is proposed, which allows to talk about a single chain (path) for each inserted edge. In the implementation, we perform the algorithm differently: First, we compute a suitable combinatorial embedding for each non-trivial block independently. Only then, we consider the C-nodes at which the blocks are joined. From the formal definition of dirty passes, we can easily deduce that C-nodes do not interact with other nodes in terms of realized embedding preferences, and hence we can independently choose which faces to embed into each other at cut vertices, after fixing the embeddings of the incident blocks.

This modification allows us to consider only two-connected graphs and SPR-trees in the following, vastly simplifying implementation details as most of the infrastructure necessary for single edge insertions can be reused.

Merging the Embedding & Repairing Dirty Passes. In [5], the formal definition of dirty passes needs to group nodes as tuples of 1–3 SPR-tree nodes and requires a tie-breaking to prohibit invalid node tuple overlaps. Yet, from the proofs it becomes clear that this is merely necessary to correctly estimate the *number* of these passes. Within the algorithm, these passes are only detected in order to identify possible flips to prohibit too many such situations. For this purpose alone, a much simpler strategy suffices: Usually, we consider one insertion path after another: We traverse its SPR-nodes and fix the embedding of each skeleton along this path as preferred. When a visited node already has a fixed embedding, we (coarsely speaking) try to flip the predecessor nodes of our current path in order to avoid dirty passes. Instead of checking the full case distinction in the dirty pass definition, it suffices to consider the case where the currently visited nodes ν and its predecessor (disregarding S-nodes) μ are P- and/or R-nodes:

We say an embedding preference at a P-node *agrees* with a fixed embedding of this node's skeleton, if the specified two edges occur clockwise neighboringly. An embedding preference of an R-node is simply a binary flag specifying whether to use a "default" planar embedding of the node's skeleton or the "mirror" (only these two embeddings exist). Now, we only have to flip μ and its predecessors along the insertion path iff μ and ν are *switching*, i.e., the new embedding preferences agree with the already fixed embedding of one of these two nodes, and agrees with the *flipped* embedding of the other node.

Doing this for all such pairs ν, μ then also repairs dirty passes on node triples, if at all possible. In all other cases of dirty passes, no flip can improve the situation anyway and hence is not necessary. It is understood that this procedure performs the same flips as the more abstract merge routine described in [5], and hence the implementation retains the approximation guarantee.

4 Experiments

Experimental Setup. We implemented all algorithms using the C++ library OGDF[1] and ran our experiments on a Linux system with an Intel Core i7 (2.67 GHz) processor and 12 GB RAM. For each instance, all edge insertion algorithms were called with the same, pre-computed maximal planar subgraph, computed via the PQ-tree based planar subgraph algorithm [17] (best of 250

---

[1] Open Graph Drawing Framework, see http://www.ogdf.net

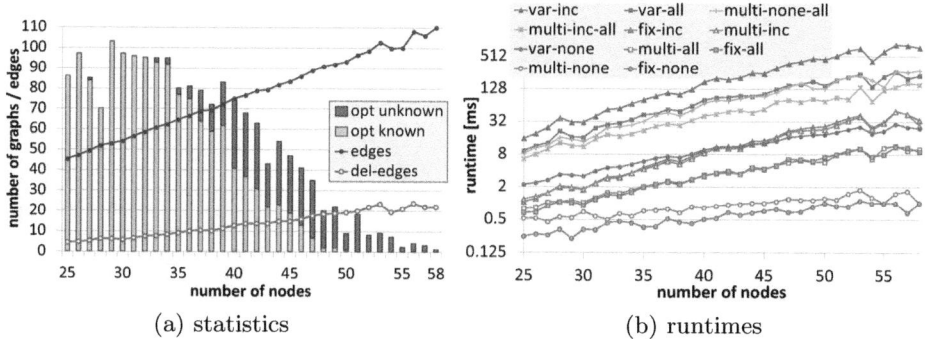

Fig. 1. *Rome* graphs

random runs, i.e., random choices of the initial *st*-edge for the numbering) and iteratively adding removed edges afterwards if they do not destroy planarity.

We consider four benchmark sets, the first two of which are the well-known *Rome* library [9] and *AT&T* graphs (available at http://graphdrawing.org/data.html). We first applied a reduction strategy that removes parallel edges, self-loops, and planar biconnected components, and reduces paths in the graph to single edges (unless this introduces parallel edges). We consider all remaining non-planar connected components with at least 25 nodes and at least two edges removed in the computed planar subgraph, which are 1843 graphs in the *Rome* set (25–58 nodes) and 311 graphs in the *AT&T* set (25–312 nodes). The *ISCA* graphs are hypergraphs taken from the ISCA'85 benchmark set of real world electrical networks, transformed into traditional graphs by substituting each hyperedge h by a new hypernode connected to all nodes contained in h, connecting all inputs (outputs) to a new node s_{in} (s_{out}, resp.), and introducing the edge $(s_{\mathrm{in}}, s_{\mathrm{out}})$. We used the same reduction and selection as described above leading to 20 graphs (25–223 nodes). Finally, the *KnownCR* graphs [11] are a collection of 1946 graphs with known crossing numbers (by proofs), consisting of generalized Petersen graphs ($P(m, 2)$, $P(m, 3)$) and products of cycles C_n, paths P_n, and 5-vertex graphs G_i ($C_m \times C_n$, $G_i \times P_n$, $G_i \times C_n$); these graphs have between 9 and 250 nodes. Our whole benchmark set can be downloaded from http://ls11-www.cs.uni-dortmund.de/people/gutweng/planexp.zip.

Rome Graphs. Fig. 1(a) gives an overview on the Rome benchmark set, displaying the number of graphs and average number of edges per node count. Furthermore, it shows the average number of edges deleted in the planar subgraphs and for how many of the graphs we know the exact crossing number from the branch-and-cut algorithm presented in [7,3].

We first consider the effect of postprocessing; see Fig. 2. We compare the results with the *best* known results (from our experiments and the branch-and-cut algorithm [3]) and show the relative difference between heuristic and best solution. Since we know the exact solutions for many of the graphs, this gives a very good impression on the actual quality of the heuristics. We note that the

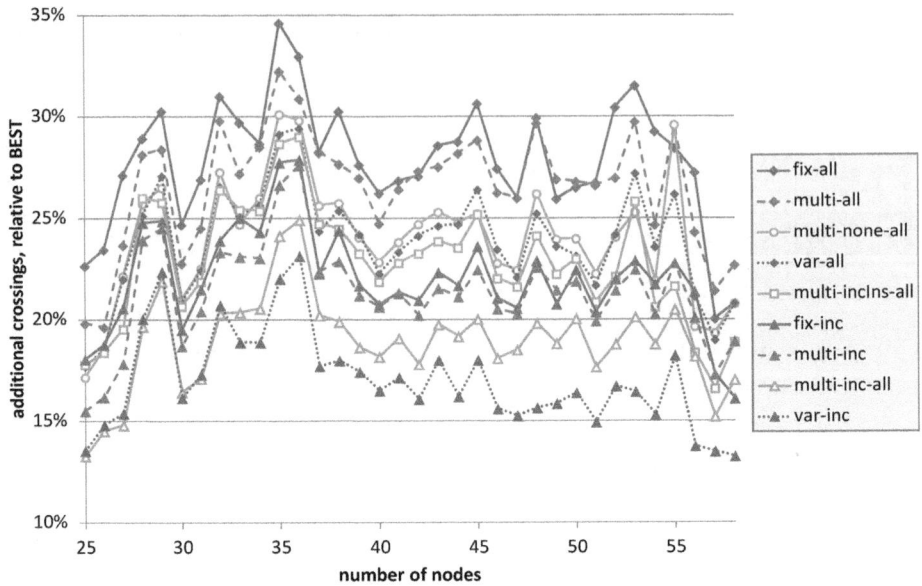

Fig. 2. Number of crossings for *Rome* graphs, relative to BEST known solutions

exact algorithm is clearly slower than any of the considered heuristics by orders of magnitudes, cf. [7,3]. As already observed in [13], postprocessing helps a lot, and this also holds for multiple edge insertion (the values without postprocessing lie between 60–70%). Our new incremental postprocessing achieves clearly better results than the previously best *all*, for all edge insertion strategies. We also observed that the advantage of *multi* over *fix* is large without postprocessing, but becomes smaller and smaller the more postprocessing is applied, since the postprocessing becomes the dominating factor and is the same for both.

Inspired by this observation, we experimented with an additional postprocessing for the *multi* strategy, where we reused the postprocessing with variable embedding; see Fig. 2. The variants *multi-none-all* and *multi-inc-all* perform *multi* with no or incremental postprocessing plus postprocessing with variable embedding afterwards; *multi-incIns-all* restricts the incremental postprocessing to the inserted edges. Whereas *multi-none-all* and *multi-incIns-all*—which retain the approximation guarantee—are about as good as *var-all*, *multi-inc-all*—which in theory does not give those guarantees—comes close to *var-inc* (for larger graphs, it lies between *var-all* and *var-inc*).

In practice, we want to obtain good solutions quickly, hence it is important to look at the runtimes; see Fig. 1(b). We can see that the overhead of *multi* compared to *fix* is small, and even becomes negligible if postprocessing is used. The *var* variants are always clearly slower, as they require a new SPR-decomposition after each edge insertion, whereas *multi* uses only a single such decomposition. The *inc* variants take about 2–4 times longer than *all*, which is acceptable regarding the achieved improvements in quality. For our postprocessing variants

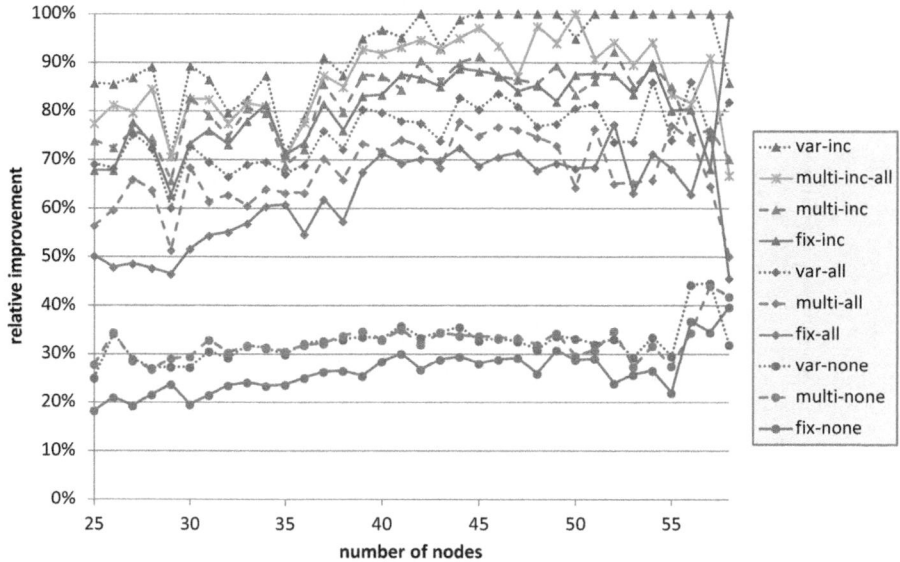

Fig. 3. Effect of permutations on number of crossings (*Rome* graphs)

for *multi* with additional *var*-postprocessing, we observe that more intensive postprocessing with fixed embedding reduces the effort required with the time-consuming *var*-postprocessing and results in smaller runtimes (i.e., *multi-inc-all* is faster than *multi-none-all*). Since *multi-inc* requires a similar runtime as *var-none* but is in quality even better than *var-all* (the previously quality-wise best known heuristic variant), it is a very good choice in practice.

Fig. 3 finally studies the effect of randomly permuting the edges to be inserted (we considered 100 permutations here). The diagram shows the relative reduction of the gap between single run and best solution (hence, 100% means that 100 permutations led to the best solution). The main message is that permutations without postprocessing are not very effective, whereas the combination of postprocessing and permutations always gets significant improvements. The incremental postprocessing variant does not only lead to best results, but is also the most effective one in combination with permutations.

KnownCR Graphs. This collection allows us to further compare the heuristic results with actual crossing numbers. Fig. 4 summarizes our findings for some selected heuristics, showing the average relative deviation from the crossing number for the different graph classes. The class $P(m, 2)$ (all whose graphs have crossing number 2 or 3) could be solved to optimality by all heuristics and we omit it in the diagram. For the classes $P(m, 3)$, $G_i \times C_n$, and $G_i \times P_n$, all heuristics perform well, being only 2-13% away from the optimum, and their order with respect to quality is as expected. The class $C_n \times C_m$ shows some unusual behavior: Without permutations, the insertion strategy seems to have only a very small influence on the solution quality; surprisingly, with 100 permutations, *fix*

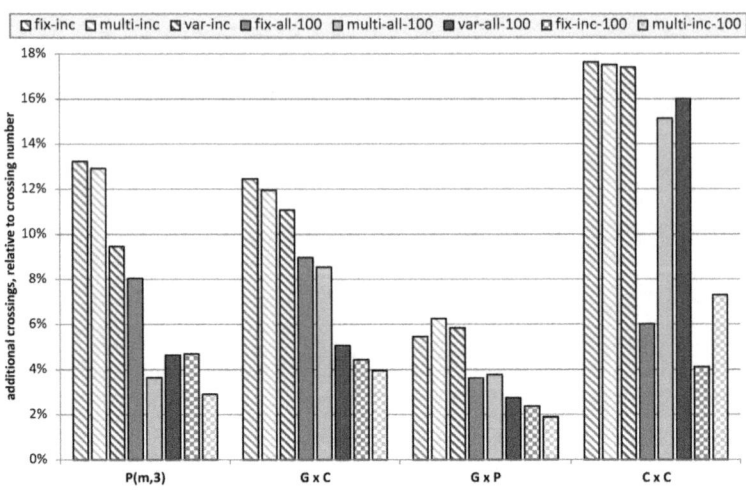

Fig. 4. Number of crossings for *KnownCR* graphs, relative to crossing number

is superior to both *multi* and *var*. Analyzing the data, we see that this happens only for a few graphs: the distinct runs of *fix* usually find slightly worse solutions than *multi* or *var*, but in some rare cases a much better solution is found. We assume that this is caused by the fact that accepting worse intermediate solutions while inserting the edges can lead to a better final solution.

AT&T Graphs. Whereas the *Rome* graphs are fairly homogeneous graphs with a simple structure and *KnownCR* consists of artificial graphs, the *AT&T* graphs are real-world graphs with quite diverse structures. For analyzing the results, we group the graphs according to the best found solutions (the first group contains graphs with 0,...,24 crossings; the last group with 700,...,799 crossings). Fig. 5 shows the relative difference between heuristic and best solution. We can confirm that incremental postprocessing clearly dominates *all* for all edge insertion strategies, and *var-inc* is by far the best strategy both without and with (*var-inc-100*) permutations. Multiple edge insertion is also slightly better than *fix*.

However, the domination of *var* comes at a price: Whereas *fix* and *multi* take about the same runtime, *var* is more than 10 times slower. Hence, *multi-inc* is again a good compromise, as it is even clearly faster (3–10 times) than *var-all*.

ISCA Graphs. We focus on the *multi* and *var* methods. Fig. 6 shows the relative difference between heuristic and best solution, for each graph in the benchmark set separately. The graphs are sorted by increasing number of edges deleted in the planar subgraph. We observe the effectiveness of postprocessing, underlining again that postprocessing is essential. We can also see that the *multi* variants come quite close to the corresponding *var* variants. This is again accompanied by a much better runtime of *multi*; in this case *multi-inc* is about 15–30 times faster than *var-inc*, and *multi-all* even about 100 times faster.

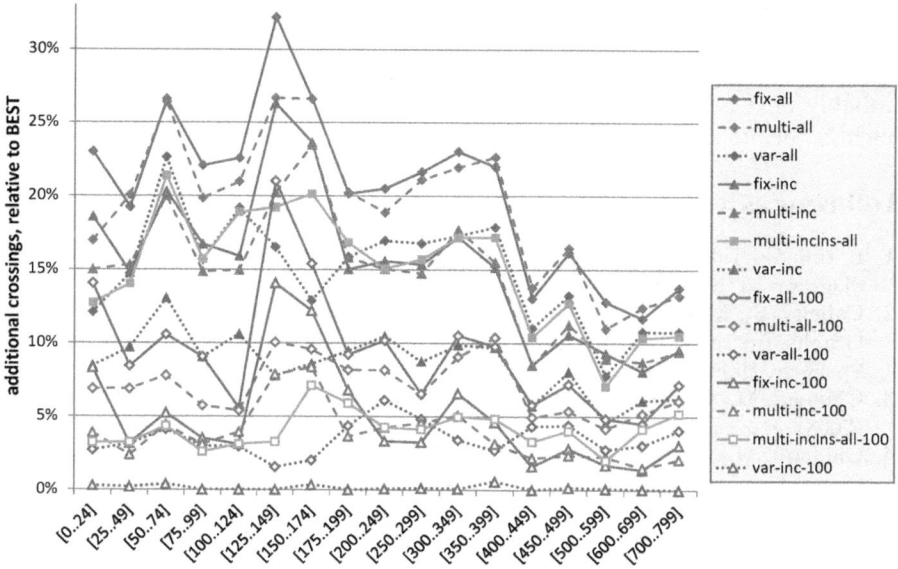

Fig. 5. Number of crossings for *AT&T* graphs, relative to best found solution

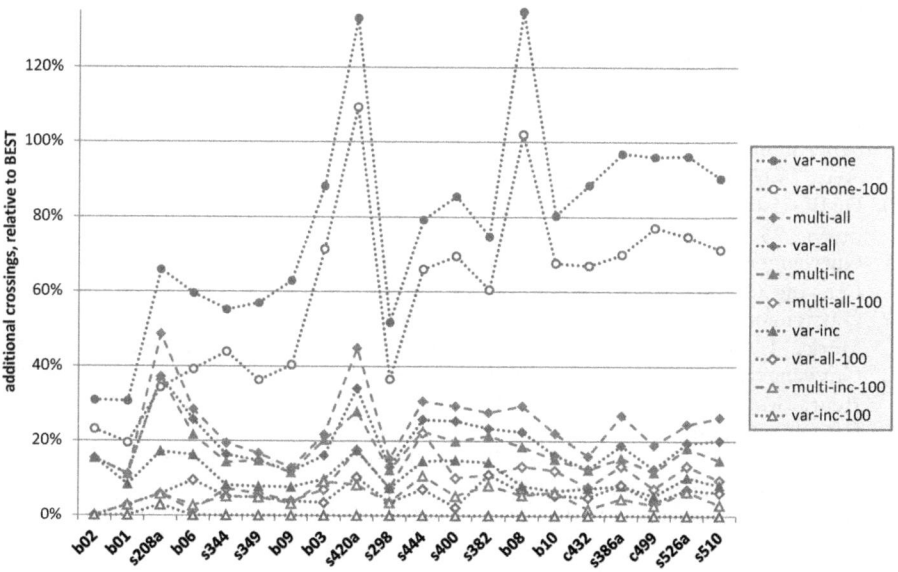

Fig. 6. Number of crossings for *ISCA* graphs, relative to best found solution

Conclusions. We presented *inc*, a new practically dominating postprocessing strategy for the planarization heuristic, and report on *multi*, the first implementation of any crossing minimization approximation algorithm for general graphs.

Both algorithms outperform any previously known heuristic in terms of solution quality, and if one cannot afford the relatively long running times for inserting all edges iteratively into a variable embedding using *inc*, the *multi* variants give the probably best balance between running time and solution quality: while being much faster, its solutions tend to be only slightly weaker than *inc*'s.

References

1. Batini, C., Talamo, M., Tamassia, R.: Computer aided layout of entity relationship diagrams. J. Syst. Software 4, 163–173 (1984)
2. Cabello, S., Mohar, B.: Crossing and Weighted Crossing Number of Near Planar Graphs. In: Tollis, I.G., Patrignani, M. (eds.) GD 2008. LNCS, vol. 5417, pp. 38–49. Springer, Heidelberg (2009)
3. Chimani, M.: Computing Crossing Numbers. PhD thesis, TU Dortmund, Germany (2008)
4. Chimani, M., Gutwenger, C., Mutzel, P., Wolf, C.: Inserting a vertex into a planar graph. In: Mathiru, C. (ed.) Proc. SODA 2009, pp. 375–383 (2009)
5. Chimani, M., Hliněný, P.: A tighter Insertion-Based Approximation of the Crossing Number. In: Aceto, L., Henzinger, M., Sgall, J. (eds.) ICALP 2011. LNCS, vol. 6755, pp. 122–134. Springer, Heidelberg (2011)
6. Chimani, M., Hliněný, P., Mutzel, P.: Vertex insertion approximates the crossing number for apex. Europ. J. Comb. (to appear, 2011)
7. Chimani, M., Mutzel, P., Bomze, I.: A New Approach to Exact Crossing Minimization. In: Halperin, D., Mehlhorn, K. (eds.) ESA 2008. LNCS, vol. 5193, pp. 284–296. Springer, Heidelberg (2008)
8. Chuzhoy, J., Makarychev, Y., Sidiropoulos, A.: On graph crossing number and edge planarization. In: Proc. SODA 2011, pp. 1050–1069. ACM Press (2011)
9. Di Battista, G., Garg, A., Liotta, G., Tamassia, R., Tassinari, E., Vargiu, F.: An experimental comparison of four graph drawing algorithms. Computational Geometry 7(5-6), 303–326 (1997)
10. Di Battista, G., Tamassia, R.: On-line planarity testing. SIAM Journal on Computing 25, 956–997 (1996)
11. Gutwenger, C.: Application of SPQR-Trees in the Planarization Approach for Drawing Graphs. PhD thesis, TU Dortmund, Germany (2010)
12. Gutwenger, C., Mutzel, P.: A Linear Time Implementation of SPQR Trees. In: Marks, J. (ed.) GD 2000. LNCS, vol. 1984, pp. 77–90. Springer, Heidelberg (2001)
13. Gutwenger, C., Mutzel, P.: An Experimental Study of Crossing Minimization Heuristics. In: Liotta, G. (ed.) GD 2003. LNCS, vol. 2912, pp. 13–24. Springer, Heidelberg (2004)
14. Gutwenger, C., Mutzel, P., Weiskircher, R.: Inserting an edge into a planar graph. Algorithmica 41(4), 289–308 (2005)
15. Hliněný, P., Salazar, G.: On the Crossing Number of Almost Planar Graphs. In: Kaufmann, M., Wagner, D. (eds.) GD 2006. LNCS, vol. 4372, pp. 162–173. Springer, Heidelberg (2007)
16. Hopcroft, J.E., Tarjan, R.E.: Dividing a graph into triconnected components. SIAM Journal on Computing 2(3), 135–158 (1973)
17. Jünger, M., Leipert, S., Mutzel, P.: A note on computing a maximal planar subgraph using PQ-trees. IEEE Trans. Comp.-Aided Design 17(7), 609–612 (1998)
18. Ziegler, T.: Crossing Minimization in Automatic Graph Drawing. PhD thesis, Saarland University, Germany (2001)

A Quantitative Comparison
of Stress-Minimization Approaches
for Offline Dynamic Graph Drawing*

Ulrik Brandes and Martin Mader

Department of Computer & Information Science, University of Konstanz
{Ulrik.Brandes,Martin.Mader}@uni-konstanz.de

Abstract. In dynamic graph drawing, the input is a sequence of graphs
for which a sequence of layouts is to be generated such that the quality
of individual layouts is balanced with layout stability over time. Qual-
itatively different extensions of drawing algorithms for static graphs to
the dynamic case have been proposed, but little is known about their
relative utility. We report on a quantitative study comparing the three
prototypical extensions via their adaptation for the stress-minimization
framework. While some findings are more subtle, the linking approach
connecting consecutive instances of the same vertex is found to be the
overall method of choice.

1 Introduction

A dynamic graph is a sequence of (static) graphs, often representing an evolving
structure at discrete times of observation. Dynamic graph drawing refers to the
problem of generating a sequence of layouts to be used either in a small multiples
representation or as frames in an animation. In the offline scenario the entire
input sequence is known in advance, whereas in the online scenario the sequence
is given one graph at a time.

Approaches to dynamic graph drawing most often augment a layout algorithm
designed for static graphs in such a way that the resulting sequence of layouts
is more stable than if each graph was drawn from scratch [5]. The motivation
for this approach is generally said to be the preservation of a viewer's mental
map [15], but it may also be interpreted as conveying the degree and location of
structural change more accurately by aligning it with layout change.

A common objective for drawing general undirected graphs is stress minimization [10,13], a special case of multidimensional scaling applied to graph-theoretic
distances. It has been found to outperform other spring embedder variants [3]
and will be the basis in this study.

The simplest (and most common) approach to add stability to an iterative
layout algorithm for static graphs is to initialize the computation for each graph

---

* This work was partially supported by DFG Research Training Group GK-1042 *Ex-
plorative Analysis and Visualization of Large Information Spaces.*

M. van Kreveld and B. Speckmann (Eds.): GD 2011, LNCS 7034, pp. 99–110, 2012.

in the sequence with the preceding layout [12,16]. The implicit assumption is that consecutive graphs are similar in general, and thus, the initial layout is not too far from a locally optimal one. The method is therefore easy to implement, more efficient than computing a layout from scratch, and applicable in both on- and offline scenarios.

However, stability is not addressed in a controlled way, hence this approach may result in excessive and unnecessary movement of vertices, and layout quality tends to degrade over the course of the sequence. Among the first to address stability directly were [1], and [17] provides a generic problem statement. The trade-off between readability and stability is formalized in [4] and a similar principle for offline scenarios is proposed in [7].

More sophisticated attempts to increase stability are typically based on one of three approaches. Maximum stability is achieved in aggregation approaches (e.g., [2,16]) where fixed vertex positions are obtained from the layout of an aggregate of all graphs in the sequence. Alternatives are based on anchoring vertices to reference positions (e.g., [4]), or linking vertices to instances of themselves that are close in the sequence (e.g., [8,9]).

Do these methods work well? Which one to implement for a given application? While the natural response to these questions appears to be a user study [18], their design may be challenging. Controlled experiments require a thorough understanding of the way in which model parameters affect outcomes. By purely algorithmic experimentation, we therefore want to provide quantitative evidence for the differential behavior of variant approaches, and thus prepare the ground for further user studies.

Our study compares aggregation, anchoring, and linking variants of stress minimization for offline dynamic graph drawing scenarios. The latter are of increasing relevance especially in longitudinal social network analysis [16], from which we hence draw some of our test cases. Our most important conclusion is that linking compares favorably with the other approaches.

After reviewing layout methods in Sect. 2, we formulate hypotheses in Sect. 3 that are based on common, though often implicit, assumptions about these methods and serve as a guideline for the experiments in Sect. 4. The experimental results are discussed in Sect. 4.3, and we conclude in Sect. 5.

2   Offline Dynamic Layout Approaches

Let $G = (V, E)$ be an undirected graph defined by a set V of n vertices, and a set E of m edges. An arbitrary pair of vertices is called *dyad*. Given a matrix D of vertex *dissimilarities* δ_{ij}, $i, j \in V$, the purpose of stress minimization is to determine positions $p_i = \langle x_i, y_i \rangle \in \mathbb{R}^2$ for every vertex $i \in V$ such that the Euclidean distances in the plane resemble the given dissimilarities as closely as possible, i.e., $\delta_{ij} \approx \|p_i - p_j\|$, where $\|\cdot\|$ denotes the Euclidean norm. For any given layout $P = (p_1, \ldots, p_n)$ this is quantified using a parameterized *stress function* stress(P),

$$\text{stress}(P) = \sum_{i<j} \omega_{ij} \left(\delta_{ij} - \|p_i - p_j\|\right)^2 , \tag{1}$$

where $W = (\omega_{ij})_{i,j \in V}$ is a weight matrix whose entries determine the contribution of each dyad. For graph drawing, lengths of shortest paths are a plausible choice for dissimilarities [10,13], and the objective is to find a layout of minimum stress. Because these distances are clearly not realizable for any non-trivial graph, weights $\omega_{ij} = \delta_{ij}^{-2}$ discount representation errors for distant pairs, thus emphasizing local accuracy.

Similar to other energy-based methods a solution can only be obtained by iterative stress reduction that yields a local minimum which may be far from an optimal layout. However, low-stress layouts can be routinely and efficiently computed using a two-step process [3]: In the first step, an initial layout is determined using *classical scaling*. In the second step, the representation of small distances is improved by iteratively and monotonically reducing stress using majorization [10].

2.1   Aggregation

Maximum stability is obtained when a vertex maintains its position throughout the entire sequence of diagrams. That is, given a sequence $G^{(1)} = (V, E^{(1)}), \ldots,$ $G^{(T)} = (V, E^{(T)})$ of T graphs with corresponding shortest-path distances $D^{(t)}$, $1 \leq t \leq T$, we are looking for one layout $\bar{P}$ for the vertices in V and let $P^{(t)} = \bar{P}$ at all times $t = 1, \ldots, T$.

We aggregate all shortest-path information by adapting input dissimilarities and weights in Eq. 1. We use $\bar{D} = (\bar{\delta}_{ij})_{i,j \in V}$, $\bar{\delta}_{ij} := \frac{1}{T} \sum_{t=1}^{T} \delta_{ij}^{(t)}$, i.e., the mean shortest-path distances, as dissimilarities, and weights $\bar{W} = (\bar{\omega}_{ij})_{i,j \in V}$ with

$$\bar{\omega}_{ij} = \frac{1}{\bar{\delta}_{ij}^2} \cdot \frac{1}{1 + \text{VAR}(\delta_{ij})} ,$$

where $\text{VAR}(\delta_{ij}) := \frac{1}{T} \sum_{t=1}^{T} \left(\delta_{ij}^{(t)} - \bar{\delta}_{ij}\right)^2$ is the variance of distances within a dyad across all observations. Thus, representation accuracy of dyads that are connected via short paths most of the time is emphasized. By additionally scaling with the variance, priority is given to structures that are relatively stable throughout the sequence. To obtain a layout we use the same algorithms as in the static case: Layout computation is initialized by classical scaling of mean distances; subsequently, $\text{stress}(\bar{P})$ is reduced via majorization. Note that, in an offline scenario, infinite distances in a dyad that might occur due to temporary disconnectedness can be handled by interpolating between the two finite distances observed previously and next for this dyad, and by adding a small constant, say 1.

2.2   Anchoring

The main idea of the anchoring approach [4] is an explicit modeling of the trade-off between layout quality as measured by an objective function, and layout stability with respect to a reference drawing as measured by a difference metric [6]. A stress function quantifying the compromise between quality of each individual graph in the sequence and deviation from reference positions is

$$\text{stress}_\alpha^A \left(P^{(t)} \right) = \underbrace{(1 - \alpha) \cdot \text{stress} \left(P^{(t)} \right)}_{\text{quality}} + \underbrace{\alpha \cdot \sum_{i \in V} \phi_i^{(t)} \left\| p_i^{(t)} - p_i \right\|^2}_{\text{stability}}, \qquad (2)$$

where $P = (p_i)_{i \in V}$ denotes the reference layout and weights $\phi_i^{(t)}$ allow for inter-vertex variation in deviation tolerance.

The stability term thus corresponds to a point-wise penalty for deviations from the reference layout, and the parameter $0 \leq \alpha \leq 1$ provides explicit control of the trade-off between quality (original stress) and stability. Note that minimizing stress_α^A for $\alpha = 0$ corresponds to regular stress minimization without control for stability, and $\alpha = 1$ yields the reference layout, since no deviation is tolerated.

For now, we use constant stability weights $\phi_i^{(t)} := 1$ for all i and t. More sophisticated choices, however, may be useful to compensate for cases with highly varying degrees or localized structural change. Before minimization of stress_α^A, we perform a Procrustes rotation [20] – an affine transformation that minimizes the sum of squared deviations from reference positions without changing relative distances – of the initial layout to the reference. After each layout of the sequence is obtained, we again apply Procrustes rotation subsequently to the whole sequence.

Depending on initialization and the type of reference, we obtain four anchoring methods. The first two are purely online, whereas the second two incorporate offline information by means of using the aggregate layout (Sect. 2.1) as reference:

APP initialize with previous layout (classical MDS for the first network), and also anchor to previous layout (no anchoring for the first network).
ACP initialize with classical scaling, anchor to previous layout (no anchoring for the first network).
APA initialize with previous layout (aggregate layout for the first network), anchor to aggregate layout.
ACA initialize with classical scaling, anchor to aggregate layout.

2.3   Linking

The main idea of the linking approach is to implicitly make use of all information about the networks of a sequence in an offline scenario. Instances of the same vertex are *linked* with each other, so as to stabilize their positions throughout the sequence. In contrast to the anchoring approach, layout calculation is not performed one after each other, but the whole system is computed simultaneously.

A general formulation of a corresponding stress function is

$$
\mathsf{stress}_\alpha^L \left(P^{(1)}, \dots, P^{(T)} \right) =
$$
$$
(1 - \alpha) \cdot \underbrace{\sum_{t=1}^{T} \mathsf{stress}\left(P^{(t)} \right)}_{\text{quality}} + \alpha \cdot \underbrace{\sum_{i \in V} \sum_{t'=1, t' \neq t}^{T} \phi_i^{(t)} \zeta(t, t') \left\| p_i^{(t)} - p_i^{(t')} \right\|^2}_{\text{stability}}, \quad (3)
$$

where $\zeta(t, t')$ is a function controlling the influence of the position at a certain time t for vertices at other time points t'. Concretely, we implemented two versions w.r.t. $\zeta(t, t')$ similar to the two alternatives stated in [8]: $\zeta_G(t, t') = e^{-\frac{1}{2}(t'-t)^2}$, a Gaussian function with mean value t and variance 1 without normalization, i.e., $\zeta_G(t, t) = 1$; and $\zeta_W(t, t') = 1$ for $|t - t'| = 1$, and $\zeta_W(t, t') = 0$ otherwise, i.e., a vertex is only linked within a time-window of size 1. Again, we use $\phi_i^{(t)} = 1$, and align all layouts in the sequence by Procrustes rotation after initialization, and after stress minimization. Depending on initialization and $\zeta(t, t')$, we obtain four linking methods:

LCG initialization by classical scaling, use ζ_G.
LAG initialization by aggregate layout as described in Sect. 2.1, use ζ_G.
LCW initialization by classical scaling, use ζ_W.
LAW initialization by aggregate layout, use ζ_W.

3   Hypotheses

Explicitly addressing stability by use of the above methods instead of simply initializing with the preceding layout implies that a better compromise between quality and stability is expected. Assessment of this claim is broken down into constituent components to structure the discussion of detailed quantitative results in Sect. 4.3.

Our first hypothesis to test is thus that the methods actually display the assumed effects at all.

H 1. Aggregation, anchoring, and linking increase dynamic stability, but reduce individual quality.

Likewise, the explicit trade-off between quality and stability should be controllable via control parameter α.

H 2. In anchoring and linking, higher values of α result in more stability and less quality.

Being an iterative method, stress minimization is known to be susceptible to poor local minima and thus to depend on good initialization [3]. As a consequence, the same caveat should be in place where the outcome is not governed by the attempt to maintain stability.

H 3. For decreasing values of control parameter α, anchoring and linking are increasingly sensitive to initialization.

And finally, the principal adaptation to the offline scenario is by either anchoring to a reference position determined from the entire sequence of graphs, or by linking with future instances. These should pay off in case there is a persistent global structure.

H 4. For dynamic graphs with persistent structure, anchoring to an aggregate layout and linking outperform online approaches.

The experiments conducted in the next section are designed to provide evidence for assessing these rather qualitative associations in detail.

4   Experiments

Instead of illustrating the approaches on selected examples, we here perform algorithmic experiments to obtain more detailed and generalizable insight into the behavior of dynamic graph drawing approaches. It is thus particularly important to use realistic input graph sequences, but we also address the issue of quantifying the output in a novel way.

4.1   Data

As mentioned above, our focal application area are longitudinal social networks. Instead of using a (necessarily small) collection of benchmark networks, though, we generate random graphs that are believed to be realistic for the application scenario, because they are obtained from the two most prevalent models in this domain.

These are exponential-family random graph models (*ERGM*, [19], modeling the characteristics of single networks) to create the initial graph of each sequence, and stochastic actor-oriented models (*SAOM*, [21], modeling the evolution between two networks) to obtain the actual sequence.[1] Both models are based on network-specific characteristics, called *effects* – such as density of the network, reciprocity of edges for directed networks, or number of triangles – and associated model parameters determining whether an effect increases or decreases the probability of a network (ERGM), or of particular network changes (SAOM). Both allow for model estimation, given networks and the desired set of effects, and for simulation of networks, given a starting network and a specified model.

---

[1] Available for the open source statistical system R (packages `ergm` and `RSiena`).

A sequence of T graphs is created in the following way: Two actual observations G_1 and G_2 of a longitudinal network serve as the basis for the creation process. Using G_1, we estimate an ERGM using basic effects,[2] from which an artificial first observation G_1^{sim} is simulated. Next, a SAOM is estimated using the real observations G_1 and G_2.[3] The thus estimated SAOM is used for the following two simulations.[4] The artificial second observation G_T^{sim} is obtained by running a simulation using G_1^{sim} and G_2 as input. Then, a simulation using G_1^{sim} and G_T^{sim} as input is performed to obtain a reliable sequence of changes leading from G_1^{sim} to G_T^{sim}, which is partitioned into $T - 1$ parts. Applying the corresponding changes to the initial observation G_1^{sim} yields a sequence of T networks.

As real input data, we use two data sets that are well studied in the social sciences.[5] The *s50* data set [14] comprises a sequence of three friendship networks of 50 female teenage pupils. We use the first and third observation as input for network sequence generation. The second real data set used is the *van de Bunt* data set [22], again, an evolving friendship network among 32 university freshmen comprising seven observations. We obtain input for network sequence generation by only considering edges with rating *best friendship* and *friendship* from the second and the seventh observation. Note also, that we removed vertex 18, since it is isolated at all time points.

In addition to the network generation process described above, we employ generation of unstructured artificial data by means of the $G(n, p)$ random graph model [11]. An initial observation is created with $n = 50$ and $p = \log(n)/n$, which produces connected graphs with high probability. Repeatedly, $k/2$ edges are formed uniformly at random and, likewise, $k/2$ edges deleted, where we do not allow deletion of edges just formed, and the resulting graph is made connected. In our experiments, we use $k = 2\sqrt{n}$ and $k = n$.

4.2   Measurements

To assess the quality and stability of layouts of a dynamic graph, we use the measures that constitute our approaches, that are, stress and sum of squared positional difference. Although it may be doubted whether these measures really capture either quality or stability, no other measures have been shown to better represent these concepts; it is therefore only reasonable to use the intrinsic measures of the approaches. Another problem is that both measures are not directly comparable across graphs of different sizes or structure. We solve this by relating both measures to the ones obtained from a common baseline method **B**, that is, we compute static layouts for each graph in the sequence as suggested

---

[2] We use effects `edges`, `mutual`, `gwodegree` and `gwesp`.

[3] We use effects for the the number of changes (`rate parameter`), outdegree, reciprocity, and transitivity (`transitive triplets`).

[4] The first `RSiena` simulation uses the unconditional method of moments, the second uses the maximum likelihood method.

[5] Publicly available at `http://www.stats.ox.ac.uk/~snijders/siena/`

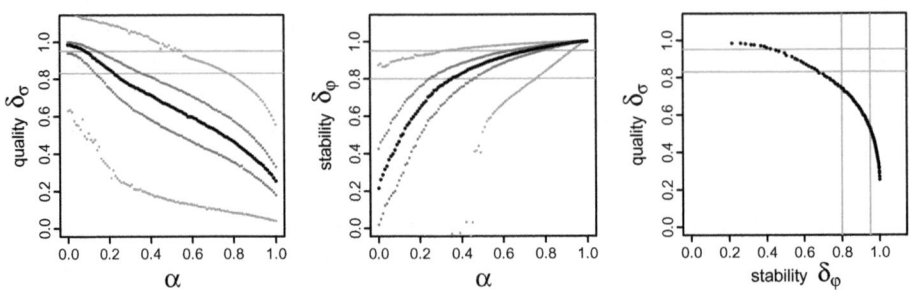

Fig. 1. Five-point summary for measurements of $\delta_\sigma^{\text{APA}}$ and $\delta_\varphi^{\text{APA}}$ subject to trade-off parameter α for 50 network sequences generated from the s50 data set

in the beginning of Sect. 2, and align the sequence by Procrustes rotation after layout calculation.

Let P_M be the layout for a graph obtained by method M. We measure quality δ_σ^M as the fraction of $\mathsf{stress}(P_B)$ and $\mathsf{stress}(P_M)$, i.e.,

$$\delta_\sigma^M = \frac{\mathsf{stress}(P_B)}{\mathsf{stress}(P_M)} .$$

Since we assume that the baseline layout is of relatively high quality, and that quality degrades when mechanisms to increase stability are employed, the range of δ_σ^M should be $[0,1)$, with δ_σ^M decreasing for higher values a.

Let $\varphi(P^{t-1}, P^t) = \sum_{i \in V} \left\| p_i^t - p_i^{t-1} \right\|^2$ be the sum of squared positional difference between two subsequent layouts P^{t-1} and P^t, where these have been aligned by Procrustes rotation, i.e., $\varphi(P^{t-1}, P^t)$ is minimal w.r.t. translation and rotation of P^{t-1} and P^t. Stability is measured as the relative decrease of positional difference w.r.t. the baseline layout, i.e.,

$$\delta_\varphi^M = 1 - \frac{\varphi(P_M^{t-1}, P_M^t)}{\varphi(P_B^{t-1}, P_B^t)} .$$

The assumption is that the baseline layout exhibits a high positional difference, that will decrease whenever mechanisms to increase stability are employed. Note that all methods presented yield the same layout for all graphs in the sequence for $\alpha = 1$, therefore δ_φ^M must be 1 for all methods in this case. Thus the range is expected to be $[0,1]$, with δ_φ^M increasing for higher values α.

For each network sequence generator, we created 50 network sequences comprising 10 graphs each. We measured δ_σ and δ_φ corresponding to trade-off parameter $\alpha \in \{0, 0.01, 0.02, \ldots, 1\}$. Thus, per generator, method, and value of α, we obtain 500 measurements of δ_σ (450 for **APP** and **ACP**, since for the first observation of each sequence in these cases $\delta_\sigma = 1$ for all α), and 450 measurements of δ_φ (not applicable to each first observation). Figure 1 shows a five point summary, i.e., minimum, first quartile, median, third quartile, and maximum,

Table 1. Median values for δ_σ and δ_φ at certain selected values of α for measurements on sequences generated from the s50 data set. Note that only measurements belonging to either the anchoring or linking approaches can be compared directly. Sequences from other generators reveal similar tendencies.

α :	0.1	0.2	0.3	0.7	0.8	0.9	0.1	0.2	0.3	0.7	0.8	0.9
APP	0.94	0.86	0.78	0.48	0.40	0.31	**0.49**	0.63	0.72	0.93	0.95	0.98
ACP	**0.97**	**0.90**	**0.80**	0.48	0.40	0.31	0.29	0.54	0.70	0.93	0.95	0.98
APA	0.94	0.86	0.77	**0.54**	**0.47**	0.37	0.46	**0.64**	**0.76**	**0.95**	**0.97**	**0.99**
ACA	0.96	0.87	0.79	0.53	**0.47**	**0.38**	0.22	0.52	0.72	**0.95**	**0.97**	**0.99**
LAG	0.96	0.91	0.86	0.68	0.62	0.54	0.51	0.69	0.78	**0.95**	0.97	0.99
LCG	**0.98**	**0.93**	**0.89**	**0.69**	**0.63**	0.54	0.35	0.59	0.72	0.94	0.97	0.99
LAW	0.95	0.90	0.85	0.68	0.62	**0.55**	**0.57**	**0.73**	**0.81**	**0.95**	0.97	0.99
LCW	0.97	0.92	0.87	0.68	**0.63**	0.55	0.43	0.64	0.75	0.94	0.97	0.99

$$\hat{\delta}_\sigma \qquad\qquad\qquad\qquad \hat{\delta}_\varphi$$

of the measurements obtained for the 50 network sequences generated from the s50 data set when applying the **APA** method. The gray horizontal lines indicate thresholds used in our experiments, that are, 5% and 20% more stress w.r.t. the baseline for quality measurements, and 80% and 95% reduction in positional difference w.r.t. to the baseline for stability measurements. Note that, for both δ_σ and δ_φ, there are outliers that contradict the intuitive assumptions regarding the range of the measures. We can only explain these by the heuristic nature of stress minimization. Still, most of the measured values are within a reasonable range around the median values, as can be observed by the inter-quartile range. Thus, we will argue about the approaches by means of the median measurements, denoted by $\hat{\delta}_\sigma$ and $\hat{\delta}_\varphi$, respectively. Table 1 shows values of $\hat{\delta}_\sigma$ and $\hat{\delta}_\varphi$ at selected levels of α for network sequences generated from the s50 data set, and Fig. 2 summarizes measurements for all methods and data sets.

4.3   Results

Figure 2 (upper row, right endpoints) shows that already a slight compromise in quality (5% additional stress compared to static baseline layouts) yields a large increase in positional stability (ranging from 24% to 82% reduction of total movement). If we allow a 20% increase of stress (left-hand side of each upper graph), all methods reduce movement by more than 50%. Across all experiments, reduction in positional difference increases very rapidly at lower ranges of α as exemplified in Fig. 1. This provides evidence that the methods are largely having the desired effects (H 1). Moreover, the actual values corroborate earlier findings that low stability mechanisms appear to be most effective [18].

The monotone behavior of median values of δ_σ and δ_φ in Table 1 support the expected dependencies on α (H 2). Also, sensitivity to initialization (H 3) is confirmed: Although small, there are noticeable differences in quality at the lower range ($\alpha \leq 0.3$) in favor of initialization with classical scaling. There are, however, large differences in stability in favor of initialization with the previous

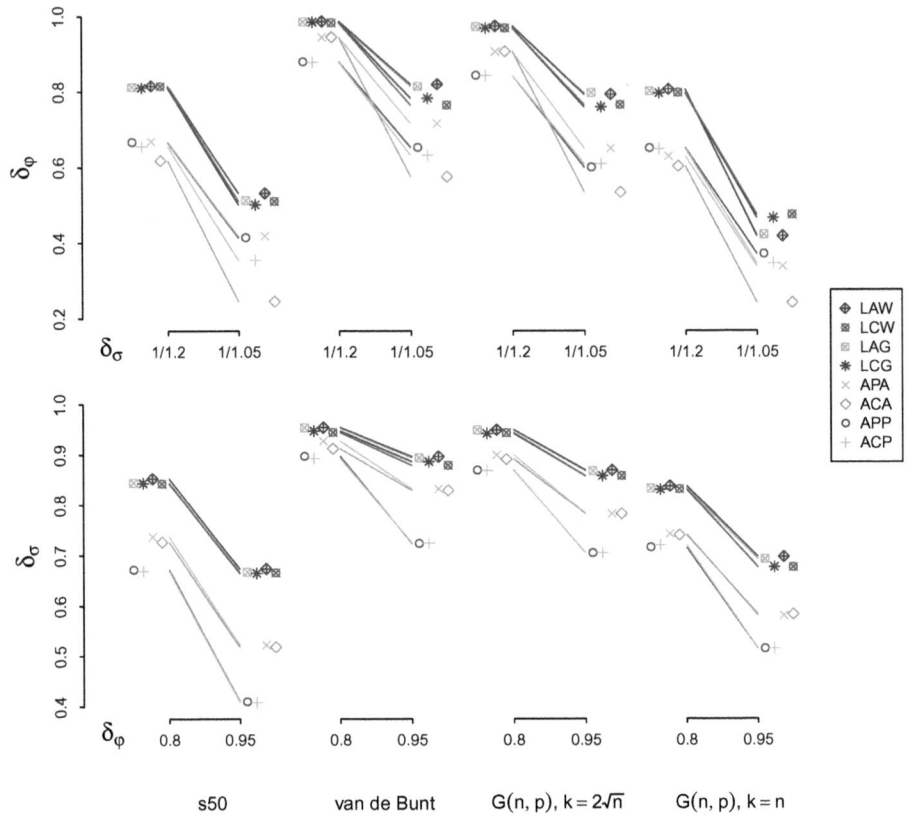

Fig. 2. Overview of median values for δ_φ at selected median values of δ_σ (upper part) and vice versa (lower part) for all measurements

layout for anchoring approaches, or the aggregate layout for linking approaches, as expected. At higher range, we observe very similar results for each pair of anchoring approaches belonging together w.r.t. the choice of reference, and generally similar results for the linking approaches, which is also visible in the lower graphs of Fig. 2. This indicates diminishing influence of initialization at higher stability requirements.

Regarding H 4, we first posit that data generated from ERGMs and SAOM learned from real data contains more persistent structures, since change in these networks is actually based on structural processes unlike the random changes in the $G(n,p)$ generator. Rather surprisingly, though, we cannot observe a general trend in Fig. 2 that **A?A** approaches perform better than **A?P** approaches at very low stability levels (stress increase of 5%). Indeed method **ACA** consistently performs worse, which indicates that initialization with classical scaling is in conflict with anchoring to the aggregate. However, the more stability is sought, the more is the difference in increase of stress between the **A?A** and the **A?P** approaches, in favor of the former, and regardless of the data set. The linking approaches perform better than the anchoring approaches throughout. Thus, our evaluation of H 4 is inconclusive, since no statement can be made for very low stability, and methods incorporating offline information apparently perform better regardless of structure for moderate, and especially, high stability requirements. To our surprise, the choice between the two functions ζ_G and ζ_W does not seem to considerably influence the results.

5 Conclusion

We compared dynamic variants of the stress-minimization approach for general undirected graphs in which vertices are at the same position throughout the sequence of layouts (aggregation), attracted by a reference position (anchoring), or attracted by positions of their copies in neighboring time slices (linking).

The comparison was based on a novel form of measurement of the trade-off between quality (in terms of stress) and stability (in terms of vertex movement): Measures were related to a baseline, determined from Procrustes aligned static layouts, to normalize over graphs of different sizes and structure.

A second novel aspect is our use of more sophisticated graph generators that eliminate reliance on small benchmark data sets and still produce application-typical data. Here, ERGMs were used for boundary observations and SAOMs for the evolution.

Our results suggest that linking is a generally preferable approach. Since it is computationally demanding, a faster alternative is anchoring to an aggregate layout initialized with the previous one in the sequence.

The present study is an attempt to move towards more precise measurement of those aspects of the performance of graph drawing algorithms that are not easily characterized analytically, but we are left with more new questions than answers to old ones. The focus on stress-minimization approaches and two particular criteria for quality and stability allowed for better comparison and more detailed insights, but different quantities may turn out important as well. Other avenues for future research include refined data generation procedures (e.g., including behavioral effects), in-depth discussion of outliers and other observations, and dependencies on specific graph structures and change sequences.

References

1. Böhringer, K.F., Paulisch, F.N.: Using constraints to achieve stability in automatic graph layout algorithms. In: Proc. of the SIGCHI Conference on Human Factors in Computing Systems (CHI 1990), pp. 43–51. ACM (1990)
2. Brandes, U., Corman, S.R.: Visual unrolling of network evolution and the analysis of dynamic discourse. Information Visualization 2(1), 40–50 (2003)
3. Brandes, U., Pich, C.: An Experimental Study on Distance-Based Graph Drawing. In: Tollis, I.G., Patrignani, M. (eds.) GD 2008. LNCS, vol. 5417, pp. 218–229. Springer, Heidelberg (2009)

4. Brandes, U., Wagner, D.: A Bayesian Paradigm for Dynamic Graph Layout. In: DiBattista, G. (ed.) GD 1997. LNCS, vol. 1353, pp. 236–247. Springer, Heidelberg (1997)
5. Branke, J.: Dynamic Graph Drawing. In: Kaufmann, M., Wagner, D. (eds.) Drawing Graphs. LNCS, vol. 2025, pp. 228–246. Springer, Heidelberg (2001)
6. Bridgeman, S.S., Tamassia, R.: Difference metrics for interactive orthogonal graph drawing algorithms. Journal of Graph Algorithms and Applications 4(3), 47–74 (2000)
7. Diehl, S., Görg, C.: Graphs, they are Changing. In: Goodrich, M.T., Kobourov, S.G. (eds.) GD 2002. LNCS, vol. 2528, pp. 23–30. Springer, Heidelberg (2002)
8. Erten, C., Harding, P., Kobourov, S., Wampler, K., Yee, G.: Graphael: Graph Animations with Evolving Layouts. In: Liotta, G. (ed.) GD 2003. LNCS, vol. 2912, pp. 98–110. Springer, Heidelberg (2004)
9. Erten, C., Kobourov, S., Le, V., Navabi, A.: Simultaneous graph drawing: Layout algorithms and visualization schemes. Journal of Graph Algorithms and Applications 9(1), 165–182 (2005)
10. Gansner, E., Koren, Y., North, S.: Graph Drawing by Stress Majorization. In: Pach, J. (ed.) GD 2004. LNCS, vol. 3383, pp. 239–250. Springer, Heidelberg (2005)
11. Gilbert, E.N.: Random graphs. The Annals of Mathematical Statistics 30(4), 1141–1144 (1959)
12. Huang, M.L., Eades, P., Wang, J.: On-line animated visualization of huge graphs using a modified spring algorithm. Journal of Visual Languages and Computing 9(6), 623–645 (1998)
13. Kamada, T., Kawai, S.: An algorithm for drawing general undirected graphs. Information Processing Letters 31, 7–15 (1989)
14. Michell, L., Amos, A.: Girls, pecking order and smoking. Social Science & Medicine 44(12), 1861–1869 (1997)
15. Misue, K., Eades, P., Lai, W., Sugiyama, K.: Layout adjustment and the mental map. Journal on Visual Languages and Computing 6(2), 183–210 (1995)
16. Moody, J., McFarland, D.A., Bender-deMoll, S.: Dynamic Network Visualization. American Journal of Sociology 110(4), 1206–1241 (2005)
17. North, S.C.: Incremental Layout with DynaDag. In: Brandenburg, F.J. (ed.) GD 1995. LNCS, vol. 1027, pp. 409–418. Springer, Heidelberg (1996)
18. Purchase, H.C., Samra, A.: Extremes are Better: Investigating Mental Map Preservation in Dynamic Graphs. In: Stapleton, G., Howse, J., Lee, J. (eds.) Diagrams 2008. LNCS (LNAI), vol. 5223, pp. 60–73. Springer, Heidelberg (2008)
19. Robins, G., Pattison, P., Kalish, Y., Lusher, D.: An introduction to exponential random graph (p*) models for social networks. social networks 29(2), 173–191 (2007)
20. Sibson, R.: Studies in the robustness of multidimensional scaling: Procrustes statistics. Journal of the Royal Statistical Society. Series B (Methodological) 40(2), 234–238 (1978)
21. Snijders, T.A.B.: The statistical evaluation of social network dynamics. Sociological Methodology 31, 361–395 (2001)
22. Van De Bunt, G.G., Van Duijn, M.A., Snijders, T.A.: Friendship networks through time: An actor-oriented dynamic statistical network model. Computational & Mathematical Organization Theory 5, 167–192 (1999)

Accelerated Bend Minimization

Sabine Cornelsen and Andreas Karrenbauer

Department of Computer & Information Science, University of Konstanz
`firstname.lastname@uni-konstanz.de`

Abstract. We present an $\mathcal{O}(n^{3/2})$ algorithm for minimizing the number of bends in an orthogonal drawing of a plane graph. It has been posed as a long standing open problem at *Graph Drawing 2003*, whether the bound of $\mathcal{O}(n^{7/4}\sqrt{\log n})$ shown by Garg and Tamassia in 1996 could be improved. To answer this question, we show how to solve the uncapacitated min-cost flow problem on a planar bidirected graph with bounded costs and face sizes in $\mathcal{O}(n^{3/2})$ time.

1 Introduction

A drawing of a planar graph is called orthogonal if all edges are non-crossing axis-parallel polylines, i.e. sequences of finitely many horizontal and vertical line segments. The intersection point of a vertical and a horizontal line segment of an edge is a bend.

If a graph has an orthogonal drawing such that the vertices are drawn as points then the degree of any vertex is at most four. Biedl and Kant [1] gave a linear-time algorithm for constructing an orthogonal drawing with at most two bends per edge of a graph with degree at most four (except for the octahedron). The problem of minimizing the number of bends in an orthogonal drawing of a planar graph with maximum degree four is $\mathcal{NP}$-complete [2] if the *embedding* of the graph, i.e. the cyclic ordering of the incident edges around each vertex, is not fixed.

Tamassia [3] considered the bend-minimization problem on *plane* graphs, i.e., on planar graphs with a fixed embedding and a fixed outer face. He showed that the problem of minimizing the total number of bends in an orthogonal drawing of a plane graph with degree at most four can be modeled by a min-cost flow problem. There are also variations of the flow-based bend minimization approach which include a restricted number of bends, vertices of degree higher than four [4,5,6,7], drawing clustered graphs [8,9], or interactive and dynamic graph drawing [10,11].

Network flows are an important topic in combinatorial optimization and we refer the interested reader to [12] and [13] for a general overview. Instead, we concentrate on the special case of planar networks in this paper. To the best of the authors knowledge, there have not been many direct contributions to compute planar min-cost flows in the past decades. One exception is a dedicated analysis of an *interior point method* [14] restricted to linear programs arising from min cost flow problems on planar graphs by Imai and Iwano [15]. In 1990,

M. van Kreveld and B. Speckmann (Eds.): GD 2011, LNCS 7034, pp. 111–122, 2012.

they proved a running time bound of $O(n^{1.594}\sqrt{\log n}\log(n\gamma))$, at which γ is an upper bound on the absolute values of costs and capacities. Much more progress has been made on important special cases such as the *shortest path* problem and the *max flow* problem, which may be used in general flow algorithms as subroutines to obtain a better running time when the input is restricted to planar graphs. This includes the famous linear-time algorithm for planar shortest path with non-negative lengths [16], near linear-time algorithms for shortest path with real lengths [17,18,19], and for max s-t-flow [20,21]. The latter problem can be solved in linear time when s and t are on the same face because of its equivalence to a shortest path problem with non-negative lengths in the dual graph shown by Hassin [22]. This result has been extended to multiple sources and sinks on the same face by Miller and Naor [23].

Garg and Tamassia [24] proved that a min-cost flow problem on a flow network with n nodes, m arcs, and the minimum cost χ of a flow can be solved in $\mathcal{O}(\chi^{3/4}m\sqrt{\log n})$ time and concluded that the bend minimization problem of an embedded planar graph with degree at most four can be solved in $\mathcal{O}(n^{7/4}\sqrt{\log n})$ time. It was posed as an important open problem in graph drawing, whether this run time could be improved [25, Problem 14].[1]

Our Contribution

In this paper, we especially exploit the fact that the flow network is planar and show how to solve the problem in $\mathcal{O}(n^{3/2})$ time. Our algorithm splits the flow network using a cycle separator. To this end, the edges on the cycle are contracted, which maintains planarity. The separator thereby shrinks to a cut node that joins two biconnected components on which the min-cost flow problem can be solved independently. The recursive solutions of the two parts are combined by expanding the separator edge by edge and adjusting the flow between the endpoints of the corresponding edge in each step.

In particular, we show that the uncapacitated min-cost flow problem on a planar bidirected graph with bounded costs and face sizes can be solved in $\mathcal{O}(n^{3/2})$ time. This result only relies on linear-time algorithms for finding cycle separators [26], and for computing max s-t-flows in (s,t)-planar graphs ([22] combined with [16]). Note that our approach combined with a result on multiple-source multiple-sink max-flow in planar graphs [27] solves the bend-minimization problem in $\mathcal{O}(n^{3/2}\log n)$ time if we additionally wish to constrain the number of bends on some edges and it yields an $\mathcal{O}(\sqrt{\chi}n\log^3 n)$ algorithm for computing a flow of minimum-cost χ on a planar flow network with n nodes and $\mathcal{O}(n)$ arcs.

The paper is organized as follows. In Section 2, we define the min-cost flow problem and briefly describe the flow model of Tamassia for bend-minimization. In Section 3, we describe the primal-dual algorithm that generally solves the min-cost flow problem. Our main result, based on the divide and conquer approach, that yields the $\mathcal{O}(n^{3/2})$ time algorithm is described in Section 4.

---

[1] The result of [15] provides a better bound, but the algorithm is not combinatorial and its correctness is hard to verify since not all details have been presented in the extended abstract. In any case, we improve w.r.t. both.

2   Bend Minimization and Flow Networks

Throughout this paper let $G = (V, E)$ be a simple undirected connected plane graph with n vertices of degree at most four and let $\mathcal{F}$ be the set of faces of a planar embedding. We consider the *vertex-face-incidence multi-graph* with node set $W_G = V \cup \mathcal{F}$ and arcs whenever the geometric intersection of two elements of W_G is non-empty in the planar embedding. Let $D_G = (W_G, A_G)$ denote the bidirected version of this graph. Let $D_{\mathcal{F}}$ be the subgraph of D_G that is induced by the face nodes only.

A *min-cost flow network* $\mathcal{N}$ consists of a directed (multi-)graph $D = (W, A)$, capacities $u : A \rightarrow \mathbb{Z}_{\geq 0} \cup \{\infty\}$, node demands $b : W \rightarrow \mathbb{Z}$, and arc costs $c : A \rightarrow \mathbb{Z}_{\geq 0}$. A map $f : A \rightarrow \mathbb{Z}_{\geq 0}$ is a *pseudo-flow* on $\mathcal{N}$ if $f(a) \leq u(a)$ for $a \in A$. A pseudo-flow f is a *flow* if the *deficiency*

$$b_f(v) = b(v) + \sum_{(w,v) \in A} f(w, v) - \sum_{(v,w) \in A} f(v, w)$$

of each node $v \in W$ is zero. The cost of a flow is $c(f) = \sum_{a \in A} c(a) f(a)$. We say that a flow problem is *uncapacitated* and with *unit costs*, respectively, if $u(a) = \infty$ and $c(a) = 1$, respectively, for all arcs $a \in A$.

The bend-minimization problem can be modeled by a min-cost flow network $\mathcal{N}_G = (D_G, u, b, c)$ [3] with the following properties.

1. $D_{\mathcal{F}}$ is planar, bidirected with infinite capacity and unit cost.
2. The degree of a face of $D_{\mathcal{F}}$ is at most 4.
3. A cycle separator of $D_{\mathcal{F}}$ is a cycle separator of D_G.
4. D_G is planar and triangulated.
5. The minimum cost of a flow in $\mathcal{N}_G$ is at most $2n + 4$ [1].

Readers to whom these properties sound familiar may safely skip the next subsection, which contains a brief presentation of Tamassia's approach [3].

Bend Minimization as Min Cost Flow

In this section, we briefly describe the approach of Tamassia [3] for constructing an orthogonal drawing of a plane graph with the minimum total number of bends. The approach consists of two phases. In the first phase an *orthogonal representation* is computed, which fixes the angle at each vertex between two consecutive adjacent edges on one hand and the number of right and left turns on an edge on the other hand. In a second step an area efficient orthogonal grid drawing is constructed from a feasible orthogonal representation. The second step can be done in linear time using topological sorting [28, page 155].

The orthogonal representation associates four labels with each edge $\{v, w\} \in E$, two for each direction. The label $1 \leq \alpha(v, w) \leq 4$ is such that $\alpha(v, w) \cdot \pi/2$ denotes the angle at vertex v between $\{v, w\}$ and the next incident edge of v in counter-clockwise direction. The label $\tau(v, w) \geq 0$ denotes the number of left-turns on $\{v, w\}$ traversed from v to w. See Fig. 1(b), for an illustration.

Let the degree deg f of a face f be the number of its incident edges where bridges count twice. Elementary geometry implies that there is an orthogonal drawing that corresponds to some given labels α and τ if and only if they imply that the sum of angles around a vertex is 2π and that the sum of angles around an inner/outer face f is $\pi \cdot (\deg(f) + \text{number of bends} \mp 2)$. The latter can be reformulated as

$$\sum_{(v,w)\in E(f)} (\alpha(v,w) + \tau(w,v) - \tau(v,w)) = 2\deg(f) \mp 4$$

where $E(f)$ denotes the arcs incident to the face f directed in counter-clockwise direction. This yields a min-cost flow formulation for finding a feasible orthogonal representation with the minimum number of bends.

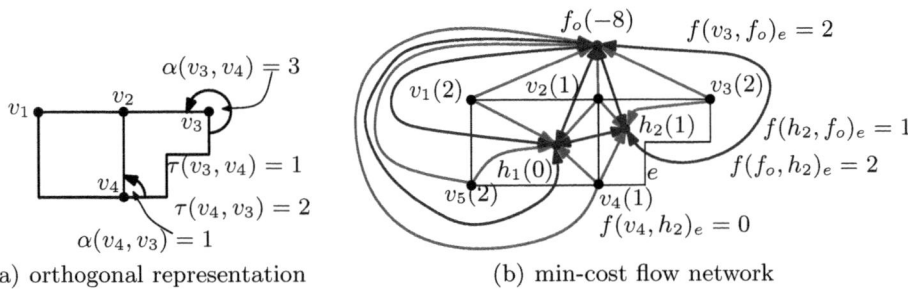

(a) orthogonal representation   (b) min-cost flow network

Fig. 1. Illustration of the approach of Tamassia [3] for solving the bend-minimization problem. In b) the directed arcs indicate the flow network. All arcs have infinite capacity, the bidirected blue arcs have cost one and the unidirected red arcs have costs zero. The node demands are indicated in brackets.

The bend minimization problem on G can be solved by the following min-cost flow network $\mathcal{N}_G$. The node set of the directed graph D_G is $W_G = V \cup \mathcal{F}$ with $b(v) = 4 - \deg(v), v \in V$, $b(h) = 4 - \deg(h)$ if $h \in \mathcal{F}$ is an inner face and $b(f_o) = -4 - \deg(f_o)$ for the outer face f_o. For each edge $e = \{v, w\} \in E$ with $(v, w) \in E(h)$ and $(w, v) \in E(g)$ the arc set A_G contains the arcs $(v, h)_e, (w, g)_e$ with costs zero and $(h, g)_e, (g, h)_e$ with costs one. All arcs have infinite capacities. Note that the index e is only used to distinguish possible multiple arcs. See Fig. 1(b), for an illustration.

Now a min-cost flow f on $\mathcal{N}_G$ corresponds to an orthogonal representation with the minimum number of bends as follows. For each edge $e = \{v, w\} \in E$ with $(v, w) \in E(h)$ and $(w, v) \in E(g)$ set $\alpha(v, w) = f(v, h)_e + 1$ and $\tau(v, w) = f(h, g)_e$.

3   The Primal Dual Algorithm

In this section, we briefly describe the primal-dual algorithm [29] for solving the min-cost flow problem.

Let $\mathcal{N} = (D = (W, A), u, b, c)$ be a min-cost flow network. An arc $a \in A$ is *saturated* by a pseudo-flow f if $f(a) = u(a)$. A *node potential* is a function $\pi : W \to \mathbb{Z}$. The *residual network* $\mathcal{N}_{f,\pi} = (D_f = (W, A_f), u_f, b_f, c_\pi)$ of the min-cost flow network $\mathcal{N}$, a pseudo-flow $f : A \to \mathbb{Z}_{\geq 0}$, and a node potential $\pi : W \to \mathbb{Z}$ is defined as follows. For each arc $a \in A$ with tail v and head w the arc set A_f contains a with $c_\pi(a) := c(a) + \pi(v) - \pi(w)$ if $u_f(a) := u(a) - f(a) > 0$. Further, if $f(a) > 0$ then A_f contains a reversed copy $-a$ from w to v with $c_\pi(-a) = -(c(a) + \pi(v) - \pi(w))$ and $u_f(-a) := f(a)$. The costs c_π are called the *reduced costs* and u_f are the *residual capacities*. The node potential is *valid* if $c_\pi(a) \geq 0$ for all $a \in A_f$. The primal-dual algorithm solves a min-cost flow problem utilizing the reduced cost optimality condition.

Lemma 1 ([12, Theorem 9.3]). *A flow has minimum cost if and only if it admits a valid node potential.*

The *primal-dual algorithm* works as follows on a min-cost flow network $\mathcal{N} = (D = (W, A), u, b, c)$. First, the equivalent min-cost max-flow network $\mathcal{N}^{st} = (D^{st} = (W \cup \{s, t\}, A^{st}), u, c, s, t)$ is constructed, i.e. a super source s and a super sink t is added to W. Note that in general this construction does not preserve planarity. However, this is not relevant for the following lemmas. For each node $v \in W$ with $b(v) > 0$ an arc (s, v) with $u(s, v) = b(v)$ and cost zero is added to A. Further, for each node $v \in W$ with $b(v) < 0$ an arc (v, t) with $u(v, t) = -b(v)$ and zero costs is added to A. The value of a flow in $\mathcal{N}^{st}$ is the sum of all flow values on the arcs incident to s. Note that $\mathcal{N}$ has a feasible flow if and only if a maximum s-t-flow of $\mathcal{N}^{st}$ saturates all arcs incident to s. Further, let f be a maximum flow with minimum costs on $\mathcal{N}^{st}$. Restricting f to A yields a min-cost flow on $\mathcal{N}$.

The primal-dual algorithm now basically augments as much flow as possible on shortest s-t-paths in the residual network. More precisely, the algorithm starts with the node potential $\pi = 0$ and the pseudo flow $f = 0$. As long as not all arcs incident to s are saturated, the algorithm adds the shortest-path distances $\text{dist}_{f,\pi}(s, v)$ in (D_f, c_π) to $\pi(v)$. Then it considers the *admissible network* $D_f^o = (W \cup \{s, t\}, A^o)$ with $A^o = \{a \in A_f^{st}; c_\pi(a) = 0\}$ and augments f by a maximum s-t-flow in (D^o, u_f). See Algorithm 1 for a pseudocode.

To analyze the number of iterations, let f_i and π_i, respectively, be the flow and potential, respectively, after the ith iteration of the primal-dual algorithm. Further, let $f_0 = 0$, $\pi_0 = 0$ be the initial flow and potential. Recall that we consider integer costs and capacities.

Lemma 2. *We have the following properties.*

1. $\pi_i(v) = \text{dist}_{f_{i-1}, \pi_0}(s, v), v \in W, i \geq 1$.
2. $\pi_i(t) < \pi_{i+1}(t), i \geq 1$.
3. $\pi_i(t) \geq i - 1$.
4. $i \leq \text{dist}_{f_i, \pi_0}(s, t)$.

Proof. 1. Let $v \in W$. If there is no $s - v$-path in $D_{f_{i-1}}$, then $\text{dist}_{f_{i-1}, \pi_0}(s, v) = \text{dist}_{f_{i-1}, \pi_{i-1}}(s, v) = \infty$, and, hence, $\pi_i(v) = \pi_{i-1}(v) + \text{dist}_{f_{i-1}, \pi_{i-1}}(s, v) = \infty$.

Algorithm 1. Primal-Dual Algorithm

Input : min-cost flow network $\mathcal{N} = (D = (W, A), u, b, c)$.
Output : min-cost max-flow f of $\mathcal{N}^{st}$ with valid node potential π,

both initialized to 0

PRIMAL-DUAL(D, u, b, c)
> **while** *there is an s-t-path in* D_f^{st} **do**
>> dist$(s, .) \leftarrow$ SINGLE-SOURCE-SHORTEST-PATH(D_f^{st}, c_π, s);
>> **for** $v \in W \cup \{t\}$ **do**
>>> $\pi(v) \leftarrow \pi(v) + \text{dist}(s, v)$;
>>
>> $f^o \leftarrow$ MAX-FLOW(D_f^o, u_f, s, t);
>> $f \leftarrow f + f^o$;
>
> **return** (f, π);

Let now $s = v_0, \ldots, v_\ell = v$ be the nodes on a shortest $s - v$-path in $(D_{f_{i-1}}, c_{\pi_i})$. Then we have that $0 = \text{dist}_{f_{i-1}, \pi_i}(s, v) = \sum_{k=1}^{\ell} c_{\pi_i}(v_{k-1}, v_k) = \sum_{k=1}^{\ell} c(v_{k-1}, v_k) + \pi_i(s) - \pi_i(v) = \text{dist}_{f_{i-1}, \pi_0}(s, v) - \pi_i(v)$, where the latter equality holds since $\pi_i(s) = 0$.

2. By definition, $\pi_{i+1}(t) = \pi_i(t) + \text{dist}_{f_i, \pi_i}(s, t)$. After augmenting a maximum s-t-flow on the arcs with zero reduced costs there is an s-t-cut on which all arcs with zero reduced costs are saturated. Hence, the residual network contains no s-t-path with zero reduced costs. Hence, $\text{dist}_{f_i, \pi_i}(s, t) > 0$.

3. $\pi_i(t) \geq i - 1$ follows immediately from $\pi_1(t) \geq 0$ and the previous item.

4. If there is no $i + 1$st iteration then $i < \infty = \text{dist}_{f_i, \pi_0}(s, t)$. Otherwise, combining the previous items, we obtain $i \leq \pi_i(t) + 1 \leq \pi_{i+1}(t) = \text{dist}_{f_i, \pi_0}(s, t)$.

□

Lemma 3. *Let there be a feasible flow on* $\mathcal{N}$, *let* χ *be the minimum cost of a flow on* $\mathcal{N}$, *and let* $i \geq 1$. *Then the primal-dual algorithm terminates after at most* $i + \chi/i$ *iterations.*

Moreover, a min-cost flow can be computed by performing at most i *max-flow computations and at most* $i + \chi/i$ *shortest path computations.*

Proof. Let $i \geq 1$. The statement is trivially true if the algorithm performs at most i iterations. So assume that the algorithm performs more than i iterations. Let $r := b_{f_i}(s)$ be the sum of the residual capacities of the arcs leaving s after iteration i. Since in each of the following iterations at least one unit of flow is sent to t it follows that the primal-dual algorithm will finish within at most $i + r$ iterations even if in the last r iterations only one unit of flow is sent from s to t along a shortest path in the residual network and thus without the need to compute any further maximum flow.

On the other hand, since there is a feasible flow on $\mathcal{N}$, all arcs incident to s have to be saturated at the end. Augmenting one unit of flow augments the total cost of a flow by at least the original cost of a shortest s-t-path in the residual network. Since $\text{dist}_{f_i, \pi_0}(s, t) = \pi_{i+1}(t) < \pi_{i+2}(t) = \text{dist}_{f_{i+1}, \pi_0}(s, t)$ it follows that the length of a shortest s-t-path increases with every step. Hence,

$\chi \geq r \cdot \mathrm{dist}_{f_i, \pi_0}(s, t) \geq r \cdot i$. Thus, at most $r \leq \chi/i$ shortest-path computations have to be performed after the ith iteration. □

Corollary 1. *Let there be a feasible flow on $\mathcal{N}$ and let χ be the minimum cost of a flow on $\mathcal{N}$. Then the primal-dual algorithm terminates after at most $2 \cdot \sqrt{\chi} + 1$ iterations.*

Proof. If $\chi = 0$ then the algorithm terminates after at most 1 iteration. Otherwise, let i be such that $i - 1 < \sqrt{\chi} \leq i$. Then the total number of iterations is bounded by $i + \chi/i < \sqrt{\chi} + 1 + \chi/\sqrt{\chi} = 2\sqrt{\chi} + 1$ iterations. □

In a network with n vertices and $\mathcal{O}(n)$ arcs the shortest-path problem can be solved in $\mathcal{O}(n \log n)$ time using the algorithm of Dijkstra [30], while the max-flow problem can be solved in $\mathcal{O}(n \log^3 n)$ time if the network is planar [27].

Remark 1. Hence, the primal-dual algorithm computes a flow with minimum cost χ on a planar min-cost flow network with n nodes and with $\mathcal{O}(n)$ arcs in $\mathcal{O}(\sqrt{\chi} n \log^3 n)$ time.

Since the number of bends in an orthogonal drawing and, hence, the cost of the flow in the corresponding min-cost flow network is in $\mathcal{O}(n)$ [1], it follows that the bend-minimization problem can be solved in $\mathcal{O}(n^{3/2} \log^3 n)$ time, even if the number of bends on some edges is restricted. In the next section, we give a divide and conquer approach that directly solves the uncapacitated bend minimization problem utilizing only less recent results.

4   A Recursive Approach

In this section, we show how to utilize a planar separator theorem to recursively solve the min-cost flow problem.

Let an assignment of non-negative weights to the vertices, faces, and edges of a plane graph G be given that sum to one. A simple cycle C of G is a *weighted cycle separator* of G if both, the weight of the interior of C and the weight of the exterior of C do not exceed $2/3$.

Miller [26] showed that every biconnected planar graph with n vertices and face degree at most d has a simple cycle separator with at most $2\sqrt{d \cdot n}$ vertices unless there is a face with weight higher than $2/3$. Moreover, such a cycle separator can be constructed in linear time. Note that the min-cost flow problem decomposes into independent subproblems for each biconnected component.

This yields the following recursive algorithm for constructing a min-cost flow on a flow-network $\mathcal{N} = (D = (W, A), u, b, c)$ where D is a plane digraph with $\mathcal{O}(n)$ nodes and arcs.

First, we find a small cycle separator $C : v_1, \ldots, v_\ell$ of D. Let W_1 be the set of nodes in the interior of C and let W_2 be the set of nodes in the exterior of C. Let A_i be the set of arcs of A that are incident to at least one node of W_i. See Fig. 2(a) for an illustration. Let $D_i = (W_i \cup \{\hat{C}\}, A_i), i = 1, 2$ be obtained from the subgraph of D induced by $W_i \cup C$ by shrinking C to a single node $\hat{C}$

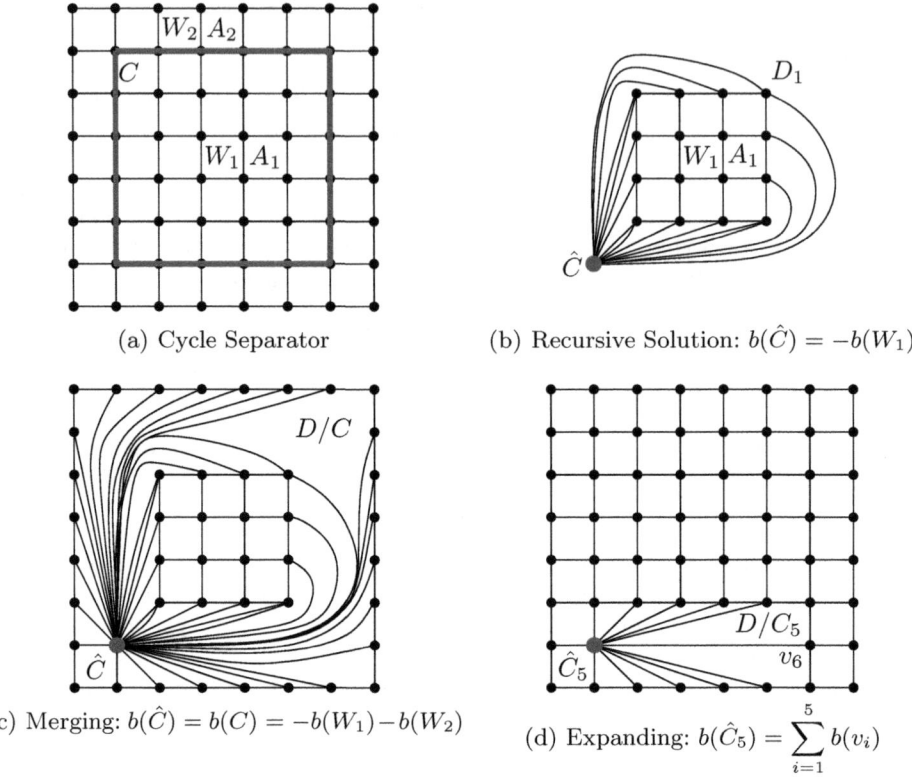

(a) Cycle Separator

(b) Recursive Solution: $b(\hat{C}) = -b(W_1)$

(c) Merging: $b(\hat{C}) = b(C) = -b(W_1) - b(W_2)$

(d) Expanding: $b(\hat{C}_5) = \sum_{i=1}^{5} b(v_i)$

Fig. 2. Illustration of Algorithm 2

maintaining all arcs between W_i and C with their respective costs and capacities. See Fig. 2(b) for an illustration. For a subset $W' \subset W$ let $b(W') = \sum_{v \in W'} b(v)$.

We now recursively solve the two min-cost flow problems

$$\mathcal{N}_i = (D_i, u|_{A_i}, \{b|_{W_i}, b(\hat{C}) = -b(W_i)\}, c|_{A_i}), i = 1, 2$$

obtaining a flow $f|_{A_i}$ with a valid node potential π_i.

Note that $\mathcal{N}_i, i = 1, 2$ has a feasible flow if $\mathcal{N}$ has a feasible flow: Let f be a feasible flow on $\mathcal{N}$. Clearly, f induces a flow on the graph D/C obtained from D by shrinking C to a single node $\hat{C}$ with demand $b(C)$. Note that D_1 is obtained from D/C by deleting W_2 and all its incident arcs. Let $f(C, W_2)$ be the amount of flow on the arcs from C to W_2 minus the amount of flow from W_2 to C. Then $f(C, W_2) = b(W_1) + b(C)$. So if we set $b(\hat{C}) = b(C) - f(C, W_2) = -b(W_1)$ then f induces a flow on D_1.

To merge the two solutions, we first set $\pi(\hat{C}) = \max\{\pi_1(\hat{C}), \pi_2(\hat{C})\}$ adjusting the potential in the respective components. Now we have a feasible flow with a valid node potential on D/C. See Fig. 2(c) for an illustration. We now expand C edge by edge assigning the nodes on C the current potential of $\hat{C}$. More precisely,

Algorithm 2. Recursive Min-Cost Flow

Input   : min-cost flow network $\mathcal{N} = (D = (W, A), u, b, c)$ admitting a flow.
Output : min-cost flow f on $\mathcal{N}$ and valid node potential π, both init. to 0.

MIN-COST-FLOW(D, u, b, c)
  $\quad (W_1, C, W_2) \leftarrow$ CYCLESEPARATOR(D);
  $\quad (f|_{A_i}, \pi_i) \leftarrow$ MIN-COST-FLOW$((W_i \cup \{\hat{C}\}, A_i), u|_{A_i}, \{b|_{W_i}, -b(W_i)\}, c|_{A_i})$;
  $\quad \pi(\hat{C}) \leftarrow \max\{\pi_1(\hat{C}), \pi_2(\hat{C})\}$;
  $\quad$ **for** $v \in W_i, i = 1, 2$ **do**
  $\quad\quad \lfloor\ \pi(v) \leftarrow \pi_i(v) - \pi_i(\hat{C}) + \pi(\hat{C})$;
  $\quad$ Let $C : v_1, \dots, v_\ell$;
  $\quad$ **for** $i = \ell, \dots, 2$ **do**
  $\quad\quad$ Expand v_i setting $\pi(v_i) \leftarrow \pi(\hat{C})$;
  $\quad\quad \lfloor\ (f, \pi) \leftarrow (f, \pi) +$ PRIMAL-DUAL$((D/\{v_1, \dots, v_{i-1}\})_f, u_f, b_f, c_\pi)$;
  $\quad$ **return** (f, π);

for $2 < i \leq \ell$ let D/C_i be obtained from D by shrinking $C_i = \{v_1, \dots, v_i\}$ to a single node $\hat{C}_i$ with demand $b(C_i)$. Assume that we have computed a flow f with a valid node potential π of D/C_i. *Expanding* v_i means extending f and π to D/C_{i-1} by setting the flow on the arcs between v_i and C_{i-1} to be zero and $\pi(v_i) = \pi(\hat{C}_{i-1}) = \pi(\hat{C}_i)$. See Fig. 2(d) for an illustration. This yields a pseudo-flow with a valid node potential, however, the deficiencies on v_i and $\hat{C}_{i-1}$ might be different from zero. To adjust the deficiencies, we run the primal-dual algorithm on the residual network. This yields a flow on $\mathcal{N}$ with a valid node potential and, hence, a min-cost flow on D. The algorithm is summarized in Algorithm 2.

Note that the max-flow within the primal-dual algorithm does only have to be performed between v_i and $\hat{C}_{i-1}$. Hence, there is no need for neither super source nor super sink and thus planarity is preserved. Moreover, v_i and $\hat{C}_{i-1}$ lie on the same face. Such a max flow computation can be done in linear time [22,16]. The same holds for the shortest path computation [16] because we maintain a valid node potential, i.e. non-negative reduced cost.

Theorem 1. *The recursive min-cost flow algorithm indicated in Algorithm 2 computes a min-cost flow on a planar bidirected uncapacitated min-cost flow network with n nodes, $\mathcal{O}(n)$ arcs, arc costs at most $c_{\max}$, and face degrees at most d in $\mathcal{O}(c_{\max}\sqrt{d}n^{3/2})$ time.*

Proof. Let m be the number of arcs in the flow network. We may assume that the network is connected and, hence, that $m \in \Theta(n)$. If the network is not biconnected, we first use the cut nodes as separators in the recursive algorithm. Since there is no expansion step, the combination of the recursive solutions of the biconnected components takes only constant time.

So assume now that the network is biconnected. Let all arcs have weight $1/m$ and let all faces and arcs have weight zero. Then the algorithm of Miller [26] constructs in linear time a cycle separator C with $\mathcal{O}(\sqrt{d \cdot n})$ nodes such that both, the interior and the exterior of C contain at most $2/3 \cdot m$ arcs. Let $c_{\max}$

be the maximum cost of an arc. Note that when expanding v_i then the only sources and sinks are v_i and $\hat{C}_{i-1}$ and there is an arc between the two of them in both directions with infinite capacity. Hence, the equivalent min-cost max-flow network remains planar and in all residual networks the length of a shortest path with respect to the original costs is at most $c_{\max}$. Hence, the primal-dual algorithm has to perform at most $c_{\max}$ max-flow operations (Lemma 2) before pushing the remaining deficiency directly over the arc incident to v_i and $\hat{C}_{i-1}$. It follows that $\mathcal{O}(c_{\max}\sqrt{d \cdot n})$ max-flow computations between two adjacent nodes of a planar graph have to be performed. Hence, each recursive step can be performed in $\mathcal{O}(c_{\max}\sqrt{d}n^{3/2}) = \mathcal{O}(c_{\max}\sqrt{d}m^{3/2})$. Hence, the run time $T(m)$ fulfills the recursion

$$T(m) \leq T(m_1) + T(m_2) + c \cdot c_{\max}\sqrt{d}m^{3/2}, \text{ with } m_1 + m_2 \leq m, \, m_1, m_2 \leq 2/3 \cdot n$$

Thus, the total running time is in $\mathcal{O}(c_{\max}\sqrt{d}m^{3/2}) = \mathcal{O}(c_{\max}\sqrt{d}n^{3/2})$. □

Note Theorem 1 remains true if the arc costs are not bounded in general and Algorithm 2 chooses separators that are not necessarily cycles but induce connected subgraphs with arc costs at most $c_{\max}$.

Corollary 2. *The bend-minimization problem on a plane graph with degree at most four and n vertices can be solved in $\mathcal{O}(n^{3/2})$ time.*

Proof. Let $G = (V, E)$ be a plane graph with n vertices and with degree at most four and let $\mathcal{N}_G = (D_G = (V \cup \mathcal{F}, A_G), u, b, c)$ be the min-cost flow network for the bend-minimization problem. Let $m = |A_G|$. Note that $m \in \Theta(n)$. For Computing the cycle separator in the recursive min-cost flow algorithm, we only consider the subgraph $D_\mathcal{F}$ induced by the face nodes. We assign each arc of $D_\mathcal{F}$ the weight $1/m$ and each face h of $D_\mathcal{F}$ the weight $\deg f/m$ while the nodes obtain zero weight. Now the cycle separator of $D_\mathcal{F}$ constructed by the algorithm of Miller [26] is a cycle separator C of the whole graph with $O(\sqrt{n})$ nodes such that both, the interior and the exterior of C contain at most $2/3 \cdot m$ arcs. Moreover the arcs on C are bidirected uncapacitated and have unit cost. Hence, each call of the primal-dual algorithm within Algorithm 2 performs one max-flow operation on two adjacent nodes and pushes the remaining deficiency over the corresponding cycle arc. Hence, each recursive step and thus, the whole algorithm can be performed in $\mathcal{O}(n^{3/2})$ time. □

If we wish to constrain the number of bends on an edge artificially, we may sacrifice a log-factor and use the result in [27] to obtain the following.

Remark 2. The bend-minimization problem on a plane graph with degree at most four and n vertices can be solved in $\mathcal{O}(n^{3/2} \log n)$ time even if the number of bends per edge is bounded by some upper bounds $u : A \to \mathbb{Z}_{\geq 0}$, provided that the bounds still admit an orthogonal drawing with a linear number of bends.

Proof. Instead of expanding the cycle separator node after node, we expand it at once. Now the nodes with deficiency other than zero are all on a path. Hence, the max-flow problem within the primal-dual algorithm is solvable in $\mathcal{O}(n\log^2 n)$ time [27]. Assume now that we perform $\sqrt{n}/\log n$ times an ordinary iteration of the primal dual algorithm. Then, by Lemma 3, at most $\mathcal{O}(n/(\sqrt{n}/\log n))$ additional shortest path computations have to be performed, each of which can be done in linear time [31]. Hence, one recursive step and thus the whole algorithm can be performed in $\mathcal{O}(\sqrt{n}/\log n \cdot n\log^2 n + \sqrt{n}\log n \cdot n) = \mathcal{O}(n^{3/2}\log n)$ time. $\qquad\square$

Acknowledgments. We are grateful to Ulrik Brandes for bringing our attention to this problem and for fruitful discussions.

References

1. Biedl, T.C., Kant, G.: A better heuristic for orthogonal graph drawings. Computational Geometry 9(3), 159–180 (1998)
2. Garg, A., Tamassia, R.: On the computational complexity of upward and rectilinear planarity testing. SIAM Journal on Computing 31(2), 601–625 (2001)
3. Tamassia, R.: On embedding a graph in the grid with the minimum number of bends. SIAM Journal on Computing 16, 421–444 (1987)
4. Fößmeier, U., Kaufmann, M.: Drawing High Degree Graphs with Low Bend Numbers. In: Brandenburg, F.J. (ed.) GD 1995. LNCS, vol. 1027, pp. 254–266. Springer, Heidelberg (1996)
5. Klau, G.W., Mutzel, P.: Quasi orthogonal drawing of planar graphs. Technical Report MPI-I-98-1-013, Max-Planck-Institut für Informatik, Saarbrücken, Germany (1998), http://data.mpi-sb.mpg.de/internet/reports.nsf
6. Tamassia, R., Di Battista, G., Batini, C.: Automatic graph drawing and readability of diagrams. IEEE Transactions on Systems, Man and Cybernetics 18(1), 61–79 (1988)
7. Bertolazzi, P., Di Battista, G., Didimo, W.: Computing orthogonal drawings with the minimum number of bends. IEEE Transactions on Computers 49(8), 826–840 (2000)
8. Brandes, U., Cornelsen, S., Fieß, C., Wagner, D.: How to draw the minimum cuts of a planar graph. Computational Geometry: Theory and Applications 29(2), 117–133 (2004)
9. Lütke-Hüttmann, D.: Knickminimales Zeichnen 4-planarer Clustergraphen. Master's thesis, Universität des Saarlandes (1999) (Diplomarbeit)
10. Brandes, U., Wagner, D.: Dynamic Grid Embedding with Few Bends and Changes. In: Chwa, K.-Y., Ibarra, O.H. (eds.) ISAAC 1998. LNCS, vol. 1533, pp. 89–98. Springer, Heidelberg (1998)
11. Brandes, U., Eiglsperger, M., Kaufmann, M., Wagner, D.: Sketch-Driven Orthogonal Graph Drawing. In: Goodrich, M.T., Kobourov, S.G. (eds.) GD 2002. LNCS, vol. 2528, pp. 1–11. Springer, Heidelberg (2002)
12. Ahuja, R.K., Magnanti, T.L., Orlin, J.B.: Network Flows. Prentice-Hall (1993)
13. Schrijver, A.: Combinatorial Optimization: Polyhedra and Efficiency. Springer, Heidelberg (2003)
14. Karmarkar, N.: A new polynomial-time algorithm for linear programming. Combinatorica 4(4), 373–395 (1984)

15. Imai, H., Iwano, K.: Efficient Sequential and Parallel Algorithms for Planar Minimum Cost Flow. In: Asano, T., Imai, H., Ibaraki, T., Nishizeki, T. (eds.) SIGAL 1990. LNCS, vol. 450, pp. 21–30. Springer, Heidelberg (1990)
16. Henzinger, M.R., Klein, P., Rao, S., Subramanian, S.: Faster shortest-path algorithms for planar graphs. Journal of Computer and System Sciences 55, 3–23 (1997); Special Issue on Selected Papers from STOC 1994
17. Fakcharoenphol, J., Rao, S.: Planar graphs, negative weight edges, shortest paths, and near linear time. J. Comput. Syst. Sci. 72, 868–889 (2006)
18. Klein, P., Mozes, S., Weimann, O.: Shortest paths in directed planar graphs with negative lengths: a linear-space $O(n \log^2 n)$-time algorithm. In: Proceedings of the Twentieth Annual ACM-SIAM Symposium on Discrete Algorithms, SODA 2009, pp. 236–245. SIAM, Philadelphia (2009)
19. Mozes, S., Wulff-Nilsen, C.: Shortest Paths in Planar Graphs with Real Lengths in $O(n \log^2 n / \log \log n)$ Time. In: de Berg, M., Meyer, U. (eds.) ESA 2010. LNCS, vol. 6347, pp. 206–217. Springer, Heidelberg (2010)
20. Weihe, K.: Maximum (s,t)-flows in planar networks in O(V log V) time. J. Comput. Syst. Sci. 55, 454–475 (1997)
21. Borradaile, G., Klein, P.: An O(n log n) algorithm for maximum st-flow in a directed planar graph. J. ACM 56, 9:1–9:30 (2009)
22. Hassin, R.: Maximum flow in (s, t) planar networks. Information Processing Letters 13(3), 107 (1981)
23. Miller, G.L., Naor, J.: Flow in planar graphs with multiple sources and sinks. SIAM J. Comput. 24, 1002–1017 (1995)
24. Garg, A., Tamassia, R.: A New Minimum Cost Flow Algorithm with Applications to Graph Drawing. In: North, S.C. (ed.) GD 1996. LNCS, vol. 1190, pp. 201–213. Springer, Heidelberg (1997)
25. Brandenburg, F.J., Eppstein, D., Goodrich, M.T., Kobourov, S.G., Liotta, G., Mutzel, P.: Selected Open Problems in Graph Drawing. In: Liotta, G. (ed.) GD 2003. LNCS, vol. 2912, pp. 515–539. Springer, Heidelberg (2004)
26. Miller, G.L.: Finding small simple cycle separators for 2-connected planar graphs. Journal of Computer and System Sciences 32(4), 265–279 (1986)
27. Borradaile, G., Klein, P., Mozes, S., Nussbaum, Y., Wulff-Nilsen, C.: Multiple-source multiple-sink maximum flow in directed planar graphs in near-linear time. In: Proceedings of the 52nd Annual Symposium on Foundations of Computer Science, FOCS 2011 (to appear, 2011)
28. Di Battista, G., Eades, P., Tamassia, R., Tollis, I.G.: Graph Drawing: Algorithms for the Visualization of Graphs. Prentice-Hall (1999)
29. Ford, L.R., Fulkerson, D.R.: Flows in Networks. Princeton University Press (1962)
30. Dijkstra, E.W.: A note on two problems in connexion with graphs. Numerische Mathematik 1, 269–271 (1959)
31. Tazari, S., Müller-Hannemann, M.: Shortest paths in linear time on minor-closed graph classes, with an application to steiner tree approximation. Discrete Applied Mathematics 157(4), 673–684 (2009)

TGI-EB: A New Framework for Edge Bundling Integrating Topology, Geometry and Importance

Quan Nguyen, Seok-Hee Hong, and Peter Eades

School of Information Technologies, University of Sydney, Australia
Capital Markets CRC, Sydney, Australia
{qnguyen,shhong,peter}@it.usyd.edu.au

Abstract. Edge bundling methods became popular for visualising large dense networks; however, most of previous work mainly relies on *geometry* to define *compatibility* between the edges.

In this paper, we present a new framework for edge bundling, which tightly integrates topology, geometry and importance. In particular, we introduce new edge compatibility measures, namely *importance compatibility* and *topology compatibility*. More specifically, we present four variations of force directed edge bundling method based on the framework: Centrality-based bundling, Radial bundling, Topology-based bundling, and Orthogonal bundling.

Our experimental results with social networks, biological networks, geographic networks and clustered graphs indicate that our new framework can be very useful to highlight the most *important topological skeletal structures* of the input networks.

1   Introduction

Overviews of large and complex networks are useful for conveying information and commonly used for extracting global patterns, such as clusters and outliers in a data set. However, visualising large and complex networks is very challenging, especially, for large dense graphs due to visual clutters which hinder human understanding and analytic tasks.

Recently, edge bundling methods became popular for visualising large dense networks, and have received much attention by the Graph Drawing community and Information Visualisation community [8, 11, 13–15]. Most of the methods are based on *geometry*, i.e., a given drawing of graphs, to define *geometry compatibility* between the edges (i.e., edges are typically polylines or splines that are bundled together if they are compatible). While those edge bundling methods reduce visual clutters and show some high level edge patterns, they may not necessarily highlight the important skeletal structure of the network.

In this paper, we present a new framework for edge bundling, which tightly integrates *topology*, *geometry* and *importance*. In particular, we introduce new measures of edge compatibility based on network analysis and topology, namely *importance compatibility* and *topology compatibility*, which are independent from the geometry of the given input drawing.

M. van Kreveld and B. Speckmann (Eds.): GD 2011, LNCS 7034, pp. 123–135, 2012.
© Springer-Verlag Berlin Heidelberg 2012

As an example to define importance compatibility, we use social network analysis methods [20]. For example, *centrality* analysis determines the relative importance of vertices and edges in a network. The *k-core* decomposition can be used to identify cohesive groups of actors within a network. As an example to define topology compatibility, we use clustered graph model.

More specifically, we present four variations of force directed edge bundling method, based on the framework:

- CenEB (Centrality-based edge bundling): tightly integrates edge centrality analysis with edge bundling.
- TopoEB (Topology-based edge bundling): tightly integrates clustered graph topology with edge bundling.
- RadEB (Radial edge bundling): tightly integrates *k*-core analysis with edge bundling.
- OrthoEB (Orthogonal edge bundling): uses orthogonal-like edge representation to produce orthogonal-like crossings.

We implemented our new framework and conducted experiments with social networks, biological networks, geographic networks and clustered graphs. Our experimental results show that our new framework can be useful to highlight the most important topological skeletal structures of the input network, and significantly improve visual analysis.

The new approach has proved very useful for the analysis on the integrated NF-κB protein-protein interaction and signalling transduction networks, clearly showing a number of significant functional groups. In fact, our visualisation guided biologists to derive new biological hypothesis, and currently laboratory experiments are being conducted.

2 Related Work

The use of attractions on control points for curved edges was first introduced by Brandes and Wagner [6] and later Finkel et al. [9], though the term "edge bundling" was coined several years later by others.

Holten [13] presented Hierarchical Edge Bundling method for hierarchical graphs using B-splines. Balzer et al. [4] proposed a multi-level compound visualisation using transparent surfaces and edge bundling for a hierarchical 3D visualisation.

Zhou et al. [21] presented a hierarchical edge clustering using Delaunay triangulation, where control points are hierarchically clustered by energy-based optimisation. Geometry Based Edge Bundling by Cui et al. [8] uses a control mesh for edge clustering, where edge bundles share the same control points on the mesh. Lambert et al. [15] generalised a control mesh to route graph edges using a shortest path algorithm and mesh edge weights are updated to encourage graph edges to share mesh edges.

Gansner et al. [11] improved circular layouts by merging splines of edges to minimise the total amount of ink needed to draw the edges. Cornelissen et al. [7]

presented a circular bundle view of the hierarchical graphs to study in software engineering, such as, the program execution traces.

Holten and van Wijk introduced a Force-Directed Edge Bundling (FDEB) algorithm [14], which models edges intuitively as flexible springs that can attract each other. The attractive force depends on the distance of the springs and the compatibility of the edges. The method achieves smoother bundles that are easy to read, although it incurs high computational complexity.

Telea et al. [19] proposed an Image-Based Edge Bundles that aims for coarse-grained edge shapes of bundled edges to further simplify visual representation of the network structure. Nachmanson et al. [16] consider edge bundling in layered drawings in which edges already routed as polylines or splines; the method preserves the topology of the original drawing and disambiguates edges.

Recently, Gansner et al. [10] introduced a multi-level method which approximates k-neighbor edge proximity graphs using kd-tree as input for their agglomerative bundling algorithm. They reported experiments on the approach up to one million edges in a few minutes.

This previous work on edge bundling reduces visual clutter and displays some high-level patterns. Yet the "bundles" are mainly based on geometry in disregard of the importance and the topology of the network. This motivates our new framework for edge bundling which integrates topology, geometry and importance, to highlight important skeletal structures of the networks.

3   Integrated Framework for Edge Bundling

This section presents our new generic framework for edge bundling which tightly integrates topology, geometry and importance. Our framework is *flexible*: one can use other measures for importance, geometry, and topology.

For our specific framework, we first use a force-directed edge bundling method as a basis, and then integrate geometry with *importance*, defined by centrality and k-core analysis. Finally, we further integrate *topology* into the model, defined by a clustered graph model.

3.1   New Edge Compatibility Measures

Existing edge bundling methods mainly use geometry to define *geometry compatibility* $G(e, e')$. For instance, several metrics are proposed in FDEB (force-directed edge bundling) method [14] to define geometry compatibility (in their paper $C(e, e')$ is used). "Angle" metric is designed to avoid bundling edges that are almost perpendicular. "Scale" metric ensures edges that differ considerably in length should not be bundled together. "Position" metric aims to avoid bundling edges that are very far apart. "Visibility" metric avoids bundling edges that are parallel and equal in length.

Importance Compatibility. Here, we introduce a new measure "*importance compatibility*" to integrate importance into geometry for edge bundling. Importance compatibility is conceptually to guide the bundling with respect to important edges and thus is independent from geometry, i.e., the given input drawing of a graph. Importance can be defined from application domain or specific analysis in analytic task.

Topology Compatibility. We now introduce another new notion of compatibility, called "*topology compatibility*". The topology compatibility can be defined from topological structure or combinatorial structure of given graph model. The topology compatibility, like importance compatibility, is independent from geometry.

3.2   The Framework

As an example of the integrated framework, we integrate our new edge compatibility measures into FDEB [14].

More specifically, the FDEB algorithm first inserts control points in each edge, and then uses a force-directed method to compute the position of the control points. Their forces depend on the "geometry compatibility" $G(e, e')$.

For a subdivision point e_i on edge e, the total force F_{e_i} exerted on e_i is a sum of the two spring forces exerted by two neighbors e_{i-1} and e_{i+1}, and the total of electrostatic forces F_s:

$$F_{e_i} = k_e(|\mathsf{p}_{e_{i-1}} - \mathsf{p}_{e_i}| + |\mathsf{p}_{e_i} - \mathsf{p}_{e_{i+1}}|) + F_s, \qquad (1)$$

where k_e is the stiffness of edge e, and $\mathsf{p}(x)$ is the location of x.

In FDEB, electrostatic force model is

$$F_s = \sum_{e' \in \mathcal{E}} G(e, e') * |\mathsf{p}_{e_i} - \mathsf{p}_{e'_i}|^{-d}, \qquad (2)$$

TGI-EB, our new general framework for edge bundling integrating topology, geometry and importance, can be described as follows. In its most general form, our electrostatic force model is

$$F_s = \sum_{e' \in \mathcal{E}} G(e, e') * I(e, e') * T(e, e') * g(|\mathsf{p}_{e_i} - \mathsf{p}_{e'_i}|), \qquad (3)$$

where $\mathcal{E}$ is the set of compatible edges of e; $G(e, e')$, $I(e, e')$ and $T(e, e')$ are geometry compatibility, importance compatibility, and topology compatibility measures for a pair of edges e and e'; and g is a function of $|\mathsf{p}_{e_i} - \mathsf{p}_{e'_i}|$, e.g., $g = |\mathsf{p}_{e_i} - \mathsf{p}_{e'_i}|^{-d}$ where d is a numeric constant.

Note that our new framework TGI-EB is very general and flexible. For example, one can derive various models by controlling the weight parameters between $G(e, e')$, $I(e, e')$ and $T(e, e')$. Furthermore, one can define different metric to define geometry compatibility, importance compatibility and topology compatibility.

3.3   Centrality Based Edge Bundling (CenEB)

As an example to define importance compatibility, we use edge centrality. Centrality is the most well-known network analysis method, which determines the relative prominence of vertices and edges in a network [5, 20]. For instance, edge centrality analysis, which finds the important edges, has been used for mesh coarsening, analyzing biological networks and community detection.

CenEB is a special case of the general model TGI-EB described in Equation 3, which integrates importance compatibility and geometry compatibility, and $T(e, e')$ is absent. We use the edge centrality metric to highlight important edges and bundle high centrality edges together.

The most general form of our electrostatic force model for CenEB is

$$F_s = \sum_{e' \in \mathcal{E}} G(e, e') * I(e, e') * g(|\mathsf{p}_{e_i} - \mathsf{p}_{e'_i}|), \qquad (4)$$

where $\mathcal{E}$ is the set of compatible edges of e, $I(e, e')$ is calculated based on the centrality values of the edges e and e'. For example, $g = |\mathsf{p}_{e_i} - \mathsf{p}_{e'_i}|^{-d}$, where d is a numeric constant, and $I(e, e')$ is defined from centrality values of e and e'.

3.4   Topology Based Edge Bundling (TopoEB)

As an example of topology compatibility, here we use a clustered graph model. A clustered graph $\mathsf{G} = (V, E)$ consists of a number of clusters $\mathsf{G}_i = (V_i, E_i)$. An edge that connects two nodes in the same cluster is called an *intra-cluster edge*, while an edge connecting two nodes from different clusters is called an *inter-cluster edge*.

Using the topology of the clustered graphs, we can define topology compatibility as follows:

- Two intra-cluster edges are not topology-compatible unless they belong to the same cluster;
- All inter-cluster edges are topology-compatible; in fact, they all belong to the *root* cluster G;
- A pair of an intra-cluster edge and an inter-cluster edge is not topology-compatible.

In fact, the benefits of topology compatibility in clustered graph model are two-fold.

- First, by using topology compatibility, the number of compatible edges $\mathcal{E}$ of an edge e can be significantly reduced, which results in faster bundling iterations.
- Second, for better flexibility, one can define a topology compatibility metric $T(e, e')$, which may allow bundling intra- and inter-cluster edges together. The metric is defined in three cases depending on whether the edges e and e' are intra-cluster edges in the same cluster (intra-intra), inter-cluster edges (inter-inter), or one inter-cluster edge and one intra-cluster edge (inter-intra).

As an example of integration for TopoEB, we now integrate topology compatibility, with the model defined above for CenEB, which combines importance compatibility and geometry compatibility. TopoEB is the special case of the general model TGI-EB in Equation 3, and can be described as follows:

$$F_s = \sum_{e' \in \mathcal{E}} G(e, e') * I(e, e') * T(e, e') * g(|\mathsf{p}_{e_i} - \mathsf{p}_{e'_i}|), \qquad (5)$$

where $I(e, e')$ is defined based on the centrality values of e and e', and $T(e, e')$ is defined from the clustered graph model. Note that, inter-cluster edges often have higher edge centralities than intra-cluster edges.

For example, the metric $T(e, e')$ can be simply defined as :

- c_{intra}: if e and e' are intra-cluster edges in the same cluster
- c_{inter}: if e and e' are inter-cluster edges
- c_{mix}: if e and e' is a pair of an intra-cluster edge and an inter-cluster edge;

where the contants c_{intra}, c_{inter} and c_{mix} are chosen from 0 to 1. It is worth-noting that when these constants are chosen equal, the value $T(e, e')$ is the same for every pair of edges and thus the method is said *topology-insensitive*. When c_{mix} is zero, there is no bundling between intra-cluster edge and inter-cluster edge. Generally, one may choose a value close to 1 for c_{intra} and c_{inter} and a small value for c_{mix}.

3.5   Radial Bundling (RadEB)

We now present another variation of edge bundling, called Radial bundling (RadEB), which uses a radial layout consisting of concentric circles for the input of edge bundling. The radial layout can be used to display hierarchy or k-core analysis of graphs. As a specific example in this paper, we used k-core analysis to define a radial layout.

An important group-level network analysis is to identify cohesive subgroups of actors with strong ties [5, 20]. A well-known example is the k-*cores* of a graph, each of which is a maximal-connected subgraph whose nodes have the induced degree at least k [5]. The k-core analysis has been used in social networks such as collaboration networks, and biological networks for analyzing PPI networks.

For a radial layout, we use forces to constrain the vertices u in each k-core to a circle of radius $r_u = f(k)$. The forces place vertices from the same k-core along the same circle.

We integrate the standard force-directed layout method with a new *radial force* for each vertex u: $F_{rad} = c_{rad}(|\mathsf{p}_u - \mathsf{p}_o| - r_u)$, where o is the center of the circles. Typically, f is a linear function, although we have also used logarithmic functions.

Clustering Constraints.   We further extended RadEB to handle clustering constraints. We introduce a *similarity clustering force*, which attracts vertices of close similarity indices together. Thus, a new attraction force $f_a(u, v)$ is applied between every pair of vertices u and v:

$$F_{clus} = \sum_{(u,v) \in E} f_a(u, v) * exp(-|\mathsf{i}_u - \mathsf{i}_v|), \qquad (6)$$

where i_u and i_v is clustering indices of u and v, respectively. The clustering index can defined based on the application: for example, functional-similarity for biological networks, and group membership for social networks.

Note that our radial layout is different from the k-core visualisation by Alvarez et al. [3], which produces a radial layout using the polar coordinates. In fact, our model is more flexible, since we can further combine clustering constraints.

After producing the radial layout for visualising k-core analysis, we apply our CenEB to the resulting layout for radial bundling.

3.6 Orthogonal Edge Bundling (OrthEB)

We also present a new variation of edge representation for edge bundling, called OrthEB, which produces orthogonal-like edge bundles. Orthogonal edge bundling can be effective to produce a bundles with right angle crossings.

More specifically, we adapt forces in CenEB using magnetic field forces [18], to produce orthogonal-like bundled edges. Figure 1a and Figure 1b show example of forces in CenEB and OrthoEB in each iteration.

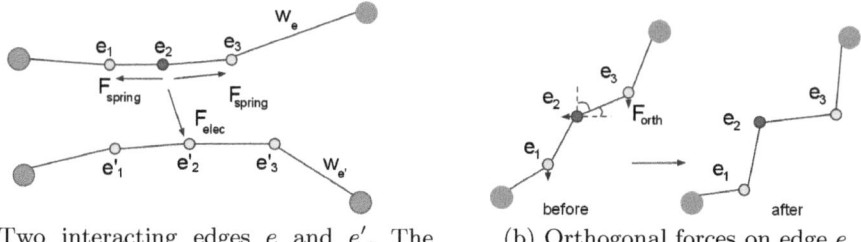

(a) Two interacting edges e and e'. The spring and electrostatic forces on a control point e_2

(b) Orthogonal forces on edge e

Fig. 1. Examples of forces in CenEB and OrthEB

The *orthogonal forces* are applied on the control points of each edge. The orthogonal force on point e_i is based on the tangent of the subsegment $e_{i-1}e_i$ of the edge, and e_i is sequentially moved towards the axis (either x-axis or y-axis) that forms smaller angle. Consequently, sub-segments are placed almost horizontally or vertically. In the final drawing, splines are used to connect the control points in each edge to achieve aesthetically pleasing bundling effects.

3.7 Time Complexity and Implementation

Like force-directed edge bundling (FDEB), our TGI-EB traverses every pair of edges to determine compatible edges, thus it takes $O(|E|^2)$ time for an iteration. Our force-directed radial layout with clustering constraints takes $O(|V|^2)$ time. Yet our experimental results show that our methods are quite fast for graphs with up to a few hundred nodes and two thousand edges. It took a few seconds to produce a nicely bundled layouts.

We have implemented our new edge bundling methods using our own implementation in Java for k-core radial layout, a prototype implementation of FDEB from the jFlowMap project [1], and various clustered graph layouts [12] implemented in GEOMI [2].

4   Experimental Results

4.1   Social Networks

As an example of a social network, we use the 2010 Graph Drawing competition data set consisting of research collaborations in Graph Drawing research papers from 2004-2010. The data set is a graph with 362 nodes and 942 edges. Our case

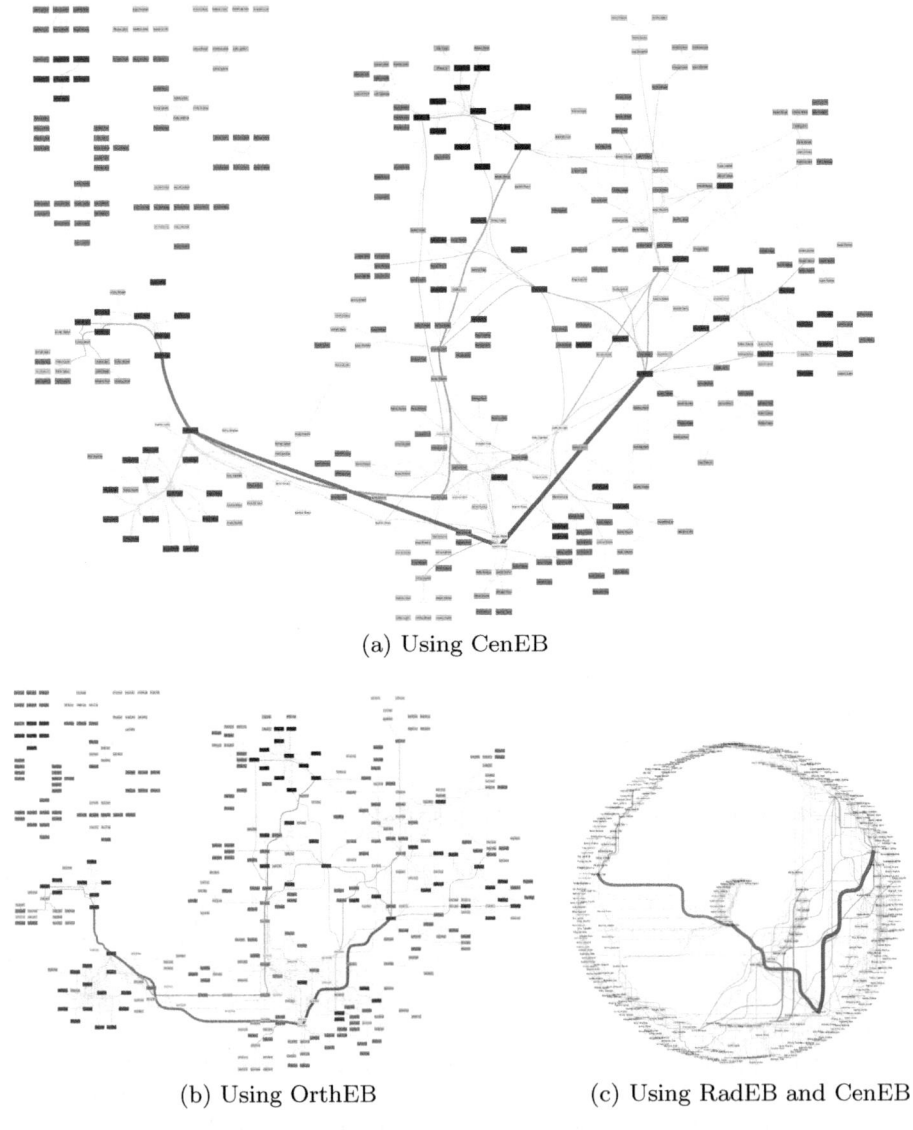

(a) Using CenEB

(b) Using OrthEB   (c) Using RadEB and CenEB

Fig. 2. Collaboration network using CenEB, OrthEB and RadEB

study on collaboration networks aims to identify important researchers, research groups and collaboration patterns.

Figure 2a and Figure 2b show visualisations using CenEB and OrthEB, respectively, with edge betweenness centrality. The figures enable the following visual analyses.

First, one can easily identify the major research groups and research collaborations between the groups. The largest group is a 13-core (red) of Spanish and German researchers; the second largest group is an 11-core (blue) of an Australian clique. Second, the drawing clearly highlights researchers with high betweenness centrality: Brandes, Brandenburg, Kaumann, Kobourov, Kratochvil, Liotta, Mutzel and Wolff. Third, one can also identify several important edges with high centrality values; for example, the collaborations between Kaufmann and Kobourov; between Kauffman, Wolff and Symvonis; between Kratochvil and Wolff; between Brandes and Dwyer; between Brandes and Symvonis; between Kobourov and Sander. Finally, one can find a clique of four people with high betweenness centrality values: Brandenburg, Kobourov, Liotta and Mutzel.

Figure 2c shows a drawing of the collaboration network produced by the integration of RadEB and OrthoEB. It shows a clearer structure of the groups within different k-core circles. The inner most circle contains the 13-core group of researchers. The next circle contains the 11-core group of researchers. The drawing also highlights the important collaboration paths between researchers. Strong collaboration paths are more visible from the orthogonal-like bundled edges.

4.2   Clustered Graphs

We have experimented with randomly generated clustered graphs with different inter-cluster edge densities: sparse and dense. We use clustered graph layouts of Ho and Hong [12] implemented in GEOMI [2]. This case study uses two clustered

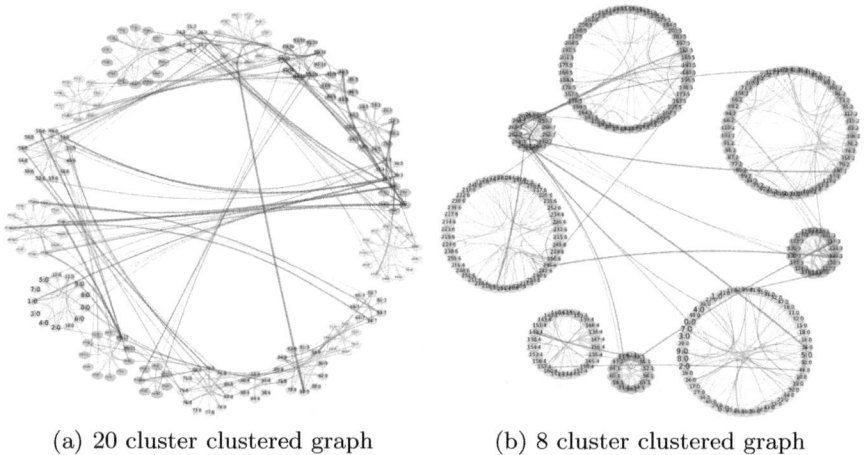

(a) 20 cluster clustered graph	(b) 8 cluster clustered graph

Fig. 3. Dense clustered graphs in Circular-Circular layouts using TopoEB

graph layouts: circular-circular layouts and circular-force directed layouts, which draw each cluster on a circle and each cluster is, respectively, drawn using a circular layout or a spring algorithm.

We found that clustered graphs with sparse inter-cluster edges have less edge bundling effects, compared to the dense inter-cluster edge instances. Thus, we present two examples with dense inter-cluster edges. Two instances were selected from randomly generated clustered graphs: one has 20 clusters consisting of 191 nodes and 2165 edges; and the other has 8 clusters consisting of 272 nodes and 2407 edges.

Figure 3a and Figure 3b show our TopoEB results on the two clustered graphs using circular-circular layout. The figures clearly show important inter-cluster and important intra-cluster edges, and the clusters from intra-cluster edge bundles. Inter-cluster edge bundling has been shown to be effective for dense clustered graphs.

4.3   Biological Networks

This case study aims to identify new important regulatory elements and structures in a protein-protein interaction (PPI) network. We use a NF-κB PPI network consisting of 778 nodes and 1868 edges, and 14 levels of coreness.

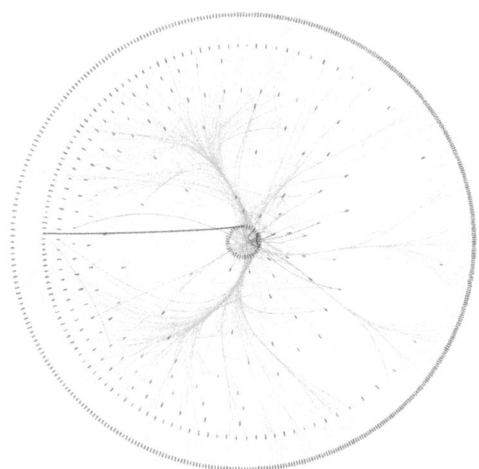

Fig. 4. NF-κB network using our RadEB and CenEB

Figure 4 shows a visualisation produced by RadEB and CenEB. It clearly shows six significant paths indicating different cell functionality. The drawing also clearly depicts several protein groups with specific functionality.

The edge centralities reflect the importance of elements. One can identify the important proteins that directly influence the translocation of the NF-κB transcription factor. Proteins that act in similar biological processes are grouped together and form network structures and motifs. In fact, our new visualisation has inspired biologists to generate a new hypothesis, based on the newly identified six important paths and important proteins around the paths. Some lab experiments are being conducted to verify the hypothesis.

4.4 Geographic Networks

For geographic networks, critical analysis include tasks, such as flight scheduling and facility allocation. We extend our case study on airlines networks to identify important airports and flights. We use the US airlines network, which contains 235 nodes and 2101 directed edges (1297 undirected edges).

Figure 5a and Figure 5b show the airlines network using CenEB and OrthEB, respectively. Airport nodes are colored with k-core values. One can identify several important flights: between SEA and DTW, between BIL and MSP, between LAX and SEA, and between BIL and ATL.

The airlines network using RadEB is shown in Figure 5c. The figure shows the most highly connected group consists of 23 airports of the 13-core around the inner most circle, including important airports, e.g., SEA, DTW, MSP, ATL, MEM and TAH. One can also identify all the important flights connecting the

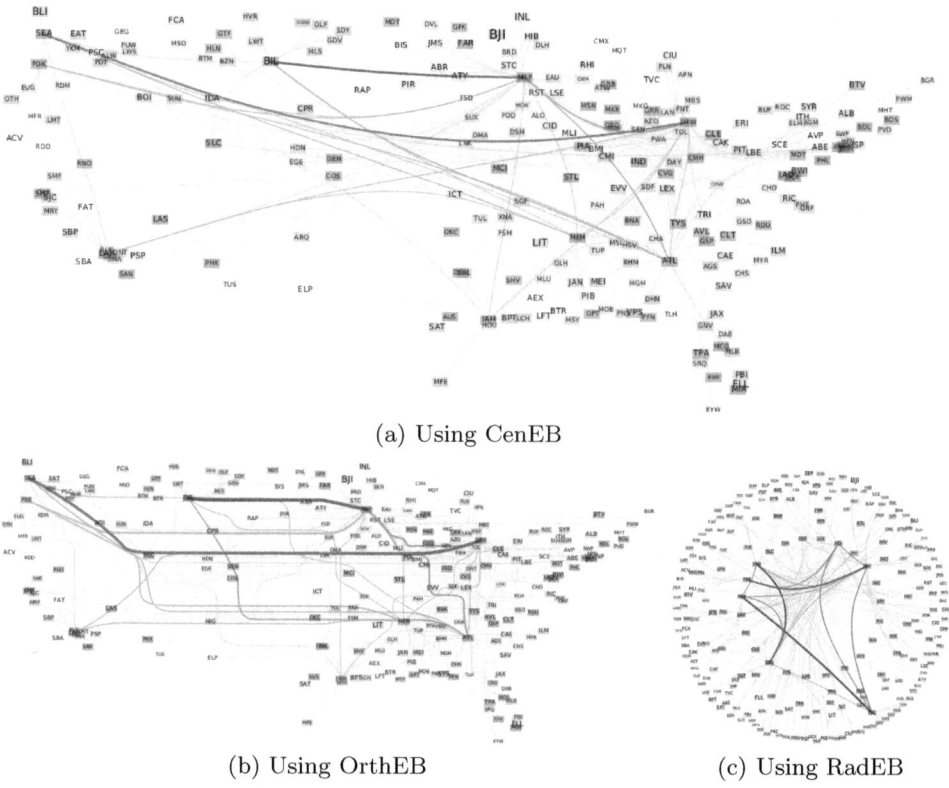

(a) Using CenEB

(b) Using OrthEB   (c) Using RadEB

Fig. 5. The US Airline network using CenEB, OrthEB and RadEB

13-core airports. Interestingly, an outlier was identified as depicted in the figure: BIL airport in a low core has several "important" flights (those connected to MEM, ATL, MSP and SEA airports). This is possibly because BIL is geographically located in the middle between MEM, ATL, MSP airports (the east) and SEA airport (the west), as shown in Figure 5a and Figure 5b.

5   Future Work

Our future work is to improve the running time to address the scalability problem for huge network instances; for example, adapting the agglomative edge bundling algorithm of Gansner et al. [10].

We also plan to design new criteria or metric to evaluate the performance of edge bundling methods. We plan to generalise the magnetic field in our orthogonal edge bundling method to handle any arbitrary angles rather than just 90 degree, similar to gradient computation in Strzodka et al. [17].

Acknowledgements. Thanks to S. J. Janowski, J. Stoye and C. Kaltschmidt from Faculty of Technology, University of Bielefeld, Germany for the PPI networks, and valuable discussions on the biological hypotheses and lab experiments.

References

1. jFlowMap (2010), http://code.google.com/p/jflowmap/
2. Ahmed, A., Dwyer, T., Forster, M., Fu, X., Ho, J., Hong, S.-H., Koschützki, D., Murray, C., Nikolov, N.S., Taib, R., Tarassov, A., Xu, K.: GEOMI: Geometry for Maximum Insight. In: Healy, P., Nikolov, N.S. (eds.) GD 2005. LNCS, vol. 3843, pp. 468–479. Springer, Heidelberg (2006),
   http://dblp.uni-trier.de/db/conf/gd/gd2005.html#AhmedDFFHHKMNTTX05
3. Alvarez, H.J.I., Dall, A.L., Barrat, A., Vespignani, A.: Large scale networks fingerprinting and visualization using the k-core decomposition. In: Advances in Neural Information Processing Systems, vol. 18, p. 41 (2006)
4. Balzer, M., Deussen, O.: Level-of-detail visualization of clustered graph layouts. In: APVIS, pp. 133–140 (2007)
5. Brandes, U., Erlebach, T.: Network analysis: methodological foundations. Springer, Heidelberg (2005)
6. Brandes, U., Wagner, D.: Using Graph Layout to Visualize Train Interconnection Data. In: Whitesides, S.H. (ed.) GD 1998. LNCS, vol. 1547, pp. 44–56. Springer, Heidelberg (1999)
7. Cornelissen, B., Zaidman, A., Holten, D., Moonen, L., van Deursen, A., van Wijk, J.J.: Execution trace analysis through massive sequence and circular bundle views. Journal of Systems and Software 81(12), 2252–2268 (2008)
8. Cui, W., Zhou, H., Qu, H., Wong, P.C., Li, X.: Geometry-based edge clustering for graph visualization. IEEE Transactions on Visualization and Computer Graphics, 1277–1284 (2008)
9. Finkel, B., Tamassia, R.: Curvilinear Graph Drawing using the Force-Directed Method. In: Pach, J. (ed.) GD 2004. LNCS, vol. 3383, pp. 448–453. Springer, Heidelberg (2005)

10. Gansner, E., Hu, Y., North, S., Scheidegger, C.: Multilevel agglomerative edge bundling for visualizing large graphs. In: IEEE PacificVis, pp. 187–194 (2011)
11. Gansner, E., Koren, Y.: Improved Circular Layouts. In: Kaufmann, M., Wagner, D. (eds.) GD 2006. LNCS, vol. 4372, pp. 386–398. Springer, Heidelberg (2007)
12. Ho, J., Hong, S.-H.: Drawing Clustered Graphs in Three Dimensions. In: Healy, P., Nikolov, N.S. (eds.) GD 2005. LNCS, vol. 3843, pp. 492–502. Springer, Heidelberg (2006)
13. Holten, D.: Hierarchical edge bundles: Visualization of adjacency relations in hierarchical data. IEEE Transactions on Visualization and Computer Graphics, 741–748 (2006)
14. Holten, D., van Wijk, J.J.: Force-directed edge bundling for graph visualization. Computer Graphics Forum 28(3), 983–990 (2009)
15. Lambert, A., Bourqui, R., Auber, D.: Winding roads: Routing edges into bundles. Computer Graphics Forum 29, 853–862 (2010)
16. Pupyrev, S., Nachmanson, L., Kaufmann, M.: Improving Layered Graph Layouts with Edge Bundling. In: Brandes, U., Cornelsen, S. (eds.) GD 2010. LNCS, vol. 6502, pp. 329–340. Springer, Heidelberg (2011)
17. Strzodka, R., Telea, A.: Generalized distance transforms and skeletons in graphics hardware. In: VisSym, pp. 221–230 (2004)
18. Sugiyama, K., Misue, K.: Graph drawing by the magnetic spring model. Journal of Visual Languages and Computing 6(3), 217–231 (1995)
19. Telea, A., Ersoy, O.: Image-Based Edge Bundles: Simplified Visualization of Large Graphs. Computer Graphics Forum 29, 843–852 (2010)
20. Wasserman, S., Faust, K.: Social network analysis: Methods and applications, 1st edn. Cambridge University Press (1994)
21. Zhou, H., Yuan, X., Cui, W., Qu, H., Chen, B.: Energy-based hierarchical edge clustering of graphs. In: IEEE PacificVis, pp. 55–61 (2008)

Edge Routing with Ordered Bundles

Sergey Pupyrev[1], Lev Nachmanson[2], Sergey Bereg[3], and Alexander E. Holroyd[4]

[1] Ural State University, Russia
spupyrev@gmail.com
[2] Microsoft Research, USA
levnach@microsoft.com
[3] University of Texas at Dallas, USA
besp@utdallas.edu
[4] Microsoft Research, USA
holroyd@microsoft.com

Abstract. We propose a new approach to edge bundling. At the first stage we route the edge paths so as to minimize a weighted sum of the total length of the paths together with their ink. As this problem is NP-hard, we provide an efficient heuristic that finds an approximate solution. The second stage then separates edges belonging to the same bundle. To achieve this, we provide a new and efficient algorithm that solves a variant of the metro-line crossing minimization problem. The method creates aesthetically pleasing edge routes that give an overview of the global graph structure, while still drawing each edge separately, without intersecting graph nodes, and with few crossings.

1 Introduction

The core components of most graph drawing algorithms are computation of positions of the nodes, and edge routing. In this paper we concentrate on the latter problem.

For many real-world graphs with substantial numbers of edges, traditional algorithms produce visually cluttered layouts. The relations between the nodes are difficult to analyze by looking at such layouts. Recently, edge bundling techniques have been developed, in which some edge segments running close are collapsed into bundles to reduce the clutter. While these methods create an overview drawing, they typically allow the edges within a bundle to cross and overlap each other arbitrarily, making individual edges hard to follow. In addition, previous approaches allowed edges to overlap nodes, thus obscuring their text or graphics.

We present a novel edge routing algorithm for undirected graphs, which we call **ordered bundles**. This algorithm produces a drawing in a "metro-line" style (see Fig. 1). The graphs for which our algorithm is best applicable are of medium size with a large number of edges, although it can process larger graphs efficiently too.

The input for our algorithm is an undirected graph with given node positions. These positions can be generated by a graph layout algorithm, or, in some applications (for instance, geographical ones) they are fixed in advance. During the algorithm the node positions are not changed. The main steps of our algorithm are similar to existing approaches, but with several innovations, which we indicate with italic text in the following description.

M. van Kreveld and B. Speckmann (Eds.): GD 2011, LNCS 7034, pp. 136–147, 2012.

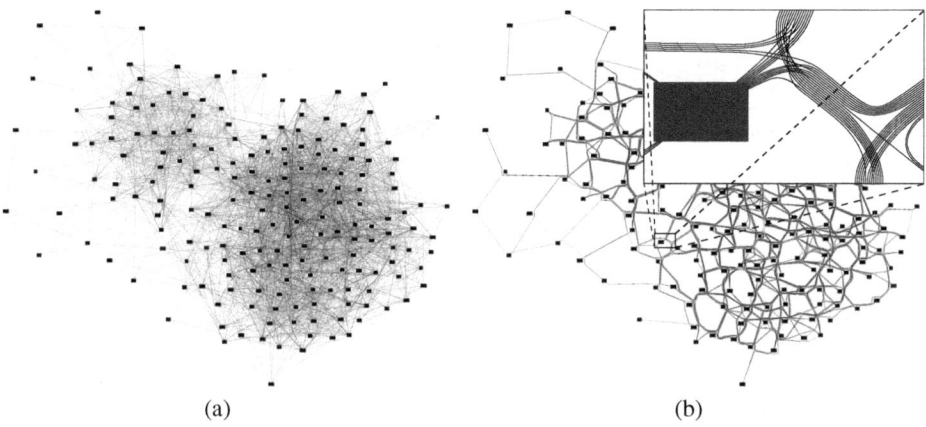

Fig. 1. Edge bundling example (jazz graph)

Edge Routing. In this step the edges are routed along paths, and the overlapping parts are organized into bundles. One approach here has been to minimize the total "ink" of the bundles. However, this often produces excessively long paths. *For this reason we introduce a novel cost function for edge routing, a weighted sum of the ink and total path length.* Minimizing this function forces the paths to share routes, creating bundles, but at the same time it keeps the paths relatively short. *We furthermore show how to route the bundles outside of the nodes.*

Edge Nudging. In this step the paths belonging to the same bundle are "nudged" away from each other. The effect of this action is that individual edges become visible, and the bundles obtain their thickness. *Contrary to previous approaches, we try to draw the edges of a bundle as parallel as possible with a given gap.* However, such a routing might not always exist because of the limited space between the node shapes. We provide a heuristic that finds a drawing with bundles of suitable thickness.

Edge Ordering. To route individual edges, an order of the edge segments inside of the bundle needs to be computed. This order minimizes the number of crossings between edges of the same bundle. The problem of finding such an order is related to a variant of the metro-line crossing minimization problem called MLCM-PA [12]. *We provide a new efficient algorithm that solves this problem exactly.*

The next section summarizes related work. In Section 3 we give a detailed explanation of our algorithm. Results of experiments are presented in Section 4. Finally we discuss some additional aspects and future work in Section 5.

2   Related Work

We believe that the first use of bundled edges in the graph drawing literature is given in [5]. The authors improve circular layouts by routing edges either on the outer or on the inner face of a circle. Edges in that paper are bundled with an algorithm that tries to minimize the total ink of the drawing. Here we follow a similar strategy, but in addition we try to keep the edges themselves short.

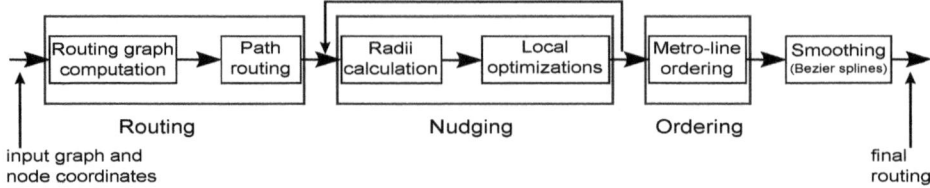

Fig. 2. Algorithm pipeline

In the hierarchical approach of [8], edges are bundled together based on an additional tree structure. Unfortunately, not every graph comes with a suitable underlying tree, and it is not clear how to extend the method to general layouts. In [13] edge bundles were computed for layered graphs. In contrast, our method applies to general graphs.

Edge bundling methods for general graphs are given in [2,4,9,10]. In the force-directed heuristic [9], edges can attract each other to organize themselves into bundles. The method is not efficient, and while it produces visually appealing drawings, they are often ambiguous in a sense that it is hard to follow an edge. The approaches [2,10] have a common feature: they both create a grid graph for edge routing. Our method also uses a special graph for edge routing, but our approach is different because we modify this graph to obtain better edge bundles. Unlike [2,4,9] we avoid edge-node overlaps.

The paper [12] inspired us to apply the technique of metro-line routing to minimize edge crossings inside bundles. Related work has been done in the VLSI community [1].

3   Algorithm

Let us establish some terminology. An undirected graph G is a pair (V, E), where V is the set of nodes and E is the set of edges, i.e. unordered pairs of nodes. A drawing of G is a representation of G in the plane in which each node $v \in V$ is drawn as a convex polygon p_v, and each edge $uv \in E$ is drawn as a simple curve connecting p_u and p_v. We will call the node polygons **obstacles** (since our focus is edge routing), and we assume that they are pairwise disjoint.

In overview, our algorithm takes the following steps (Fig. 2). We generate a routing graph $\widetilde{G}$ with straight-line edges that avoid the obstacles. We route the edges of G through $\widetilde{G}$. We will refer to an edge of G as a **path** in $\widetilde{G}$. Following [5,13], we define the **ink** I of a set of paths on $\widetilde{G}$ as the sum of the lengths of the edges of $\widetilde{G}$ used in these paths; if an edge is used by several paths its length is still counted only once.

A set of paths sharing the same edge of $\widetilde{G}$ is called a **bundle**. After the initial routing, the paths of the same bundle overlap on its edge. We estimate the space required to draw the paths separately, and we modify the positions of $\widetilde{G}$'s nodes, thus changing the paths' geometry. Then we order each bundle, and draw the paths individually with gaps, according to this order. To complete the drawing, we smooth the paths by fitting Bezier segments into the path corners. Next we give a detailed description of the steps.

3.1   Edge Routing

As we mentioned before, minimizing ink only often leads to extremely long paths that are difficult to follow. Therefore, we try to minimize a novel cost function of path routes,

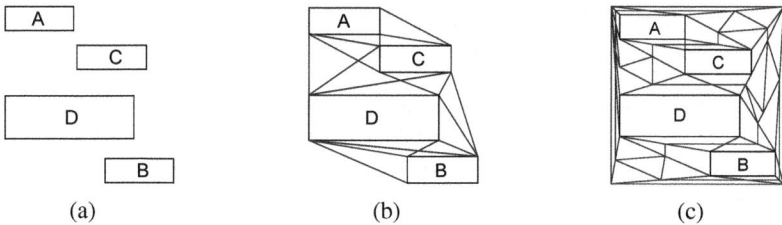

Fig. 3. (a) The obstacles. (b) A visibility graph. (c) A refinement of the Delaunay triangulation.

which takes both ink and path lengths into account: *routing cost* $= \alpha I + \beta \sum_{uv \in E} \ell_{uv}$. Here ℓ_{uv} is the length of path uv in the routing. The non-negative real constants α and β determine the contribution of the terms to the function. In our default settings $\alpha = 1$ and $\beta = 500$.

Problem 1 (Edge Routing). Given the graph G with fixed node positions, route the edges of E in the plane so that *routing cost* is minimized.

Problem 1 is NP-hard, because its instance with $\alpha = 1$, $\beta = 0$, and G being a complete graph is equivalent to the Geometric Steiner Tree problem, which is known to be NP-hard [6]. Therefore, we use an approximate solution, which starts with the construction of a routing graph $\widetilde{G}$.

Generation of the Routing Graph. We consider two approaches to the routing graph construction. The first one follows [3], which builds a sparse visibility graph (Fig. 3(b)). The second approach is a refinement of the Delaunay triangulation, in which for each two sides of a triangle that are not node polygon sides, we add the segment connecting their midpoints (Fig. 3(c)). This creates routes in the middle of the "channels" between the nodes. Both methods work in $O(n \log n)$ time, and create $\widetilde{G}$ with $O(n)$ edges, where n is the number of node polygon corners. In both cases the edges do not intersect the node interiors. For each node $v \in V$, we add edges connecting the center of polygon p_v to its corners. We denote the set of nodes of $\widetilde{G}$ by W, and the set of edges of $\widetilde{G}$ by U. We write $|W| = n$, and assume that $\widetilde{G}$ contains $O(n)$ edges. We will discuss the effect of the routing graph choice to the final routing later. Next we route the paths on $\widetilde{G}$.

Path Routing. We try to route the paths on $\widetilde{G}$ with the minimal *routing cost*. The problem can be formulated as follows.

Problem 2 (Path Routing). Given the graph $\widetilde{G}$ and a set of pairs of nodes $(a_i, b_i) \in W^2$, find paths between a_i and b_i for all i so that *routing cost* is minimized.

Here a_i and b_i correspond to the centers of obstacles connected by the edges of E. We stress that Problem 2 is different from Problem 1 in that the paths are now constrained to be routed through the edges of $\widetilde{G}$. The Problem 2 is again NP-hard, because its instance with $\beta = 0$ is a Steiner Forest Problem [6]. Therefore we solve an easier task, where some paths are already known, and we need to route the next path. We will route it by minimizing an *additional cost*, which is the increment of the *routing cost* associated with this path. For a path uv we define *additional cost* $= \alpha \, \Delta I + \beta \, \ell_{uv}$. Here ΔI is the increment in the ink, which equals the sum of edge lengths of uv that were not part of any previous path.

Problem 3 (Single Path Routing). Given the graph $\widetilde{G}$ and a set of already routed paths, find a path from a to b so that *additional cost* is minimized.

Let us assign the following weights to edges of $\widetilde{G}$: the weight of edge e is equal to $\alpha \, \delta_e + \beta \, \ell_e$, where δ_e is ℓ_e if e is not taken by a previous path, and 0 otherwise. It can be seen that the minimum *additional cost* is achieved by a shortest path from a to b according to these weights. We can thus apply the Dijkstra algorithm to find a path solving the Problem 3.

To solve Problem 2 approximately, we organize the edges of E in a sequence (a_1, b_1), $\ldots, (a_m, b_m)$, and iteratively solve Problem 3 for already routed paths (a_i, b_i), $i < k$, and $a = a_k$, $b = b_k$ for $k = 1, \ldots, m$. The routing of a single path takes $O(n \log n)$ time with the Dijkstra algorithm in our settings, because the number of edges in $\widetilde{G}$ is $O(n)$. All steps take $O(|E|n \log n)$ time.

Can we do better in this iterative approach than routing one path at a time? It turns out that we can route optimally a set of paths with a common end, which will be an improvement in some settings. We define an *additional cost* of a set of paths by analogy with the *additional cost* of a path.

Problem 4 (Multiple Path Routing). Given the graph $\widetilde{G}$ with some paths already routed, find paths for $(a^*, b_1), \ldots, (a^*, b_k)$ so that *additional cost* is minimized.

We can solve this problem by a dynamic programming approach. We first fix a set of pre-existing paths in $\widetilde{G}$; *additional cost* will always be with respect to these paths. Let us call a **state** a pair (v, P), where v is a node of $\widetilde{G}$, and P is a subset of $\{b_1, \ldots, b_k\}$. We need to solve our problem for the state $(a^*, \{b_1, \ldots, b_k\})$. We reduce the problem to solving it for "smaller" states, that are the states with fewer elements in P. For a state (v, P) we define its cost $f(v, P)$ as the minimal *additional cost* of a set of paths $\{(v, b), b \in P\}$. A set of paths giving the minimal $f(v, P)$ is called an **optimal** set for state (v, P). Let us clarify the structure of an optimal set of paths.

By the subgraph generated by a set of paths in $\widetilde{G}$ we mean the subgraph of $\widetilde{G}$ comprising all edges and nodes in the paths.

Lemma 1. *For each state there exists an optimal set of paths that generates a tree.*

Proof. Let Π be any optimal set of paths for state (v, P), and G' be the graph generated by Π, and note that it is connected. Let T be a shortest path tree of G', rooted at v, with respect to ordinary edge lengths. Let Π' be the set of paths connecting v to the points of P in T. The *additional cost* of Π' is at most that of Π. Indeed, the increment in I is no greater because T is a subgraph of G'. Each path of Π' is shortest in G' and thus no longer than the corresponding path of Π. Hence, Π' is an optimal set for (v, P).   □

Lemma 1 leads us to the following formula.
$$f(v, P) = \min \begin{cases} f(u, P) + \alpha \, \delta_{uv} + |P| \, \beta \, \ell_{uv}, & \text{for } u \in W \text{ adjacent to } v, \\ f(v, P') + f(v, P - P'), & \text{for } P' \text{ with } \emptyset \subset P' \subset P \end{cases}$$
The minimum is taken over both expressions on the right as u and P' vary. To verify this, we consider some optimal set of paths for (v, P) that form a tree, and split into two cases. The first line corresponds to the case where u is the only neighbor of v in the

tree. The second line is the case where v has at least two neighbors, thus the paths can be partitioned into two proper subsets with no common edges.

Now we describe how to compute $f(v, P)$. Let us assume, that f is known for all states (u, P'), where P' is a proper subset of P. To compute $f(v, P)$, a new graph H is constructed with $\widetilde{G}$ as a subgraph. An edge e of $\widetilde{G}$ has weight $\alpha\, \delta_e + |P|\, \beta\, \ell_e$ in H. We add a new node h to H and connect it with all nodes of $\widetilde{G}$. For every new edge hu we assign weight $\min_{P'} f(u, P') + f(u, P - P')$, where P' varies over proper non-empty subsets of P. One can see that the required value $f(v, P)$ is the length of a shortest path from v to h in graph H. We can compute it with the Dijkstra algorithm.

To solve Problem 4 we work bottom-up. We compute all $f(v, P)$ with $|P| = 1$ and v is a node of $\widetilde{G}$, by the algorithm for Problem 3, where we find a path with the minimal *additional cost*. Then we compute the values $f(v, P)$ for each v and $|P| = 2, \ldots, k$ by creating the corresponding graphs H. The answer for the problem is $f(a^*, \{b_1, \ldots, b_k\})$.

Running time. The main steps of the algorithm are the construction of graph H and finding a shortest path on it with the Dijkstra algorithm for each state (v, P). Luckily, graph H depends only on the P component of a state. The construction of graph H for a fixed set P takes $O(2^{|P|}n)$ time. We execute the Dijkstra algorithm only once per P starting from h to compute $f(v, P)$ for all $v \in W$. Thus, finding $f(v, P)$ for a known H and for all $v \in W$ takes $O(n \log n)$ time. Summing over all possible sets P produces

$$O\left(\sum_P (2^{|P|}n + n \log n)\right) = O(3^k n + 2^k n \log n).$$

To utilize the method solving Problem 4, we organize the paths into a sequence of subsets of paths having a common end. We route the paths of the first element of the sequence with the minimal *additional cost*, solving Problem 4. Then, using this routing we solve Problem 4 for the second subset, using an updated *additional cost* function, and so on. To avoid a long running time we need to keep the path subsets small. We experimented with $k = 5, 10$, and the results are shown in Section 4.

In practice, we set *routing cost* $= \alpha I + \beta \sum_{uv \in E} \frac{\ell_{uv}}{d_{uv}}$, where d_{uv} is the Euclidean distance between the nodes u and v. This way we penalize the relative growth of path lengths to avoid long paths for short edges.

3.2   Local Adjustments and Spline Routing

To save space we omit some details in this section. Routing the paths through $\widetilde{G}$ defines the bundles. In the final drawing we would like to draw the paths of a bundle in a particular order, as will be explained in Section 3.3, while keeping them at a predefined distance from each other, and outside the obstacles. For this we need to have some free space around the edges of $\widetilde{G}$. To provide the free space, we surround each node of $\widetilde{G}$ by a circle, called a **hub**, with the center at the node position, and each edge by a rectangle, that are disjoint from the obstacles (Fig. 4(b)). In the final drawing every path is represented as a sequence of line segments and cubic Bezier segments, where each line segment is contained in a rectangle, and each cubic Bezier segment is contained in a hub (Fig. 4(c)). Such a path does not intersect the obstacles. To draw a Bezier segment inside of a hub, we place each control point of the segment inside of the hub; since a circle is convex, and a Bezier segment is contained in the convex hull of its control points, this keeps the segment inside the hub.

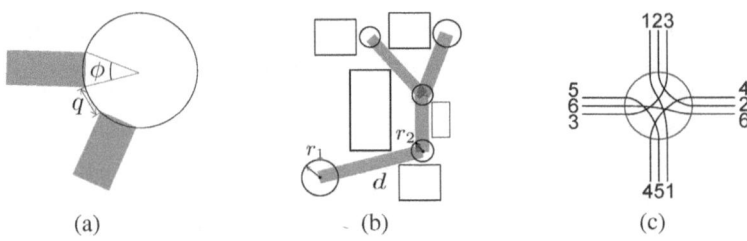

Fig. 4. (a) Desired radius (b) Allowed radius (c) Paths intersecting at a node

Hub Radii Calculation. The radius of a hub is defined as the minimum of two radii: a desired one, and an allowed one. The calculation of the desired radius can be explained with Fig. 4(a). We would like a hub to be large enough to accommodate an incoming bundle by keeping angle ϕ at most $\pi/4$. We also would like to keep two bundles separated before entering a hub by having distance q at least the given edge separation. The allowed radius is explained with a help of Fig. 4(b). To keep two connected hubs separated, we require (a) that $r_1 + r_2 \leq \gamma d$, for some $0 < \gamma < 1$, and (b) that each hub does not intersect the obstacles.

Local Optimization. To be able to route thick bundles without overlapping the obstacles, we first apply a heuristic preprocessing step before the local optimizations, in which each node of $\widetilde{G}$ participating in a path and belonging to an obstacle is moved away from the obstacles. The *routing cost* usually becomes larger after this step, but we obtain the necessary space around the paths. In order to minimize *routing cost* locally, we next iteratively adjust the position of each node of $\widetilde{G}$ by moving it in a random direction and trying to diminish *routing cost*. We also try to glue some of the nodes of $\widetilde{G}$ together, if it is beneficial. During these transformations, we do not modify the positions of the nodes corresponding to the corners and polygon centers and we pay attention to preserve conditions (a) and (b) mentioned above. Let m be the number of the obstacles, and c be the time required to find out if a circle or a rectangle intersects an obstacle. Using an R-tree [7] on the obstacles, one can find out if a circle or a rectangle intersects the obstacles in $O(c \log m)$ time. The number of edges in $\widetilde{G}$ is $O(n)$, therefore, a pass locally optimizing the position of every node of $\widetilde{G}$ can be done in $O(cn \log m)$ time. This is the most expensive stage of the algorithm, since we proceed iterating as long as we diminish *routing cost*, and we do not have a good upper bound for this step.

3.3   Ordering Paths

At this point the routing is completed and the bundles have been defined. We draw the paths of a given bundle parallel to the corresponding edge, therefore two paths may need to cross at a node as shown in Fig. 4(c). The order of paths in bundles affects crossings of paths. Let P be the set of paths in $\widetilde{G}$ computed by path routing. We address the following problem.

Problem 5. Given the graph $\widetilde{G}$ and a set of paths P, find an ordering of paths for each edge of $\widetilde{G}$ that minimizes the number of crossings.

In our setting, the paths terminate at the nodes of $\widetilde{G}$ corresponding to the centers of the obstacles, and these nodes cannot be intermediate points of paths. Thus we have:

Path Terminal Property. No node is both an endpoint of some path and an intermediate point of some path.

We call nodes that are endpoints of paths **terminal** nodes. In this section we will need to assume the following additional property of P.

Path Intersection Property. The common nodes and edges of any two paths form a path (which may be empty, or a single node).

The paths produced by our algorithm in Section 3.1 do *not* necessarily have the path intersection property. However, any set of paths may be modified so as to satisfy the property via the following algorithm. Let H be the graph formed by the union of the original paths, and assign each edge a label. The labels may be independent uniformly random real numbers in $[0, 1]$, or distinct integers. Now for each path, re-route it along the *shortest* path in H between its endpoints according to the original edge lengths, but breaking ties between paths of equal length via the sums of the real labels, or via lexicographic ordering of the sets of integer labels on the two paths.

Ideally, every two paths either do not cross or cross one time if needed (Fig. 5). An ordering of paths is **consistent** if any two paths cross at most once at a node. Clearly, if two paths cross once in some ordering, they must cross in every ordering. Hence consistent orderings have the minimum number of crossings. However, consistent orders are clearly not necessarily unique, and the choice of a particular one may greatly influence the quality of final drawing. The following property might appear desirable. A consistent order of paths is **nice** if, for any two paths, their order along all their common edges is the same (i.e. they may cross only at an endpoint of their common subpath). Unfortunately, we found an example of $(\widetilde{G}, P)$ having no nice consistent order. On the other hand, we prove that

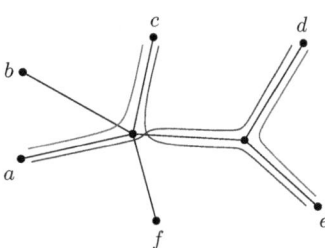

Fig. 5. Paths ac and de do not cross, while paths ad and ce cross once

a consistent ordering always exists, and we provide an efficient algorithm to construct one. We consider the following problem.

Problem 6 (Path Ordering). Given the graph $\widetilde{G}$ and a set of simple paths P satisfying the path terminal and intersection properties, compute a consistent ordering of paths for all edges of $\widetilde{G}$.

Algorithm. A basic step of our algorithm is the deletion of a node of $\widetilde{G}$ (Fig. 6). For every non-terminal node v do the following. Let P_v be the set of paths passing through v. Number the edges incident to v as $e_1, e_2, \ldots, e_t$ in clockwise order, and let $v_1, \ldots, v_t$ be the corresponding nodes adjacent to v. For every path $\pi \in P_v$ using edges e_a and e_b, represent it by pair (a, b). For each pair (a, b), add a new edge (v_a, v_b). Assign the paths labeled by (a, b) to this edge. The new edges incident to v_a should be inserted into $v'_a s$ clockwise order in the position previously occupied by e_a, in the order determined by the positions of v_b. Delete node v from the graph and the paths.

After all non-terminal nodes have been deleted, we reverse the process and undo the deletions, adding orders to the edges. Consider the deletion of v. The new order along edge e_a is obtained by concatenating the orders along the edges $\{(v_a, v_b) : b \neq a\}$.

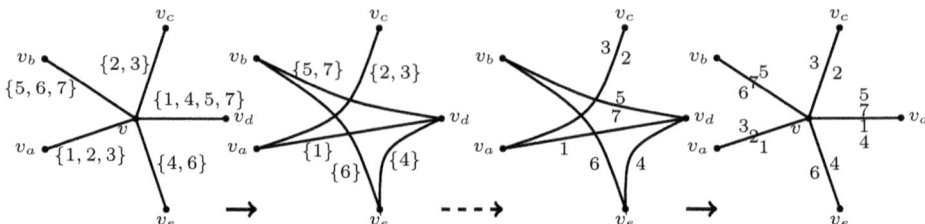

Fig. 6. Removal of node v; intervening steps; re-insertion of v. Paths of P_v are numbered $1, \ldots, 7$, and are shown in braces when unordered, and then with placement indicating their orders.

Implementation. First, create a new graph H. Initially, H is the subgraph of $\widetilde{G}$ generated by all the paths of P. Every path in P is stored as a list of nodes in H. For every edge e, L_e is a list of paths containing e. We assume that, for every node v of H, the list of edges incident to v in clockwise order is given. Note that these lists are dynamic since H undergoes deletions of nodes. We keep track of the deletions in a forest F. Initially, F contains isolated nodes corresponding to the edges of H. When a node is deleted and an edge e is replaced by new edges, we add them in F as children of the node corresponding to e. For instance, $L_{d,v}$ contains paths 1,4,5, and 7 in Fig. 6. When node v is processed, list $L_{d,v}$ is split into new sublists $L_{d,e}$, $L_{d,a}$, and $L_{d,b}$. The (clockwise) order of sublists is important. Then we replace edge dv by edges de, da, and db in graph H and in order of edges around d. The first phase finishes, when all non-terminals are deleted from H. In the second phase, we process each tree in F in bottom-up order. The list of a node is simply the concatenation of the lists of its children. The leaves of F correspond (one-to-one) to the paths of P.

Theorem 1. *Given the graph $\widetilde{G} = (W, U)$, a set of simple paths P in $\widetilde{G}$ satisfying the path terminal and intersection properties, and a clockwise order of the edges around each node, an ordering of paths along edges of $\widetilde{G}$ minimizing the number of crossings can be computed in $O(|W| + |U| + L)$ time, where L is the total length of paths in P.*

Proof. Correctness. We need to show that the edge (v_a, v_b) added in Phase 1 is new. Indeed, if edge (v_a, v_b) already existed, there would be a path passing from v_a to v_b and a path passing v_a, v, v_b, which contradicts the path intersection property.

The ordering of paths is consistent since the split of paths P_v makes only necessary crossings. Two paths π_1 and π_2 will produce a crossing only when the last node of their common subpath is deleted and the clockwise order of the nodes around v is $\cdots v_a \cdots v_b \cdots v_{a'} \cdots v_{b'} \cdots$, where $\pi_1 = \cdots v_a v v_{a'} \cdots$ and $\pi_2 = \cdots v_b v v_{b'} \cdots$.

Running time. The time for processing node v (the deletion of v) is $O(1 + d_{v,H} + s_v)$, where $d_{v,H}$ is the degree of v in H at the current step and s_v is the number of paths passing through v. The theorem follows since $d_{v,H} \le d_{v,\widetilde{G}} + s_v$. $\qquad\qquad\square$

Table 1. The percentage of routing cost improvement, $1 - (routing\ cost$ of bundled graph$)/$ $(routing\ cost$ of straight edges$)$. For the cells with "–" running time exceeds 10 hours.

Graph	Visibility graph					Delaunay triangulation				
	$|W|$	$|U|$	$k=1$	$k=5$	$k=10$	$|W|$	$|U|$	$k=1$	$k=5$	$k=10$
tail	105	348	8.4	16.9	18.8	257	638	13.3	16.0	18.5
airlines	1175	5297	61.3	62.4	63.3	2825	7076	61.3	62.1	62.7
jazz	955	4478	32.2	33.2	33.6	2297	5798	32.2	34.3	36.0
protein	7290	32585	16.2	16.2	17.3	17501	43676	15.5	16.6	17.1
power grid	24705	109779	1.0	1.5	–	59297	148280	0.1	0.1	–
Java	7690	32712	32.0	34.9	–	18461	46350	30.0	34.1	–
migrations	8575	41451	74.5	75.3	–	20585	51510	74.4	75.9	–

Overall, the complexity of the Ordering step is $O(|E|n + n \log D)$, where $|E|$ is the number of edges of the original graph G (the number of paths), n is the number of nodes in $\widetilde{G}$, and D is the maximum degree of $\widetilde{G}$.

4   Experimental Results

We implemented our algorithm in MSAGL tool [11]. Edge bundling was applied for synthetic graph collections and several real-world graphs (see [13] for a detailed description of our dataset). Unless node coordinates are available, we used the tool to position the nodes. All our experiments were run on a 3.1 GHz quad-core machine with 4 GB of RAM. Tables 1 and 2 give measurements of the method on some test cases.

The quality analysis of ink minimization heuristics is given in Table 1. We compare the routing cost gain of our algorithm with different settings. An iterative approach with routing paths one by one corresponds to a $k = 1$ case. The results of routing multiple paths at a time are shown for groups of size 5 and 10.

The variant of the algorithm that routes multiple paths with the same endpoint produces routings with smaller *routing cost*, while its running time is much longer (e.g. 4 seconds with $k = 1$, 2 minutes with $k = 5$, and 1.5 hours with $k = 10$ for airlines graph). The approaches with different routing graphs are quite similar in both *routing cost* minimization and running time. Moreover, we could not identify significant differences in the quality of final drawings. We believe it is a result of Local Optimization step of our algorithm in which edge routes are shortened and smoothed. Overall, we chose $k = 1$ with sparse visibility graph as a default settings for our routing.

Table 2 shows the CPU times of algorithm steps. As can be seen, ordered bundles can be constructed for graphs with several thousand of nodes and edges in less than a minute. The most expensive steps are Local Optimization and Edge Routing.

We now demonstrate the algorithm on real-world examples. A migration graph used for comparison of edge bundling algorithms is shown in Fig. 7. In our opinion, on a global scale ordered edge bundles are aesthetically as pleasant as other drawings of the graph (see e.g. [2,4,9,10]). On a local scale, our result outperforms previous approaches by arranging edge intersections. Another advantage of our routing scheme is shown in Fig. 8. Multiple edges are visualized separately making them easier to discover (compare the edge between nodes Editor and Application on both drawings).

Table 2. Performance of the algorithm (in seconds). The results are given for $k = 1$ with sparse visibility graph.

Graph	$|V|$	$|E|$	source	Visibility	Routing	Radii	Optimizations	Ordering	Overall
tail	21	68	[13]	0.11	0.02	0.04	0.13	0.01	0.34
airlines	235	1297	[2]	0.16	0.31	0.17	2.50	0.15	3.32
jazz	191	2732	[14]	0.14	0.42	0.23	3.02	0.18	4.04
protein	1458	1948	[14]	0.57	0.88	0.48	11.45	0.10	13.52
power grid	4941	6594	[14]	1.86	6.11	1.15	16.85	0.18	26.31
Java	1538	7817	GD'06 Contest	0.59	3.49	1.37	28.80	0.96	35.35
migrations	1715	6529	[2]	0.50	3.39	1.38	29.19	1.16	35.75

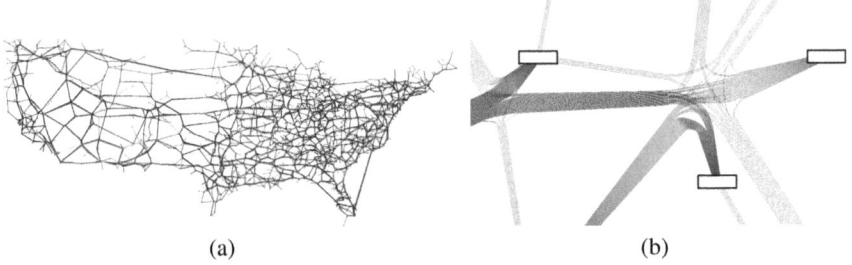

(a)   (b)

Fig. 7. Migration graph. (a) Overview. (b) Detail.

5   Conclusions and Future Work

We have presented a new edge routing algorithm based on ordered bundles that improves the quality of single edge routes when compared to existing methods. Our technique differs from classical edge bundling, in that the edges are not allowed to actually overlap, but are run in parallel channels. The algorithm ensures that the nodes do not overlap with the bundles and that the resulting edge paths are relatively short.

In our opinion, the novel cost function can be considered as a quality measure for different bundling heuristics. In a future, it would be interesting to verify if layouts with smaller *routing cost* correspond to subjectively better images. We are also exploring a possible extensions of the function to control the curvature of the resulting edges.

An important contribution of the paper is an efficient algorithm that finds an order of edges inside of bundles with minimal number of crossings. As mentioned above, this order is not unique. We left the question of choosing the best order as future research.

Our method splits the overlapped edge segments. The main limitation of our technique is that the routing and nudging steps are performed independently. A minimal *routing cost* might correspond to a routing, where edges can not be drawn with ideal thickness. In contrast, the nudging step moves bundles, thus, increasing ink and edge lengths. We plan next to combine these two steps.

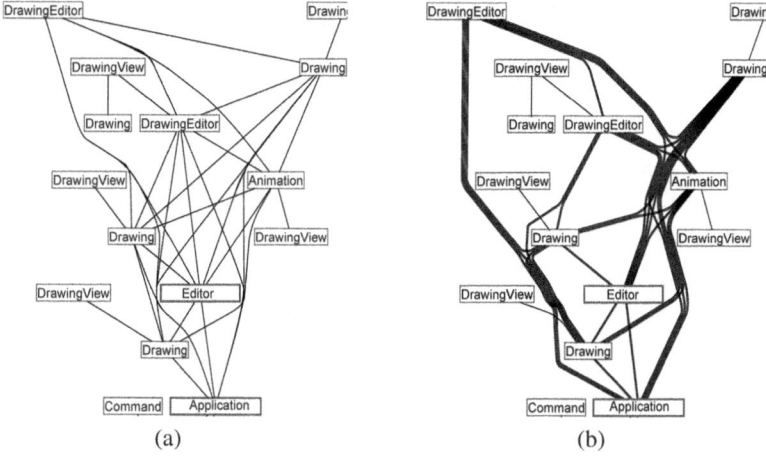

Fig. 8. `Tail` graph. (a) Original. (b) Bundled.

References

1. Chen, H.-F.S., Lee, D.T.: On crossing minimization problem. IEEE Transactions on Computer-aided Design of Integrated Circuits and Systems 17, 406–418 (1998)
2. Cui, W., Zhou, H., Qu, H., Wong, P.C., Li, X.: Geometry-based edge clustering for graph visualization. IEEE Trans. on Visualization and Computer Graphics 14(6), 1277–1284 (2008)
3. Dwyer, T., Nachmanson, L.: Fast Edge-Routing for Large Graphs. In: Eppstein, D., Gansner, E.R. (eds.) GD 2009. LNCS, vol. 5849, pp. 147–158. Springer, Heidelberg (2010)
4. Gansner, E., Hu, Y., North, S., Scheidegger, C.: Multilevel agglomerative edge bundling for visualizing large graphs. In: Proc. IEEE Pacific Visualization Symposium (to appear, 2011)
5. Gansner, E.R., Koren, Y.: Improved Circular Layouts. In: Kaufmann, M., Wagner, D. (eds.) GD 2006. LNCS, vol. 4372, pp. 386–398. Springer, Heidelberg (2007)
6. Garey, M.R., Johnson, D.S.: Computers and Intractability: A Guide to the Theory of NP-Completeness. W.H. Freeman and Company, New York (1979)
7. Guttman, A.: R-trees: A dynamic index structure for spatial searching. In: Proc. Int. Conf. on Management of Data, pp. 47–57 (1984)
8. Holten, D.: Hierarchical edge bundles: Visualization of adjacency relations in hierarchical data. IEEE Transactions on Visualization and Computer Graphics 12(5), 741–748 (2006)
9. Holten, D., van Wijk, J.J.: Force-directed edge bundling for graph visualization. Computer Graphics Forum 28(3), 983–990 (2009)
10. Lambert, A., Bourqui, R., Auber, D.: Winding Roads: Routing edges into bundles. Computer Graphics Forum 29(3), 853–862 (2010)
11. Nachmanson, L., Robertson, G., Lee, B.: Drawing Graphs with GLEE. In: Hong, S.-H., Nishizeki, T., Quan, W. (eds.) GD 2007. LNCS, vol. 4875, pp. 389–394. Springer, Heidelberg (2008)
12. Nöllenburg, M.: An Improved Algorithm for the Metro-Line Crossing Minimization Problem. In: Eppstein, D., Gansner, E.R. (eds.) GD 2009. LNCS, vol. 5849, pp. 381–392. Springer, Heidelberg (2010)
13. Pupyrev, S., Nachmanson, L., Kaufmann, M.: Improving Layered Graph Layouts with Edge Bundling. In: Brandes, U., Cornelsen, S. (eds.) GD 2010. LNCS, vol. 6502, pp. 329–340. Springer, Heidelberg (2011)
14. Gephi dataset, http://wiki.gephi.org/index.php?title=Datasets

Right Angle Crossing Graphs and 1-Planarity*

Peter Eades[1] and Giuseppe Liotta[2]

[1] School of Information Technologies, University of Sydney
peter@it.usyd.edu.au
[2] Università degli Studi di Perugia, Italy
liotta@diei.unipg.it

Abstract. A Right Angle Crossing Graph (also called RAC graph for short) is a graph that has a straight-line drawing where any two crossing edges are orthogonal to each other. A 1-planar graph is a graph that has a drawing where every edge is crossed at most once. We study the relationship between RAC graphs and 1-planar graphs in the extremal case that the RAC graphs have as many edges as possible. It is known that a maximally dense RAC graph with $n > 3$ vertices has $4n - 10$ edges. We show that every maximally dense RAC graph is 1-planar. Also, we show that for every integer i such that $i \geq 0$, there exists a 1-planar graph with $n = 8 + 4i$ vertices and $4n - 10$ edges that is not a RAC graph.

1   Introduction

A *drawing* of a graph G maps each vertex u of G to a distinct point p_u in the plane, each edge (u, v) of G to a Jordan arc connecting p_u and p_v and not passing through any other vertex, and is such that any two edges have at most one point in common. A *1-planar drawing* is a drawing of a graph where every edge can be crossed by at most one other edge. A *1-planar graph* is a graph that has a 1-planar drawing. A *straight-line drawing* is a drawing of a graph such that every edge is a straight-line segment. A *Right Angle Crossing drawing* (or *RAC drawing*, for short) is a straight-line drawing where any two crossing edges form right angles at their intersection point. A *Right Angle Crossing graph* (or *RAC graph*, for short) is a graph that has a RAC drawing.

Pach and Tóth prove that 1-planar graphs with n vertices have at most $4n - 8$ edges, which is a tight upper bound [9]. Korzhik and Mohar prove that recognizing 1-planar graphs is NP-hard [8]. Suzuki studies the combinatorial properties of the so-called *optimal 1-planar* graphs, i.e. those n-vertex 1-planar graph having $4n - 8$ edges [10]. A limited list of additional papers on 1-planar graphs includes [4,7]. Didimo et al. show that a RAC graph with $n > 3$ vertices has at most $4n - 10$ edges and that this bound is tight [5]. Argyriou at al. prove that recognizing RAC graphs is NP-hard [2]. For recent references about RAC graphs and their variants see also [1,3,6,11].

This paper studies the relationship between RAC graphs and 1-planar graphs in the extremal case that the RAC graphs are as dense as possible. A RAC graph is *maximally dense* if it has $n > 3$ vertices and $4n - 10$ edges. While, at a first glance, one might think that, in order to maximize the number of edges in a RAC graph, a good strategy is that each edge should be crossed many times, we prove the following.

---

* Work supported in part by MIUR of Italy under project AlgoDEEP prot. 2008TFBWL4 and by an IVFR Grant of the Australian Government.

M. van Kreveld and B. Speckmann (Eds.): GD 2011, LNCS 7034, pp. 148–153, 2012.

Theorem 1. *Every maximally dense RAC graph is 1-planar. Also, for every integer i such that $i \geq 0$, there exists a 1-planar graph with $n = 8 + 4i$ vertices and $4n - 10$ edges that is not a RAC graph.*

We observe that the first part of Theorem 1 is trivially true if the maximally dense RAC graph has exactly 4 vertices. Namely, the maximally dense RAC graph with 4 vertices is K_4 which is planar and hence 1-planar. We prove that a maximally dense RAC graph with at least 5 vertices is also 1-planar by showing that all RAC drawings with $4n - 10$ edges are such that no edge is crossed twice. For reasons of space, some proofs are omitted or sketched in this abstract.

2   Red-Blue-Green Coloring of Maximally Dense RAC Graphs

Let G be a maximally dense RAC graph and let D be any RAC drawing of G. Let E be the set of the edges of D. In [5] the following 3-coloring of the edges of D (and hence of G) is described. Every edge of D is either a *red edge* or a *blue edge*, or a *green edge*. An edge is red if and only if it is not crossed by any other edge; a blue edge is only crossed by green edges, and a green edge is only crossed by blue edges. We call this 3-coloring of the edges of D a *red-blue-green coloring* of D and denote it as Π_{rbg}. Let $D_{rb} = (V, E_r \cup E_b)$ be the sub-drawing of D consisting of the red and blue edges and let G_{rb} be the corresponding subgraph of G. We call G_{rb} the *red-blue subgraph* of G induced by Π_{rbg} and we call D_{rb} the *red-blue sub-drawing* of D induced by Π_{rbg}. Note that, by construction, D_{rb} has no crossing edges and thus G_{rb} is a planar graph. We will always consider G_{rb} as a planar embedded graph, where the planar embedding is given by D_{rb}. Analogously, we define the *red-green* subgraph of G induced by Π_{rbg} denoted as G_{rg}, and the *red-green* sub-drawing of D induced by Π_{rbg} denoted as D_{rg}. Also G_{rg} has the planar embedding of D_{rg}, and thus G_{rg} and G_{rb} have the same external face.

The next lemmas will particularly focus on the size and the coloring of some specific faces of the red-blue graph G_{rb}. We will consider its external face, denoted as f_{ext}, and its *fence faces*, defined as those internal faces that share at least one edge with f_{ext}. In the proofs that follow, we denote with m_r the number of red edges, with m_b the number of blue edges, and with m_g the number of green edges. Without loss of generality, we will assume from now on that our red-blue-green coloring is such that $m_b \geq m_g$. Also, we denote with f_{rb} the number of faces of G_{rb} and with n the number of its vertices.

Lemma 1. *[5] Every internal face of G_{rb} has at least two red edges. Also, all edges of f_{ext} are red.*

All remaining lemmas of this section assume that the maximally dense RAC graph G has at least 5 vertices.

Lemma 2. *Face f_{ext} is a 3-cycle.*

Sketch of Proof: By Lemma 1, every internal face of G_{rb} has at least two red edges and all edges of f_{ext} are red. Hence, denoting with $|f_{ext}|$ the number of edges of f_{ext}, we have $m_r \geq (f_{rb} - 1) + \frac{|f_{ext}|}{2}$. Since G_{rb} is a planar graph, Euler's formula implies that

$m_r + m_b \leq n + f_{rb} - 2$. It follows $m_b \leq n - 1 - \frac{|f_{ext}|}{2}$. Since also the red-green subgraph of G is planar and it has the same external face of G_{rb}, by Euler's formula we also have that $m_r + m_g \leq 3n - 3 - |f_{ext}|$. It follows that $m_r + m_b + m_g \leq 4n - 4 - \frac{3|f_{ext}|}{2}$. Observe that $|f_{ext}| \geq 5$ would imply $m_r + m_b + m_g < 4n - 10$, which is impossible because G is a maximally dense RAC graph. We now show that the external face of G_{rb} cannot be a 4-cycle either. By contradiction, assume that $|f_{ext}| = 4$. Consider first the case that some fence face of G_{rb} has more than 3 edges: Since $|f_{ext}| = 4$ and a fence face has size at least 4, we have $m_r + m_b \leq 3n - 8$. By the inequalities above, we also have $m_r \geq f_{rb} + 1$ and $m_b \leq n - 3$. Since G is maximally dense, we have $m_r + m_b + m_g = 4n - 10$. It follows that $m_r + m_g \geq 3n - 7 > m_r + m_b$, which is however impossible because we are assuming $m_b \geq m_g$. Lastly, consider the case that $|f_{ext}| = 4$ and all fence faces are 3-cycles. Note that there must be four fence faces: If there were only three fence faces there would be a vertex of degree at most three in G, which is impossible in a maximally dense graph with at least 5 vertices. Since in every RAC drawing of G each fence face is drawn as a triangle, for at least one of these four triangles the angle opposite to the edge that belongs to f_{ext} must be larger than or equal to $\frac{\pi}{2}$. This observation, together with Lemma 1, implies that at least one of the fence faces consists of all red edges in any red-blue-green coloring. We therefore have the following: $m_r \geq (f_{rb} - 2) + \frac{|f_{ext}|}{2} + \frac{3}{2} = f_{rb} + \frac{3}{2}$. Since m_r is an integer, we have $m_r \geq f_{rb} + 2$. By $m_r + m_b \leq n + f_{rb} - 2$ we obtain $m_b \leq n - 4$, and by $m_r + m_b + m_g = 4n - 10$ we obtain $m_r + m_g \geq 3n - 6$. However, G_{rg} is a planar graph and it has the same external face as G_{rb}, that has size 4; so, G_{rg} cannot be a maximal planar graph, a contradiction. It follows that f_{ext} must be a 3-cycle. □

Lemma 3. *Graph G_{rb} is biconnected.*

Lemma 4. *Graph G_{rb} has three fence faces. Also, each fence face of G_{rb} is a 3-cycle.*

Lemma 5. *G_{rb} and G_{rg} are both maximal planar graphs.*

Sketch of Proof: By Lemmas 2 and 4, f_{ext} is a 3-cycle consisting of red edges and the three fence faces are all 3-cycles. By simple geometric arguments it follows that in any red-blue-green coloring of a RAC drawing of G, at least two of the triangles representing these fence faces consist of red edges. We therefore have: $m_r \geq (f_{rb} - 3) + \frac{|f_{ext}|}{2} + \frac{3}{2} + \frac{3}{2}$, which implies $m_r \geq f_{rb} + 2$. By $m_r + m_b \leq n + f_{rb} - 2$, we obtain $m_b \leq n - 4$. By $m_r + m_b + m_g = 4n - 10$ we have $m_r + m_g \geq 3n - 6$. Since G_{rg} is a planar graph, it has exactly $3n - 6$ edges and so does G_{rb} because $m_b \geq m_g$. It follows that G_{rb} and G_{rg} are are both maximal planar graphs. □

3   Proof of Theorem 1

The following lemma is the key for proving the first part of Theorem 1.

Lemma 6. *Every RAC drawing of a maximally dense RAC graph is also a 1-planar drawing.*

Proof. The proof is immediate if the maximally dense RAC graph has 4 vertices. Let G be a maximally dense RAC graph with at least 5 vertices, let D be a RAC drawing of G and consider any red-blue-green coloring of the edges of D. Let e be a blue edge of D. By Lemma 5, every blue edge $e = (u, v)$ of G_{rb} is shared by two internal triangular faces, that we denote as f and f'. Let u, v, w be the vertices of f and u, v, w' be the vertices of f'. Since by Lemma 1 every face of G_{rb} has two red edges, we have that edges (u, w) and (w, v) are not crossed by any other edge; similarly, edges (u, w') and (w', v) of f' are both red. Since every blue edge is crossed by some green edges, we have that there can be only one green edge crossing e, namely edge (w, w'). It follows that the RAC drawing D is also a 1-planar drawing. □

To show the second part of Theorem 1, we describe an infinite family of 1-planar graphs that have the same edge density as the maximally dense RAC graphs but are not maximally dense RAC graphs. Consider first the graph G_0 of of Figure 1 (a). Clearly it is 1-planar; also, it has $n = 8$ vertices and $4n - 10 = 22$ edges.

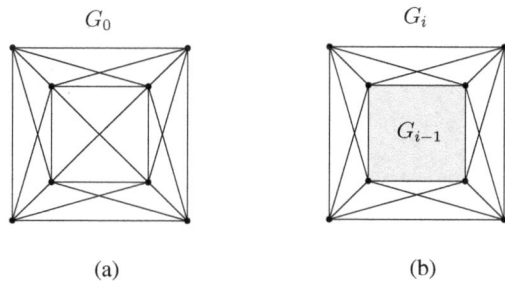

G_0 G_i

G_{i-1}

(a)   (b)

Fig. 1. (a) Graph G_0; (b) Constructing graph G_i from G_{i-1}

Lemma 7. *Graph G_0 is not a RAC graph.*

Proof. Observe that G_0 has the following properties: (1) Every vertex of G_0 has degree at least five and at most six; (2) For every 3-cycle of G_0 with vertices u, v, w, there exists a fourth vertex z such that the subgraph induced by u, v, w, z is the complete graph K_4; (3) There is a 4-cycle through the remaining four vertices of G_0, i.e. the vertices that do not form this K_4.

Suppose, for a contradiction, that G_0 had a RAC drawing D_0. By Lemma 2, the external face of D_0 is a triangle; let u, v, w be the vertices of this external face. Let z be the vertex such that the sub-drawing of D induced by vertices u, v, w, z is a planar representation of K_4. Let f_0, f_1, and f_2 be the three internal faces of this sub-drawing. Let v_0, v_1, v_2, v_3 be the remaining four vertices of G_0. They can be either all inside the same face, or they can be in two faces, or they can be in three faces. The three cases are illustrated in Figure 2.

Assume that v_0, v_1, v_2, v_3 are all in a same face, say f_0. Refer to Figure 2 (a). By Lemma 4, D_0 has three fence faces and these faces are triangles. As discussed in the proof of Lemma 5, in any red-blue-green coloring of D the edges of at least two of these three triangles are red. Since f_1 and f_2 are both fence faces, either (w, z) is a red edge or (u, z) is a red edge. Assume, w.l.o.g. that (w, z) is red. Since vertex v has degree at least five and (w, z) is red, there must be at least two edges that connect v to one of the vertices inside f_0; both such edges must cross (u, z) (see the dotted edges in Figure 2 (a)). However, by Lemma 6, D_0 is also a 1-planar drawing and (u, z) cannot be crossed twice; a contradiction.

Assume that v_0, v_1, v_2 are in f_0 and v_3 is in f_2. Refer to Figure 2 (b). Since there is a cycle with vertices v_0, v_1, v_2, v_3, there are at least two edges incident to v_3 that cross the boundary of f_2. If both these edges cross edge (u, z), then the same argument as in the previous case applies. If one of these edges crosses (v, z), it must also cross (w, z) to reach any one of v_0, v_1, v_2 (see for example the dotted edge (v_2, v_3) in Figure 2 (b)). But this would violate Lemma 6, a contradiction.

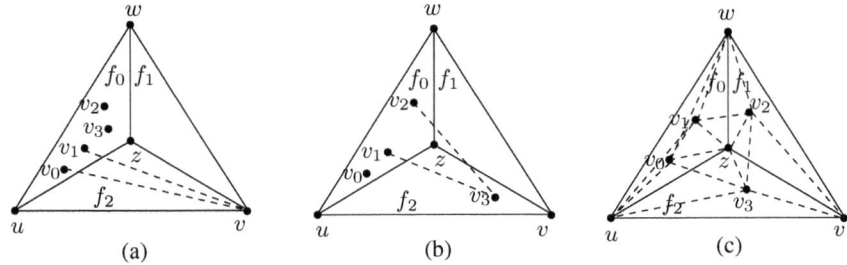

Fig. 2. The three cases in the proof of Lemma 7. (a) v_0, v_1, v_2, v_3 are all in f_0, (w, z) is a red edge, and two dotted edges cross (u, z); (b) v_3 is in f_2 and edge (v_2, v_3) violates the 1-planarity condition; (c) v_3 is in f_2, v_2 in f_1 and z has degree seven.

Finally, assume that v_0, v_1 are in f_0, v_2 is in f_1 and v_3 is in f_2, as depicted in Figure 2 (c). Since there is a 4-cycle with vertices v_0, v_1, v_2, v_3, there is an edge of this cycle crossing (u, z), one crossing (v, z), and one crossing (w, z). Again by Lemma 6, neither (u, z), nor (v, z), nor (w, z) can be crossed by any other edge. In order to guarantee that every vertex of G_0 has degree at least five, we must have that v_0 and v_1 are adjacent to all vertices of f_0, v_2 is adjacent to all vertices of f_1, and v_3 is adjacent to all vertices of f_3 (see the dotted edge v_2, v_3) in Figure 2 (c)). This implies that z has degree seven, which is however impossible because every vertex of G_0 has degree at most six. The statement of the lemma follows. $\qquad\square$

Lemma 8. *For every integer i such that $i \geq 0$, there exists a 1-planar graph with $n = 8 + 4i$ vertices and $4n - 10$ edges that is not a RAC graph.*

Proof. Let $\mathcal{G}$ be a family of graphs defined as follows. G_0 is a graph of $\mathcal{G}$. Graph G_i of $\mathcal{G}$ is obtained from G_{i-1} by adding four vertices to the external face of G_{i-1} and 16 edges as described in Figure 1 (b). Observe that every graph in $\mathcal{G}$ is 1-planar and it has $n = 8 + 4i$ vertices and $4n - 10$ edges. Suppose that G_i had a RAC drawing D_i. Since any sub-drawing of a RAC drawing is RAC drawing too, the sub-drawing of D_i representing graph G_0 should also be a RAC drawing of G_0, contradicting Lemma 7. It follows that no graph of $\mathcal{G}$ is a RAC graph, which proves the lemma. $\qquad\square$

Lemma 6 and Lemma 8 prove Theorem 1. Note that Theorem 1 does not hold if we drop the requirement of maximal density. Consider, for example, the graph G formed from K_5 by adding 11 paths of length 3 between every pair of vertices. It can be proved that G is a RAC graph, but it is not 1-planar.

4   Open Problems

1. Establish whether recognizing maximally dense RAC graphs is computationally as difficult as recognizing RAC graphs in the general case [2].
2. Characterize 1-planar graphs that admit a RAC drawing.

References

1. Ackerman, E., Fulek, R., Tóth, C.D.: On the Size of Graphs that Admit Polyline Drawings with Few Bends and Crossing Angles. In: Brandes, U., Cornelsen, S. (eds.) GD 2010. LNCS, vol. 6502, pp. 1–12. Springer, Heidelberg (2011)
2. Argyriou, E.N., Bekos, M.A., Symvonis, A.: The Straight-Line RAC Drawing Problem is NP-Hard. In: Černá, I., Gyimóthy, T., Hromkovič, J., Jefferey, K., Královič, R., Vukolić, M., Wolf, S. (eds.) SOFSEM 2011. LNCS, vol. 6543, pp. 74–85. Springer, Heidelberg (2011)
3. Arikushi, K., Fulek, R., Keszegh, B., Morić, F., Tóth, C.D.: Graphs that Admit Right Angle Crossing Drawings. In: Thilikos, D.M. (ed.) WG 2010. LNCS, vol. 6410, pp. 135–146. Springer, Heidelberg (2010)
4. Borodin, O.V., Kostochka, A.V., Raspaud, A., Sopena, E.: Acyclic colouring of 1-planar graphs. Discrete Applied Mathematics 114(1-3), 29–41 (2001)
5. Didimo, W., Eades, P., Liotta, G.: Drawing Graphs with Right Angle Crossings. In: Dehne, F., Gavrilova, M., Sack, J.-R., Tóth, C.D. (eds.) WADS 2009. LNCS, vol. 5664, pp. 206–217. Springer, Heidelberg (2009)
6. Didimo, W., Eades, P., Liotta, G.: A characterization of complete bipartite rac graphs. Inf. Process. Lett. 110(16), 687–691 (2010)
7. Fabrici, I., Madaras, T.: The structure of 1-planar graphs. Discrete Mathematics 307(7-8), 854–865 (2007)
8. Korzhik, V.P., Mohar, B.: Minimal Obstructions for 1-Immersions and Hardness of 1-Planarity testing. In: Tollis, I.G., Patrignani, M. (eds.) GD 2008. LNCS, vol. 5417, pp. 302–312. Springer, Heidelberg (2009)
9. Pach, J., Tóth, G.: Graphs drawn with few crossings per edge. Combinatorica 17(3), 427–439 (1997)
10. Suzuki, Y.: Optimal 1-planar graphs which triangulate other surfaces. Discrete Mathematics 310(1), 6–11 (2010)
11. van Kreveld, M.: The Quality Ratio of RAC Drawings and Planar Drawings of Planar Graphs. In: Brandes, U., Cornelsen, S. (eds.) GD 2010. LNCS, vol. 6502, pp. 371–376. Springer, Heidelberg (2011)

Pinning Balloons with Perfect Angles and Optimal Area

Immanuel Halupczok and André Schulz

Institut für Mathematische Logik und Grundlagenforschung,
Universität Münster, Germany
{ihalu_01,andre.schulz}@uni-muenster.de

Abstract. We study the problem of arranging a set of n disks with pre-scribed radii on n rays emanating from the origin such that two neighboring rays are separated by an angle of $2\pi/n$. The center of the disks have to lie on the rays, and no two disk centers are allowed to lie on the same ray. We require that the disks have disjoint interiors, and that for every ray the segment between the origin and the boundary of its associated disk avoids the interior of the disks. Let $\widetilde{r}$ be the sum of the disk radii. We introduce a greedy strategy that constructs such a disk arrangement that can be covered with a disk centered at the origin whose radius is at most $2\widetilde{r}$, which is best possible. The greedy strategy needs $O(n)$ arithmetic operations.

As an application of our result we present an algorithm for embedding unordered trees with straight lines and perfect angular resolution such that it can be covered with a disk of radius $n^{3.0367}$, while having no edge of length smaller than 1. The tree drawing algorithm is an enhancement of a recent result by Duncan et al. [Symp. of Graph Drawing, 2010] that exploits the heavy-edge tree decomposition technique to construct a drawing of the tree that can be covered with a disk of radius $2n^4$.

1 Introduction

When a graph is drawn in the plane, the vertices are usually represented as small dots. From a theoretical point of view a vertex is realized as a point, hence as an object without volume. In many applications, however, it makes sense to draw the vertices as disks with volume. The radii of the vertices can enhance the drawing by visualizing associated vertex weights [2,5]. This idea finds also applications in so-called *bubble drawings* [8], and *balloon drawings* [9,10].

Two important quality measures for aesthetically pleasant drawings are the *area* of a drawing and its *angular resolution*. The area of a drawing denotes the area of the smallest disk that covers the drawing with no edge lengths smaller than 1. The angular resolution denotes the minimum angle between two neighboring edges emanating at a vertex. Unfortunately, drawings of planar graphs with bounded angular resolution require exponential area [11]. On the other hand, by a recent result of Duncan et al. [6], it is possible to draw any unordered tree as plane straight-line graph with *perfect angular resolution*, that is the edges

M. van Kreveld and B. Speckmann (Eds.): GD 2011, LNCS 7034, pp. 154–165, 2012.
© Springer-Verlag Berlin Heidelberg 2012

incident to a vertex v are separated by an angle of at least $2\pi/\mathrm{degree}(v)$, and polynomial area. In the same paper it was observed that an ordered tree drawn with perfect angular resolution requires exponential area. Surprisingly, even ordered trees can be drawn in polynomial area with perfect angular resolution when the edges are drawn as circular arcs [6].

The following sub-problem appears naturally in tree drawing algorithms. Suppose we have drawings of all subtrees of the children of the root. How can we group the subtrees around the root, such that the final drawing is densely packed? Often one assumes that every subtree lies exclusively in some region, say a disk. Hence, at its core, a tree drawing algorithm has to arrange disjoint disks "nicely" around a new vertex. Furthermore this task is also a fundamental base case for bubble drawing algorithms or for algorithms that realize vertices as large disks. In the paper we show how to layout the balloons with perfect angular resolution and optimal area.

More formally, let $\mathcal{B} = \{B_1, B_2, \ldots, B_n\}$ be a set of n disks. To distinguish the disks B_i from other disks we call them *balloons*. The balloon B_i has radius r_i, and the balloons are sorted in increasing order of their radii. We are interested in layouts, in which the balloons of $\mathcal{B}$ have disjoint interiors and are evenly angularly spaced. In particular, we draw for every balloon a *spoke*, that is a line segment from the origin to the balloon center. The spokes have to avoid the interior of the other balloons and two neighboring spokes are separated by an angle of $2\pi/n$. Furthermore the drawing should require only small area. We measure the area of the balloon layout by the radius of the smallest disk that is centered at the origin and covers all balloons.

Results. We show how to locate the balloons with perfect angular resolution such that the drawing can be covered with a disk of radius $2\tilde{r}$, for $\tilde{r}$ being the sum of the radii. This is clearly the best possible result in the worst case, since when $|\mathcal{B}| = 1$, the area of the best balloon layout is clearly $2r_1$. We also study a modified version of the balloon layout problem that finds application in a tree drawing algorithm. Here, one and two spokes may remain without balloon, but the angle between the two unused spokes has to be at least $2\pi/3$. In this setting we obtain a balloon drawing that can be covered with a disk of radius $(1+\sqrt{2-2/\sqrt{5}})\tilde{r} \approx 2.0514\tilde{r}$. The induced algorithm draws unordered trees with perfect angular resolution and with area smaller than $n^{3.0367}$.

Related work. Without explicitly stated, Duncan et al. [6] studied the balloon layout problem (with one or two unused spokes) as part of their drawing algorithm for unordered trees and obtained a bound of $4\tilde{r}$ for the area. The induced tree drawing algorithm produces drawings with area smaller than $2n^4$. For the special case of orthogonal straight-line drawings of ternary trees (they automatically guarantee perfect angular resolution) Frati [7] provided an algorithm whose drawings require $O(n^{1.6131})$ area; the drawing of the complete ternary tree requires $O(n^{1.262})$ area. Bachmaier et al. obtained a drawing of the complete 6-regular tree with perfect angular resolution with area $O(n^{1.37})$ [1].

In contrast to our setting the so-called balloon drawings [9,10] place all balloons at the same distance. Also related are the (non-planar) *ringed circular layouts* [13]. Without the perfect angular resolution constraint trees can be drawn with area $O(n \log n)$ [4].

Conventions. We normalize the radii of the balloons such that they sum up to 1. In intermediate stages of the drawing algorithm a spoke may be without a balloon. In this case we consider the spoke as a ray emanating from the origin that fulfills the angular resolution constraint. When we say that "we place balloon B on s at distance x" we mean that the balloon B is placed on a spoke s (that had no associated balloon yet) such that its center lies on s at Euclidean distance x from the origin. In the remainder of the paper all disks covering the balloons are considered as centered at the origin.

2   The Greedy Strategy

In the following section we introduce the greedy strategy for placing $\mathcal{B}$ with perfect angles. To keep things simple we assume for now that the number of balloons n is a power of two. The general case is discussed later.

We place the balloons in increasing order of their radii. Thus we start with the smallest balloon and end with the largest balloon. The placement of the balloons is carried out in *rounds*. In every round we locate half of the balloons that have not been placed yet. Thus, we "consume" a certain number of spokes in each round. Let S be the list of spokes that are available in the beginning of a round in cyclic order. In every round we select every other spoke as a spoke on which a balloon is placed in the current round. This ensures that consecutive spokes that receive a balloon in round i are separated by an angle of $\alpha_i := 2^{i+1}\pi/n$. For every round we define the *safe disk* SD_i centered at the origin with radius safe_i. The safe disk is the smallest disk covering all balloons that were placed in previous rounds. In round i we place all balloons such that they avoid the interior of the safe disk SD_i. Thus, the best we can hope for is to place the balloons such that they touch SD_i. Whenever this is possible we speak of a *contact situation*, depicted in Figure 1(a). The safe disks ensure that balloons placed in the current round will not intersect the interior of the balloons that were placed in previous rounds. However, we have to guarantee that balloons placed in the same round will also not interfere with the remaining spokes. Suppose that B_j is assigned to the spoke s_k. We enforce B_j to lie inside a wedge with opening angle α_i centered at s_k. This wedge is named W_k. Since the spokes that are used in round i are separated by α_i, the wedges of round i have disjoint interiors. Whenever a balloon touches the boundary of its associated wedge we speak of a *wedge situation*, as shown in Figure 1(b).

The greedy strategy tries first to place B_j at its spoke s_k, such that it touches SD_i. If this would imply that B_j is not contained inside W_k, we move the center of B_j on s_k away from the origin, until B_j touches the boundary of W_k. In case a wedge situation occurs, we can compute the location of the center of B_j with help of the following lemma, whose proof can be found in Duncan et al. [6].

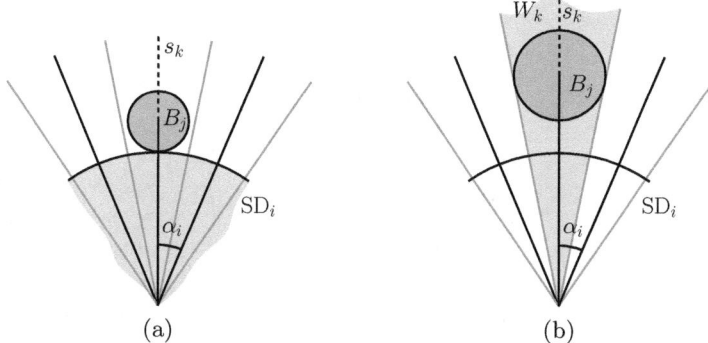

Fig. 1. In a contact situation (a) we place B_i such that it touches SD_i. In contrast, in a wedge situation (b), we place B_i such that it touches the boundary of W_k (when it is placed on s_k).

Lemma 1. *Let W be a wedge with opening angle φ centered at a spoke s. Further let B be a balloon with radius r that is placed such that (1) its center lies on s, and (2) it touches the boundary of W. Then B is contained inside a disk centered at the origin with radius*

$$(1 + \sin (\varphi/2))/(\sin (\varphi/2)) \cdot r.$$

In the remainder of the paper we use as notation

$$\alpha(\varphi) := (1 + \sin (\varphi/2))/(\sin (\varphi/2)). \qquad (1)$$

Notice that when a wedge situation occurs in round i, then in particular a wedge situation has to occur for the last balloon that is added in round i, since the balloons are sorted by increasing radii. All balloons placed in round i are sandwiched between SD_i and SD_{i+1}. We call the region $SD_{i+1} \setminus SD_i$ the i-th layer L_i.[1] The *width* of layer L_i is defined as $\mathrm{safe}_{i+1} - \mathrm{safe}_i$. When a wedge situation occurs in round i, the layer L_i is called a *wedge layer*, otherwise a *contact layer*. An example of a wedge layer is shown in Figure 2.

2.1   Splitting the Set of Spokes

We come now back to the case where n is not necessarily a power of two. In this setting there might be an odd number of spokes in some round. In such a round we place only $\lfloor k/2 \rfloor$ balloons, such that no two of them are assigned to consecutive spokes. This however has two drawbacks: First, the angles might not split evenly, and second, the layers will be filled with less balloons.

We can always pick $\lfloor k/2 \rfloor$ spokes such that in the remaining set of spokes at most two separating angles are smaller than the others, which are all equal.

---

[1] By convention $SD_1 = \emptyset$, and for i being the last round, $SD_{i+1} = $ smallest disk covering all balloons.

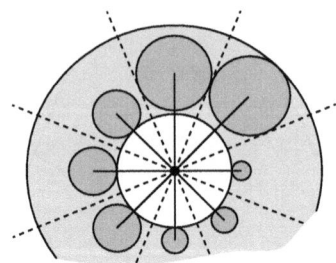

Fig. 2. A wedge layer (shaded) that had been filled with balloons by the greedy strategy

Moreover, the two smaller angles are each at least half as big as the remaining angles. We call every set of spokes for which this property holds *well-separated*. Furthermore we assume that a well-separated set of spokes is ordered such that the two smaller angles are realized between the first and second, and between the second and third spoke. Algorithm 1 describes a strategy that picks $\lfloor k/2 \rfloor$ of the spokes and ensures that the remaining set of spokes is still well-separated if the original set was well-separated.

Algorithm 1. SplitSpokes(S)

> **Input** : S set of spokes
> **Output**: (T, T'), such that T' are the spokes that will be used in the current
> round, $T = S \setminus T'$.
> 1 $T' \leftarrow$ every spoke of S with even index
> 2 $T \leftarrow S \setminus T'$
> 3 reorder T by putting the last spoke in front
> 4 **return** (T, T')

Lemma 2. *Let S be a well-separated set of at least three spokes and let φ denote the size of the big angles in S. Let (T, T') be the return value of Algorithm 1.*

(1) If $|T| > 2$, then T is well-separated.
(2) If $|T| = 2$, then the smaller angle between the two spokes is at least $2\pi/3$.
(3) The wedge with angle φ centered at the first spoke in T' contains no spoke of S in its interior.
(4) A wedge with angle 2φ centered at a spoke in T' that is not the first spoke contains no spoke of S in its interior.

Proof. Let the angle between the first and second spoke in S be γ_1, and let the angle between the second and third spoke in S be γ_2. Since S is well-separated, we have $\varphi/2 \leq \gamma_1, \gamma_2 \leq \varphi$. Hence the wedge centered at the second spoke of S with angle φ does not contain any other spoke of S in its interior, which proves (3). Property (4) is due to the fact that every spoke in S with even index larger than 2 is separated from its neighboring spokes by an angle of φ.

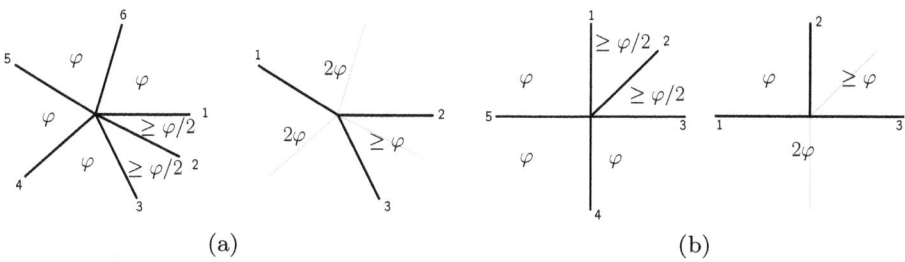

Fig. 3. Merging the angles as implied by Algorithm 1. In case we have an even number of spokes (a), and in case we have an odd number of spokes (b). The spoke numbers are shown as small numbers.

After line 2 of Algorithm 1, the angle between the first and second spoke of T equals $\gamma_1 + \gamma_2 \geq \varphi$. In case S has an even number of spokes the remaining angles of size φ are grouped pairwise and therefore the corresponding angles in T are all 2φ, which proves property (1) for this case. If the set S contains an odd number of spokes, the additional angle between the last spoke in T and the first spoke in T is also φ. Hence after reordering, the new set T is again well-separated, and (1) follows. Figure 3 illustrates the outcome of Algorithm 1.

To see that (2) is true, notice the following. T contains two spokes, if S contains three or four spokes. In case S contains 4 spokes, the sum of the two small angles is at least $2\pi/3$. In case S contains three spokes, the sum of the two small angles between the spokes is at least π. The large angle between the spokes in S is at least $2\pi/3$. This angle appears also between the spokes in T. □

To ensure that the balloons of each layer cannot interfere with each other and with the remaining spokes, we place them inside the wedges defined by Lemma 2(3–4). All wedges have the same opening angle, say φ, except the first wedge, whose opening angle is at least $\varphi/2$. The balloon with the smallest radius in each round is placed inside the wedge with the (possible) smaller opening angle.

2.2   The Final Layer

It is important to analyze the situation where the greedy strategy has to stop. In every round we reduce the number of spokes from k to $\lceil k/2 \rceil$. If we subdivide the spokes in this fashion we will come to a point where exactly two spokes are left. The final two balloons are placed in the last round as follows: (1) The balloon B_n will be placed such that it touches the safe disk. (2) The balloon B_{n-1} will be placed such that it is contained inside a wedge with opening angle $\pi/3$, centered at its spoke, while avoiding the interior of the current safe disk.

Lemma 3. *When the balloons are placed as discussed in the previous paragraph, then one of the following is true:*

1. *The width of the last layer is $2r_n$.*
2. *All balloons can be covered with a disk of radius two.*

Proof. Let φ be the smaller of the two angles between the spokes in the final round i. Due to Lemma 2, φ is at least $2\pi/3$. The tangent of B_n at its intersection with SD_i separates B_n from the spoke of B_{n-1}. Since the angle between this tangent and the spoke of B_{n-1} is at least $\varphi - \pi/2 \geq \pi/6$ it is safe to place B_{n-1} inside a wedge centered at its spoke with opening angle $\pi/3$. Thus, either B_{n-1} touches SD_i, or it is contained inside a disk of radius $\alpha(\pi/3)r_{n-1} = 3r_{n-1}$. In the former case the width of the layer is $2r_n$, in latter case the radius of the covering disk is at most $\max\{2r_n, 3/2\}$ (recall that $r_{n-1} \leq 1/2$). $\qquad\square$

Due to Lemma 3 we can assume that the width of the last layer equals $2r_n$. Thus even if B_{n-1} defines a wedge situation we consider the last layer as contact layer. We summarize the discussion in Algorithm 2.

Algorithm 2. GreedyBalloon(S)

    Input : S: spokes in cyclic order.

```
1  k ← 0                                        // number of balloons placed so far
2  safe ← 0                                     // radius of the current safe disk
3  while |S| > 2 do
4  │   (T, T') ← Splitspokes(S)
5  │   width ← 0                                // width of the current layer so far
6  │   for i ← k + 1 to k + |T'| do
7  │   │   s ← (i − k)-th spoke of T'
8  │   │   φ ← 2(minimal angle between s and one of its neighboring spokes in S)
9  │   │   c ← max {α(φ)rᵢ − rᵢ, safe + rᵢ}     // center of Bᵢ
10 │   │   place Bᵢ on s at distance c
11 │   │   width ← max{width, c + rᵢ − safe}
12 │   end
13 │   safe ← safe + width
14 │   k ← k + |T'|
15 │   S ← T
16 end
17 let s₁, s₂ be the spokes in S
18 place Bₙ on s₁ at distance safe + rₙ
19 place Bₙ₋₁ on s₂ at distance max{2rₙ₋₁, safe + rₙ₋₁}
```

2.3   Quality of the Greedy Strategy

We denote by R the radius of the smallest disk that covers all balloons. In order to determine R we have to consider only certain radii.

Lemma 4. *The radius of the smallest disk R that covers all balloons drawn with Algorithm 2 can be determined with the knowledge of*

1. *the number of spokes,*
2. *the radius of the largest and smallest balloon in the outermost wedge layer,*
3. *the radii of the largest balloons in each of the contact layers following the outermost wedge layer.*

Proof. Suppose the last wedge situation occurs in round i. Then the radius of SD_{i+1} is determined by a balloon that touches its wedge. All wedges have the same opening angle, except maybe the first wedge. Since the smallest balloon is placed inside the first wedge, the wedge situation that defines the radius of SD_{i+1} depends on the possible wedge situation of the largest and smallest balloon only. The following layers are all contact layers. Their width is determined by the diameter of the largest balloons in each layer. The radius R equals therefore the radius of SD_{i+1} with the addition of the widths of the following contact layers. □

Since we are interested in a worst case bound for R we make the following assumptions to simplify the analysis of the algorithm.

Lemma 5. *Let r_w be the radius of the balloon, whose wedge situation determined the width of the last wedge layer L_k. The radius R of the smallest covering disk is maximized when*

$$r_w = r_{w+1} = r_{w+2} = \cdots = r_{n-1}, \, and$$
$$r_1 = r_2 = r_3 = \cdots = r_{w-1} = 0.$$

Proof. We consider the radii as resources that we want to spend to make R as large as possible. Since no radius of a balloon with smaller index than w matters for R, we set these radii to zero to save resources. If B_w is the smallest balloon in its layer, all radii of balloons in L_k have the same radius in the worst case. Otherwise we could shrink some of these balloons without changing the width of L_k and spent the resources to increase r_n and therefore R.

Only the balloon added last in each contact layer determines the width of its layer. We select the radii of the other balloons in contact layers as small as possible, i.e., as large as the radius of the largest balloon in the previous layer. If any of these radii would be larger we could make such a radius smaller and increase r_n instead, which would increase R.

Assume we have at least two contact layers following L_k. Let B_c be the largest balloon in the contact layer L_{k+1}, that is the balloon last added in L_{k+1}. Due to the discussion in the previous paragraph we can assume that the balloon B_{c+1} in the next layer has radius r_c. If $r_c > r_w$, we could lower the radius by $r_c - r_w$ for B_c and B_{c+1} each. By this we can increase r_n by $2(r_c - r_w)$. As a consequence the radius R increases by $r_c - r_w$. Therefore in the worst case all radii in layer L_{k+1} equal r_w. By an inductive argument the radii in the last contact layers are all r_w. The only exception is the largest balloon B_n. □

Theorem 1. *Algorithm 2 constructs a drawing of balloons with disjoint interiors and spokes that intersect only the interior of their associated balloon that can be covered with a disk of radius two, which is best possible.*

Proof. We define as $\bar{L}_i$ the i-th last layer such that $\bar{L}_1$ is the last layer. Suppose there were ℓ spokes left, before the last wedge layer was filled. We denote the number of contact layers that follow the last wedge layer by k. By Algorithm 1 the number k is given by a function $k = f(\ell)$, which is defined as follows

$$f(\ell) := \begin{cases} 1 & \text{if } 3 \le \ell \le 4, \\ 1 + f\left(\frac{\ell}{2}\right) & \text{if } \ell > 4, \text{ even}, \\ 1 + f\left(\frac{\ell+1}{2}\right) & \text{if } \ell > 4, \text{ odd}. \end{cases} \tag{2}$$

By induction, $f(\ell) \le \log(\ell - 1)$. The radius of the covering disk R equals the radius of $\bar{L}_k$'s safe disk plus the width of the last k contact layers. Let B_w be the balloon that determined safe$_k$. By Lemma 5 we can assume that all balloons following B_w have radius r_w, except B_n. All other radii are zero.

As previously discussed, the balloon B_w is either the first or the last balloon in the last wedge layer. We discuss the two possibilities by case distinction. Let us first assume that B_w is the last balloon of layer $\bar{L}_{k+1}$. By construction the last balloon is placed inside the wedge with largest opening angle (in this round). Therefore its opening angle φ is minimized, when the angles between all pairs of neighboring spokes are equal. We have ℓ spokes in $\bar{L}_{k+1}$, and therefore two spokes are separated by $2\pi/\ell$ and $\varphi = 4\pi/\ell$. Furthermore, we have $k-1$ layers of width $2r_w$, and one layer of width $2r_n$ following $\bar{L}_{k+1}$. In layer $\bar{L}_{k+1}$ we place no more than $\ell/2$ balloons and therefore in the last k layers we have at least $\ell/2$ balloons in total. Since there is one balloon in $\bar{L}_{k+1}$ with radius r_w and only one balloon in the last k layers with radius different from r_w, we get $r_n \le 1 - r_w\ell/2$. This leads to

$$R \le \alpha(\varphi)r_w + 2(k-1)r_w + 2r_n \le 2 + [\alpha(4\pi/\ell) + 2\log(\ell-1) - \ell - 2]\, r_w.$$

The last wedge layer must contain at least three spokes. Since $\alpha(4\pi/\ell) + 2\log(\ell - 1) - \ell - 2$ is decreasing[2] for $\ell \ge 4$ and negative for $\ell = 3, 4$, we get $R \le 2$.

We assume now that B_w was placed first in $\bar{L}_{k+1}$. Again, let φ be the angle of the wedge that contains B_w centered at its spoke. Due to Lemma 2 the angles between two neighboring spokes are all of size ψ except two angles, which are at least $\psi/2$ (the small angles). The angle φ is twice the minimum of the two small angles, and hence minimized when one of the small angles has size ψ and the other has size $\psi/2$. In this case we have $\ell - 1$ angles of size ψ and one angle of size $\psi/2$. Since all angles sum up to 2π, we have $\psi = 2\pi/(\ell - 1/2)$, which is a lower bound for φ. Notice that all balloons in $\bar{L}_{k+1}$ have now radius r_w, hence we have $\ell - 1$ balloons of radius r_w, and therefore $r_n \le 1 - (\ell - 1)r_w$. We conclude with

$$R \le \alpha(\varphi)r_w + 2(k-1)r_w + 2r_n \le 2 + [\alpha(2\pi/(\ell - 1/2)) + 2\log(\ell-1) - 2\ell]\, r_w.$$

For $\ell \ge 2$ the expression $\alpha(2\pi/(\ell - 1/2)) + 2\log(\ell - 1) - 2\ell$ is negative and decreasing and the theorem follows.   □

---

[2] The estimation of this expression and of similar following expressions was obtained by computer algebra software.

3   Drawing Unordered Trees with Perfect Angles

The greedy strategy can be used to construct drawings of unordered trees with perfect angular resolution and small area. In fact, the balloon layout problem studied in Section 2 is a subproblem of the drawing algorithm of Duncan et al. [6], where it is used to draw depth-1 trees. With the help of the so called heavy edge tree-decomposition (see Tarjan [12]) these trees are combined to the original tree. Since our proposed strategy uses significantly smaller area, it implies an improvement for the area of the tree drawing.

We start with a brief review of the heavy edge tree-decomposition. Let u be a non-leaf of the rooted tree T. We denote by T_u the subtree of T rooted at u. Let v be the child of u such that T_v has the largest number of nodes (compared to the subtrees of the other children of u), breaking ties arbitrarily. We call the edge (u, v) a *heavy edge*, and the edges to the other children of u *light edges*. The heavy edges induce a decomposition of T into paths, called *heavy paths*, and light edges; see Figure 4 on the left.

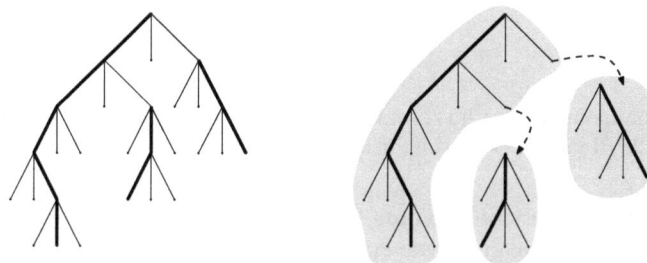

Fig. 4. An example of a heavy-edge tree-decomposition

In order to draw a tree with perfect angular resolution we have to modify the greedy strategy such that (1) one or two spokes can remain without balloon, and (2) the separating angle between the two unused spokes is at least $2\pi/3$. This however, comes at a cost, we have to make the disk that covers the drawing slightly larger. Thus instead of a disk with radius 2 we might need a disk of radius κ, where $\kappa = (1 + \sqrt{2 - 2/\sqrt{5}}) \approx 2.0514$. Notice that we use an additional construction, which can be found in the full paper, to make the balloon packing slightly denser.

By construction, every non-leaf tree node lies on exactly one heavy path. Let C be the union of the heavy path that is incident to the root with its incident light edges. By deleting C the original tree splits into subtress. Assume that we have constructed the drawings for these subtrees by recursion. We are left with drawing C such that the leaves of C are drawn as disjoint disks. The disk radii are chosen such that each drawing of a subtree fits inside its associated disk. For every node on the heavy path we apply the greedy strategy to draw the associated disks as balloons while leaving the heavy edges as free spokes without a disk. These drawings of depth-1 trees can be combined by the strategy of

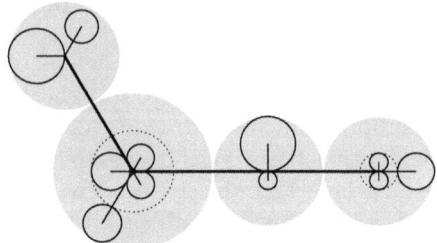

Fig. 5. Drawing of the root-heavy path with incident light edges and *safe regions* for the missing subtrees

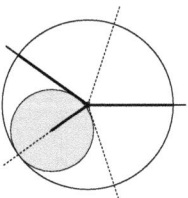

Fig. 6. Three balloons with radius $\varepsilon, \varepsilon, 1-2\varepsilon$ and 5 spokes. Separating the unused spokes by an angle $\geq 2\pi/3$ yields a covering disk with radius $\alpha(2\pi/5) = \kappa$. when $\varepsilon \to 0$.

Duncan et al.[6, full version, Lemma 2.3]. Figure 5 illustrates this construction. Notice that for every layer of the recursion we have to scale the intermediate drawing by a factor of 2κ. However, since by construction every root-leaf path in the original tree visits at most $\log n$ light edges, the recursion depth is logarithmic. A more detailed analysis, which can be found in the full version of the paper, proves the following theorem.

Theorem 2. *Using Algorithm 2 in the framework of Duncan et al. produces a drawing of an unordered tree with n nodes that has perfect angular resolution and that can be covered with a disk of radius $n^2 \cdot n^{\log \kappa} < n^{3.0367}$, while having no edge with length smaller 1.*

4   Concluding Remarks

The algorithm presented in this paper runs in linear time. Notice that even when the set $\mathcal{B}$ is not ordered by radii we can obtain a running time of $O(n)$. In fact, Algorithm 2 works correctly when $\mathcal{B}$ is *weakly ordered* by radii, that is, (1) the median and the smallest element are in the "right" position, (2) every radii between median and smallest element is not larger than the median, and (3) the sequence of elements larger than the median are weakly ordered. Since the median and the smallest radius can be found in $O(n)$ time [3], the recursive definition implies that $\mathcal{B}$ can be weakly ordered in $O(n)$ time.

The only case, where we obtain strict inequalities in the proof of Theorem 1, is when $|\mathcal{B}| = 1$. By placing all balloons slightly inside the wedges, resp., slightly outside the safe disks we can therefore modify all constructions such that no balloons touch.

As a final remark we point out that Theorem 1 can be generalized such that it holds for one or two unused spokes, while guaranteeing that the whole balloon drawing can be covered with a disk of radius 2. However, as depicted in Figure 6, the slightly worse bound of κ cannot be avoided if one has to guarantee that the smaller angle between the two unused spokes is at least $2\pi/3$.

References

1. Bachmaier, C., Brandenburg, F.-J., Brunner, W., Hofmeier, A., Matzeder, M., Unfried, T.: Tree Drawings on the Hexagonal Grid. In: Tollis, I.G., Patrignani, M. (eds.) GD 2008. LNCS, vol. 5417, pp. 372–383. Springer, Heidelberg (2009)
2. Barequet, G., Goodrich, M.T., Riley, C.: Drawing Graphs with Large Vertices and Thick Edges. In: Dehne, F.K.H.A., Sack, J.-R., Smid, M.H.M. (eds.) WADS 2003. LNCS, vol. 2748, pp. 281–293. Springer, Heidelberg (2003)
3. Blum, M., Floyd, R.W., Pratt, V.R., Rivest, R.L., Tarjan, R.E.: Time bounds for selection. J. Comput. Syst. Sci. 7(4), 448–461 (1973)
4. Crescenzi, P., Battista, G.D., Piperno, A.: A note on optimal area algorithms for upward drawings of binary trees. Computational Geometry: Theory & Application Geom. 2, 187–200 (1992)
5. Duncan, C.A., Efrat, A., Kobourov, S.G., Wenk, C.: Drawing with fat edges. Int. J. Found. Comput. Sci. 17(5), 1143–1164 (2006)
6. Duncan, C.A., Eppstein, D., Goodrich, M.T., Kobourov, S.G., Nöllenburg, M.: Drawing Trees with Perfect Angular Resolution and Polynomial Area. In: Brandes, U., Cornelsen, S. (eds.) GD 2010. LNCS, vol. 6502, pp. 183–194. Springer, Heidelberg (2011), http://arxiv.org/pdf/1009.0581v1
7. Frati, F.: Straight-Line Orthogonal Drawings of Binary and Ternary Trees. In: Hong, S.-H., Nishizeki, T., Quan, W. (eds.) GD 2007. LNCS, vol. 4875, pp. 76–87. Springer, Heidelberg (2008)
8. Grivet, S., Auber, D., Domenger, J.P., Melancon, G.: Bubble tree drawing algorithm. In: International Conference on Computer Vision and Graphics, pp. 633–641. Springer, Heidelberg (2004)
9. Lin, C.-C., Yen, H.-C.: On balloon drawings of rooted trees. Journal of Graph Algorithms and Applications 11(2), 431–452 (2007)
10. Lin, C.-C., Yen, H.-C., Poon, S.-H., Fan, J.-H.: Complexity analysis of balloon drawing for rooted trees. Theor. Comput. Sci. 412(4-5), 430–447 (2011)
11. Malitz, S.M., Papakostas, A.: On the angular resolution of planar graphs. SIAM J. Discrete Math. 7(2), 172–183 (1994)
12. Tarjan, R.E.: Linking and cutting trees. In: Data Structures and Network Algorithms, ch. 5, pp. 59–70. SIAM (1983)
13. Teoh, S.T., Ma, K.-L.: RINGS: A Technique for Visualizing Large Hierarchies. In: Goodrich, M.T., Kobourov, S.G. (eds.) GD 2002. LNCS, vol. 2528, pp. 268–275. Springer, Heidelberg (2002)

Approximate Proximity Drawings

William Evans[1], Emden R. Gansner[2], Michael Kaufmann[3], Giuseppe Liotta[4],
Henk Meijer[5], and Andreas Spillner[6]

[1] University of British Columbia, Canada
`will@cs.ubc.ca`
[2] AT&T Research Labs, US
`erg@research.att.com`
[3] Universität Tübingen, Germany
`mk@informatik.uni-tuebingen.de`
[4] Università degli Studi di Perugia, Italy
`liotta@diei.unipg.it`
[5] Roosevelt Academy, The Netherlands
`h.meijer@roac.nl`
[6] Universität Greifswald, Germany
`andreas.spillner@uni-greifswald.de`

Abstract. We introduce and study a generalization of the well-known region of influence proximity drawings, called $(\varepsilon_1, \varepsilon_2)$-*proximity drawings*. Intuitively, given a definition of proximity and two real numbers $\varepsilon_1 \geq 0$ and $\varepsilon_2 \geq 0$, an $(\varepsilon_1, \varepsilon_2)$-proximity drawing of a graph is a planar straight-line drawing Γ such that: (i) for every pair of adjacent vertices u, v, their proximity region "shrunk" by the multiplicative factor $\frac{1}{1+\varepsilon_1}$ does not contain any vertices of Γ; (ii) for every pair of non-adjacent vertices u, v, their proximity region "blown-up" by the factor $(1 + \varepsilon_2)$ contains some vertices of Γ other than u and v. We show that by using this generalization, we can significantly enlarge the family of the representable planar graphs for relevant definitions of proximity drawings, including Gabriel drawings, Delaunay drawings, and β-drawings, even for arbitrarily small values of ε_1 and ε_2. We also study the extremal case of $(0, \varepsilon_2)$-proximity drawings, which generalizes the well-known weak proximity drawing model.

1 Introduction and Overview

Proximity drawings are straight-line drawings of graphs where any two adjacent vertices are deemed to be close according to some proximity measure, while any two non-adjacent vertices are far from one another with respect to the same measure. Different definitions of proximity give rise to different types of proximity drawings. In the *region of influence* based proximity drawings two vertices u and v are adjacent if and only if some regions of the plane, defined by using the coordinates of u and v, are *empty*, i.e. they do not contain any vertices of the drawing other than, possibly, u and v. Throughout this paper we shall always assume that the proximity regions are closed sets; hence if a vertex is on the boundary of the proximity region of u and v, the region is not empty.

For example, the *Gabriel disk* of two points u and v in the plane is the disk having u and v as their antipodal points and a *Gabriel drawing* (also called a *Gabriel graph*) is a

M. van Kreveld and B. Speckmann (Eds.): GD 2011, LNCS 7034, pp. 166–178, 2012.
© Springer-Verlag Berlin Heidelberg 2012

planar straight-line drawing such that any two vertices are connected by an edge if and only if their Gabriel disk is empty A generalization of the Gabriel disk is the so-called β-*region of influence*: For a given value of β such that $1 \leq \beta \leq \infty$, the β-region of influence of two vertices u and v having Euclidean distance $d(u, v)$ is the intersection of the two disks of radius $\frac{\beta d(u,v)}{2}$, centered on the line through u and v, one containing u and touching v, the other containing v and touching u (hence the β-region for $\beta = 1$ is the Gabriel disk). Given a value of β, a straight-line drawing is a β-*drawing* (also called a β-*skeleton*) if and only if for any edge (u, v) the β-region of influence of u and v is empty. *Delaunay drawings* use a definition of proximity that extends the one used for Gabriel drawings. Namely, the *Delaunay disks* of two vertices u and v are the disks having $\overline{uv}$ as a chord (the Gabriel disk is therefore a particular Delaunay disk). In a *Delaunay drawing* (also called a *Delaunay graph*) an edge (u, v) exists if and only if at least one of the Delaunay disks of u and v is empty.

As is not hard to imagine, by changing the definition of region of influence, the combinatorial properties of those graphs that admit a certain type of proximity drawing can change significantly. For example, it is known that not all trees having vertices of degree four admit a Gabriel drawing [4] while they have a β-drawing for $1 < \beta \leq 2$ [12]. Unfortunately, the adoption of region of influence based proximity rules seems to dramatically restrict the family of representable graphs. Also, despite the many papers published on the topic, full combinatorial characterization of proximity drawable graphs remains an elusive goal for most types of regions of influence. The interested reader is also referred to [6,11,14] for references and results on these topics.

1.1 Problem and Results

In this paper, we want to compute planar straight-line drawings of graphs where adjacent vertices are relatively close to each other while non-adjacent vertices are relatively far apart. In order to overcome the restrictions on the families of representable graphs imposed by region of influence based proximity drawings, we study graph visualizations that are "good approximations" of these proximity drawings. The idea is to use slightly smaller regions of influence to justify the existence of an edge and slightly larger regions of influence to justify non-adjacent vertices.

More formally, let D be a disk with center c and radius r, and let ε_1 and ε_2 be two non-negative real numbers. The ε_1-*shrunk disk of* D is the disk centered at c and having radius $\frac{r}{1+\varepsilon_1}$; the ε_2-*expanded disk of* D is the disk centered at c and having radius $(1 + \varepsilon_2)r$. An $(\varepsilon_1, \varepsilon_2)$-*proximity drawing* is a planar straight-line proximity drawing where the region of influence of two adjacent vertices is defined by using ε_1-shrunk disks, while the region of influence of two non-adjacent vertices uses ε_2-expanded disks. In the next sections we study $(\varepsilon_1, \varepsilon_2)$-*Gabriel drawings*, $(\varepsilon_1, \varepsilon_2)$-*Delaunay drawings*, and $(\varepsilon_1, \varepsilon_2)$-$\beta$-*drawings*.

It is immediate to observe that all planar graphs with at least one edge or at least three vertices have an $(\varepsilon_1, \varepsilon_2)$-proximity drawing for sufficiently large values of $\varepsilon_1, \varepsilon_2$. For example, every planar straight-line drawing Γ with at least one edge is a (∞, ∞)-Gabriel drawing since an ∞-shrunk Gabriel disk reduces to a point (and thus the ∞-shrunk disk of every edge in Γ is empty) and an ∞-expanded Gabriel disk is the whole plane (and thus the ∞-expanded disk of every pair of non-adjacent vertices of Γ is never empty). At the other extreme, a $(0, 0)$-Gabriel drawing is a Gabriel drawing, since a 0-shrunk Gabriel disk is a Gabriel disk and so is a 0-expanded Gabriel disk. Hence, not all planar graphs admit a $(0, 0)$-Gabriel drawing [4].

Based on this observation, our main target is to establish values of ε_1 and of ε_2 that make it possible to compute $(\varepsilon_1, \varepsilon_2)$-proximity drawings for meaningful families of planar graphs. Our results are as follows:

- We prove that every embedded planar graph admits, for any $\varepsilon_1 > 0$ and any $\varepsilon_2 > 0$, an $(\varepsilon_1, \varepsilon_2)$-Gabriel drawing, an $(\varepsilon_1, \varepsilon_2)$-Delaunay drawing, and an $(\varepsilon_1, \varepsilon_2)$-$\beta$-drawing (for all $1 \leq \beta \leq \infty$) that preserve the given embedding. (See Theorems 1, 4, and 5.)
- We show that the above results are, in a sense, tight by exhibiting embedded planar graphs that do not have an embedding preserving $(\varepsilon_1, \varepsilon_2)$-proximity drawing with either $\varepsilon_1 = 0$ or $\varepsilon_2 = 0$. (See again Theorems 1, 4, and 5.)
- We study $(0, \varepsilon_2)$-proximity drawings which, as explained in the next section, make it possible to express different proximity conventions in a unified framework. In particular, we study $(0, \varepsilon_2)$-Gabriel drawings of outerplanar graphs, extending previous results of [7,12]. (See Theorems 2 and 3.)

We emphasize that the main contribution of this paper is in introducing the concept of $(\varepsilon_1, \varepsilon_2)$-proximity drawing and in proving the existence of $(\varepsilon_1, \varepsilon_2)$-proximity drawings for relevant families of graphs. Hence, we shall not spend words on the time complexities of our algorithms; it is not hard to see, however, that our drawing techniques all require polynomial time when adopting the real RAM model of computation. For reasons of space, some proofs are sketched or omitted.

1.2   Related Work

Several generalizations, variants, and relaxations of proximity drawings have been defined in the literature such as, for example, k-localized Delaunay triangulations, approximate minimum spanning trees, and witness proximity drawings. The interested reader can, for example, use [2,10,13] as starting points to study these topics.

Although each proximity drawing mentioned above would deserve some special attention, in this introduction we can just spend a few words on *weak proximity drawings* [7] that are more closely related with $(\varepsilon_1, \varepsilon_2)$-proximity drawings. In a weak proximity drawing, the region of influence of any pair of adjacent vertices must be empty, while no condition is given for the non-adjacent pairs. Hence, weak proximity drawings guarantee visual closeness of groups of edge-related vertices but do not ensure that unrelated vertices are far apart. In contrast, $(\varepsilon_1, \varepsilon_2)$-proximity drawings guarantee some relative closeness of the adjacent pairs of vertices and some relative separation of the non-adjacent pairs for any finite values of ε_1 and ε_2.

Note that a weak proximity drawing is a $(0, \infty)$-proximity drawing and that a proximity drawing in the traditional sense is a $(0, 0)$-proximity drawing. Therefore, $(0, \varepsilon_2)$-proximity drawings make it possible to study proximity drawability in a unified framework: as the value of ε_2 increases, $(0, \varepsilon_2)$-proximity drawings approach weak proximity drawings. Several questions can be asked within this unifying framework.

For example, not all trees have a Gabriel drawing, while all trees have a weak Gabriel drawing. What is the minimum threshold value such that if ε_2 is larger than this threshold all trees are drawable? Theorem 2 answers this question.

2   Approximate Gabriel Drawings

Let Γ be a planar straight-line drawing of a graph and let $\varepsilon_1, \varepsilon_2$ be two non-negative numbers. Let u, v be any two vertices of Γ and let $D(u, v)$ be the Gabriel disk of u, v (that is, the disk having u and v as the end-points of its diameter). We say that Γ is an $(\varepsilon_1, \varepsilon_2)$-*Gabriel drawing* if: (i) for every edge (u, v) of Γ the ε_1-shrunk disk of $D(u, v)$ is empty (i.e. it does not contain any vertex of Γ other than, possibly, u and v); and (ii) for every pair of non-adjacent vertices u, v of Γ, the ε_2-expanded disk of $D(u, v)$ is not empty (i.e. it contains some vertex w of Γ other than u and v). Note that a Gabriel graph is a special case of an $(\varepsilon_1, \varepsilon_2)$-Gabriel drawing, namely the one in which $\varepsilon_1 = \varepsilon_2 = 0$.

Fig. 1 is an example of an $(\varepsilon_1, \varepsilon_2)$-Gabriel drawing for $\varepsilon_1 = 0$ and $\varepsilon_2 = 0.7$. The drawing is not a Gabriel drawing; for example, the dotted disk in the figure is a Gabriel disk, while the solid one is its 0.7-expanded version. Note that *no* Gabriel drawing exists for the tree in Fig. 1 [4].

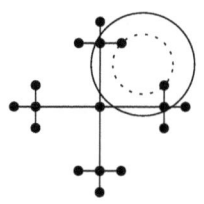

Fig. 1. A $(0,0.7)$-Gabriel drawing of a tree that does not have a $(0, 0)$-Gabriel drawing

In order to establish values of $\varepsilon_1, \varepsilon_2$ that allow an $(\varepsilon_1, \varepsilon_2)$-Gabriel drawing of every planar graph, we start by considering the extremal cases that either $\varepsilon_1 = 0$ and $\varepsilon_2 > 0$ or $\varepsilon_1 > 0$ and $\varepsilon_2 = 0$. An *embedded planar graph* is a planar graph together with a planar topological embedding. A planar straight-line drawing Γ of an embedded planar graph G *maintains* (or *preserves*) the embedding of G if Γ and G have the same set of faces and for every vertex v the circular order of the edges around v is the same in G and in Γ; in this case we shall also sometimes say that Γ is an *embedding preserving drawing* of G.

The next two lemmas study the relationship between embedding preserving $(\varepsilon_1, \varepsilon_2)$-Gabriel drawings with either $\varepsilon_1 = 0$ or $\varepsilon_2 = 0$ and Gabriel graphs. We say that an embedded planar graph is *Gabriel drawable* if there exists a Gabriel graph Γ such that Γ is an embedding preserving drawing of G.

Lemma 1. *Let G be an embedded maximal planar graph, that is, a triangulation, and let ε_2 be any given real number such that $\varepsilon_2 \geq 0$. G has an embedding preserving $(0, \varepsilon_2)$-Gabriel drawing if and only if G has an embedding preserving Gabriel drawing.*

Proof. If G has an embedding preserving Gabriel drawing Γ, then Γ is also a $(0, 0)$-Gabriel drawing of G and therefore a $(0, \varepsilon_2)$-Gabriel drawing for any $\varepsilon_2 > 0$. If G has an embedding preserving $(0, \varepsilon_2)$-Gabriel drawing Γ, let V be the vertex set of Γ. Let $GG(V)$ be the Gabriel graph having V as its vertex set. Since Γ is a $(0, \varepsilon_2)$-Gabriel drawing, for every pair u, v of adjacent vertices in Γ the disk having u and v as antipodal points does not contain any other element of V. Hence, every edge of Γ is also an edge in $GG(V)$. Since the Gabriel graph of a point set is a planar geometric graph [11] and G is a triangulation, it follows that $GG(V)$ coincides with Γ.   □

It is immediate to verify that every embedded planar triangulation with a separating three-cycle does not have a Gabriel drawing. Therefore, Lemma 1 implies the following.

Corollary 1. *There exist embedded planar graphs that do not have an embedding preserving $(0, \varepsilon_2)$-Gabriel drawing, for any $\varepsilon_2 \geq 0$.*

The proof of the next lemma is omitted. It can be established using a similar argument to the one in the proof of Lemma 1, but focusing on pairs of non-adjacent vertices. The key observation is that, for any embedding preserving $(\varepsilon_1, 0)$-Gabriel drawing Γ of an embedded tree T, any edge of the Gabriel graph $GG(V)$ of the points in the vertex set V of Γ must also be an edge of Γ. But this implies, since $GG(V)$ is connected [11], that $GG(V)$ coincides with Γ.

Lemma 2. *Let T be an embedded tree and let ε_1 be any given real number such that $\varepsilon_1 \geq 0$. T has an embedding preserving $(\varepsilon_1, 0)$-Gabriel drawing if and only if T has an embedding preserving Gabriel drawing.*

Lemma 2 and the characterization of which trees admit a Gabriel drawing in [4] immediately imply the following.

Corollary 2. *There exist embedded planar graphs that do not have an embedding preserving $(\varepsilon_1, 0)$-Gabriel drawing, for any $\varepsilon_1 \geq 0$.*

Motivated by Corollaries 1 and 2, we move our attention to $(\varepsilon_1, \varepsilon_2)$-Gabriel drawings where both $\varepsilon_1 > 0$ and $\varepsilon_2 > 0$. We prove that one can compute a drawing that approximates a Gabriel drawing for (almost) every planar graph, provided that the Gabriel region is scaled down for the edges and is scaled up for the non-adjacent pairs of vertices by any arbitrarily small chosen amount. Note, however, that for small values of ε_2 any $(\varepsilon_1, \varepsilon_2)$-Gabriel drawing must contain at least one edge, namely between a pair of vertices with minimum distance.

Lemma 3. *Let $\varepsilon_1, \varepsilon_2$ be any two real numbers such that $\varepsilon_1 > 0$ and $\varepsilon_2 > 0$. Every embedded planar graph with at least one edge has an embedding preserving $(\varepsilon_1, \varepsilon_2)$-Gabriel drawing.*

Proof. Let G be a planar graph with a given planar embedding. Consider some maximal planar supergraph G' of G together with a planar embedding of G' that respects the given embedding of G. We choose G' in such a way that there is a canonical ordering $v_1, \ldots, v_n$ of the vertices of G' so that (v_1, v_2) is an edge of G. Let G_i be the subgraph of G induced by $V_i = \{v_1, v_2, \ldots, v_i\}$. We show how to construct a drawing Γ_i of G_i by induction so that, for all $i \geq 2$, (a) Γ_i is an embedding preserving $(\varepsilon_1, \varepsilon_2)$-Gabriel drawing of G_i; (b) all vertices in V_i that lie on the outer face of Γ_i are horizontally visible from the right; and (c) vertices $v_2, v_3, \ldots, v_i$ have y-coordinates $n, n-1, \ldots, n-i+2$. Clearly we can satisfy these properties for $i = 2$, since (v_1, v_2) is an edge of G, by drawing v_1 and v_2 at points $(0, 1)$ and $(0, n)$, respectively.

Next, assuming we have Γ_i for some $i \geq 2$, we show how to construct Γ_{i+1}. We will place vertex v_{i+1} at y-coordinate $n - i + 1$ far enough to the right so that for every $v_j, v_k \in V_i$, (i) an edge from v_{i+1} to v_j is permitted by the ε_1-shrunk Gabriel disk

$D(v_{i+1}, v_j)$ (i.e., the shrunken disk is empty); (ii) if (v_{i+1}, v_j) is *not* an edge in G then the ε_2-expanded Gabriel disk $D(v_{i+1}, v_j)$ prevents the edge (i.e., the expanded disk contains a vertex); and (iii) v_{i+1} does not lie in the ε_1-shrunk Gabriel disk $D(v_j, v_k)$.

Let D be the smallest disk centered on y-coordinate $n - i + 1$ that encloses Γ_i. Let c be the center of D and r be the radius of D. Let ℓr be the (still to be determined) distance of v_{i+1} from the rightmost point of D. We choose ℓ so that for every $p \in D$, if C is the disk with diameter $\overline{pv_{i+1}}$ (in fact, C can be any disk with chord $\overline{pv_{i+1}}$), (I) the ε_1-shrunk C does not intersect D (implying Property (i)), and (II) the ε_2-expanded C contains D (implying Property (ii)). Let b be the center of C. Since b is on the perpendicular bisector of $\overline{pv_{i+1}}$, $d(b, p) \geq (\ell/2)r$. Refer to Fig. 2.

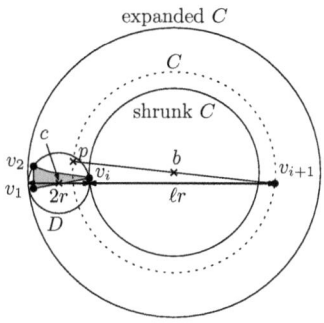

Fig. 2. Proof of Lemma 3

Property (I) is equivalent to $\frac{d(b,p)}{d(b,c)-r} < 1 + \varepsilon_1$.
By the triangle inequality, $d(b, p) \leq d(b, c) + d(c, p) \leq d(b, c) + r$. Thus, $\frac{d(b,p)}{d(b,c)-r} \leq \frac{d(b,p)}{d(b,p)-2r}$
$= 1 + \frac{2r}{d(b,p)-2r} \leq 1 + \frac{2r}{(\ell/2)r-2r} = 1 + \frac{4}{\ell-4}$.
Property (II) is equivalent to $\frac{d(b,c)+r}{d(b,p)} < 1 + \varepsilon_2$.
By the triangle inequality, $d(b, c) \leq d(b, p) + d(p, c) \leq d(b, p) + r$. Thus, $\frac{d(b,c)+r}{d(b,p)} \leq \frac{d(b,p)+2r}{d(b,p)}$
$= 1 + \frac{2r}{d(b,p)} \leq 1 + 4/\ell$. If we choose ℓ large enough so that $4/(\ell - 4) < \varepsilon_1$ and $4/\ell < \varepsilon_2$ then we satisfy both Properties (i) and (ii). Property (iii) is immediate since the ε_1-shrunk disks $D(v_j, v_k)$ for edges (v_j, v_k) with $v_j, v_k \in V_i$ are contained in D and v_{i+1} lies outside D. In addition, we choose ℓ large enough so that no edge from v_{i+1} to v_j, $j \leq i$, crosses any (already drawn) edge in Γ_i. Thus, we ensure that Γ_{i+1} respects the given embedding and is a $(\varepsilon_1, \varepsilon_2)$-Gabriel drawing of G_{i+1}. □

The results in this section can be summarized as follows.

Theorem 1. *Let G be an embedded planar graph with at least one edge. For any given values of $\varepsilon_1, \varepsilon_2$ such that $\varepsilon_1 > 0$ and $\varepsilon_2 > 0$, G admits an embedding preserving $(\varepsilon_1, \varepsilon_2)$-Gabriel. Also, there exist embedded planar graphs that do not have an embedding preserving $(0, \varepsilon_2)$-Gabriel drawing and embedded planar graphs that do not have an embedding preserving $(\varepsilon_1, 0)$-Gabriel drawing.*

Theorem 1 naturally gives rise to two research directions. One is about extending the set of proximity regions that make it possible to compute $(\varepsilon_1, \varepsilon_2)$-proximity drawings for all planar graphs and for any arbitrarily small positive values of ε_1 and ε_2. The second is about studying subfamilies of planar graphs that admit an $(\varepsilon_1, \varepsilon_2)$-Gabriel drawing in the extremal cases that either $\varepsilon_1 = 0$ or $\varepsilon_2 = 0$. The next two sections study these questions; as for the extremal case, we shall focus on $(0, \varepsilon_2)$-Gabriel drawings because, as explained in the introduction, they generalize the notion of weak Gabriel drawings.

3 $(0, \varepsilon_2)$-Gabriel Drawings

This section studies $(0, \varepsilon_2)$-Gabriel drawings. Observe that this family of approximate proximity drawings generalizes *weak Gabriel drawings*, which are equivalent to $(0, \infty)$-Gabriel drawings. Di Battista et al. [7] proved that all biconnected outerplanar graphs and all trees have a $(0, \infty)$-Gabriel drawing, while Bose et al. [4] proved that not all trees have a $(0, 0)$-Gabriel drawing. The next two lemmas and Theorem 2 establish a tight threshold value for ε_2 for the $(0, \varepsilon_2)$-Gabriel drawability of embedded trees.

Lemma 4. *For any real number $\varepsilon_2 < 2$, there exists a tree that does not admit a $(0, \varepsilon_2)$-Gabriel drawing.*

Proof. Let $0 \le \varepsilon_2 < 2$ be a real number. Consider the star tree S_d with central vertex v of degree d. We show that if d is sufficiently large then S_d has no $(0, \varepsilon_2)$-Gabriel drawing. To this end, consider an arbitrary drawing Γ of S_d and assume for a contradiction that Γ is a $(0, \varepsilon_2)$-Gabriel drawing. Select two distinct leaves u and w of S_d such that in Γ the angle α between $\overline{uv}$ and $\overline{vw}$ is minimal. Note that, for d sufficiently large, we have $\alpha < \pi/4$.

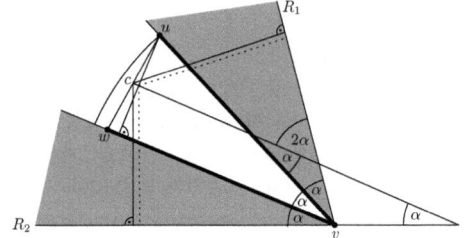

Fig. 3. Proof of Lemma 4

We assume without loss of generality that $d(v, w) \le d(u, v) = 1$ holds. Let c denote the midpoint of $\overline{uw}$. The situation is depicted in Fig. 3. Note that, since $D(u, v)$ does not contain any vertices other than u and v, we have $d(v, w) \ge \cos \alpha$. This implies $\sin \alpha \le d(u, w) \le 2 \sin \frac{\alpha}{2}$.

Since α is minimal, the shaded area in Fig. 3, that is, the wedges with apex v and aperture angle α adjacent to the wedge defined by u, v and w, cannot contain any vertex in their interior. Hence, to obtain a lower bound on the minimum value by which $D(u, w)$ must be expanded to contain a vertex other than u and w, it suffices to consider the minimum of $\frac{2d(c,v)}{d(u,w)}$, $\frac{2d_1}{d(u,w)}$, and $\frac{2d_2}{d(u,w)}$, where d_1 and d_2 denote the distance of c from the rays R_1 and R_2, respectively (see Fig. 3).

Now, we have $\frac{2d(c,v)}{d(u,w)} \ge \frac{\cos \alpha}{\sin(\alpha/2)}$ which tends to $+\infty$ as d tends to $+\infty$ and, thus, α tends to 0. It also follows, using simple geometric arguments, that

$$\frac{2d_1}{d(u, w)} \ge \left(\frac{\cos \alpha}{2} + \frac{1}{4 \cos \alpha} \right) \frac{\sin(2\alpha)}{\sin(\alpha/2)} \quad \text{and}$$

$$\frac{2d_2}{d(u, w)} \ge \left(\frac{\cos \alpha}{2} + \frac{1}{4 \cos \alpha} + \frac{\cos(2\alpha)}{4 \cos \alpha} + \frac{\sin(2\alpha)}{4 \sin \alpha} \right) \frac{\sin \alpha}{\sin(\alpha/2)}$$

hold. It is routine to check that the right hand sight in both inequalities above tends to 3 as d tends to $+\infty$. But this implies that, for sufficiently large d, the ε_2-expanded disk $D(u, w)$ does not contain any vertices other than u and w, a contradiction.   □

Lemma 5. *Let T be an embedded tree. Then T admits an embedding preserving $(0, \varepsilon_2)$ -Gabriel drawing for any real number $\varepsilon_2 \geq 2$.*

Proof. (Sketch) Root T at an arbitrary vertex t. First, draw t at an arbitrary point. In general, let W and $E(W)$ denote the set of vertices and edges, respectively, already drawn. In a single step of our algorithm we consider an arbitrary $v \in W$ such that the set U of its children are all undrawn. If no such v exists, we will have drawn T.

Let r be the minimum of $\frac{1}{2} \min\{d(v, w) : w \in W \setminus \{v\}\}$ and $\frac{1}{2} \min\{d(v, D(w, w')) :$ $(w, w') \in E(W)$ and $w, w' \neq v\}$, where $d(v, D)$ is the distance from v to disk D (if v is t then set $r = 1$). Draw the vertices in U equally spaced on the semicircle of radius r centered at v whose base side is perpendicular to the line segment between v and v's parent, so that the given embedding of T is maintained (cf. Fig. 4(a) where the semicircle is drawn shaded).

Note that, for all $u \in U$, (a) the disk $D(v, u)$ is empty, (b) the ε_2-expanded disk of $D(u, w)$ contains v for all $w \in W \setminus \{v\}$, (c) the edge (v, u) does not cross any edge in $E(W)$, and (d) for every edge $(w, w') \in$ $E(W)$, $D(w, w')$ does not contain u.

It remains to show that, for every pair $u, u' \in U, u \neq$ u', the ε_2-expanded disk of $D(u, u')$, denoted in the following by D', contains a vertex in $W \cup U \setminus \{u, u'\}$. If u and u' are not consecutive on the semicircle then a vertex in U between them lies in D'. Otherwise, let $\alpha = \angle uvu'$. If $\alpha \geq \pi/4$ then v is in D'. If $\alpha \leq \pi/5$ then the vertex $u'' \in U$ that follows u' is in D (see Fig. 4(b)).

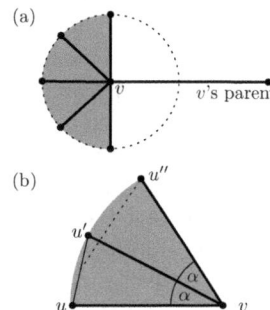

Fig. 4. Proof of Lemma 5

Lemmas 4 and 5 can be summarized as follows.

Theorem 2. *Every embedded tree has an embedding preserving $(0, \varepsilon_2)$-Gabriel drawing for any given value of ε_2 such that $\varepsilon_2 \geq 2$. Also, for each value of ε_2 such that $0 \leq \varepsilon_2 < 2$, there exists a tree T such that T does not have a $(0, \varepsilon_2)$-Gabriel drawing.*

We now consider outerplanar graphs with cycles. Lenhart and Liotta [12] proved that all biconnected outerplanar graphs with a given outerplanar embedding have a $(0, 0)$-Gabriel drawing that maintains the embedding, while a connected outerplanar graph where a cut vertex is shared by more that four biconnected components is not $(0, 0)$-Gabriel drawable. The next theorem shows that this upper bound on the number of components sharing a cutvertex can be removed in $(0, \varepsilon_2)$-Gabriel drawings, provided that the input graph does not have any degree-one vertices. In the statement, by embedded outerplanar graph we mean an outerplanar graph with a planar embedding where all vertices are on the external face. The proof is omitted.

Theorem 3. *Let G be an embedded outerplanar graph that does not have vertices of degree one or zero. G has a $(0, \varepsilon_2)$-Gabriel drawing that maintains the embedding for any given value of ε_2 such that $\varepsilon_2 > 0$.*

4   Approximate β-Drawings and Delaunay Drawings

In this section we extend Theorem 1 to other families of $(\varepsilon_1, \varepsilon_2)$-proximity drawings. Subsection 4.1 studies an infinite family of $(\varepsilon_1, \varepsilon_2)$-proximity drawings that includes the $(\varepsilon_1, \varepsilon_2)$-Gabriel drawings as a special case. Subsection 4.2 introduces and studies approximations of the Delaunay drawings.

4.1 $(\varepsilon_1, \varepsilon_2)$-$\beta$-Drawings

Let $\varepsilon_1, \varepsilon_2$ be any two non-negative numbers and let β be any real number such that $\beta \geq 1$. Let Γ be a planar straight-line drawing of a graph and let u, v be any two vertices of Γ. The β-region of influence of u and v, denoted as $\beta(u, v)$, is the intersection of two disks D_u and D_v such that: (i) both D_u and D_v have the center along the line through u, v; (ii) both D_u and D_v have radius $\frac{\beta d(u,v)}{2}$, where $d(u, v)$ is the Euclidean distance between u and v; D_u contains v and D_v contains u; and (iii) the circumference of D_u contains u and the circumference of D_v contains v. The ε_1-*shrunk β-region of influence* of u and v is defined as the intersection of the ε_1-shrunk disk of D_u with the ε_1-shrunk disk of D_v. Similarly, the ε_2-*expanded β-region of influence* of u and v is the intersection of the ε_2-expanded disks of D_u and D_v.

We say that Γ is an $(\varepsilon_1, \varepsilon_2)$-$\beta$-drawing if: (i) for every edge (u, v) of Γ the ε_1-shrunk β-region of influence of u and v is empty; and (ii) for every pair of non-adjacent vertices u, v of Γ, the ε_2-expanded β-region of influence of u and v is not empty.

Not all embedded planar graphs have a $(0, 0)$-β-drawing [4]. Also, by definition, an $(\varepsilon_1, \varepsilon_2)$-$\beta$-drawing with $\beta = 1$ is a $(\varepsilon_1, \varepsilon_2)$-Gabriel drawing. Hence, by Corollaries 1 and 2, it follows that not all embedded planar graphs admit an $(\varepsilon_1, \varepsilon_2)$-$\beta$-drawing that respects the given embedding, when either ε_1 or ε_2 is set to 0. On the other hand, we can extend Lemma 3 to all values of $\beta > 1$. The proof technique is similar to the one in Lemma 3. Therefore, the proof is omitted.

Lemma 6. *Let $\varepsilon_1, \varepsilon_2$ be any two real numbers such that $\varepsilon_1 > 0$ and $\varepsilon_2 > 0$ and let β be any real number such that $\beta \geq 1$. Every embedded planar graph with at least one edge has a $(\varepsilon_1, \varepsilon_2)$-$\beta$-drawing that maintains the given embedding.*

We can summarize the discussion of this section as follows.

Theorem 4. *Let G be an embedded planar graph with at least one edge. For any given values of $\varepsilon_1, \varepsilon_2$ such that $\varepsilon_1 > 0$ and $\varepsilon_2 > 0$ and for any value of β such that $\beta \geq 1$, G admits an embedded $(\varepsilon_1, \varepsilon_2)$-$\beta$-drawing. Also, there exist embedded planar graphs that do not have a $(0, \varepsilon_2)$-β-drawing and planar graphs that do not have a $(\varepsilon_1, 0)$-β-drawing that maintain the given embedding.*

4.2 $(\varepsilon_1, \varepsilon_2)$-Delaunay drawings

Let Γ be a planar straight-line drawing of a graph and let $\varepsilon_1, \varepsilon_2$ be any two non-negative numbers. Let u, v be any two vertices of Γ and let $\mathcal{D}(u, v)$ be the set of all disks in the plane that have $\overline{uv}$ as a chord. Let $\mathcal{D}_{\varepsilon_1}(u, v)$ be the set of the ε_1-shrunk disks of

$\mathcal{D}(u, v)$ and let $\mathcal{D}_{\varepsilon_2}(u, v)$ be the set of the ε_2-expanded disks of $\mathcal{D}(u, v)$. The drawing Γ is an $(\varepsilon_1, \varepsilon_2)$-*Delaunay drawing* if: (i) for any two adjacent vertices u, v of Γ, there exists at least one empty disk in $\mathcal{D}_{\varepsilon_1}(u, v)$; and (ii) for any two non-adjacent vertices u, v of Γ, all disks of $\mathcal{D}_{\varepsilon_2}(u, v)$ contain some vertex of Γ other than u and v. Note that a Delaunay drawing is a special case of $(\varepsilon_1, \varepsilon_2)$-Delaunay drawings, namely the one in which $\varepsilon_1 = \varepsilon_2 = 0$. Fig. 5 is an example of an $(\varepsilon_1, \varepsilon_2)$-Delaunay drawing for $\varepsilon_1 = 0.25$ and $\varepsilon_2 = 0.2$. In this figure, two Delaunay disks are described (dotted) and their corresponding ε_1-shrunk and ε_2-expanded counterparts (solid) are depicted. The graph with the planar embedding of Fig. 5 does not admit an embedding preserving Delaunay drawing [8].

Recall that, in the context of Delaunay drawings, a point set P is degenerate if either four or more co-circular points in P define a circle that does not contain another point in P in its interior, or there are three or more collinear points in P on the boundary of the convex hull of P (see, e.g. [9]). Note that, for any $(0, 0)$-Delaunay drawing Γ, if the point set P representing the vertices in Γ is non-degenerate, Γ coincides with the well-known *non-degenerate Delaunay triangulation* of P. We say that an embedded maximal planar graph G is *Delaunay drawable* if the exists a set of points such that the

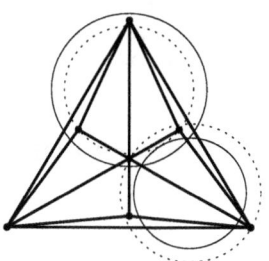

Fig. 5. An $(\varepsilon_1, \varepsilon_2)$-Delaunay drawing for $\varepsilon_1 = 0.25$ and $\varepsilon_2 = 0.2$ of a planar embedded graph that does not have an embedding preserving Delaunay drawing

Delaunay graph of this point set is an embedding preserving drawing of G. Since Delaunay triangulations are among the most studied graphs in computational geometry, we start by investigating the relationship between $(\varepsilon_1, \varepsilon_2)$-Delaunay drawability and Delaunay drawability of maximal planar graphs. The proof of the next lemma is again omitted.

Lemma 7. *There exist embedded maximal planar graphs that do not admit an embedding preserving $(\varepsilon_1, 0)$-Delaunay drawing, for any value $\varepsilon_1 \geq 0$.*

While Lemma 7 considers $(\varepsilon_1, 0)$-Delaunay drawings of maximal planar triangulations, one can wonder what happens with the other extreme, that is, with $(0, \varepsilon_2)$-Delaunay drawings. With arguments similar to those in the proof of Lemma 1 and the results in [9], the following lemma and corollary can be proved.

Lemma 8. *Let G be an embedded maximal planar triangulation and let ε_2 be any given real number such that $\varepsilon_2 \geq 0$. G has an embedding preserving $(0, \varepsilon_2)$-Delaunay drawing if and only if G is Delaunay drawable.*

Corollary 3. *There exist embedded planar graphs that do not have an embedding preserving $(0, \varepsilon_2)$-Delaunay drawing, for any $\varepsilon_2 \geq 0$.*

By using again similar arguments as those in the proof of Lemma 3 we can prove the following.

Lemma 9. *Let $\varepsilon_1, \varepsilon_2$ be any two real numbers such that $\varepsilon_1 > 0$ and $\varepsilon_2 > 0$. Every planar graph with at least one edge has a $(\varepsilon_1, \varepsilon_2)$-Delaunay drawing.*

The discussion of this section is summarized in the following theorem, which establishes that one can compute a drawing that approximates a Delaunay drawing for every planar graph, provided that the Delaunay disks are scaled down for the edges and are scaled up for the non-adjacent pairs of vertices by any arbitrarily small chosen amount.

Theorem 5. *Let G be an embedded planar graph. For any given values of $\varepsilon_1, \varepsilon_2$ such that $\varepsilon_1 > 0$ and $\varepsilon_2 > 0$, G admits an embedding preserving $(\varepsilon_1, \varepsilon_2)$-Delaunay drawing. Also, there exist embedded planar graphs that do not have an embedding preserving $(0, \varepsilon_2)$-Delaunay drawing and embedded planar graphs that do not have an embedding preserving $(\varepsilon_1, 0)$-Delaunay drawing.*

5   Conclusions and Open Problems

In this paper we have introduced an approximate version of the well-studied proximity drawings. In comparison with the standard definition of region of influence based proximity drawing, our drawings consider a slightly smaller region of influence for the adjacent pairs of vertices and a slightly larger region for the non-adjacent pairs. The amount by which the region of influence can be scaled up or down depends on two non-negative real numbers ε_1 and ε_2; the resulting straight-line drawing is called an $(\varepsilon_1, \varepsilon_2)$-proximity drawing. Intuitively, the smaller these parameters are the closer an $(\varepsilon_1, \varepsilon_2)$-proximity drawing is to the standard proximity drawing.

The paper has investigated the approximation of three well-known proximity drawings, namely Gabriel drawings, Delaunay drawings, and β-drawings. For each of these types of proximity drawings, we showed that every planar graph has a planar straight-line drawing that can be made arbitrarily close to satisfy the usual proximity rule. This contrasts with well-known results that only restricted subfamilies of planar graphs have a (standard) Gabriel drawing, or a Delaunay drawing, or a β-drawing. Also extremal cases which generalize and extend the notion of weak proximity have been investigated. A first natural direction for future research is therefore the following.

Question 1. Extend the study of approximate proximity to other classical or emerging families of proximity drawings, such as the rectangle of influence drawings and/or the witness Delaunay drawings. A good starting point for this question may be, for example [3].

We remark that the major contribution of this paper is in analyzing to what extent the class of representable graphs can vary if the standard definition of proximity is approximate in the manner described above. Based on the presented results, we believe that the proposed definition of approximate proximity may be effectively adopted in practice to represent planar graphs where proximity constraints need to be maintained. However, in order to do so, relevant questions about the area and the bit complexity of the computed drawings must be addressed.

As an example, we recall a recent paper by Angelini et al. [1] proving that drawing a tree of maximum degree five as a Euclidean minimum spanning tree may require exponential area. Since the family of β-drawable trees for $\beta = 2$ is the family of trees having maximum degree five and since a 2-drawing of a tree is also a Euclidean minimum spanning tree, it follows that $(0, 0)$-2 drawings may require exponential area. On the other hand, every straight-line planar drawing is an $(\varepsilon_1, \varepsilon_2)$-proximity drawing for a sufficiently small value of ε_1 and a sufficiently large value of ε_2. In fact, every planar graph has a $(0, \infty)$-proximity drawing with integer coordinates and in polynomial area (see, e.g., [5]). This discussion leads to the following research direction.

Question 2. Study *polynomial area approximation schemes*, that is, for any fixed ε_1, ε_2, the size of the computed drawing is bounded by a polynomial in the number of vertices of the given graph. Similar studies have been described in [10] in the context of drawing a tree as a minimum spanning tree approximation.

Acknowledgments. This research was initiated during the *Bici BWGD 2011: Bertinoro Workshop on Graph Drawing*. The authors are thankful to the workshoppers for useful discussions and also to the anonymous referees for their helpful comments. Research supported in part by MIUR of Italy under project AlgoDEEP prot. 2008TFBWL4 and NSERC of Canada, as well as by EUROGIGA project GraDR 10-EUROGIGA-OP-003.

References

1. Angelini, P., Bruckdorfer, T., Chiesa, M., Frati, F., Kaufmann, M., Squarcella, C.: On the Area Requirements of Euclidean Minimum Spanning Trees. In: Dehne, F., Iacono, J., Sack, J.-R. (eds.) WADS 2011. LNCS, vol. 6844, pp. 25–36. Springer, Heidelberg (2011)
2. Aronov, B., Dulieu, M., Hurtado, F.: Witness (Delaunay) graphs. Comput. Geom. 44(6-7), 329–344 (2011)
3. Aronov, B., Dulieu, M., Hurtado, F.: Witness Rectangle Graphs. In: Dehne, F., Iacono, J., Sack, J.-R. (eds.) WADS 2011. LNCS, vol. 6844, pp. 73–85. Springer, Heidelberg (2011)
4. Bose, P., Lenhart, W., Liotta, G.: Characterizing proximity trees. Algorithmica 16(1), 83–110 (1996)
5. de Fraysseix, H., Pach, J., Pollack, R.: How to draw a planar graph on a grid. Combinatorica 10(1), 41–51 (1990)
6. Di Battista, G., Lenhart, W., Liotta, G.: Proximity Drawability: a Survey. In: Tamassia, R., Tollis, I.G. (eds.) GD 1994. LNCS, vol. 894, pp. 328–339. Springer, Heidelberg (1995)
7. Di Battista, G., Liotta, G., Whitesides, S.: The strength of weak proximity. J. Discrete Algorithms 4(3), 384–400 (2006)
8. Dillencourt, M.B.: Realizability of Delaunay triangulations. Inf. Process. Lett. 33(6), 283–287 (1990)
9. Dillencourt, M.B., Smith, W.D.: A Simple Method for Resolving Degeneracies in Delaunay Triangulations. In: Lingas, A., Carlsson, S., Karlsson, R. (eds.) ICALP 1993. LNCS, vol. 700, pp. 177–188. Springer, Heidelberg (1993)
10. Di Giacomo, E., Didimo, W., Liotta, G., Meijer, H.: Drawing a Tree as a Minimum Spanning Tree Approximation. In: Cheong, O., Chwa, K.-Y., Park, K. (eds.) ISAAC 2010, Part II. LNCS, vol. 6507, pp. 61–72. Springer, Heidelberg (2010)

11. Jaromczyk, J.W., Toussaint, G.T.: Relative neighborhood graphs and their relatives. Proc. IEEE 80(9), 1502–1517 (1992)
12. Lenhart, W., Liotta, G.: Proximity Drawings of Outerplanar Graphs. In: North, S.C. (ed.) GD 1996. LNCS, vol. 1190, pp. 286–302. Springer, Heidelberg (1997)
13. Li, X.: Applications of computational geometry in wireless networks. In: Cheng, X., Huang, X., Du, D.-Z. (eds.) Ad Hoc Wireless Networking, pp. 197–264. Kluwer Academic Publishers (2004)
14. Liotta, G.: Proximity drawings. In: Tamassia, R. (ed.) Handbook of Graph Drawing and Visualization. CRC Press (to appear)

Generalizing Geometric Graphs*

Edith Brunel, Andreas Gemsa, Marcus Krug, Ignaz Rutter, and Dorothea Wagner

Faculty of Informatics, Karlsruhe Institute of Technology (KIT), Germany
`firstname.lastname@kit.edu`

Abstract. Network visualization is essential for understanding the data obtained from huge real-world networks such as flight-networks, the AS-network or social networks. Although we can compute layouts for these networks reasonably fast, even the most recent display media are not capable of displaying these layouts in an adequate way. Moreover, the human viewer may be overwhelmed by the displayed level of detail. The increasing amount of data therefore requires techniques aiming at a sensible reduction of the visual complexity of huge layouts.

We consider the problem of computing a generalization of a given layout reducing the complexity of the drawing to an amount that can be displayed without clutter and handled by a human viewer. We take a first step at formulating graph generalization within a mathematical model and we consider the resulting problems from an algorithmic point of view. Although these problems are NP-hard in general, we provide efficient approximation algorithms as well as efficient and effective heuristics. At the end of the paper we showcase some sample generalizations.

1 Introduction

As a natural consequence of the increasing amount of available data we are facing large and even huge networks such as road and flight networks, the AS-network and social networks with millions of vertices. Visualization of these networks is a key to assessing the inherent graph-based information. There are several methods for computing layouts of huge graphs with millions of vertices within a few minutes [21,23,19].

But, how do we display such layouts? Modern HD displays feature roughly 2 Mio pixels and a standard A4 page allows roughly 8.7 Mio dots at a resolution of 300 pixels per inch. Even if we require only a minimal distance of 10 pixels or dots between the vertices of the graph, then we can display only several thousand vertices, and not too many edges. If we additionally seek to display graph structure and keep visual clutter low, the number of vertices we can display degrades even further.

Even worse, the human perception is not capable of extracting detailed information from huge layouts with millions of vertices. Since, by a simple counting argument, there are incompressible adjacency matrices, a graph with only 1 Mio vertices may encode incompressible information of up to 125 Gigabytes. This exceeds by a factor of 3.6 the average daily information consumption of an American, estimated at 34 (highly compressible) Gigabytes [4]. It is thus apparent that even an æsthetically pleasing layout of a huge graph may not be suited for displaying information to a human viewer.

---

* Research was partially supported by EUROGIGA project GraDR 10-EuroGIGA-OP-003.

M. van Kreveld and B. Speckmann (Eds.): GD 2011, LNCS 7034, pp. 179–190, 2012.

Related Work. Known approaches to coping with the huge amount of data by allowing for some kind of abstraction can be categorized into *structural* and *geometric* methods. While structural methods create a new layout for the data typically using a clustering of the graph, geometric methods are applied to a given layout maintaining the user's mental map [26].

Eades and Feng [13] describe a multilevel visualization method for clustered graphs. A force-directed layout algorithm based on a hierarchical decomposition of the graph is given by Quigley and Eades [28]. This method allows for visualizing the graph at different levels of abstraction. Abello el al. [2] discuss graph sketches for very large graphs based on mapping clusters of the graph to certain regions of the screen.

Fisheye visualizations [15,30], on the other hand, apply a distortion to a given layout to emphasize the layout in the area of interest. The resolution of the drawing deteriorates towards the boundary of the drawing and parts of the drawing in this area are usually densely cluttered. Abello et al. [1] study the visualization of large graphs with compound-fisheye views and treemaps, employing hierarchical clustering and a treemap representation of this clustering. Edge Bundling techniques [31,22] aim at reducing the complexity of layouts by bundling similar edges.

Generalization has received considerable attention in cartography [25]. Mackaness and Bear [24] highlight the potential of graph theory for map generalization. Saalfeld states the map generalization problem as a straight-line graph drawing problem [29] and formulates a number of challenges resulting from this perspective. Among others, he asks for a rigorous mathematical model for graph-based generalizations and provable guarantees. We are not aware of any work aiming at assessing this problem in general.

Our Contribution. We take a first step towards establishing a mathematical model for the problem of generalizing geometric graphs. Our model is based on the fact that vertices have a fixed size and edges have a fixed width on the screen. *Visual clutter* refers to an agglomeration of overlapping visual features in a limited area resulting in indistinguishable features. Our goal is to either avoid or reduce visual clutter. We identify three types of clutter. *Vertex-Clutter* occurs when two or more vertices are too close to each other. It may render the drawing unusable due to hidden edge information; see Fig. 1. *Edge-Clutter* occurs when too many edges cross a limited area. Even if vertices are far enough apart, edge clutter may lead to indistinguishable edge information; see Fig. 2. *Vertex-Edge-Clutter* occurs when a vertex is too close to an edge. In this case, we are unable to tell, whether the vertex is incident to the edge or not; see Fig. 3.

We devise a framework that allows for assessing all types of clutter in an incremental way by modeling the elimination or reduction of each type of clutter as an optimization problem, which we analyze in terms of complexity. We show that these problems are NP-hard in general and we provide approximation algorithms as well as effective and efficient heuristics that can be applied to huge graphs within reasonable time.

Preliminaries. A *geometric graph* is a pair $G = (P, E)$ such that $P \subseteq \mathbb{R}^2$ is a finite set of n points in the plane and E is a set of m straight-line segments with endpoints in P. If not otherwise stated, graph refers to a geometric graph throughout this paper. For $p \in P$ and a non-negative number $r \in \mathbb{R}_0^+$, we denote by $B(p, r)$ the disk with center p and radius r. We model the finite resolution of a screen by assuming that each point p occupies the

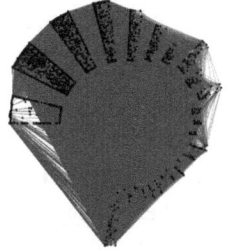
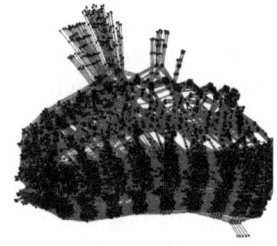

Fig. 1. Vertex-Clutter   **Fig. 2.** Edge-Clutter   **Fig. 3.** Vertex-Edge-Clutter

locus of points whose distance to p is bounded by $s \in \mathbb{R}_0^+$ and, similarly, each edge e occupies the locus of points whose distance to e is bounded by $w \in \mathbb{R}_0^+$.

A *generalization* of G is a pair (H, φ) where $H = (Q, F)$ is a geometric graph with $Q \subseteq P$ such that $\varphi: P \to Q$ maps vertices of G to vertices of H and F is a subset of edges resulting from a contraction of G according to φ. Since the subgraph induced by $\varphi^{-1}(q)$ is contracted into a single vertex, we call this subgraph the *cluster* of q, denoted by C_q. Given $Q \subseteq P$, we denote by $\nu: P \to Q$ the *Voronoi mapping* which maps $p \in P$ to its closest neighbor in Q with respect to the Euclidean metric. We call the corresponding clusters *Voronoi clusters*. Throughout the paper distance refers to the Euclidean metric.

Organization of the Paper. In Section 2 we consider the problem of eliminating vertex-clutter. We discuss our model for the generalization of the vertex set and show NP-hardness of the corresponding optimization problem. We further show that the size of the generalized pointset can be approximated and we devise an efficient heuristic for further optimization. In Section 3 we study the reduction of edge-clutter. We show that it is in general NP-hard to find a sparse or short subset of the edges maintaining monotone tendencies. When the original graph is complete, however, or if we are not restricted to use edges of the original graph, we can efficiently compute a sparse graph approximately representing monotone tendencies of the edges. In Section 4 we model the problem of reducing vertex-edge clutter and we show how to compute a drawing that allows for unambiguously deciding whether an edge is incident to a vertex or not, thus effectively eliminating vertex-edge clutter. We showcase some sample generalizations and conclude with a short discussion as well as open problems in Section 5. We omit some details due to space constraints; for full proofs and additional sample generalizations we refer the reader to the long version of this paper [6].

2 Generalizing the Vertex Set without Vertex-Clutter

In this section we consider the problem of computing a generalization (H, φ) without vertex clutter for a geometric graph $G = (V, E)$, where $H = (Q, F)$. We focus on the case that φ is the Voronoi-mapping assigning each vertex in P to its nearest neighbor in Q. In order to avoid vertex-clutter we require a minimal distance $r \in \mathbb{R}_0^+$ between the vertices of a generalized geometric graph. Hence for each vertex $p \in Q$ in the generalized graph the disk $B(p, \rho(p))$ with $\rho: P \to \mathbb{R}_0^+$ and $\rho(p) \geq r$ may not contain any other point from Q. We call a pointset Q with this property a ρ-*set of P*. This prerequisite, however, must be balanced with additional quality measures such as the size of the ρ-set, the clustering induced by φ and the distribution of the points in Q in order to avoid trivial solutions such as a single vertex.

Choosing $\rho \equiv r$ uniformly for all points $p \in P$ may have a severe effect on the distribution of the points when maximizing the size of a ρ-set, since the distances to the nearest neighbors in an inclusion-maximal ρ-set tend to be uniformly distributed regardless of the original distribution. However, it may be more appropriate to approximate the distribution of the original pointset. In order to approximate this distribution by an inclusion-maximal ρ-set we can choose ρ as follows. Let p_0 be the point that maximizes the number of points in $B(p,r) \cap P$ over all $p \in P$ and let $k = |B(p_0,r) \cap P| - 1$. For each $p \in P$ let $d_k(p) \geq r$ denote p's distance to its k-nearest neighbor in P. By choosing $\rho(p) = d_k(p) \geq r$ any inclusion-maximal pointset will have approximately the same distribution as the original pointset since for each point in the generalized pointset we discarded the same amount of points from the original graph.

Since, in general, it is not clear which behavior is more appropriate, we introduce a parameter $\alpha \in [0,1]$ and let the user decide by setting $\rho(p) := \max\{r, \alpha d_k(p)\}$. That is, the user can choose between retaining as many points in areas with low clutter as possible ($\alpha = 0$) and approximating the distribution of the pointset ($\alpha = 1$) as well as interpolations between the two extremes.

We consider two measures to assess the quality of a ρ-set Q. While the size of Q is a measure of the amount of data that is retained, the quality of the clustering induced by φ is a measure for the amount of data that is lost due to the contraction of the vertices. There are several established ways of assessing the quality of clusterings, such as coverage, performance, conductance [16], and modularity [5]. We consider a measure similar to coverage, which we adapt to our purpose as follows. For each cluster C_q let n_q denote the number of vertices and m_q denote the number of edges in C_q, respectively. We define the *local coverage* of a cluster C_q by $\mathrm{lcov}(C_q) = 2m_q/(n_q(n_q - 1))$, i.e., as the amount of intra-cluster coherence that is explained by the intra-cluster edges. The local coverage of the generalization is defined as $\mathrm{lcov}(H, \varphi) = \min_{q \in Q} \mathrm{lcov}(\varphi^{-1}(q))$.

We consider the following multi-objective optimization problem. Given a geometric graph $G = (P,E)$, a non-negative radius $r \in \mathbb{R}_0^+$ and $\alpha \in [0,1]$ the LOCAL COVERAGE CLUSTER PACKING (LCCP) problem is to compute a ρ-set $Q \subseteq P$ and a mapping $\varphi \colon P \to Q$ that maximizes both $|Q|$ and $\mathrm{lcov}(H, \varphi)$.

2.1 Complexity

The problem of computing a ρ-set of maximum size for $\alpha = 0$ can be reduced to the problem of computing a maximum independent set in the intersection graph of the disks with radius $r/2$ centered at the points in P. Clark et al. [9] prove that this problem is NP-hard in unit-disk graphs, even if the disk representation of the graph is given.

Corollary 1. *Maximizing the size of a ρ-set is NP-hard for $\alpha = 0$.*

Next, we show that it is also NP-hard to maximize the local coverage in the induced clusters of a ρ-set as well as the total size of the generalization obtained by choosing a ρ-set if the clustering is obtained by the Voronoi mapping induced by the points in Q.

Theorem 1. *Maximizing* lcov(H,v) *of a generalization* (H,v) *is NP-hard for* $\alpha = 0$.

The proof is by reduction from the NP-hard problem PLANAR MONOTONE 3-SAT [11]. Given an instance of this problem we construct a geometric graph G composed of gadgets acting as variables, literals and clauses, respectively, such that for constant ρ G contains a ρ-set with local coverage 1 if and only if the corresponding planar monotone 3-sat-formula is satisfiable. A full proof can be found in [6].

2.2   Approximating the Maximum Size of a Generalization

Although it is unlikely that we can efficiently compute a ρ-set with maximum size, we show that we can approximate the size of a maximum ρ-set.

Theorem 2. *Let G be a geometric graph and let $r \in \mathbb{R}_0^+$ and $\alpha \in [0,1]$ be given. In $O(kn + n\log^5 n(\log\log n)^2)$ time we can compute a generalization $\mathcal{H}$ of G that approximates the maximum number of vertices of a generalization by a factor of $(7k+2)/3$, where $k = \max_{p\in P}|B(p,\rho(p))\cap P| - 1$.*

In order to prove Theorem 2 we use the following auxiliary lemma, whose proof can be found in [6].

Lemma 1. *Let p_0 be a point in the plane and let $k \in \mathbb{N}$. Then there are at most $6k$ points Q such that p_0 is among the k closest points for each of the points $q \in Q$.*

Proof (Proof of Theorem 2). Let H be the graph on the set of points such that pq is a (directed) edge if and only if $q \in B(p,\rho(p))$. The graph H contains an independent set of size s if and only if G contains a ρ-set of this size. Each independent set in H corresponds to a ρ-set in G since each point in H is connected to all points that are closer than $\rho(p)$ and it is connected to all points q such that p is in the $\rho(q)$ disk around q. On the other hand, each ρ-set in G induces an independent set due to this construction.

By choice of ρ, each vertex has out-degree bounded by $k = \max_{p\in P}|B(p,r)\cap P| - 1$ for any value of α. There is an ingoing edge from q into p if and only if p is among the k closest neighbors of q. By Lemma 1 there are at most $6k$ points such that p is among the closest k points for each of these points. Hence, the in-degree of each vertex is bounded by $6k$. In total, each vertex has degree at most $7k$. Hence, by a result due to Halldórsson and Radhakrishnan [20] we can approximate the maximum size of an independent set by a factor of $(7k+2)/3$. The algorithm greedily chooses the minimum degree vertex in each step and can be implemented to run in time $O(kn)$, given the graph H.

In order to compute H we locate the points in a closed disk by a circular range query in $O(\log n + k)$ time using $O(n\log^5 n(\log\log n)^2)$ preprocessing time [7]. Hence, the total running time is $O(kn + n\log^5 n(\log\log n)^2)$.   □

Based on this approximation, we heuristically compute a ρ-set Q balancing both the size of Q and the local coverage of the Voronoi clustering induced by Q. For $p \in P$ let $\widetilde{m}(p)$ be the number of edges whose endpoints are both contained in $B(p,\rho(p)/2)$ and let $\widetilde{n}(p)$ be the number of points in $B(p,\rho(p))$. We show the following.

Lemma 2. *Let Q be an inclusion-maximal ρ-set and let $\alpha = 0$. Further, let $H = (Q,F)$ be the generalization obtained from $G = (P,E)$ by the Voronoi-mapping v. Then the value $\min_{q\in Q} 2\widetilde{m}(q)/(\widetilde{n}(q)(\widetilde{n}(q)-1))$ is a lower bound for* lcov(H,v).

Proof. For $\alpha = 0$ we have $\rho \equiv r$. Whenever p is chosen as a cluster center in Q, the points in $B(p, r/2)$ are closer to p than to any other point in Q, since the closest point to p in Q has distance to p at least r. Hence, the edges in $B(p, r/2)$ are intra-cluster edges of C_p. On the other hand, the number of points in each of the clusters is bounded by $\tilde{n}(p)$ whenever $\alpha = 0$ and Q is an inclusion-maximal ρ-set. To see this, consider any vertex q that is not contained in $B(p, r)$, but closer to p than to any other cluster center. Then q is contained in none of the disks centered in the cluster centers and, thus, q must be a cluster center itself, since Q is inclusion-maximal. Hence, the claim holds.   □

Based on Lemma 2 we propose a heuristic, called GREEDY WEIGHT HEURISTIC, that operates as follows. First we compute an estimate of $2\tilde{m}(q)/(\tilde{n}(q)(\tilde{n}(q) - 1))$ for each $p \in P$. Subsequently we sort the points according to these estimates in $O(n \log n)$ time and iteratively consider the points in this order. If the current vertex is not covered by a previous vertex, then it is chosen for the ρ-set, otherwise it is discarded.

Instead of computing $\tilde{m}(p)$ and $\tilde{n}(p)$ exactly, we estimate these numbers by counting the number of vertices and edges in the bounding boxes of the disks $B(p, \rho(p)/2)$. To count the number of edges we use a 4-dimensional range searching query on a data structure containing tuples of points corresponding to edges in E with query time $O(\log^3 m)$ [8]. We use the 2-dimensional counterpart to locate points. Further, we use a data structure for dynamic nearest neighbor queries with $O(\log^2 n)$ query time [3], into which we insert the selected points to decide whether the current point is covered by a previously selected point. The total running time is $O((n + m) \log^3 m + n \log^2 n)$.

3   Minimizing Edge-Clutter

In order to reduce the clutter resulting from an excess of edges in certain areas we must filter out some of the edges without destroying the visual appearance of the graph. The total length of the edges seems to be a good measure for the clutteredness of the graph since it is proportional to the ink used for the drawing. While a minimum spanning tree will minimize this quantity, it is unlikely to preserve the visual appearance of the graph. We therefore require that monotone tendencies of the edges are preserved in order to best maintain the mental map of the adjacencies between vertices of the graph.

Let ℓ be a line in the plane and let $S = (p_1, \ldots, p_k)$ be a sequence of points. We say that S is ℓ-*monotone* if the order of the orthogonal projections of $p_1, \ldots, p_k$ onto ℓ is the same as the order of the points in S. Let $G = (P, E)$ be a geometric graph and let (H, φ) be a generalization of G such that $H = (P, F)$, i.e., $F \subseteq E$. We say that H is a *monotone generalization of* G if for every edge $e \in E$ with endpoints p and q there is a p-q-path π_e in H such that π_e is ℓ_e-monotone, where ℓ_e is the line defined by the endpoints of e. Given $G = (P, E)$ the SHORTEST GEODESIC SUBGRAPH (SGS) problem asks for a monotone generalization H of G minimizing the total length of H.

Theorem 3. SHORTEST GEODESIC SUBGRAPH *is NP-hard.*

Proof. We reduce from monotone 3-SAT, an NP-complete variant of 3-SAT where each clause contains either only positive or only negative literals [17]. Let φ be an instance of monotone 3-SAT with variables $x_1, \ldots, x_n$ and clauses $C_1, \ldots, C_m$. We construct the

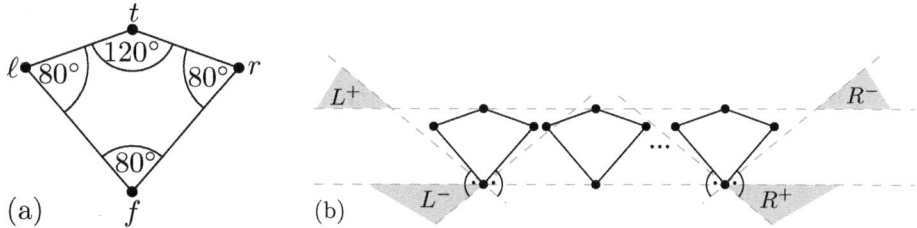

Fig. 4. Overview of the reduction from 3-SAT to SHORTEST GEODESIC SUBGRAPH. A kite with foot point f, top point t and left and right points r and ℓ (a), and the arrangement of the kites in the reduction with the corresponding regions for clause vertices (b).

following instance G_φ of SHORTEST GEODESIC SUBGRAPH. For each variable x we create a kite as shown in Figure 4. Note that the angles at f, ℓ and r are strictly less than $90°$, and the angle at t is strictly greater than $90°$. The two edges incident to the top vertex t are the *top edges*, the edges $f\ell$ and fr are the left and right *side edges*, respectively. We place the kites so that their foot points f lie evenly spaced on the x-axis and the kites are disjoint. The region R^+ (resp. L^+) is the region below the x-axis (resp. above the horizontal line defined by the top points of the kites) and to the right (resp. left) of the line through the bottom point of the rightmost (resp. leftmost) kite that is perpendicular to the right sides of the kites. We define R^- and L^- analogously, but with lines orthogonal to the left sides of the kites.

It follows immediately from the construction that a path that is monotone in the direction from a point in R^+ to a point in L^+ may not contain any right edge of a kite as this would mean a turn of more than $90°$, which is not monotone. Analogously, monotone paths from L^- to R^- may not contain left edges of kites. In our reduction the kites will act as variables, and edges from R^+ to L^+ (from L^- to R^-) will act as clauses with only positive (only negative) literals.

For each clause C_i consisting of only positive literals, we add a *clause vertex* c_i^1 into R^+ and a clause vertex c_i^2 in L^+. We add *connector edges* that connect c_i^1 to the foot points of all kites that correspond to variables that occur in C_i and that connect c_i^2 to all the left points of kites that correspond to variables that occur in C_i. Finally, we add the *clause edge* $c_i^1 c_i^2$. We treat the clauses consisting of only negative literals analogously, except that we place the new vertices in L^- and R^-, respectively, and we connect the new vertices in R^- to the right kite points instead to the left.

This completes our construction, and we claim that an optimal solution of this instance allows us to decide whether the initial formula φ was satisfiable. We will make this more precise in the following. A subset of edges of G_φ is called *tight* if it contains both top edges of each kite, all connector edges, and exactly one of the two side edges of each kite. The proof relies on two claims; full proofs are in [6].

Claim. Any feasible solution contains a tight edge set.
Claim. There exists a tight set that is feasible if and only if φ is satisfiable.

Note that the total length L is the same for all tight edge sets. The first claim shows that any geodesic subgraph has length at least L. And thus, the second claim implies that φ

is satisfiable if and only if G_φ admits a geodesic subgraph of length at most L. Since the construction can easily be performed in polynomial time this concludes the proof.   □

As we have seen, the restriction to edges from the input graph makes it difficult to construct short monotone subgraphs. One possibility is thus to drop this constraint and to allow arbitrary edges. Additionally, we would like to control the distance of the monotone path π_e and the edge it is approximating in terms of monotonicity. This is motivated by the observation that the shortest monotone generalization of a clique whose vertices are arranged equidistantly on a circle is the convex hull of the pointset. Given a line segment s with length ℓ_s and a point p with distance d_p from s we call the ratio d_p/ℓ_s the *drift of p from s*. The *drift of a path* π_e with endpoints pq is defined as the maximum drift of any point on π_e from the segment pq. Given a geometric graph $G = (P,E)$ and a non-negative real number $\delta \in \mathbb{R}_0^+$ the SPARSE GEODESIC NETWORK (SGN) problem asks for a geometric graph $H = (P,F)$ with minimum total length such that for each edge e in E there is an ℓ_e-monotone path π_e with drift at most δ, where ℓ_e denotes the line defined by the endpoints of e. We show the following.

Lemma 3. *Given a (complete) geometric graph* $G = (P,E)$, *the Delaunay graph* $\mathscr{D}(P)$ *contains for each edge* $e \in E$ *an* ℓ_e-*monotone path* π_e *with drift at most* $1/2$.

Proof. Let P be a set of points and let $p,q \in P$. Without loss of generality we assume that p and q are on the x-axis such that $x(p) < x(q)$. According to Dobkin et al. [12] we can construct an x-monotone path in the Delaunay graph $\mathscr{D}(P)$ of P as follows. Let $\mathscr{V}(P)$ denote the Voronoi diagram of P and let $p_1,\ldots,p_k$ be the ordered points corresponding to the Voronoi cells that are traversed when following the line from p to q. Then the path $p,p_1,\ldots,p_k,q$ is an x-monotone path in the Delaunay graph. Further, all points p_i are contained within the disk with radius $d(p,q)/2$ centered in the midpoint of the segment pq. Hence, the drift is at most $1/2$.   □

Although the Delaunay graph seems to be well suited to represent monotone tendencies, this result also shows the limitations of allowing arbitrary edges. In the following we therefore focus on subgraphs of the original graph and describe a greedy heuristic for computing a monotone generalization with bounded drift δ and short total length, which we call MONOTONE DRIFT HEURISTIC. Given a geometric graph $G = (P,E)$ and a maximal drift δ we sort the edges of G with respect to increasing length in $O(m\log m)$ time. Then we consider the edges $e_1,\ldots,e_m$ in this order and iteratively construct a sequence of graphs $H_0,\ldots,H_m$, where $H_0 = (P,\emptyset)$. We insert the edge e_i into H_{i-1} whenever there is no ℓ_{e_i}-monotone path with drift at most δ in H_{i-1}. This can be tested by performing a modified depth-first search exploring only monotone subpaths in $O(n+m)$ time. Hence, the total running time of this approach is $O(nm + m^2)$.

4   Vertex-Edge-Clutter

Vertex-edge-clutter is the most complicated type of clutter since it involves both vertices and edges and the selection of these features cannot be handled independently as in the previous sections. On the other hand, this type of clutter may be considered as the least annoying type of clutter. While vertex-edge clutter is caused by edges that are close to a vertex resulting in the difficulty to determine correct incidences,the human perception is rather good at determining whether a line passes a disk through the center or not. For instance, it is easy to see that the leftmost line in Fig. 5 is not incident to the vertex although it crosses the vertex. Additionally, the human perception is also good at determining whether a line has a bend or not, which is illustrated in Fig. 5.

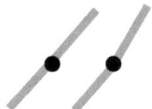

Fig. 5. Line perception

Hence, as long as there is neither vertex-clutter nor edge-clutter and as long as no pair of edges incident to a common vertex form a 180°-angle, we will be able to unambiguously tell whether an edge is incident to a vertex or not. In order to attack vertex-edge clutter we therefore propose the following optimization problem. For a pair of edges incident to a common vertex p we define the *angular straight-line deviation* as the smaller of the two angles that is enclosed by the lines defined by the two edges, respectively. The angular straight-line deviation of p is then defined as the minimum angular straight-line deviation over all pairs of edges incident to p and the angular straight-line deviation of a geometric graph G is the minimum angular straight-line deviation over all vertices of G. Note, that the angular straight-line deviation is maximized if all angles are close to a right angle. Given a geometric graph $G = (P,E)$ and a non-negative value $r \in \mathbb{R}^+$, the OPTIMAL ANGLE ADJUSTMENT problem is to find a new position for each vertex p inside $B(p,r)$ minimizing the angular straight-line deviation of the resulting geometric graph.

We tackle this problem by maximizing the vertices' distances from the lines defined by the edges incident to their neighbors. Let $G = (P,E)$ be a geometric graph and let $v \in P$ be a vertex. Let $N(v)$ denote its neighbors in G. Further, let $E(v)$ denote the edges incident to v and let $F(v)$ denote the set of edges incident to the vertices in $N(v)$ but not to v. By moving v we change the angles formed by pairs of edges in $E(v)$ as well as the angles formed by pairs of edges (e,f) such that $e \in E$ and $f \in F$, respectively. Let $L_F(v)$ be the set of lines defined by the edges in $F(v)$ and let $L_E(v)$ be the set of lines defined by all pairs of vertices in $N(v)$. Note, that there will be an angle of 180 degrees involving an edge incident to v if and only if v is placed on one of the lines in $L_E(v) \cup L_F(v)$. Given $p \in \mathbb{R}^2$ we denote by $\mu_v(p)$ the minimum distance of p to the lines in $L_E(v) \cup L_F(v)$. We prove the following.

Theorem 4. *Given a graph $G = (P,E)$, a vertex $v \in P$ and a positive radius $r \in \mathbb{R}^+$ we can compute a new position p^* for v in $B(v,r)$ such that $\mu_v(p^*) > 0$ and such that p^* maximizes $\mu_v(p)$ over all $p \in B(v,r)$ in $O(t^3\alpha(t))$ time where $t = \min\{\Delta^2,m\}$, Δ denotes the maximum degree of G and $\alpha(\cdot)$ denotes the inverse Ackermann function.*

Proof. First, we compute the set of edges $L_F(v)'$ incident to v's neighbors, but not to v, that intersect $B(v,r)$ as well as the set of lines $L_E(v)'$ defined by all pairs of v's neighbors intersecting $B(v,r)$. Let $L = L_E(v)' \cup L_F(v)'$. We compute the arrangement of lines in L in $O(|L|^2)$ time. Over each of the resulting faces C we compute the lower envelope of the hyperplanes defining the distance to the boundaries of the faces and project the graph $\mathcal{G}_C$ defined by the resulting 3-dimensional polytope onto the plane.

The lower envelope of a set of n hyperplanes can be computed in $O(n^2\alpha(n))$ time where $\alpha(\cdot)$ denotes the inverse of the Ackermann function [14]. Hence the lower envelopes can be computed in time $O(|L|^2 \alpha(|L|))$ for each face, resulting in a total complexity of $O(|L|^3 \alpha(|L|))$. For each face C we inspect the vertices of $\mathscr{G}_C$ in $B(v,r)$ as well as its intersection with $B(v,c)$ and thus compute the point p^* maximizing μ_v in $B(v,r)$. Since L is bounded by $\max\{\Delta^2, m\}$ we obtain the claimed time complexity. Further, since $r > 0$ and therefore $B(v,r)$ is non-degenerate, there must be a non-degenerate face in the arrangement containing a point p^* in its interior such that $\mu(p^*) > 0$.   □

Using Theorem 4 we can incrementally compute a new position for each vertex v such that none of the edges incident to v encloses an angle of 180 degrees with any other edge. Since the angles between pairs of edges that are not incident to v are not affected by this operation, we can iteratively apply Theorem 4 to the vertices one after another to obtain a drawing with strictly positive angular straight-line deviation. At the same time this approach heuristically maximizes this deviation.

Note that we may assume that we apply the angle adjustment to a generalized graph whose complexity tends to be significantly lower than the complexity of the original graph, i.e., both m and Δ should be considerably smaller.

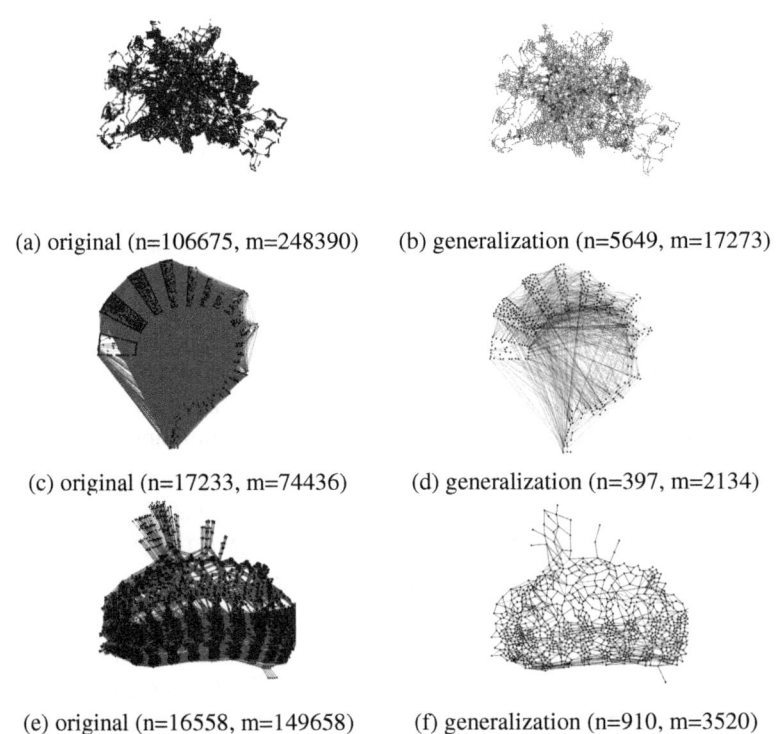

(a) original (n=106675, m=248390)   (b) generalization (n=5649, m=17273)

(c) original (n=17233, m=74436)   (d) generalization (n=397, m=2134)

(e) original (n=16558, m=149658)   (f) generalization (n=910, m=3520)

Fig. 6. Sample Generalizations computed by implementations of GREEDY WEIGHT HEURISTIC and MONOTONE DRIFT HEURISTIC: (a)–(b) OSM Streetmap Data of Berlin [27], (c)–(d) LunarVis Layout of the AS-Graph [18], (d)–(f) Netlib PDS10 problem graph from the University of Florida sparse matrix collection [10].

5   Conclusion and Open Problems

We have undertaken a first step at studying the problem of generalizing geometric graphs within a rigorous mathematical model. We formalized the problem by considering an incremental framework modeling the elimination or reduction of different types of clutter as optimization problems, which we analyzed in terms of complexity. Since these problems turned out to be NP-hard in general, we also devised efficient approximation algorithms as well as efficient heuristics. We showed how to heuristically eliminate vertex-clutter in $O((n+m)\log^3 m + n\log^2 n)$ time and how to reduce edge clutter in $O(nm + m^2)$ time considering geometric features such as point distributions and geodesic tendencies. After the elimination of vertex-clutter and edge-clutter we can expect the graph to be much smaller than the original graph. Hence, even larger complexities may scale accordingly. Thus, even the relatively high complexity of our heuristic for reducing vertex-edge clutter may be practical.

Even without this step, however, the resulting generalizations exhibit considerably less clutter and are easier to analyze. We showcase some generalizations produced by our heuristics in Figure 6 as well as in [6]. We conclude by listing some open problems.

- Is it possible to approximate both the local coverage and the size of a ρ-set in the vertex generalization step?
- What is the complexity of the LOCAL COVERAGE CLUSTER PACKING problem for different type of mappings?
- Is it possible to approximate the size of a shortest geodesic subgraph, possibly in the presence of a limited drift?
- What is the complexity of the optimal angle adjustment problem?
- How can the generalization problem be adapted to a dynamic scenario, where consistency issues play an additional role.

Acknowledgments.   We thank Robert Görke for the helpful discussion and for providing the LunarVis layout.

References

1. Abello, J., Kobourov, S.G., Yusufov, R.: Visualizing Large Graphs with Compound-Fisheye Views and Treemaps. In: Pach, J. (ed.) GD 2004. LNCS, vol. 3383, pp. 431–441. Springer, Heidelberg (2005)
2. Abello, J., Korn, J., Finocchi, I.: Graph sketches. In: Proceedings of the IEEE Symposium on Information Visualization 2001 (INFOVIS 2001), p. 67. IEEE Computer Society (2001)
3. Bentley, J.L., Saxe, J.B.: Decomposable searching problems I. static-to-dynamic transformation. Journal of Algorithms 1(4), 301–358 (1980)
4. Bohn, R.E., Short, J.E.: How much information? 2009 Report on American consumers. Global Information Industry Center, University of California, San Diego (2009)
5. Brandes, U., Delling, D., Gaertler, M., Görke, R., Hoefer, M., Nikoloski, Z., Wagner, D.: On modularity clustering. IEEE Trans. Knowledge and Data Engineering 20, 172–188 (2008)
6. Brunel, E., Gemsa, A., Krug, M., Rutter, I., Wagner, D.: Generalizing Geometric Graphs. Technical Report 27, Karlsruhe Institute of Technology (2011)

7. Chazelle, B., Cole, R., Preparata, F.P., Yap, C.: New upper bounds for neighbor searching. Information and Control 68(1-3), 105–124 (1986)
8. Chazelle, B.: Functional approach to data structures and its use in multidimensional searching. SIAM J. Comput. 17, 427–462 (1988)
9. Clark, B.N., Colbourn, C.J., Johnson, D.S.: Unit disk graphs. Discrete Mathematics 86(1-3), 165–177 (1990)
10. Davis, T.A.: University of florida sparse matrix collection. NA Digest 92 (1994)
11. de Berg, M., Khosravi, A.: Optimal Binary Space Partitions in the Plane. In: Thai, M.T., Sahni, S. (eds.) COCOON 2010. LNCS, vol. 6196, pp. 216–225. Springer, Heidelberg (2010)
12. Dobkin, D., Friedman, S., Supowit, K.: Delaunay graphs are almost as good as complete graphs. Discrete & Computational Geometry 5, 399–407 (1990)
13. Eades, P., Feng, Q.-W.: Multilevel Visualization of Clustered Graphs. In: North, S.C. (ed.) GD 1996. LNCS, vol. 1190, pp. 101–112. Springer, Heidelberg (1997)
14. Edelsbrunner, H., Guibas, L., Sharir, M.: The upper envelope of piecewise linear functions: Algorithms and applications. Discr. & Comp. Geometry 4, 311–336 (1989)
15. Furnas, G.W.: Generalized fisheye views. SIGCHI Bull. 17, 16–23 (1986)
16. Gaertler, M.: Clustering. In: Brandes, U., Erlebach, T. (eds.) Network Analysis. LNCS, vol. 3418, pp. 178–215. Springer, Heidelberg (2005)
17. Garey, M.R., Johnson, D.S.: Computers and Intractability. A Guide to the Theory of NP-Completeness. W. H. Freeman and Company (1979)
18. Görke, R., Gaertler, M., Wagner, D.: Lunarvis - Analytic Visualizations of Large Graphs. In: Hong, S.-H., Nishizeki, T., Quan, W. (eds.) GD 2007. LNCS, vol. 4875, pp. 352–364. Springer, Heidelberg (2008)
19. Hachul, S., Jünger, M.: Drawing Large Graphs with a Potential-Field-Based Multilevel Algorithm. In: Pach, J. (ed.) GD 2004. LNCS, vol. 3383, pp. 285–295. Springer, Heidelberg (2005)
20. Halldórsson, M., Radhakrishnan, J.: Greed is good: Approximating independent sets in sparse and bounded-degree graphs. Algorithmica 18, 145–163 (1997)
21. Harel, D., Koren, Y.: Graph Drawing by High-Dimensional Embedding. In: Goodrich, M.T., Kobourov, S.G. (eds.) GD 2002. LNCS, vol. 2528, pp. 207–219. Springer, Heidelberg (2002)
22. Holten, D., van Wijk, J.J.: Force-directed edge bundling for graph visualization. In: Proc. of the 11th Eurographics/IEEE-VGTC Symp. on Vis, pp. 983–990 (2009)
23. Koren, Y., Carmel, L., Harel, D.: Drawing huge graphs by algebraic multigrid optimization. Multiscale Modeling and Simulation 1, 645–673 (2003)
24. Mackaness, W.A., Beard, K.M.: Use of graph theory to support map generalization. Cartography and Geographic Information Science 20, 210–221 (1993)
25. Mackaness, W.A., Ruas, A., Sarjakoski, L.T. (eds.): Generalisation of Geographic Information. Cartographic Modelling and Applications. Elsevier B.V. (2007)
26. Misue, K., Eades, P., Lai, W., Sugiyama, K.: Layout adjustment and the mental map. Journal of Visual Languages & Computing 6(2), 183–210 (1995)
27. Openstreetmap database (2011), http://www.openstreetmap.de/
28. Quigley, A., Eades, P.: Fade: Graph Drawing, Clustering, and Visual Abstraction. In: Marks, J. (ed.) GD 2000. LNCS, vol. 1984, pp. 197–210. Springer, Heidelberg (2001)
29. Saalfeld, A.: Map Generalization as a Graph Drawing Problem. In: Tamassia, R., Tollis, I.G. (eds.) GD 1994. LNCS, vol. 894, pp. 444–451. Springer, Heidelberg (1995)
30. Sarkar, M., Brown, M.H.: Graphical fisheye views of graphs. In: Proceedings of the SIGCHI Conference on Human Factors in Computing Systems, CHI 1992, pp. 83–91. ACM, New York (1992)
31. Telea, A., Ersoy, O.: Image-based edge bundles: Simplified visualization of large graphs. Computer Graphics Forum 29(3), 843–852 (2010)

How to Visualize the K-Root Name Server (Demo)*

Giuseppe Di Battista[1], Claudio Squarcella[1], and Wolfgang Nagele[2]

[1] Dipartimento di Informatica e Automazione, Università Roma Tre, Italy
{gdb,squarcel}@dia.uniroma3.it
[2] RIPE NCC, Amsterdam, The Netherlands
wnagele@ripe.net

Abstract. We present a system that visualizes the evolution of the service provided by one of the most popular root name servers, called K-root, operated by the RIPE Network Coordination Centre (RIPE NCC) and distributed in several locations (instances) worldwide. The system can be used either to monitor what happened during a prescribed time interval or to observe the status of the service in near real-time. The system visualizes how and when the clients of K-root migrate from one instance to another, how the number of clients associated with each instance changes over time, and what are the instances that contribute to offer the service to a selected Internet Service Provider. In addition, the visualization aims at distinguishing usual from unusual operational patterns. This helps not only to improve the quality of the service but also to spot security-related issues and to investigate unexpected routing changes.

1 Introduction

A computer that needs to know the IP address which corresponds to a *domain name* sends a query to a *name server*. Hence, all Internet Server Providers (ISPs) make one or more name servers available to their customers in order to answer their requests.

A name server that receives a query executes a *resolution* process. The resolution computes an answer to the query by iteratively querying other name servers and quite often requires to send a query to special name servers called *root name servers* or simply *root servers*. For this reason root servers are a critical part of the Internet. They receive hundreds of thousands of queries per second and must answer immediately. Currently, there are 13 root servers, identified by a letter from A to M and operated by different organizations, e.g. A by VeriSign, B by USC, and C by Cogent. A name server selects its favorite root servers according to its query optimization policies.

For resiliency and efficiency reasons, each root server is implemented with computers spread across several locations distributed worldwide. Each location is an *instance*. Currently each root server has from 1 to 70 instances, e.g., A has 6 instances, F has 49, and K has 18. While a name server can freely select a root server for each of its queries, it cannot select the specific instance that will answer it. The instance is selected by a widely adopted mechanism called *anycast*, which leaves the responsibility of choosing the topologically nearest instance to the current status of the Internet routing. Hence, a

---

* Partially supported by the ESF project 10-EuroGIGA-OP-003 GraDR "Graph Drawings and Representations" and by the MIUR of Italy, under project AlgoDEEP, prot. 2008TFBWL4.

M. van Kreveld and B. Speckmann (Eds.): GD 2011, LNCS 7034, pp. 191–202, 2012.

name server of a certain provider that sends a sequence of queries to a root server, say K-root, can have that each of the queries is answered by a different instance, according to the current status of the routing. This has consequences both from the point of view of the name server and from the point of view of the root server. The first can experience fluctuations of the elapsed service time, while the latter can suffer changes in the distribution of the workload among the instances.

The purpose of this work is to visualize the evolution of the service provided by one of the most popular root servers, called K-root. It is operated by the RIPE Network Coordination Centre (RIPE NCC) that is one of five Regional Internet Registries providing Internet resource allocations, registration services and coordination activities that support the operation of the Internet globally. The visualization can be activated either to monitor what happened during a prescribed time interval or to check the status of the service in near real-time. We visualize how and when name servers (that are the clients of the root server) *migrate* from one instance to another, how the number of clients associated with each instance changes over time, and which is the status of the service offered to a certain ISP. In addition, the visualization aims at distinguishing usual from unusual operational patterns. This helps not only to improve the quality of the service but also to spot security-related issues and to investigate unexpected routing changes.

The paper is organized as follows. In Section 2 we discuss the adopted visualization metaphor. In Section 3 we present the layout algorithm used in our system. In Section 4 we give technical details and briefly address user feedbacks. In Section 5 we compare our system with the existing literature. Concluding remarks are in Section 6.

2   Selecting a Metaphor

The choice of the visualization metaphor and of the interaction features of our system comes from an intensive discussion with the RIPE NCC, aimed at collecting the most important visualization requirements for the K-root service. It comes out that the crucial need is to visualize how and when clients *migrate* from one instance to another. The concept of migration can be defined as follows. Let u, v be a pair of instances. We say that a client *migrates* from u to v during interval t', t'' ($t' < t''$) if its last request of service before time t' is asked to u and its last request of service before time t'' is asked to v. To give an idea of the migration phenomenon, in 24 hours of normal operation about $50,000$ clients issue a service request to more than one instance.

Further, migrations are not all the same. There are pairs u, v of instances such that migrating from u to v or vice versa is considered *usual* by the operators. For example, u and v are placed in network locations that have high connectivity between them, or the Internet routing frequently oscillates moving clients form u to v or vice versa. There are other migrations that are considered *unusual*, like for example those involving pairs of instances in places with very poor connectivity between them. Unusual migrations can put in evidence suspicious activities, misconfigurations, or large-scale faults. Given a pair of instances u, v, deciding if u, v is subject to usual or unusual migration is knowledge that comes from the RIPE NCC experts and is subject to change over time, with a frequency that is much lower than the one of the migrations.

An example of chart currently used by RIPE NCC to visualize the distribution of queries to the instances of the K-root service is in Fig. 1.a. Note that it does not provide any migration information, while the operators need to perceive unusual migration patterns keeping in the background usual migration behaviors. For this reason we define a *migration graph*. Its vertices are the instances and there is an edge (u, v) if migrating from u to v or vice versa is considered usual. Hence, given a pair u, v of instances subject to a migration, if (u, v) is an edge, then the migration is considered *usual*, otherwise it is considered *unusual*.

A second requirement is to visualize the relative weight of instances, giving an immediate perception of the number of clients they serve. Also, it is required to visualize how this weight is related to the migrations. More generally, there is the need to visualize the evolution of the service over time, both within a prescribed time interval, for ex-post analysis, and in near real-time. This requirement is extremely challenging, not only from the visualization perspective, but also because of the very high volumes of involved data: K-root receives about $20,000$ queries per second.

Finally, there is the need of understanding what happens to the queries issued by a specific Internet Service Provider (ISP). As an example, consider the main name servers that provide the name resolution service to the customers of an ISP. It is interesting to observe which K-root instances answer their queries and how this evolves over time.

Motivated by the above requirements we adopt a geographic map metaphor, which is quite appropriate for describing migrations. The service offered by K-root is represented as a map. Each instance is a country and its size is roughly proportional to the current number of its clients. Two countries are adjacent if the corresponding instances are usually exchanging clients, i.e. are adjacent in the migration graph. The map changes over time as follows: 1. countries change their size according to the fluctuations in the number of their clients, 2. usual migration flows are pictured as bubbles traversing the boundaries of adjacent countries, and 3. unusual migration flows are highlighted with impact graphics as bridges across the countries.

One might object that a few migration graphs can be represented with our metaphor. In fact, if: 1. vertices are represented as planar regions with disjoint interiors, 2. vertices are adjacent in the graph iff they share a point in the map, and 3. *no four regions meet at a point*, then only planar graphs can be represented. However, following [6], we remove from our metaphor the emphasized condition. Hence, we can represent a much wider class of graphs as migration graphs. Such graphs are called *planar map graphs* in [6], and they can contain up to $27n$ maximal cliques in a graph with n vertices.

Fig. 1 shows the system in action. Fig. 1.b shows the map at a certain instant. The green, light blue, orange, red, and yellow countries (corresponding to the main instances) share a point, and hence they are a clique. The other figures show several snapshots of an animation: Fig. 1.c shows how countries have different sizes in different instants. Fig. 1.d shows how we represent usual migration flows. Fig. 1.e shows the bridges that appear to emphasize unusual migration flows. Fig. 1.f shows how the name servers of a well known ISP are distributed among the instances. Each circle represents a group of name servers of that ISP with size proportional to their number.

Of course, the selected metaphor is not the only possible choice. Alternatives have been investigated and screened out for different reasons. As a first example, we could visualize the service on a real geographical map, since the actual coordinates of each instance are known. We discard this choice for several reasons: 1. Even if instances have a geographical location, the choice done by the clients is largely independent on the geography, 2. The (usual and unusual) migration patterns are also largely independent on the geography, and 3. Combining the geographical data with the (usual and unusual) migration patterns leads to information cluttering. Alternatively, we could visualize the

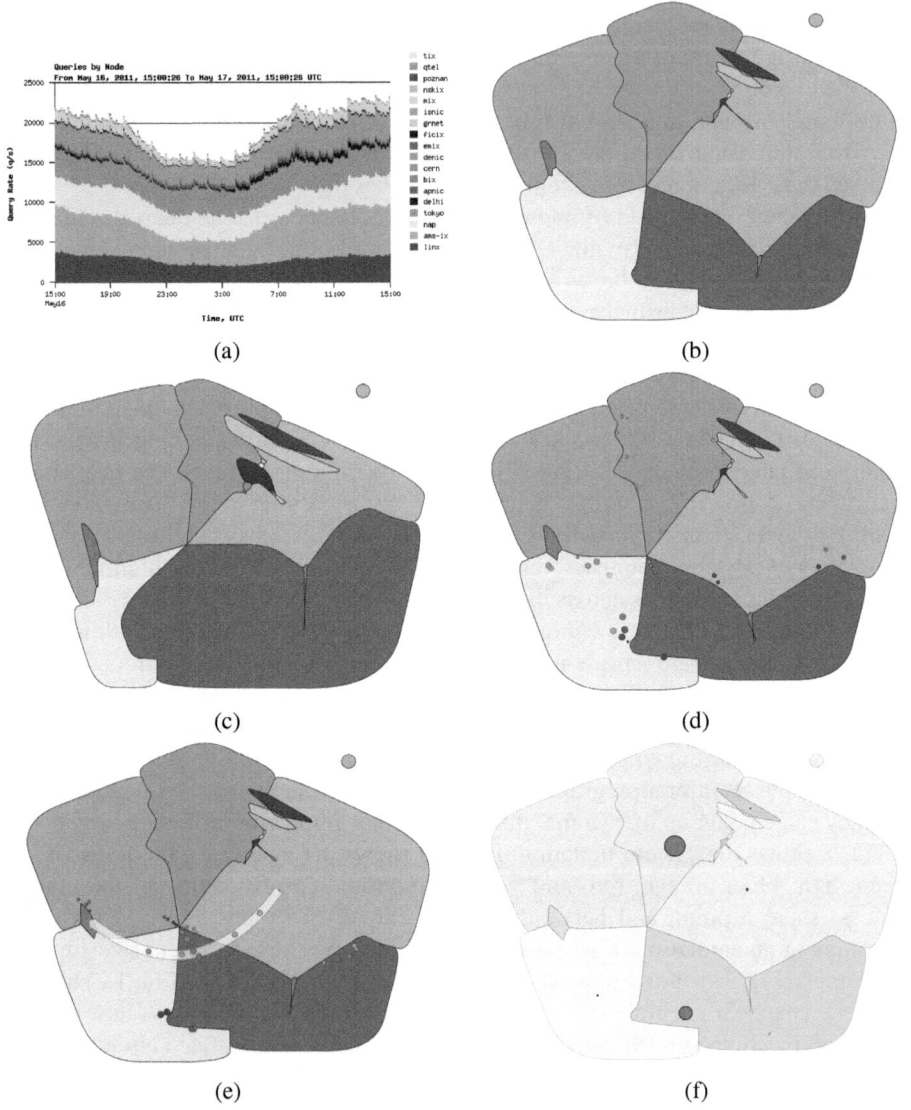

Fig. 1. (a) Distribution of the queries among the K-root instances. (b) The map at a certain instant. The animation: (c) Countries have different sizes in different instants. (d) Usual migration flows. (e) Unusual migration flows. (f) The name servers of a well known ISP.

migration graph with vertices and edges. Although it would be easy to give to each vertex an area that is proportional to its importance (see a historical example in [5]), it would be difficult to give different emphasis to usual and unusual migrations. Further, the reason behind edges having different lengths would result obscure and misleading for the user.

3   The Algorithm

The algorithm that supports our visualization framework takes as input a migration graph G and a sequence of time instants $t_1, \ldots, t_k$, where t_1, t_k is the time interval of interest and $t_2, \ldots, t_{k-1}$ depend on the adopted sampling unit. We assume that G is connected. If not, each connected component is considered separately. The algorithm constructs an animation describing the behavior of the clients in the sequence of time instants $t_1, \ldots, t_k$. We denote by $c_t(v)$ the number of clients whose last request of service before time t is asked to instance v. Given a time interval t', t'', the number of *migrants* associated with u, v at t', t'', denoted $m_{t',t''}(u, v)$, is the number of distinct clients that migrate from u to v during t', t''. We denote the *flow* between u and v as $f_{t',t''}(u, v) = \max(0, m_{t',t''}(u, v) - m_{t',t''}(v, u))$.

The algorithm is composed by two phases: the Preprocessing and the Animation, that is repeated for each t_i. The Preprocessing is composed of three steps:

1. Check if G is a map graph. If yes, then construct its *backbone*, i.e. a planar graph obtained from G by substituting some of its cliques with stars. If not, edges are removed until G is a map graph. Compute a planar topology for the backbone.
2. Find a straight-line drawing of the backbone preserving its planar topology, such that each vertex v has a surrounding "free area" that is roughly proportional to the average of $c_t(v)$ in $t_1, \ldots, t_k$.
3. Construct a constrained Delaunay triangulation, called *skeleton*, of the drawing found in the previous step. The skeleton will be used as the underlying graph during the entire animation.

The animation is performed for each interval t_i, t_{i+1} and is composed of two steps:

4. Draw the skeleton: i.e. construct a planar straight-line drawing of the skeleton preserving its topology, such that for each vertex v its incident faces can be split to determine an area surrounding v roughly proportional to $c_{t_{i+1}}(v)$.
5. Draw the map: Construct a drawing of the map at time t_{i+1} and compute the animation from t_i to t_{i+1}.

In Step 1 we check if $G(V, E)$ is a map graph. If yes, then we construct a planar embedded backbone. The backbone is obtained from G by removing the edges of a suitable set of cliques and substituting the edges of each of such cliques with a star connecting a new vertex to the vertices of the clique. More formally, let $v_1, \ldots, v_k$ be the vertices of a clique whose edges are removed. Such edges are replaced with a new vertex c and by edges $(v_1, c), \ldots, (v_k, c)$. An example where the map graph is the one of Fig. 1.b is presented in Fig. 2.a and Fig. 2.b. In [25] it is shown that testing if a graph is a map

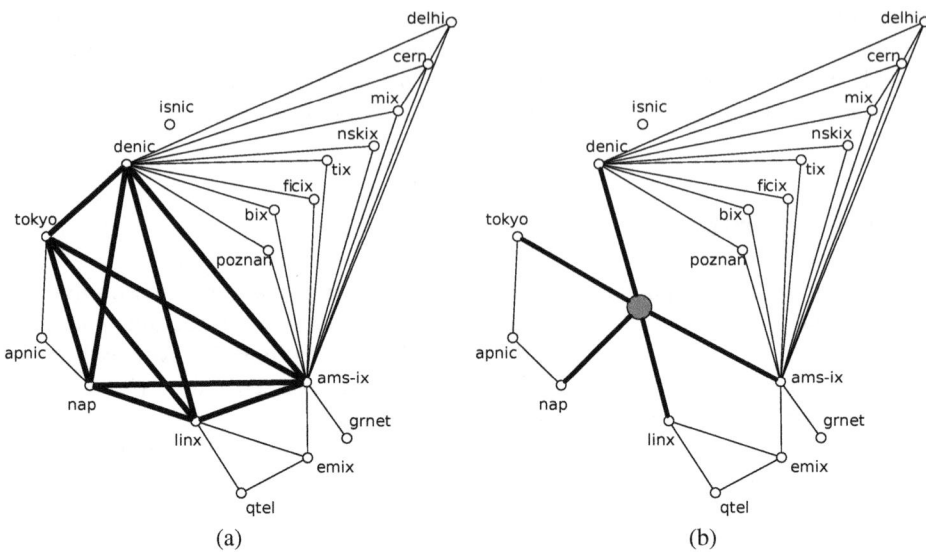

Fig. 2. (a) Migration graph for K-root. A clique of size 5 is highlighted with thick edges. (b) Backbone obtained substituting the clique with a star, centered at the grey vertex.

graph can be done in polynomial time. However, in [7] it is argued that the exponent of the polynomial bounding its running time from above is about 120. This makes it impractical to use the algorithm in [25]. Hence, we use a simple heuristic that works as follows. We first check if G is planar. If yes, we are done. Otherwise, we look in G for a maximal clique with the algorithm in [4], that is known to be efficient in practice. Then, we replace the clique with a star and perform again the planarity testing. This is repeated until either the obtained graph is planar or until no clique is found. If we are not able to find a backbone for G, then we remove the edge (u, v) with the smallest number of migrations in the given time interval, i.e. such that $\sum_{i=1}^{k-1} f_{t_i,t_{i+1}}(u, v) + f_{t_i,t_{i+1}}(v, u)$ is minimized, and repeat the process. The removed edges correspond to migration patterns that we can consider less interesting. It is also possible to involve RIPE NCC experts in this process, identifying and discarding less interesting migration patterns with their help.

Step 2 is devoted to find a straight-line drawing of the backbone, such that each vertex has a surrounding free area that is roughly proportional to the average area it will have during the animation. To perform this step we use a spring embedder [26] that preserves the given planar topology (see, e.g., [11]). Each vertex v has a positive charge $w(v)$ equal to $\frac{\sum_{i=1}^{k} c_{t_i}(v)}{k}$ and each edge (u, v) is a spring with preferred length equal to $\frac{\sqrt{w(u)}+\sqrt{w(v)}}{\sqrt{\pi}}$, that is the sum of the radii of two circles of area $w(u)$ and $w(v)$.

Step 3 adds an additional set of edges E' to the drawing of the backbone, transforming it into a maximally triangulated planar drawing. Such edges are needed to easily morph the geographical map in Step 5. All edges in the subset $A = E' \setminus E$ are marked as *additional*. For this purpose we use a constrained Delaunay triangulation, in order

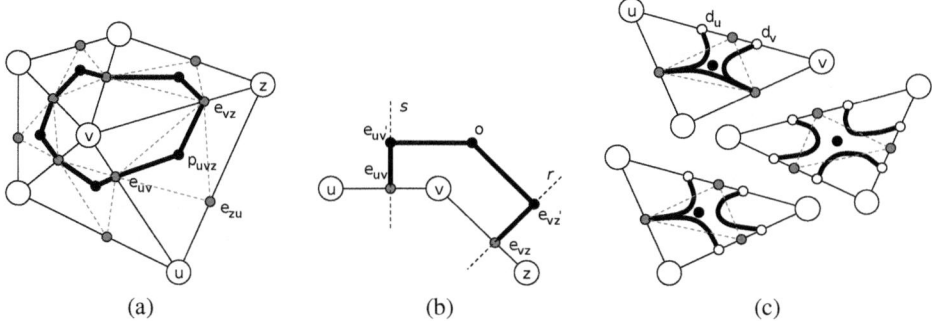

(a)   (b)   (c)

Fig. 3. (a) Costruction of the country border for a vertex that is not on the convex hull. Big white circles represent vertices of the skeleton. For each edge (u, v), a small grey circle represents the point e_{uv}. For each triangle $\Delta(u, v, z)$, a small black circle represents the point p_{uvz}. (b) Costruction of the country border for a vertex on the convex hull. (c) Three possible cases of construction of the country border with additional edges. For each additional edge (u, v), two small white circles represent the points d_u and d_v.

to maximize the angles between adjacent edges in the resulting graph. This is useful to give more degrees of freedom to the spring embedder used in Step 4.

In Step 4 the layout of the skeleton is modified to make it suitable for the construction of the map at any instant t of $t_1, \ldots, t_k$. To achieve this, a spring embedder is used where charges and preferred spring lengths change over time (see, e.g., [12]). Its initial setting is similar to the one explained for Step 2: each vertex v has a positive charge $w(v)$ that is equal to $c_t(v)$, while each edge (u, v) is a spring with preferred length equal to $\frac{\sqrt{w(u)} + \sqrt{w(v)}}{\sqrt{\pi}}$. The layout evolves with an additional constraint: consider the angle $\widehat{uvz}$ that is spanned in the external face by each triplet of vertices u, v, z that are consecutive on the convex hull. The condition $\widehat{uvz} > \pi$ is ensured. Moreover, positive charges (vertices) and springs (edges) are constantly updated to increase the precision of the map. Each triangle $\Delta(v_1, v_2, v_3)$ with area denoted by $A(\Delta(v_1, v_2, v_3))$ is split such that each of its vertices v_i is assigned an area denoted by $A(\Delta(v_1, v_2, v_3), v_i) = A(\Delta(v_1, v_2, v_3)) \frac{c_t(v_i)}{c_t(v_1) + c_t(v_2) + c_t(v_3)}$. Hence, given the set of triangles F_v with a common vertex v, the positive charge of v is regularly updated with $w(v)' = \frac{\alpha}{2\pi} \frac{c_t(v)^2}{\sum_{i \in F_v} A(i, v)}$, where α is the angle spanned by F_v (which is different from 2π only for the vertices of the external face). Spring lengths are updated accordingly with $\frac{\sqrt{w(u)'} + \sqrt{w(v)'}}{\sqrt{\pi}}$.

In Step 5 the map is computed, based on the skeleton. Each edge (u, v) is split at a point e_{uv} such that $\overline{ue_{uv}}/c_t(u) = \overline{e_{uv}v}/c_t(v)$. Then, for each triangle $\Delta(u, v, z)$ a point p_{uvz} is found such that the polygons $(u, e_{uv}, p_{uvz}, e_{zu})$, $(v, e_{vz}, p_{uvz}, e_{uv})$ and $(z, e_{zu}, p_{uvz}, e_{vz})$ have areas respectively proportional to $c_t(u)$, $c_t(v)$ and $c_t(z)$. It is easy to prove that p_{uvz} always lies inside triangle $\Delta(e_{uv}, e_{vz}, e_{zu})$.

For each vertex v that is not on the convex hull, consider the related set of triangles $F_v = \Delta(u_1, v, u_2), \Delta(u_2, v, u_3), \ldots, \Delta(u_{last}, v, u_1)$ surrounding v in clockwise order.

The country border for v is the closed polygon $(e_{u_1v}, p_{u_1vu_2}, e_{u_2v}, p_{u_2vu_3}, \ldots, e_{u_{last}v}, p_{u_{last}vu_1})$. See Fig. 3.a for details.

Vertices on the convex hull are handled in a different way. Note that, for graphs with at least three vertices, each of such vertices v has two neighbors u and z on the convex hull. We call $F_v = \Delta(z, v, u_1), \Delta(u_1, v, u_2), \ldots, \Delta(u_{last}, v, u)$ the set of triangles surrounding v. The angle $\widehat{uvz}$ that is spanned in the external face is always greater than π, as explained in Step 4. As a consequence, v can get an area on the external face that is only bounded by the line s orthogonal to (u, v) passing through e_{uv} and the line r orthogonal to (v, z) passing through e_{vz}. Given the area value $R = c_t(v) - \sum_{i \in F_v} A(i, v)$, we build the polygon $(v, e_{uv}, e'_{uv}, o, e'_{vz}, e_{vz})$ whose area is R, where e'_{uv} lies on line s, e'_{vz} lies on line r and o lies on the external face. Hence, the country border for v is the closed polygon $(e_{vz}, p_{zvu_1}, e_{u_1v}, p_{u_1vu_2}, \ldots, e_{u_{last}v}, p_{u_{last}vu}, e_{uv}, e'_{uv}, o, e'_{vz})$. See Fig. 3.b for an illustration. Finally, connected graphs with less than 3 vertices are easily converted into maps assigning circle-like country borders to each vertex.

Once all the country borders have been computed, the animation is performed. The geographical map evolves from its previous state with a linear morphing preserving adjacencies at any time. Usual migrations between countries are represented as bubbles traversing the border at randomly chosen points. Unusual migrations are represented as bridges connecting two countries, with bubbles traversing them. The size of bubbles and bridges reflects the amount of clients flowing from one country to another.

Apart from the main algorithm described above, a number of expedients are implemented to obtain a map that looks better and fully represents the underlying data. First, country borders are represented with Bézier curves where possible. This helps to give a natural look to the map. Second, at the end of Step 3, each vertex v in the skeleton that represents an instance and has degree $\delta(v)$ greater than a threshold T_δ is replaced with a path of $m = \lceil \frac{\delta(v)}{T_\delta} \rceil$ consecutive vertices. Each of them is assigned $\frac{c_t(v)}{m}$ clients and retains a fraction of the original adjacencies, with degree lower than T_δ. This helps finding better layouts for the skeleton graph in Step 4. The country border for such a path of vertices is computed as the symmetric difference between their borders. Finally, edges added in Step 3 and marked as *additional* are later handled in a different way. In particular, the spring embedder used in Step 4 assigns a fixed additional length D to springs representing additional edges. During the construction of the map (Step 5), two points d_u and d_v are found on each additional edge (u, v) together with e_{uv}, such that $\overline{ud_u}/c_t(u) = \overline{d_vv}/c_t(v)$ and $\overline{ud_u} + \overline{d_vv} + D = \overline{uv}$. Then the construction of the border is slightly different with respect to the one explained in Step 5. For each edge (u, v) marked as additional, the two vertices u and v respectively choose d_u and d_v as boundary points, instead of e_{uv}. In this way countries that are not adjacent in the graph do not share boundary points in the map. Note that for each triangle $\Delta(u, v, z)$ the point p_{uvz} is still shared by country borders for vertices u, v and z. This inconsistency is removed in practice using Bézier curves. See Fig. 3.c for an illustration.

4   Technical Aspects and User Feedback

Our visualization framework has been implemented as a Web application, composed of a Javascript front-end and a server written in Java. It relies on Google Web Toolkit, a framework for the creation of Web applications. It makes use of a cross-browser Javascript library for vector graphics called Raphaël, based on the Scalable Vector Graphics format. This implies that images and snapshots can be zoomed and exported without loss of quality (see Fig. 1 for an example). A demo of the application is available online at `http://dia.uniroma3.it/~squarcel/visual-k/`.

An associative map is kept in memory to store the current state of each client, including the instance that answered its last query. At regular time intervals all the new queries received by K-root are analyzed to detect usual and unusual migrations. These are translated into an animation step and are later used to update the associative map.

We performed a stress test using a full trace of queries received by all the K-root instances during a 48-hour time window. Even focusing on the minimal amount of information needed, in the form of a triplet $(timestamp, client\_id, instance\_id)$, the volume of data is impressive (around 200 Gigabytes) and poses a challenge for the creation of a scalable system. We ran the stress test on a laptop with a 2.4 GHz Intel Core 2 Duo processor and 4 GB of RAM. The results show that our framework can handle an update rate between 10 and 15 seconds, including both the analysis of query data and the generation of the corresponding geographical map. The result is of course expected to improve on more powerful hardware. Hence, although such an approach is not strictly real-time, it represents an approximation that satisfies the operational needs.

The system has two types of potential users: (1) the RIPE NCC DNS Services staff and (2) the vast audience of ISPs that could benefit of knowing what instances serve their clients. The system has been designed cooperating with the users of type (1). Their participation to the design process allowed to precisely focus on the requirements. As an example, during the interactions the users gave a negative evaluation of a first version of the migration graph where both usual and unusual migration patterns were represented using country adjacencies. This allowed to devise the current version of the graph. It is particularly interesting to use the system together with BGPlay [8]: once an unusual migration is spotted, BGPlay can be used to check if there is a correlation with some routing change. About users of type (2), the possibility of putting the system at their disposal depends on the future policies of the RIPE NCC. In fact, the logs of queries are strictly confidential and an anonymization policy is currently being discussed.

5   State of the Art

The problem of using geographical maps to visualize non-geographical information has been extensively studied. In this section we provide a brief overview of the literature, focusing on similarities and differences with our approach.

A methodological reference is provided by the cognitive study in [13]. It identifies four semantic primitives to be used when representing information entities with a geographical metaphor. *Boundaries*: discontinuities in the information space can be represented with borders. *Aggregate*: homogeneous zones preferably represent homogeneous entity types. We use aggregate and boundaries to group clients using the same instance and to separate such groups, respectively. *Loci*: information items preferably have a meaningful location in the information space. We put side-by-side instances that are expected to share clients. *Trajectories*: semantic relationships between information entities at different locations can be shown with paths or routes. We exploit different types of trajectories to represent migrations.

There are at least two systems whose features are similar to ours: GMap and BGPlay Island. GMap [19] visualizes clustered graphs with geographical maps. After determining the layout of the graph with a force directed approach, clusters of nodes are detected according to their relative distance. A cluster is represented with one or more geographical regions. GMap produces maps that look very similar to our maps. However, its target is quite different from ours: 1. if two vertices are connected by an edge it is not guaranteed that they have a common boundary, 2. if two vertices have a common boundary is not guaranteed that they are connected by an edge, and 3. GMap is not meant to visualize maps whose borders evolve over time. Using the terminology of [13] we can say that in [19] the aggregate primitive prevails over the others. BGPlay Island [9] extends the widely used BGPlay routing visualization system [8] and uses a topographic metaphor to show hierarchies of Internet Service Providers (ISPs). However, BGPlay Island uses the metaphor of a terrain map rather than the one of a political map and the most stressed primitive of [13] is the locus one.

Other related literature is the one on *cartograms*. Area cartograms are drawings derived from standard geographical maps, where each country is deformed so that its area is proportional to a variable specific of that country, e.g. its population. The deformation process should preserve the original shape as much as possible. The idea behind cartograms is very close to our map metaphor, which in fact can be seen as an area cartogram derived from an imaginary world. Many algorithms for computing area cartograms are available in the literature (see, for example, [15,17,21]). However, their attempt to preserve the original shape is irrelevant in our setting, since our countries do not have a prescribed shape. Also, they have high computational time, which makes them unsuited for a real-time monitoring tool. In [21] the latter issue is tackled with an algorithm that can be parallelized, but, unfortunately, results are exposed to inaccuracy (e.g. overlap between countries). Recent approaches [27,10,18,22,1,3,2] for the computation of area cartograms tend to keep the countries in their original locations but give them a regular shape, like a rectangle or a "T" or and "L". However, the more regular the shapes are, the less graphs can be represented. Further, none of the above results takes into account scenarios that include planar map graphs. Finally, the computed layouts are sometimes hard to read and therefore not suitable for an intuitive visualization.

Voronoi diagrams represent an option for partitioning information spaces into separate regions. In [23] the authors introduce an adaptive version of the multiplicatively weighted Voronoi diagram [20], where each vertex in a graph is assigned a closed region with prescribed area. Similarly to Voronoi diagrams, however, adjacencies between regions depend on geometric proximity. Hence the solution is not compatible with the notion of adjacency graph.

In a recent work [14] it is shown that planar graphs can be represented with adjacent convex hexagons. Such shapes could be a valid alternative for our scope. We think that it is possible to modify the proposed algorithm to represent also planar map graphs, using polygons with more sides and loosing the convexity. However, the problem of assigning prescribed areas to the shapes seems difficult to be addressed.

A previous attempt at visualizing the activity of Internet services, including K-root, is in [16]. Sets of clients sending requests to the same instance are located on a real geographical map and a coordinate centroid is computed. Then a circle is displayed, centered at the centroid and composed of wedges that represent the amount, distribution and latency of clients. Such a tool differs from our approach, in that it is meant to visualize static snapshots of the service, not focusing on the migrations of clients.

6   Conclusions and Future Work

We have presented a system for the visualization of the behaviour of the K-root DNS name server. It relies on a map metaphor that uses an animation to show the migration of clients among the instances that compose the server.

While in [24] it is argued that animations are not generally suitable to convey information on trends in data visualization, it is also argued that they are quite useful to create a visualization that is appealing to the user. At the same time, a real-time monitoring tool necessarily deals with the evolution of the underlying data. In our framework we find a reasonable balance between the two needs, using graphical elements that are independent on the animation. A static snapshot of each step of the animation contains all the information we want to visualize, as Fig. 1 clearly shows. The animation is only needed to gracefully link two consecutive steps, helping the user to focus on the context.

There are several future research directions that can be undertaken. One would be to deploy our system to other root servers. Such a step is technically easy, but it has drawbacks from the organizational point of view, since logs of queries are strictly confidential and dealing with them requires an adequate agreement. Another interesting possibility would be to apply the same techniques to other Internet services based on *anycast*. One possible example, mostly interesting nowadays, is the IPv6 6to4 Relay Routing Service, devised to facilitate the transition between IPv4 and IPv6.

References

1. Akbari Jokar, M., Shoja Sangchooli, A.: Constructing a block layout by face area. The International Journal of Advanced Manufacturing Technology 54, 801–809 (2011)
2. Biedl, T., Ruiz Velázquez, L.: Orthogonal Cartograms with Few Corners Perface. In: Dehne, F., Iacono, J., Sack, J.-R. (eds.) WADS 2011. LNCS, vol. 6844, pp. 98–109. Springer, Heidelberg (2011)
3. Biedl, T., Velázquez, L.E.R.: Drawing planar 3-trees with given face-areas (2010)
4. Bron, C., Kerbosch, J.: Algorithm 457: finding all cliques of an undirected graph. Commun. ACM 16, 575–577 (1973)
5. Carpano, M.-J.: Automatic display of hierarchized graphs for computer-aided decision analysis. IEEE Transactions on Systems, Man and Cybernetics 10(11), 705–715 (1980)
6. Chen, Z.-Z., Grigni, M., Papadimitriou, C.H.: Planar map graphs. In: Proceedings of the Thirtieth Annual ACM Symposium on Theory of Computing, STOC 1998. ACM (1998)
7. Chen, Z.-Z., Grigni, M., Papadimitriou, C.H.: Recognizing hole-free 4-map graphs in cubic time. Algorithmica 45(2), 227–262 (2006)
8. Colitti, L., Di Battista, G., Mariani, F., Patrignani, M., Pizzonia, M.: Visualizing interdomain routing with BGPlay. Journal of Graph Algorithms and Applications, Special Issue on the 2003 Symposium on Graph Drawing, GD 2003 9(1), 117–148 (2005)

9. Cortese, P.F., Di Battista, G., Moneta, A., Patrignani, M., Pizzonia, M.: Topographic visualization of prefix propagation in the internet. IEEE Transactions on Visualization and Computer Graphics 12(5), 725–732 (2006)

10. de Berg, M., Mumford, E., Speckmann, B.: Optimal BSPs and rectilinear cartograms. In: Proceedings of the 14th Annual ACM International Symposium on Advances in Geographic Information Systems, GIS 2006, pp. 19–26. ACM, New York (2006)

11. Didimo, W., Liotta, G., Romeo, S.A.: Topology-Driven Force-Directed Algorithms. In: Brandes, U., Cornelsen, S. (eds.) GD 2010. LNCS, vol. 6502, pp. 165–176. Springer, Heidelberg (2011)

12. Erten, C., Harding, P., Kobourov, S., Wampler, K., Yee, G.: GraphAEL: Graph Animations with Evolving Layouts. In: Liotta, G. (ed.) GD 2003. LNCS, vol. 2912, pp. 98–110. Springer, Heidelberg (2004)

13. Fabrikant, S.I., Skupin, A.: Cognitively plausible information visualization. Exploring Geovisualization, 667–690 (November 2005)

14. Gansner, E., Hu, Y., Kaufmann, M., Kobourov, S.: Optimal Polygonal Representation of Planar Graphs. In: López-Ortiz, A. (ed.) LATIN 2010. LNCS, vol. 6034, pp. 417–432. Springer, Heidelberg (2010)

15. Gastner, M.T., Newman, M.E.J.: Diffusion-based method for producing density-equalizing maps. Proceedings of the National Academy of Sciences of the United States of America 101(20), 7499–7504 (2004)

16. Huffaker, B., Fomenkov, M. Claffy, K.: Influence maps - a novel 2-d visualization of massive geographically distributed data sets. Internet Protocol Forum (October 2008)

17. Inoue, R., Shimizu, E.: A new algorithm for continuous area cartogram construction with triangulation of regions and restriction on bearing changes of edges. Cartography and Geographic Information Science 33(2), 115–125 (2006)

18. Kawaguchi, A., Nagamochi, H.: Orthogonal Drawings for Plane Graphs with Specified Face Areas. In: Cai, J.-Y., Cooper, S.B., Zhu, H. (eds.) TAMC 2007. LNCS, vol. 4484, pp. 584–594. Springer, Heidelberg (2007)

19. Mashima, D., Kobourov, S., Hu, Y.: Visualizing Dynamic Data with Maps. In: Proc. 4th IEEE Pacific Visualization Symposium (March 2011)

20. Okabe, A., Boots, B., Sugihara, K., Chiu, S.N.: Spatial tessellations: Concepts and applications of Voronoi diagrams, 2nd edn. Probability and Statistics. Wiley, NYC (2000)

21. Ouyang, M., Revesz, P.Z.: Algorithms for cartogram animation. In: Proceedings of the 2000 International Symposium on Database Engineering & Applications, IDEAS 2000, pp. 231–235. IEEE Computer Society, Washington, DC, USA (2000)

22. Rahman, M. S., Miura, K., Nishizeki, T.: Octagonal drawings of plane graphs with prescribed face areas. Comput. Geom. Theory Appl. 42, 214–230 (2009)

23. Reitsma, R., Trubin, S.: Information space partitioning using adaptive voronoi diagrams. Information Visualization 6, 123–138 (2007)

24. Robertson, G., Fernandez, R., Fisher, D., Lee, B., Stasko, J.: Effectiveness of animation in trend visualization. IEEE Transactions on Visualization and Computer Graphics 14, 1325–1332 (2008)

25. Thorup, M.: Map graphs in polynomial time. In: Proceedings of the 39th Annual Symposium on Foundations of Computer Science, FOCS 1998 (1998)

26. Tollis, I.G., Di Battista, G., Eades, P., Tamassia, R.: Graph Drawing: Algorithms for the Visualization of Graphs. Prentice Hall (1998)

27. van Kreveld, M., Speckmann, B.: On rectangular cartograms. Comput. Geom. Theory Appl. 37, 175–187 (2007)

Optimizing a Radial Layout of Bipartite Graphs for a Tool Visualizing Security Alerts

Maxime Dumas[1], Michael J. McGuffin[1], Jean-Marc Robert[1],
and Marie-Claire Willig[2]

[1] ETS
Montréal, Canada
{michael.mcguffin,jean-marc.robert}@etsmtl.ca
[2] EESTIN, UHP Nancy 1
Vandoeuvre Lès Nancy, France

Abstract. Effective tools are crucial for visualizing large quantities of information. While developing these tools, numerous graph drawing problems emerge. We present solutions for reducing clutter in a radial visualization of a bipartite graph representing the alerts generated by an IDS protecting a computer network. Our solutions rely essentially on (i) unambiguous edge bundling to reduce the number of edges to display and (ii) the minimization of the total sum of the edge lengths.

Keywords: IDS alerts, bipartite graph layout, edge bundling.

1 Introduction

Intrusion Detection Systems (IDSs) are important tools for protecting enterprise networks. Unfortunately, they generate large quantities of information that are challenging to analyze. Few visualization tools have been proposed to ease the effort of network defence analysts [1,5]. These tools either do not use any graph layouts or do not focus on optimizing graph layouts.

Recently, the authors developed a new tool for IDS visualization named Alert-Wheel [4] that employs a radial overview visualization with a novel form of edge bundling, and incorporates features for filtering and drilling down on IDS alerts. IDSs such as SNORT [7] generate alerts when abnormal traffic flows are detected. The information in each alert identifies the category of the malicious behaviour (e.g., *network-scan*, *web-application-attack*, etc.) and the origin of the flow (the source IP address of the packets, from which an *Autonomous System* (AS) *Number* can be computed). Hence, these alerts can be visualized as the edges of a bipartite graph, where each node is either the AS node of the source, or the alert category. (The use of AS rather than IP addresses greatly reduces the number of source nodes in the bipartite graph.)

AlertWheel relies on a new way of drawing bipartite graphs that is visually clearer than the status quo (see Fig. 1). The inner circle corresponds to a limited number (up to 32) of alert categories, and the outer circle corresponds to AS nodes (see Fig. 5 and 7). In the development of this tool, multiple graph drawing

M. van Kreveld and B. Speckmann (Eds.): GD 2011, LNCS 7034, pp. 203–214, 2012.

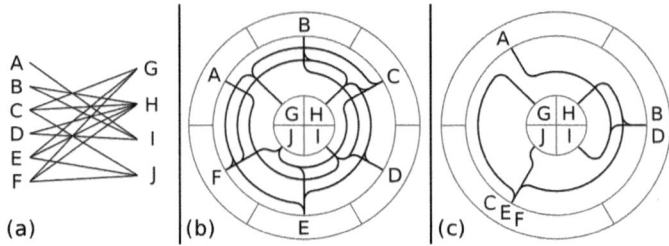

Fig. 1. The same bipartite graph shown in different ways. (a) Status-quo approach. (b) An improved way of drawing edges: each edge starting on an outer node leads to a circular bundle, and each circular bundle leads to a single inner node. (c) A further improvement that groups together nodes having the same neighbors.

problems were encountered. The aim of this paper is to present these abstract problems and the heuristics used to solve them. The companion paper [4] focuses on the overall tool and its interactive features.

2   Problem Statement

AlertWheel displays a radial visualization of a bipartite graph composed of an inner circle on which there are n interior points $i_1, \cdots, i_n$ (representing categories) and an outer circle on which there are m exterior points $o_1, \cdots, o_m$ (representing AS source nodes). Each edge between an interior and exterior point represents an observed alert. The positions of the interior points are determined by a central pie chart (Fig. 5 and 7). The interior points are sorted such that i_1 is connected to the minimum number of exterior points and i_n is connected to the maximum.

Our objective is to layout the numerous edges connecting the points as efficiently as possible, to ease reading and interpretation. This is done by

- Grouping edges into bundles [6,8];
- Reducing the sum of the edge lengths.

The edge bundling is illustrated in Fig. 1 and 2. Edges are layered on concentric circles where each concentric circle corresponds to one interior point. Edges can share either radial or circular segments with other edges, without introducing any ambiguity. Grouping together nodes with the same neighbors (Fig. 1c) further reduces clutter, and is very useful for visualizing security alerts. During outbreaks of very virulent malware, many infected computers may knock at the door of a given network and generate alerts of the same categories.

In the development of AlertWheel, the following graph drawing problems had to be addressed:

Problem I.   Choose an assignment of concentric circles to interior points.
Problem II.   Choose the position of a single exterior point minimizing the sum of lengths of edges to its neighbors.

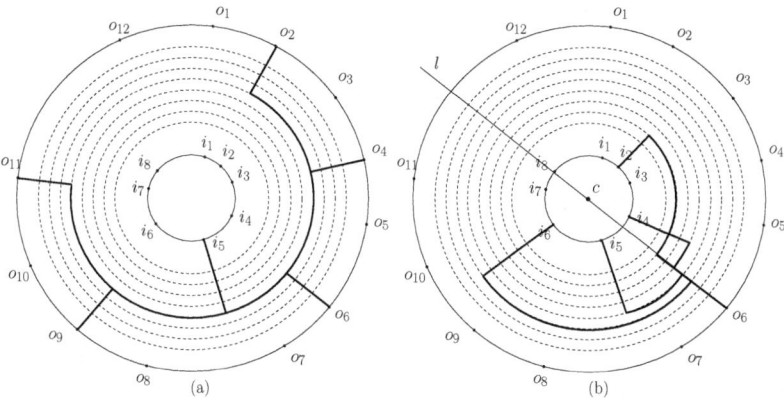

Fig. 2. Edge bundling. (a) Connecting multiple exterior points to one interior point – e.g., computers from five AS nodes generating alerts of the same category. (b) Connecting multiple interior points to one exterior point – e.g., a single AS node generating alerts of four categories.

Problem III. Choose the positions of all exterior points minimizing the total sum of the edge lengths.

In the next sections, solutions to these problems are presented. Exact solutions are only given for the first two problems, and heuristics are given for the last one. These heuristics are compared with each other and with the non-achievable lower bound derived from the second problem. Problem III is the core layout problem in AlertWheel. Efficient solutions to this problem are thus crucial.

3   Related Work

Network defence analysts continuously monitor their computer networks to detect any malicious activities. They have to analyze the data buried in numerous log files. Without appropriate tools, they cannot manage all this information.

   The ultimate goal for security visualization tools is to present the data as simply as possible. To achieve this ambitious objective, numerous graph drawing problems have to be addressed [10]. These generally involve optimizing some aesthetic properties of the graph layout such as the number of edge crossings, the number of edge bends, etc.

   Purchase [9] asserts that minimizing the number of edge crossings in a graph layout is the most important issue to deal with. Unfortunately, numerous variants of this problem have been shown to be NP-complete [11,2]. Edge bundling can be used to mitigate the effects of edge crossings. By merging edges into bundles [6,8], the number of edge crossings is significantly reduced. In such a case, reducing the length of the bundled edges remains an important goal. The rest of this paper addresses this goal.

4 Assignment of Concentric Circles to Interior Points

Concentric circles are used to layout the edges of the bipartite graph. These are particularly useful to bundle the edges connected to a given interior point (as in Fig. 2a). Assume that some set of radii r_k, $k = 1, \dots, n$ of concentric circles has been chosen, for example with equal spacing. The first problem is to find an optimal assignment of each radius to a unique interior point, i.e., an assignment minimizing the sum of the edge lengths. Unfortunately, the total length of edges depends on the assignment of points on the outer circle. Thus, in this section, we simplify the problem and seek an assignment of radii to interior points that minimizes the total length of the *radial* components of the edges.

Let E_k be the subset of exterior points connected to an interior point i_k using the concentric circle with radius r_k (as in Fig. 2a) and let e_k be its cardinality. The sum of the radial edge lengths l_k^R is given by $e_k(r_{n+1} - r_k) + (r_k - r_0)$ where r_0 and r_{n+1} are the radii of the inner and outer circles, respectively.

The following lemma shows that the optimal assignment depends only on the number of edges using each circle. It can be proved easily by contradiction.

Lemma 1. *Suppose that the numbers of edges e_k are such that $e_1 \leq \cdots \leq e_n$. If the radii are such that $r_1 \leq \cdots \leq r_n$, then $S^R = \sum_{k=1}^{n} l_k^R$ is minimum.*

5 Optimally Connecting an Exterior Point to Interior Ones

The next problem is to find the optimal position of an exterior point to minimize the sum of its edge lengths. We assume that the assignment of radii is fixed and given by Lemma 1. This leaves only the *circular* components to be minimized. (Note that this heuristic does not guarantee that the *total* length of edges is minimized.) Then, the objective can be restated as finding the optimal position of an exterior point minimizing the total length of the circular components of its edges, given the assumed assignment of radii to interior points.

Consider an exterior point o connected to k interior points $i_1, i_2, \cdots, i_k$ (as in Fig. 2b). Let l be the line passing through the center c of the circles and the exterior point o. This line partitions the interior points into three sets: the points lying above l ($\mathcal{A}$), the points lying below l ($\mathcal{B}$) and the points lying on l ($\mathcal{O}$).

Let Θ_j be the angle in radians defined by the points o, c and i_j, defined s.t. $0 \leq \Theta_j \leq \pi$. The sum of the circular edge components is

$$S^C = \sum_{i_j \in \mathcal{A}} \Theta_j \cdot r_j + \sum_{i_j \in \mathcal{B}} \Theta_j \cdot r_j + \sum_{i_j \in \mathcal{O}} \Theta_j \cdot r_j \tag{1}$$

The following lemma characterizes the optimal solutions minimizing the sum of the circular edge components for a given exterior point o.

Lemma 2. *There is an optimal position for the exterior point o minimizing the sum of the circular edge components s.t. the line l defined by c and o passes through an interior point i_j between c and o.*

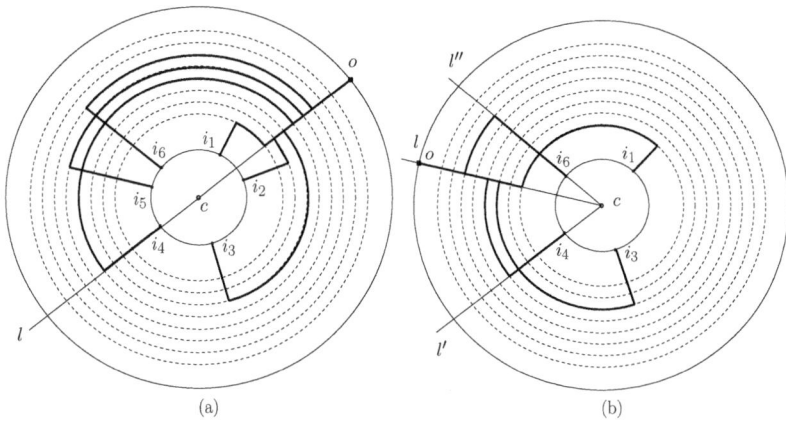

Fig. 3. Optimally positioning an exterior point connected to interior points

Proof. Suppose there is an optimal position for o which does not satisfy the criteria as in Fig. 3a. (Note that, in this case, there could still be a point in $\mathcal{O}$, located on the "other side" of o (Fig. 3a).) Let S^C be the corresponding sum of the lengths as given by Eq. 1. W.l.o.g. suppose that $\sum_{i_j \in \mathcal{A}} r_j > \sum_{i_j \in \mathcal{B}} r_j$ as in Fig. 3a. By rotating l counter-clockwise by a small angle $\epsilon > 0$, the point in $\mathcal{O}$ (if it exists) would be located above l. The sum of lengths would then be

$$S' = \sum_{i_j \in \mathcal{A}} (\Theta_j - \epsilon) \cdot r_j + \sum_{i_j \in \mathcal{B}} (\Theta_j + \epsilon) \cdot r_j + \sum_{i_j \in \mathcal{O}} (\Theta_j - \epsilon) \cdot r_j$$

$$= S^C - \epsilon \left[\sum_{i_j \in \mathcal{A} \cup \mathcal{O}} r_j - \sum_{i_j \in \mathcal{B}} r_j \right] < S^C.$$

This contradicts the optimality hypothesis of S^C.

In the special case $\sum_{i_j \in \mathcal{A}} r_j = \sum_{i_j \in \mathcal{B}} r_j$ and $\mathcal{O} = \emptyset$, the line l can still be rotated onto either of two interior points (yielding l' and l'') without worsening the sum of lengths (Fig. 3b). □

This lemma gives a straightforward linear time algorithm to find an optimal solution once the interior points have been fixed and assigned radii. The algorithm simply has to sweep through the finite number of candidate solutions.

As the point o moves around the circle, the sum of the circular edge lengths can reach numerous local minima (Fig. 4). Hence, binary search algorithms based solely on local decisions could lead to non-optimal solutions.

6   Connecting Multiple Exterior Points to Multiple Interior Points

The last problem to consider is choosing the positions of exterior points minimizing the total sum of the circular edge lengths. This is the core layout problem

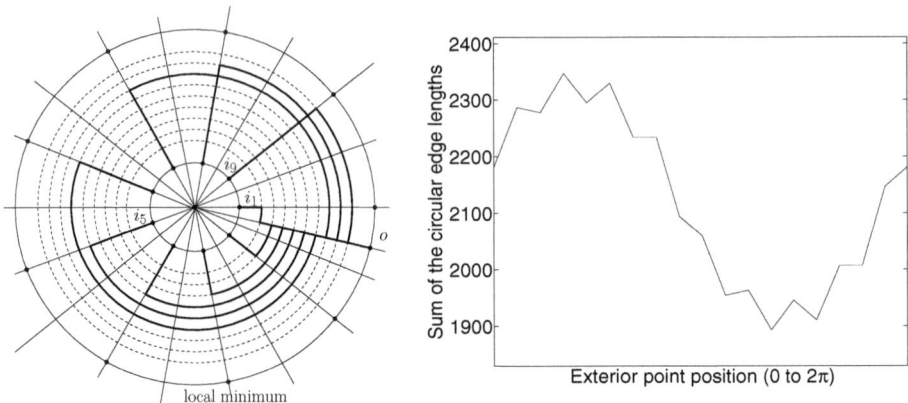

Fig. 4. Multi-modality of the function giving the sum of lengths. (Left) The construction of an example with 9 points – the radius r_k is given by $r_0 + \delta + (k-1) \times \frac{r_{n+1} - r_0 - 2\delta}{n-1}$. (Right) The corresponding sum of the circular edge lengths.

of the AlertWheel visualization tool. Due to the nature of the problem, the positions of the interior points are fixed. Hence, changing the positions of exterior points is the only way to optimize the layout.

Two heuristics are presented to solve this optimization problem. These heuristics are compared with the naive solution of ordering the points on a first-come, first-served basis. A lower bound can be derived from the algorithm presented in the previous section. However, this solution may not be achievable since it allows many exterior points to coincide.

Let us first introduce some notation. Let $I_k \subseteq \{i_1, \cdots, i_n\}$ be the subset of the interior points connected to the exterior point o_k. These points represent the *hyperedge* associated with o_k. Also, let $\mathcal{I} = \{I_k | 1 \leq k \leq m\}$ be the set of the hyperedges to be laid out.

6.1 Heuristic I: The Minimum Perfect Matching

The first heuristic is an algorithm distributing the exterior points evenly on the outer circle. The eases point labeling in the visualization tool but does not guarantee that the obtained optimal solution is globally optimal.

The algorithm is based on the minimum perfect matching problem [3]. It constructs a complete bipartite graph $K_{m,m}$ representing the cost of associating each hyperedge I_k to each potential layout position. A minimum-weight perfect matching would give a one-to-one correspondence between the set of hyperedges and the set of layout positions which minimizes the total sum of the hyperedge lengths.

A more formal description of this heuristic is presented in Algorithm 1. The running time of this algorithm is $O(m^2 n + m^3) = O(m^3)$, assuming $n < m$.

1 Let $\mathcal{I} = \{I_k | 1 \leq k \leq m\}$ be the hyperedges associated to the m exterior points.

2 Let $\mathcal{P} = \{p_k | 1 \leq k \leq m\}$ be a set of evenly distributed positions on the outer circle.

3 **for** $i \leftarrow 1$ **to** m **do**

4   **for** $j \leftarrow 1$ **to** m **do**

5   Compute the length $w_{i,j}$ of the hyperedge I_i at the position p_j.

6 Compute the minimum perfect matching of the complete bipartite graph defined by the sets $\mathcal{I}$, $\mathcal{P}$ and $\mathcal{W} = \{w_{i,j} | 1 \leq i, j \leq m\}$

Algorithm 1. Minimum perfect matching heuristic

In Fig. 5, the performance of this algorithm is compared with the naive algorithm ordering the points on a first-come, first-served basis. As expected, the heuristic yields a better result on this example than the naive algorithm. A more thorough comparison is presented in Section 6.3.

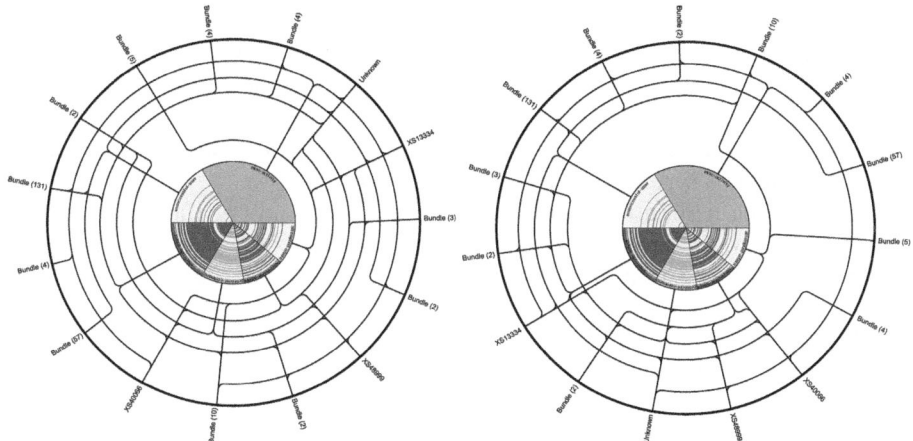

Fig. 5. AlertWheel tool: Comparing the naive algorithm (left) and the perfect matching heuristic (right). In the former case, the sum of the circular edge lengths is 22036 units. In the latter case, the sum is 17158, or 22% less.

6.2 Heuristic II: The Anchor Algorithm

The second heuristic is based on the idea of finding the optimal position of each exterior point and placing it as close as possible to this position, which we call an anchor. Lemma 2 determines the anchors for the exterior points. Based on the computed anchors, the exterior points are partitioned into sets containing points competing for the same optimal anchors. Then, the optimal local positioning of the exterior points around each anchor is determined.

Unfortunately, this algorithm does not guarantee an optimal solution. If an exterior point is placed too far from its optimal anchor, it could be better off at another anchor. Nevertheless, this heuristic yields good results, as we will see.

Let us introduce some notation. Let $C_p \subseteq \{o_1, \cdots, o_m\}$ be the points competing for the anchor p. To simplify the notation, assume that $C_p = \{o_1, \cdots, o_t\}$. The anchor p and center c determine a line l that divides the interior points into the points lying above l ($\mathcal{A}$), the points lying below l ($\mathcal{B}$) and the points lying on l. To simply the argument, we assume that only the anchor point lies on l. For any exterior point $o_i \in C_p$ connected to the hyperedge I_i, we define

$$a_i = \sum_{\substack{i_j \in I_i \\ i_j \in \mathcal{A}}} r_j \quad \text{and} \quad b_i = \sum_{\substack{i_j \in I_i \\ i_j \in \mathcal{B}}} r_j.$$

Finally, let r^* be the radius associated with the interior point defining the anchor p. By Lem. 2, this interior point must be in I_i.

Let $A = \{o_i \in C_p | a_i \geq b_i\}$ and let $B = \{o_i \in C_p | a_i < b_i\}$. Now, suppose that $o_i \in A$. As the exterior point o_i moves away from the anchor p by an angle $\Theta > 0$, the total sum of the circular edge lengths increased by

$$(a_i + r^* - b_i)\Theta > 0 \quad \text{if } o_i \text{ is moving away clockwise}$$
$$(b_i + r^* - a_i)\Theta > 0 \quad \text{if } o_i \text{ is moving away counter} - \text{clockwise}.$$

Both expressions must be positive. Otherwise, the point p would not represent an optimal anchor for o_i. This follows from Lem. 2.

The following lemma characterizes the optimal layouts of the exterior points in A. Intuitively, these points should be moved away from the anchor counter-clockwise to reduce the impact on the sum of the circular edge lengths.

Lemma 3. *Let $0 \leq i \leq |A|$. There are only two optimal layouts of the exterior points in A around the anchor p s.t. i points of A move away clockwise.*

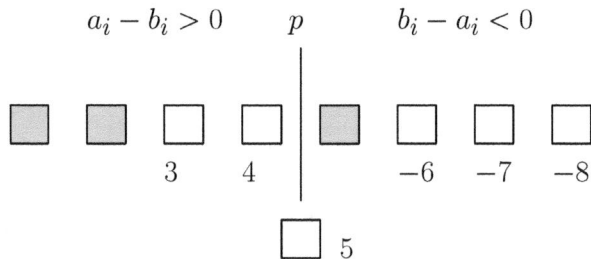

Fig. 6. Ordered layout of the exterior points in A for $i = 2$. The values represent the values of $a_i - b_i > 0$ (on the left of p) and the values of $b_i - a_i < 0$ (on the right of p). The shaded boxes represent the positions of the points in B.

Proof. W.l.o.g., suppose that $A = \{o_1, \cdots, o_r\}$ is such that the values of $a_i - b_i$ are sorted in increasing order. In order to minimize the total sum of the circular edge lengths, the points must be laid out as follows:

- $o_1, \cdots, o_i$ clockwise w.r.t. the anchor and $o_{i+1}, \cdots, o_r$ counter-clockwise w.r.t. the anchor – and according to the sorted order;
- $o_1, \cdots, o_i$ clockwise w.r.t. the anchor, o_{i+1} on the anchor and $o_{i+2}, \cdots, o_r$ counter-clockwise w.r.t. the anchor – and according to the sorted order.

Let us consider only the simpler alternative since the same argument applies to both. First, let us prove the optimality of the layout of the points which have been moved counter-clockwise (i.e., at the right of p in Fig. 6). Suppose there is an optimal layout which does not respect the increasing order. Let S^* be the sum of the circular edge lengths of this optimal solution. Suppose there are two consecutive points o_{i*} and o_{j*} counter-clockwise w.r.t. the anchor point s.t. $a_{i*} - b_{i*} > a_{j*} - b_{j*} > 0$. Thus, o_{i*} and o_{j*} have been moved counter-clockwise by an angle of $inc \times \Theta$ and $(inc + 1) \times \Theta$, respectively. The value inc is the incremental angular difference between adjacent exterior points. The weight of the these two consecutive points in S^* is

$$(b_{i*} + r^* - a_{i*}) \times inc \times \Theta + (b_{j*} + r^* - a_{j*}) \times (inc + 1) \times \Theta.$$

By permuting the two points, a smaller weight can be obtained. This contradicts the optimality of the solution.

Finally, suppose there is a point $o_{i*} \in B$ which has been also moved counter-clockwise (a shaded box in Fig. 6). Since $(b_{i*} - a_{i*}) > 0$, this point must be closer to the anchor than any other point in A in any optimal layout. Otherwise, by permuting these points, a smaller sum of the circular edge lengths would be obtained. Similar arguments can be used to prove the optimality of the layout of the points which have been moved clockwise (i.e. at the left of p in Fig. 6). □

Based on this characterization of the optimal layouts, a more formal description of this heuristic is presented in Algorithm 2. The running time of this algorithm is in $O(mn + m^2)$.

Figure 7 presents one example showing that the anchor heuristic out performs the naive algorithm, as expected. A better comparison is presented in Sect. 6.3.

6.3   Empirical Comparison

To perform a more thorough comparison of the proposed heuristics, the different algorithms were applied to the same random bipartite graphs, generated as follows. There are ten points which have been fixed on the inner circle. There are n points which have to be laid out on the outer circle. These points have to be connected to the inner points as follow:

- 10% of the exterior nodes have 5 edges.
- 20% of the exterior nodes have 4 edges.

1 Find the n anchors $p_1, \cdots, p_n$ determined by the interior points (by Lem. 2).
2 **for** *each exterior point o_i* **do**
3   Find the optimal anchor p_{j_i}.
4   Add o_i to the anchor list C_{j_i}.
5 **for** *each anchor list C_{j_i}* **do**
6   Find the optimal position of the $|C_{j_i}|$ points around p_{j_i}.
7   Let A and B be the set of points as defined in Lem. 3
8   **for** $i \leftarrow 0$ **to** $|A|$ **do**
9   **for** $j \leftarrow 0$ **to** $|B|$ **do**
10   Find the optimal solutions with i points of A and j points of B
                which have been moved clockwise (by Lem. 3).
11   Find the optimal solutions among all the solutions in the previous step.

Algorithm 2. Anchor heuristic

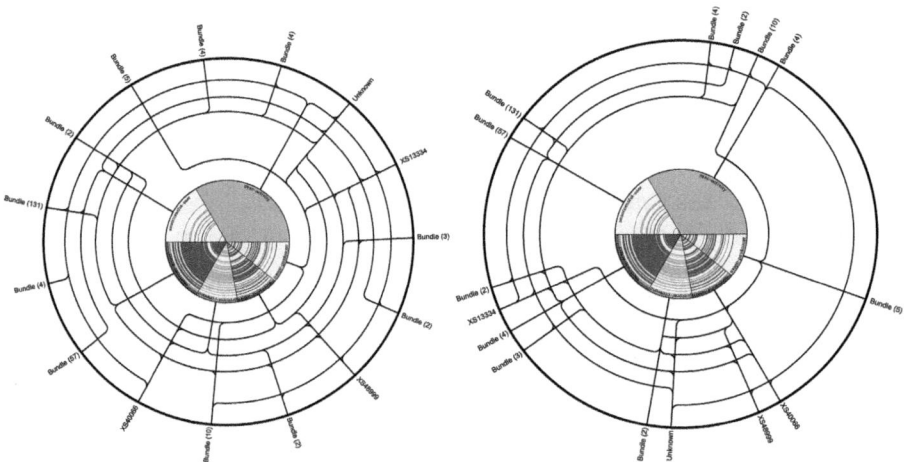

Fig. 7. AlertWheel tool: Comparing the naive algorithm (left) and the anchor heuristic (right). In the former case, the sum of the circular edge lengths is 22036 units. In the latter case, the sum is 17184, or 22% less.

- 30% of the exterior nodes have 3 edges.
- 20% of the exterior nodes have 2 edges.
- 20% of the exterior nodes have only one edge.

For each of these exterior points, their connected neighbors are randomly selected among the ten interior points.

The results of the experiment are presented in Fig. 8. For each number of nodes, 20 bipartite graphs have been generated. The figure shows the average of the total sum of circular edge lengths for each algorithm.

The performance of the algorithms can be compared with a theoretical lower bound. The lower bound is found by finding the optimal position of each exterior point (as given by Lem. 2) and allowing exterior points to coincide. Thus, this lower bound is unachievable in practice. As expected, the anchor heuristic gives very good results. If the number of competing exterior points for a given anchor is small, each point should be very close to its optimal solution. This should yield a solution that is close to the globally optimal solution.

To conclude, it should be mentioned that the anchor heuristic has a disadvantage for the AlertWheel visualization tool. Because the exterior points are non uniformly distributed on the outer circle, a radial labelling of the nodes has to be used instead of a circular labelling (as in Fig. 5 and 7).

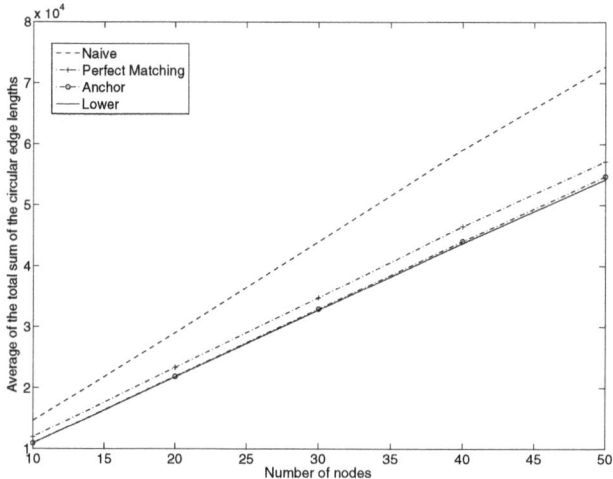

Fig. 8. Comparing the performance of the different algorithms on random bipartite graphs with $n = 10, 20, 10, 40$ and 50 nodes

7   Conclusions

In addition to the user interaction problems that had to be solved during the development of AlertWheel, multiple bipartite graph drawing problems also had to be addressed. In all cases, the naive approaches to laying out the graph without any optimization give poor results. The large quantities of information to deal with have necessitated finding good heuristics to reduce the clutter in the drawing of the radial representation of the bipartite graph representing the observed security alerts of an IDS. One of these heuristics gives very good results which are close the globally optimal solution.

References

1. Abdullah, K., Lee, C., Conti, G., Copeland, J.A., Stasko, J.: IDS RainStorm: visualizing IDS alarms. In: IEEE Workshop on Visualization for Computer Security, pp. 1–10 (2005)
2. Bachmaier, C.: A radial adaptation of the Sugiyama framework for visualizing hierarchical information. IEEE Transactions on Visualization and Computer Graphics 13, 583–594 (2007)
3. Cormen, T.H., Leiserson, C.E., Rivest, R.L., Stein, C.: Introduction to Algorithms, 3rd edn. MIT Press (2009)
4. Dumas, M., Robert, J.M., McGuffin, M.J., Willig, M.C.: AlertWheel: Radial bipartite graph visualization applied to intrusion detection system alerts (submitted for publication)
5. Foresti, S., Agutter, J., Livnat, Y., Moon, S., Erbacher, R.: Visual correlation of network alerts. IEEE Comput. Graph. Appl. 26, 48–59 (2006)
6. Gansner, E., Koren, Y.: Improved Circular Layouts. In: Kaufmann, M., Wagner, D. (eds.) GD 2006. LNCS, vol. 4372, pp. 386–398. Springer, Heidelberg (2007)
7. Northcutt, S., Novak, J.: Network Intrusion Detection, 3rd edn. New Riders, Indianapolis (2002)
8. Pupyrev, S., Nachmanson, L., Kaufmann, M.: Improving Layered Graph Layouts with Edge Bundling. In: Brandes, U., Cornelsen, S. (eds.) GD 2010. LNCS, vol. 6502, pp. 329–340. Springer, Heidelberg (2011)
9. Purchase, H.C.: Which Aesthetic has the Greatest Effect on Human Understanding? In: Di Battista, G. (ed.) GD 1997. LNCS, vol. 1353, pp. 248–261. Springer, Heidelberg (1997)
10. Tamassia, R., Palazzi, B., Papamanthou, C.: Graph Drawing for Security Visualization. In: Tollis, I.G., Patrignani, M. (eds.) GD 2008. LNCS, vol. 5417, pp. 2–13. Springer, Heidelberg (2009)
11. Zheng, L., Song, L., Eades, P.: Crossing minimization problems of drawing bipartite graphs in two clusters. In: Asia-Pacific Symposium on Information Visualisation, pp. 33–37 (2005)

Visual Community Detection: An Evaluation of 2D, 3D Perspective and 3D Stereoscopic Displays

Nicolas Greffard, Fabien Picarougne, and Pascale Kuntz

KoD Research Team - LINA - Polytech'Nantes, rue Christian Pauc
BP50609 France 44306 Nantes Cedex 3
{Nicolas.Greffard,Fabien.Picarougne,Pascale.Kuntz}@univ-nantes.fr

Abstract. 3D drawing problems of the 90s were essentially restricted on representations in 3D perspective. However, recent technologies offer 3D stereoscopic representations of high quality which allow the introduction of binocular disparities, which is one of the main depth perception cues, not provided by the 3D perspective. This paper explores the relevance of stereoscopy for the visual identification of communities, which is a task of great importance in the analysis of social networks. A user study conducted on 35 participants with graphs of various complexity shows that stereoscopy outperforms 3D perspective in the vast majority of the cases. When comparing stereoscopy with 2D layouts, the response time is significantly lower for 2D but the quality of the results closely depend on the graph complexity: for a large number of clusters and a high probability of cluster overlapping stereoscopy outperforms 2D whereas for simple structures 2D layouts are more efficient.

1 Introduction

Long after the pionnering work of Kolmogorov [1], 3D drawings knew a phase of great interest in the mid-90's in the graph drawing community. Besides the beauty of the theoretical questions, this interest was mostly motivated both by the availability of new 3D display hardware, and by the exploration of new applications which emerged in particular in VLSI design (e.g. [2]). The most studied models included orthogonal grid drawings, convex and straight line drawings (e.g. [3], [4], [5]). And, the most common aesthetic criteria were the bounding area volume, the minimization of edge length and bends. NP-completeness proofs were deduced from 2D for the different criteria, and several theoretical bounds were highlighted in different cases. And, different algorithms and tools (e.g. GIOTTO3D, GEM-3D, 3D CUBE) were developed.

However, despite all these efforts, the 3D phase rapidly declined and it is sometimes considered as a prejudicial epiphenomenon in the graph drawing community (see Eades's invited talk at GD'10[1]). The main criticsm concerns the lack of layout lisibility often illustrated by the paradigmatic 3D drawing of K_7. Some

---

[1] http://www.graphdrawing.org/gd2010/invited.html

M. van Kreveld and B. Speckmann (Eds.): GD 2011, LNCS 7034, pp. 215–225, 2012.
© Springer-Verlag Berlin Heidelberg 2012

encouraging results for the multilayer layouts have nevertheless continued to punctually draw the attention. But as said by Eades they "use 3D with a 2D attitude"; the third dimension is added for representing a parameter (e.g. time).

Nevertheless, we believe that the 3D "trial" is essentially due to a non proper definition of 3D, and consequently to an inappropriate choice of the aesthetics. In the works previously quoted, 3D drawings are 2D representations of a perspective view, and the aesthetics are directly re-used from 2D without specific analysis of their adequation in 3D.

Recent cheap technologies offer 3D stereoscopic representations of high quality which allow the introduction of binocular disparities, which is one of the main depth perception cues, not provided by the 3D perspective. Stereoscopy may also be combined with virtual reality (e.g. [6]) but this approach is far beyond the scope of this paper. Although the general question of the benefits offered by stereoscopy over 2D still remains open, investigations on the 3D expressiveness are attracting a growing community of computer scientists (e.g. for an overview, [7]).

In graph drawing, recent works have shown the interest of stereoscopic 3D representations of node-link layouts of graphs for local analysis : in particular, [8] have experimentally confirmed the power of 3D, previously hightlighted by [9], to trace out short paths between close (distance 2 or 3) vertices in limited size graphs (with less than 150 nodes).

In this paper, we explore the relevance of stereoscopy for a higher level task: the identification of communities (i.e. vertex subsets strongly connected to each other) . This task is of great importance in the analysis of social networks where visualization knows an increasing interest. Most often, communities are first identified with a clustering approach for which various algorithms have been proposed (see [10] for a recent state-of-the art), and visual representations of clustered graphs are then used. But, the community detection suffers from a major problem: in many real life situations, communities do not form a non ambiguous partition of the graph and several overlappings are present. In order to tackle this difficulty, alternative strategies have been developed: e.g. give a particular place to some pre-defined vertices ("central actors") which are members of different communities ([11]), or duplicate vertices which belong to different communities (e.g. [12]). Other representations than node-link diagrams have also been proposed but we here restrict ourselves to the latter which is by far the most popular visual representation of social networks, and the only one to have been investigated in 3D.

Here, we do not propose an $n + 1th$ strategy, but we analyse the resort to stereoscopy for detecting communities in a "crude" representation of the whole graph - obtained with the Fruchterman-Reingold algorithm. Roughly speaking, the question we are asking in our experiments is "How different is stereoscopic representation from 2D, and 3D perspective views for identifying communities in medium size graphs with different complexities ?" Beyond its popularity, the choice of the algorithm, which has been shown to be outperformed in 2D by other approaches, is here justified by two reasons: it directly computes a layout which highlights communities without pre- or post- processings, and it is directly applicable to both 2D, 3D perspective and stereoscopy which consequently limits bias in the comparisons. For pseudo-random graphs associated with different parameters such as the cluster size and the density of intra and inter cluster links, we conducted a user study: we asked participants the number of communities they could detect and we measured the time required to answer.

The rest of this paper is organized as follows. Section 2 briefly recalls some psycho-visual generalities on human 3D perception which guide our research. Section 3 details the experimental procedure for the comparisons. Finally, the results are analyzed in Section 4.

2   Stereoscopic Perception

We live in a 3D space and a long period of evolution has endowed us with organs that allow us to perceive a three dimensional space from visual information. A huge amount of research in cognitive sciences has been dedicated to the mechanisms implied in the perception of 3D environment (e.g. [13]). In addition to those associated with the shape and object detection, the biological mechanisms, which govern the perception of a distance in the optical axis -the depth-, play a crucial role. Depth perception is certainly a combination of several perceptive mechanisms, and various studies have been carried out in the last decades to try to measure the relative performance of different functions and visual cues used by the human brain to perceive depth (e.g. [14], [15], [16]). Roughly speaking, it seems that partial occlusion (an object partially in front of another one) is one of the most important factors whatever the distance between the person and the object. For limited distances (less than 40m), binocular disparities associated with stereoscopic vision and perspective motions also have an important effect, whereas for larger distances -not considered in this paper- other factors like aerial perspective come into play ([17], [18]).

The comparison of the relative effects of stereoscopy and motion cues is still widely discussed. They seem to be equivalent or complementary in various tasks (e.g. [19], [20]). More precisely, motion itself associated with a 3D perspective may give a faithful depth reproduction: object rotations jointly act with the spatial memory to form a 3D mental representation of the observed object ([15]). But motion is also useful both in perspective and in stereoscopy to detect objects hidden in a particular vision axis. Consequently, measuring the interdependence between these two cues is a very difficult task. This paper restricts itself to the comparison of stereoscopy and 3D perspective both combined with basic motion possibilities at a macroscopic level. However the analysis of the specific effects of motion are in our short term plans.

3   Experimental Design

Three viewing methods were employed during the experiment:

– 2D: a 2D graph layout was computed with the Fruchterman-Reingold algorithm and displayed on a 2D surface. The users were allowed to zoom in/out on the graph using the mousewheel and to apply a z-axis rotation on the layout by moving the mouse.
– 3D Perspective (3D persp): a 3D graph layout was computed with the Fruchterman-Reingold algorithm and displayed on a 2D surface with a perspective projection. Along with the zoom in/out and the z-axis rotation, the user was allowed to spin the viewpoint around the graph (x/y-axis rotation).
– 3D Stereoscopy (3D stereo): the same 3D layout with perspective projection as case 2 was used but with two viewpoints -computed in real time- to introduce the binocular disparity: one viewpoint for each eye, with a slight shift on the horizontal viewing axis to mimic the actual separation between the human eyes. The same interactions than in case 2 were allowed.

3.1   Apparatus

The visualization system ran on an Intel Core 2 Duo (3.00 Ghz) E8400 processor, with 4 GB of RAM and an NVidia Quadro FX 3800 GPU. All graphs were displayed in shades of white on a black background, using Gouraud shading (without projective shadows) and an anti aliasing algorithm to improve the quality of the display. The visualization was displayed on a white painted wall by an ACER H5360 3D projector $(2,30 \times 1,30 \text{ m}^2$ screen) with a resolution of 1280×720 pixels (view angle of 0.05 degrees for a pixel in the center of the screen). Our system uses active stereoscopy with Nvidia 3D Vision Shutter glasses. By using these glasses we decrease the perceived luminosity. To avoid any bias and ensure the same level of luminosity for each viewing method, participants had to wear the glasses through all the experiment. Participants could also

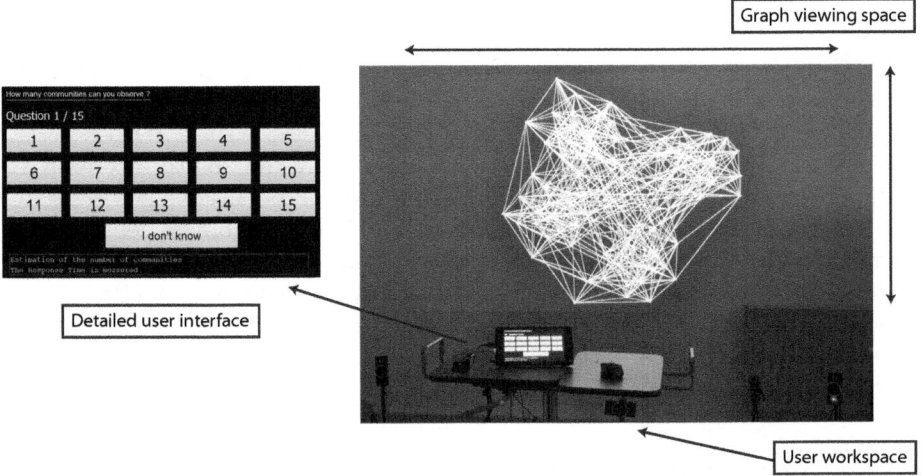

Fig. 1. Photograph of the experimental setup, along with a snapshot of the user interface of the tablet pc

interact with the system through a wireless mouse. The answers were entered using a touch screen tablet PC: different numbers were proposed (between 1 and 15, plus "don't know") and participants just had to touch the corresponding number. The experimental set up can be seen on Fig. 1.

3.2 Graph Database

In order to analyse the stereoscopic viewing for different topologies we have generated graphs with a classical pseudo-random model (e.g. [21]). The generic model $G\left(k, nv, \frac{p_{int}}{p_{ext}}\right)$ depends on four parameters: the number k of *a priori* clusters, the number of vertex per cluster nv, the probability p_{int} (resp. p_{ext}) of edge between two vertices belonging to the same cluster (resp. different clusters). We have generated 480 graphs with parameters ranges specified as $k \in \{4, 5, ..., 11\}$, $nv \in \{10, 20, 30, 40\}$, $\frac{p_{ext}}{p_{int}} \in \{\frac{0.02}{0.8}, \frac{0.02}{0.7}, \frac{0.03}{0.8}, \frac{0.03}{0.7}, \frac{0.03}{0.6}, \frac{0.04}{0.7}, \frac{0.03}{0.5}, \frac{0.05}{0.8}, \frac{0.05}{0.7}, \frac{0.05}{0.6},$ $\frac{0.065}{0.6}, \frac{0.07}{0.6}, \frac{0.1}{0.8}, \frac{0.08}{0.6}, \frac{0.1}{0.7}\}$. The parameters p_{int} and p_{ext} were empirically determined during a previous study by two confirmed users. Furthermore, the largest graphs were discarded to avoid any performance issue. A few examples of such graphs are shown on Fig. 2.

3.3 Participants

35 participants (25 males, 10 females) carried out the experiment. Aged from 20 to 50, 30 of them were computer science students or researchers. Three of the subjects were left-handed with a right-handed use of the mouse. Only two subjects had never visualized any stereoscopic material, and eleven out of the 35 participants were not familiar with 3D software such as video games.

3.4 Experimental Procedure

To limit the experimentation duration, 15 layouts per viewing method were successively presented to each participant. To avoid any potential bias, the 3 viewing methods appeared in a random order (no interleaving). The 3×15 layouts were randomly picked from the database -without duplication within the same viewing method. This process aims at avoiding any learning bias in the community detection task. The maximal task completion time was 28 minutes (15 minutes on average).

Before the experiment, a few questions were asked to the participants to gather their experience with graph theory, graph visualization, stereoscopic displays and 3D software. A description sheet was handed out to briefly explain the experimental process, and a quick demo presented the three viewing methods with an easily readable graph (3 clusters, 20 nodes per cluster and a high p_{int} and small p_{ext}). Then, the participants had to go through a training session to get familiar with the system; the training session consisted in 3 layouts of increasing complexity for each viewing method.

Participants were asked to estimate the number of communities displayed as fast as possible, and told that the experimenter couldn't help them. If a participant felt unable to detect communities, he/she was told to skip the layout

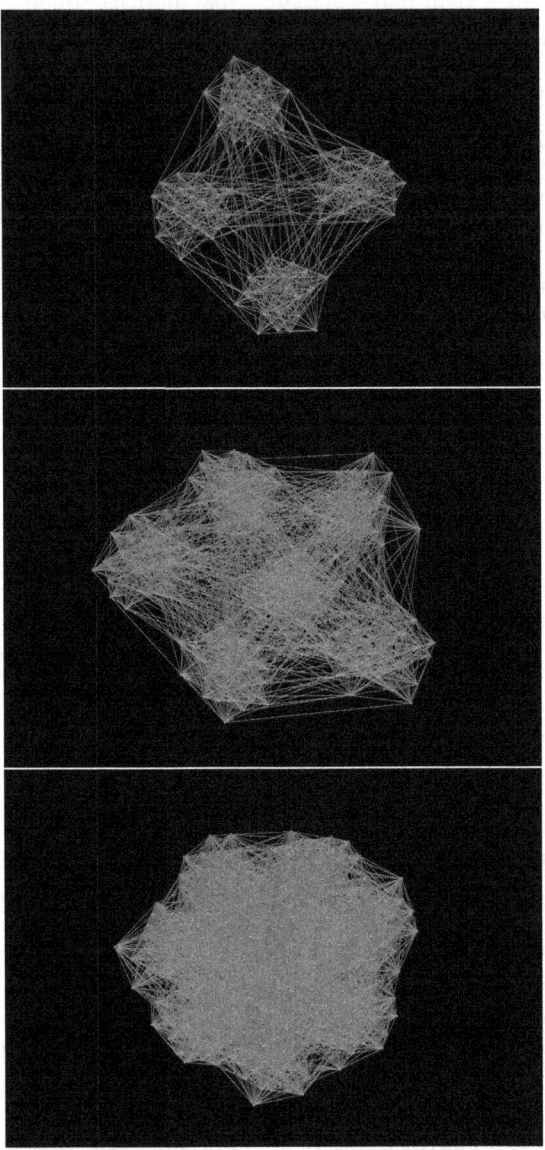

Fig. 2. Snapshots of the drawings of graphs of increasing complexity. $G_1 = G(k = 4, nv = 20, \frac{p_{ext}=0.05}{p_{int}=0.8})$; $G_2 = G(k = 6, nv = 30, \frac{p_{ext}=0.05}{p_{int}=0.7})$; $G_3 = G(k = 8, nv = 30, \frac{p_{ext}=0.07}{p_{int}=0.8})$.

by pressing a button labelled "I don't know". Once all the layouts of a viewing method had been presented, the following method started with a notice of the experimenter ("you have finished the xth series, we move on to the next series"). At the end of the experiment, the participants were asked to state their visualization preferences (easiest and hardest cases), and their estimation (cases with the best and the worst performance).

4   Results

4.1   Quality of Community Detection

Let X be the set of answers given by all participants and I a subset of such that all the instances i of I share the same ratio range p_{ext}/p_{int}. Let I_{ans}^i be the number of communities proposed by a participant for an instance i, and I_{act}^i the a priori number of communities in the model (section 3.2). For each viewing method vm, $vm = \{2D; 3Dpersp; 3Dstereo\}$, the error is measured by the average of the differences :

$$Error_{vm}(I) = \frac{1}{card(I)} \sum_{i \in I} \left(\left| I_{ans}^i - I_{act}^i \right| \right)$$

Note that the answer "I don't know" was counted as an error (i.e. the detection of only 1 community), and that for high values of p_{ext} and low values of p_{int} the term "error" is not perfect since the detection may be very ambiguous. But in these cases, the quality measurement of the community detection remains an open difficult question (see Fortunato, 2010).

The results depend on the complexity of the graph structuration here measured by the ratio p_{ext}/p_{int}: for small values, the communities are easy to identify, whereas for high values important overlappings and small community densities may make the detection difficult. Table 1 shows that 2D is significantly better for a small complexity (two way ANOVA: $p = 0.01$): communities are well-separated on the layout and are easily detected on a plane. For larger complexities, stereoscopy is slightly better (two way ANOVA: $p = 0.1$) in particular for large k

Table 1. Error for community detection. For each viewing method vm and for each interval of structure complexity (p_{ext}/p_{int}) I, mean error $Error_{vm}(I)$, along with the cardinality of each interval.

Complexity (p_{ext}/p_{int})	Cardinality	2D	3D persp	3D stereo
[0.02; 0.04]	369	**0.10**	0.37	0.27
]0.04; 0.06]	406	1.62	1.64	**1.40**
]0.06; 0.11]	405	3.27	3.22	**2.78**
]0.11; 0.15]	260	3.47	3.71	**2.99**

Table 2. Error for community detection. For each viewing method vm and for I, mean error $Error_{vm}(I)$ depending on k.

Complexity (p_{ext}/p_{int})	viewing method	$k = 4$	$k = 5$	$k = 6$	$k = 7$	$k = 8$	$k = 9$	$k = 10$	$k = 11$
	2D	0.0	0.0	0.1	**0.05**	**0.06**	0.16	**0.13**	**0.67**
[0.02; 0.04]	3D persp	0.0	0.0	0.13	0.08	0.25	0.35	0.37	1.56
	3D stereo	0.0	0.0	**0.0**	0.07	0.16	**0.11**	0.42	1.67
	2D	0.0	0.14	**0.0**	**0.19**	1.27	2.65	**3.35**	6.94
]0.04; 0.06]	3D persp	0.0	0.0	0.08	0.58	1.45	1.92	4.13	5.93
	3D stereo	0.0	0.0	0.13	0.9	**0.63**	**0.69**	3.81	**5.5**
	2D	0.0	0.0	0.5	1.35	2.24	3.95	6.15	7.9
]0.06; 0.11]	3D persp	0.0	0.13	0.32	**1.07**	**1.63**	4.44	6.12	7.61
	3D stereo	0.0	0.0	**0.21**	1.6	1.9	**3.42**	5	**6.88**
	2D	**0.28**	**0.88**	**2**	**4.69**	6.14	7.13	9	10
]0.11; 0.15]	3D persp	0.73	1.23	2.3	5.81	5.44	**6**	8.75	8.75
	3D stereo	0.93	1.35	2.06	4.75	**5.25**	8	**6.6**	**6**

Table 3. Standard deviation of the error for community detection for each viewing method and for a threshold of structure complexity (p_{ext}/p_{int})

Complexity (p_{ext}/p_{int})	2D	3D persp	3D stereo
< 0.06	1.28	1.16	1.67
≥ 0.06	3.63	3.69	3.38

values. Table 2 shows that for $k > 7$ stereoscopy is significantly better than 2D for a complexity greater than 0.06 (two way ANOVA: $p = 0.02$). The additional perceptive dimension combined with the motion seems to help to distinguish the aggregates even in presence of "noise" (overlappings). The situation is different for 3D perspective for which occlusions partly explain the debased results. Nevertheless, let us note that, whatever the viewing method, the error variation is important as soon as the complexity increases (Table 3). We have observed that this variation is similar for any value of k. Moreover, in our experimental sample group, it can not be explained by the non-familiarity with the 3D software but complementary experiments are required to reject this hypothesis.

4.2   Response Time

For each viewing method vm, the response time is the average time of response $Time_{vm}(I)$ of participants.

Table 4 shows that the response time for 2D is significantly smaller than for 3D whatever the graph complexity. And, the response times of 3D perspective and stereoscopy are very similar. However, as discussed in the conclusion, a more precise analysis of the participant behavior (mouse motion) seems to establish that the time exploitation is different in the two cases.

Table 4. Response time for community detection. For each viewing method vm and for each interval of structure complexity (p_{ext}/p_{int}), mean time $Time_v m(I)$ in seconds.

Complexity (p_{ext}/p_{int})	2D	3D persp	3D stereo
[0.02; 0.04]	**7.3**	14.2	12.3
]0.04; 0.06]	**11.1**	17.9	17.7
]0.06; 0.11]	**12.4**	22.5	24.7
]0.11; 0.15]	**13.1**	21.9	21.1

4.3 Participant Perception

Table 5 underlines participant preferences for the stereoscopy, and difficulties felt with both 2D and 3D perspective. We are aware that a bias could exist: the experimentation by itself shows our interest in 3D to the participants who may be unconsciously inclined to share the researcher's enthusiasm. Nevertheless, part of the subjectivity is corroborated by the experimentations: among the participants who estimated that their best results were obtained with stereoscopy, 54% had their intuition confirmed by the results (whereas only 15.5% of them obtained their best results with the 3D perspective). Consequently, the "reject" of 3D perspective recalled in the introduction is verified. But, the comparisons show that this subjective perception is significantly different from stereoscopy.

Table 5. Subjective perception of the participants. For each viewing method vm, percentage of participants who answered that the case vm is the easiest (resp. the hardest) and the one for which they believe to obtain the best (resp. worst) performances. (NA: don't know)

Answer	2D	3D persp	3D stereo	NA
easiest	14.2	0	**68.6**	17.2
hardest	37.1	43	**5.7**	14.2
best performances	11.3	0	**74.3**	14.4
worst performances	43	34.3	**5.7**	17

5 Conclusion

As far as we know, this paper presents a pionnering research in the use of stereoscopy for a visualization problem which has known an increasing interest in the last decade: the detection of communities in large graphs. Our first experiments highlight an important difference between stereoscopy and classical 3D perspective which has been widely critized by the graph drawing community. Moreover, even if the debate remains widely open, experimental results seem to show the interest of stereoscopy against 2D for complex structures with numerous clusters of variable density and many overlappings. Obviously, additional experiments are needed to confirm these results on larger populations and real life databases; and to better understand the observed differences.

In the near future, our objective is to go beyond the measurement of errors by apprehending more precisely the role played by motion in 3D environment. To that aim, during the experiments we stored the mouse movements for each participant. A first superficial analysis highlights important use differences between 3D perspective and stereoscopy. This research could lead to investigating new optimization criteria in the graph drawing community which take into account not only the aesthetics of the layouts but also the handling for their interpretation and use.

References

[1] Kolmogorov, A., Barzdin, Y.: Abour realization of sets in 3-dimensional space, problems cybernet (1967)

[2] Rosenberg, A.L.: Three-dimensional vlsi: a case study. J. ACM 30, 397–416 (1983)

[3] Di Battista, G., Patrignani, M., Vargiu, F.: A Split&Push Approach to 3D Orthogonal Drawing. In: Whitesides, S.H. (ed.) GD 1998. LNCS, vol. 1547, pp. 87–101. Springer, Heidelberg (1999)

[4] Eades, P., Symvonis, A., Whitesides, S.: Three-dimensional orthogonal graph drawing algorithms. Discrete Applied Mathematics 103(1-3), 55–87 (2000)

[5] Wood, D.R.: Optimal three-dimensional orthogonal graph drawing in the general position model. Theor. Comput. Sci. 299, 151–178 (2003)

[6] Halpin, H., Zielinski, D., Brady, R., Kelly, G.: Exploring Semantic Social Networks using Virtual Reality. In: Sheth, A.P., Staab, S., Dean, M., Paolucci, M., Maynard, D., Finin, T., Thirunarayan, K. (eds.) ISWC 2008. LNCS, vol. 5318, pp. 599–614. Springer, Heidelberg (2008)

[7] Teyseyre, A., Campo, M.: An overview of 3D software visualization. IEEE Trans. on Visualization and Computer Graphics 15(1), 114–135 (2009)

[8] Ware, C., Mitchell, P.: Visualizing graphs in three dimensions. ACM Transactions on Applied Perception 5, 2–15 (2008)

[9] Belcher, D., Billinghurst, M., Hayes, S., Stiles, R.: Using augmented reality for visualizing complex graphs in three dimensions. In: ISMAR, pp. 84–92 (2003)

[10] Fortunato, S.: Community detection in graphs. Physics Reports 486, 75–174 (2010)

[11] Auber, D., Chiricota, Y., Jourdan, F., Melancon, G.: Multiscale visualization of small world networks. In: INFOVIS 2003: Proceedings of the IEEE Symposium on Information Visualization (INFOVIS 2003), pp. 75–81 (2003)

[12] Henry, N., Bezerianos, A., Fekete, J.D.: Improving the readability of clustered social networks using node duplications (2008)

[13] Landy, M.S., Maloney, L.T., Young, M.J.: Psychophysical estimation of the human depth combination rule. vol. 1383, pp. 247–254. SPIE (1991)

[14] Hubona, G.S., Wheeler, P.N., Shirah, G.W., Brandt, M.: The relative contributions of stereo, lighting, and background scenes in promoting 3D depth visualization. ACM Trans. Comput.-Hum. Interact. 6, 214–242 (1999)

[15] van Schooten, B.W., van Dijk, E.M.A.G., Zudilova-Seinstra, E., Suinesiaputra, A., Reiber, J.H.C.: The effect of stereoscopy and motion cues on 3D interpretation task performance. In: Proceedings of the International Conference on Advanced Visual Interfaces, AVI 2010, pp. 167–170. ACM, New York (2010)

[16] Ware, C., Mitchell, P.: Reevaluating stereo and motion cues for visualizing graphs in three dimensions. In: Proceedings of the 2nd Symposium on Applied Perception in Graphics and Visualization, APGV 2005, vol. 95. ACM (2005)

[17] Cutting, J.: How the eye measures reality and virtual reality. Behavior Research Methods, Instrumentation, and Computers 29, 29–36 (1997)

[18] Saracini, C., Franke, R., Blümel, E., Belardinelli, M.: Comparing distance perception in different virtual environments. Cognitive Processing 10, 294–296 (2009)

[19] Ware, C., Franck, G.: Evaluating stereo and motion cues for visualizing information nets in three dimensions. ACM Transactions on Graphics 15, 121–139 (1996)

[20] Domini, F., Caudek, C., Tassinari, H.: Stereo and motion information are not independently processed by the visual system. Vision Res. 46, 1707–1723 (2006)

[21] Garbers, J., Promel, H.J., Steger, A.: Finding clusters in vlsi circuits. In: Proceedings of ICCAD 1990, pp. 520–523 (1990)

Evaluating Partially Drawn Links
for Directed Graph Edges

Michael Burch, Corinna Vehlow, Natalia Konevtsova, and Daniel Weiskopf

VISUS, University of Stuttgart
Allmandring 19, 70569 Stuttgart, Germany

Abstract. We investigate the readability of node-link diagrams for directed graphs when using partially drawn links instead of showing each
link explicitly in its full length. Providing the complete link information
between related nodes in a graph can lead to visual clutter caused by
many edge crossings. To reduce visual clutter, we draw only partial links.
Then, the question arises if such diagrams are still readable, understand-
able, and interpretable. As a step toward answering this question, we
conducted a controlled user experiment with 42 participants to uncover
differences in accuracy and completion time for three different tasks:
identifying the existence of a direct link, the existence of an indirect
connection with one intermediate node, and the node with the largest
number of outgoing edges. Furthermore, we compared tapered and traditional edge representations, three different graph sizes, and six different
link lengths. In all configurations, the nodes of the graph were placed according to the force-directed layout by Fruchterman and Reingold. One
result of this study is that the characteristics of completion times and
error rates depend on the type of task. A general observation is that
partially drawn links can lead to shorter task completion times, which
occurs for nearly all graph sizes, tasks, and both tapered and traditional
edge representations. In contrast, there is a tendency toward higher error
rates for shorter links, which in fact is task-dependent.

1   Introduction

Visualizing graph data as node-link diagrams can lead to visual clutter [17].
This problem is most pronounced for dense graphs causing a huge number of
edge crossings. Therefore, typical graph layout algorithms for node-link diagrams
follow certain aesthetic criteria for graph drawing where the reduction of edge
crossings is ranked very high. Other important aesthetic criteria include the
minimization of edge lengths, the maximization of angles at link intersections,
and the preservation of symmetries.

In our work, we ask the question if node-link graph visualizations are still
useful, readable, and interpretable when reducing visual clutter by drawing links
partially instead of showing each link explicitly in its full length and applying
some sophisticated layout algorithm. To visually encode directed edges, we draw
partial links beginning at the start vertex and pointing to the target vertex

M. van Kreveld and B. Speckmann (Eds.): GD 2011, LNCS 7034, pp. 226–237, 2012.
© Springer-Verlag Berlin Heidelberg 2012

instead of ending exactly there. By doing this, many explicit link intersections are avoided, a fact that definitely reduces visual clutter but increases ambiguities. We speculate that this may lead to more graph misinterpretations but we are unsure how error rates and completion times behave when solving graph-related tasks. We also speculate that there is an optimal range of link lengths that balances the goals of reducing visual clutter on the one hand and minimizing error rates and completion times on the other hand.

We conducted a controlled user experiment with 42 participants to find out if the partial link visualization strategy for directed graphs has any benefits over the traditional complete link representations and, if so, what the best-suited link lengths would be. For the experiments, we used artificial data sets with constant characteristics: randomly generated graph data that follows the Barabási-Albert [1] model for scale-free networks where the degrees form a power-law distribution. The graphs were laid out with the force-directed algorithm by Fruchterman and Reingold [5], which meets relevant aesthetic criteria for graph drawing.

We used the following relevant independent variables in our study: edge style (tapered straight links according to Holten and Van Wijk [8] and straight links as used in traditional approaches [3]), varying number of vertices to reflect different sizes of graphs, and varying length of links to test for partially drawn links. We employed three different tasks in our study: (1) identifying the existence of a direct link between two highlighted nodes, (2) identifying the existence of an indirect connection with one intermediate node between two highlighted nodes, and (3) the detection of the node with the highest number of outgoing edges. The user study collected accuracies and completion times for those tasks to identify performance for the different settings of the independent variables.

2 Related Work

Graph visualization techniques aim at producing graph layouts that are readable, interpretable, and look aesthetically pleasing to the viewer; see Di Battista et al. [3] for an overview of graph visualization. We focus on the issue of visual clutter in graph layouts, which is getting more and more prominent with increasing data set size.

One approach to reducing clutter relies on partially drawn links. Early work in this direction is due to Becker et al. [2], who visualized graphs with half-links (called half-lines in their paper), i.e., a directed link is connected with its start vertex and points to its target vertex but cut at halfway. With line-shortening, they even used links with further reduced length. This visualization strategy reduces visual clutter by reducing the number of explicit link crossings; however, Becker et al. did not provide any user study to evaluate the effectiveness of their visualization approach. In recent work, Rusu et al. [18] investigated another variant of partially drawn links in diagrams of undirected graphs. They introduced short breaks in full links (instead of one piece of a short link in our case), relying on the Gestalt principle of closure to perceive the whole link; see

Koffka [12] for background information on Gestalt psychology. They provided a preliminary user-based evaluation; they conducted a subjective study, whereas we focus on a task-based evaluation with accuracies and completion times.

Graph drawing aesthetics also aim at reducing visual clutter for good readability. Rosenholtz et al. [17] developed a measurement technique for (generic) display clutter, based on color and luminance contrast features. They demonstrated that their measure can be used in an automated way to make design suggestions for drawing properties such as the location of an item. In general, the graph layout strongly affects the extent of clutter. There are many corresponding node-link graph layout algorithms; many of them employ a force-directed node placement, e.g., Eades' spring-embedder model [4], the Kamada-Kawai model [11], or the Fruchterman-Reingold model [5]. We base the graph layouts in our study on the Fruchterman-Reingold model because it aims at meeting several aesthetic criteria for graph drawing.

Purchase et al. [14,15,16] conducted several empirical studies on the aesthetics of graph layouts and discovered that the layout significantly affects user preferences and task performances. In their first study [16], they investigated effects of three common aesthetics criteria on the readability of graphs: symmetry, link crossings, and bends. They reported that minimization of bends and link crossing improves task performance, where the latter was identified as most important factor on graph reading performance [14].

Ware et al. [19] found out that not only edge crossing but also continuity is an important factor for aesthetic considerations. They indicated that clutter rather depends on the number of edges that cross a path itself, than the total number of edge crossings in the diagram. They reported that the angle between crossings affects readability. Their results were also supported by eye-tracking studies of Huang et al. [9,10] that showed that small angles cause slow eye movements. Holten et al. [7,8] performed several studies to evaluate the performance and preference of different directed edge representations. Their results showed a significant performance advantage for tapered and non-compressed animation representations compared to standard arrowheads. However, their study did not include the half-links of Becker et al. [2]. To close this gap, we focus our study on partially drawn straight links and less on the style of edge representation.

To evaluate the performance of graph layouts or edge representations, it is critical to choose adequate tasks. Lee et al. [13] suggested a list of low-level tasks and complex tasks to allow the generalization of experimental results. We picked three relevant tasks from their category of topology-based tasks for our study.

Finally, matrix visualization is another approach to graph visualization, substantially different from node-link diagrams. We restrict ourselves to evaluating node-link diagrams. For a user study on comparing matrix and node-link diagrams, we refer to Ghoniem et al. [6]. Amongst other results, they reported that tasks connected to finding paths were supported more effectively and efficiently in node-link diagrams than in matrix visualizations.

3   Graph Generation and Layout

We base our directed graph data used throughout the study on the Barabási-Albert graph model [1]. The graphs are laid out by the Fruchterman-Reingold algorithm [5]. We compare tapered [8] and traditional edge representations; and we vary the link lengths as well as the graph sizes.

3.1   Graph Model

Graphs are randomly generated by using the Barabási-Albert model [1], which produces scale-free networks following a power-law distribution for node fan-in and fan-out. By doing this, we guarantee that all graphs have similar statistical properties throughout the study. Based on this graph generation model, we implemented a Java program that generates directed graph data on demand.

3.2   Graph Layout

The graph data is then represented by applying the Fruchterman-Reingold algorithm [5], implemented in our Java-based study software. Figure 1 shows visualizations of example graphs generated by the Barabási-Albert model for three different graph sizes. Here, links are represented in tapered style that was also used by Holten et al. [7]. Please note that they restricted their study to links of full length.

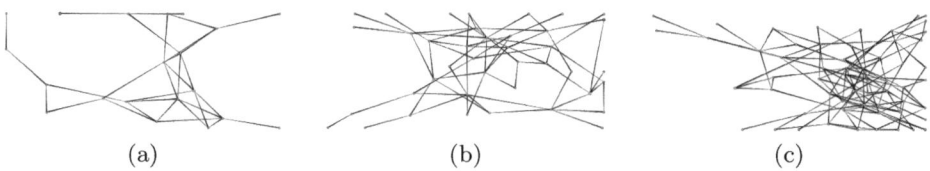

(a)   (b)   (c)

Fig. 1. Directed graph data produced by applying the Barabási-Albert model [1], laid out by the Fruchterman-Reingold algorithm [5], and displayed using tapered links of full length. Three different graph sizes are shown: (a) small graph with 20 nodes, (b) medium sized graph with 40 nodes, (c) large graph with 60 nodes.

3.3   Edge Representation

We use two different styles for representing directed graph edges:

- **Tapered Straight Links.** A needle-like shape that originates with its thicker end from the start vertex and points with its thinner end to the target vertex.
- **Traditional Straight Links.** An equally thick line that originates from the start vertex and heads to the target vertex.

Figure 2 compares tapered style and traditional style.

(a)   (b)

Fig. 2. Different edge styles: (a) traditional link, (b) tapered link. Both links are partial links with 75 percent of full link length.

3.4   Link Length

We vary the link lengths for both tapered and traditional edge representation styles. We use 100, 90, 75, 50, 25, and 12.5 percent of the length the link would have when drawn completely. For the traditional representation, we omit the 100% link length because otherwise the direction of a link could not be recognized by the viewer. Figures 3 (a)–(f) show examples of a graph consisting of 5 nodes and 10 links in tapered style; all 6 variations of link lengths are used. Figures 3 (g)–(k) show the same graph in the same layout, but with the traditional edge representation style; all variations of link lengths are used, except for the 100% link length.

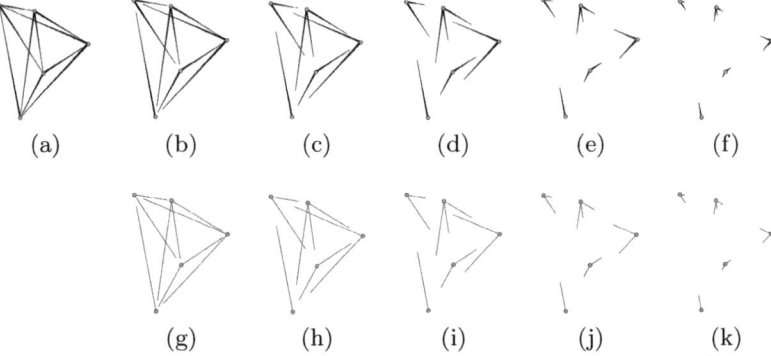

(a)   (b)   (c)   (d)   (e)   (f)

(g)   (h)   (i)   (j)   (k)

Fig. 3. Different link lengths are used in the study for both tapered and traditional edge styles: (a) tapered with 100% link length, (b) tapered 90%, (c) tapered 75%, (d) tapered 50%, (e) tapered 25%, (f) tapered 12.5%, (g) traditional 90%, (h) traditional 75%, (i) traditional 50%, (j) traditional 25%, and (k) traditional 12.5%.

3.5   Graph Size

Another variable that we vary is graph size, i.e., the number of vertices a graph contains. We choose 20 (small), 40 (medium), and 60 vertices (large) to test if there is any impact of graph size on accuracy and completion time. Figure 1 compares the three graph sizes for the example of full-length tapered edge representations. The graph density depends on the edges produced by the Barabási-Albert model.

4 User Experiment

We conducted a controlled user experiment with 42 participants to address the following research questions.

4.1 Research Questions

Since partially drawn links reduce visual clutter on the one hand, but increase the degree of ambiguity on the other hand, we considered the following research questions as relevant:

- **Research Question 1.** Can the tasks be answered more quickly with decreasing link length? In contrast, does the error rate increase due to more and more ambiguities for the target vertices?
- **Research Question 2.** Are the effects of Research Question 1 (i.e., decreasing completion time and increasing error rate) more pronounced for large graphs due to higher levels of overall visual clutter?
- **Research Question 3.** Are there any differences between tapered and traditional edge representations? Tapered links need more pixels to be drawn on screen. Hence, visual clutter is reduced more than in the traditional edge representation the shorter visible links are. Therefore, completion times should decrease more in the tapered style than in the traditional style. The error rates should stay similar since ambiguities occur equally in both styles.

4.2 Design

A repeated-measures design was used with three relevant independent within-subjects variables:

- **Edge Style.** Two possible edge representations: tapered straight links [8] and traditional straight links.
- **Number of Vertices.** Three graph sizes: 20 (small), 40 (medium), and 60 (large) vertices per graph.
- **Length of Links.** Six (five) different lengths of links, as percentage relative to the corresponding complete link: 100 (only for tapered), 90, 75, 50, 25, and 12.5 percent of the complete link.

We checked each of the three tasks in a separate block to reduce cognitive load from task switching. The three blocks were permuted to compensate learning and fatigue effects. Inside each task block, we randomized and balanced the graph sizes and link lengths, alternating between the two edge representation styles. This led to 3 [for graph sizes] × 3 [for tasks] × (5 [link length for traditional] + 6 [length for tapered]) = 99 configurations. Each of the 42 subjects performed each configuration twice (i.e., two repetitions), leading to 198 trials per subject and 8,316 trials in total.

We used a continue-on-demand study design, i.e., participants could decide when the next graph was represented by pressing a "Next" button. Participants were encouraged to take a longer break between the two repetition blocks.

4.3   Participants

We had 42 participants, 16 of whom were female and 26 male. The average age was 24.0 years; the youngest participant was at the age of 20 and the oldest at the age of 30 years. The participants were students of our university, except for one participant that had recently graduated. 16 of the participants were students of computer science or software engineering. All participants had normal or corrected-to-normal color vision, as confirmed by an Ishihara test and a Snellen chart; 15 of them wore glasses and 7 of them contact lenses. 7 participants claimed that they were familiar with graphs, 35 reported that were not (before the study). However, even the latter group was able to read node-link diagrams after a short introduction, as checked by asking graph-specific questions before the main test runs. Participants were compensated with EUR 10. Each experiment took between 44 to 100 minutes, depending on the speed of the participant. The average experiment time was 66 minutes.

4.4   Study Procedure

Participants were first asked to fill out a questionnaire about age, field of study, and prior knowledge in graph visualization techniques. Next, they read a short manual on the different graph diagrams, followed by test questions to check if they were able to read the node-link diagrams and solve the given tasks. Serving as a practice run-through, the initial test phase was conducted with a different set of stimuli data than the real experiment. Then, the actual experiment consisted of two larger blocks of trials (two repetitions as described in Section 4.2). During the experiment, subjects were sitting in front of a TFT screen with a resolution of 1920 × 1200 pixels at a distance of approximately 60 centimeters.

There was a "Give Up" option clearly present throughout the study; however, it was not used by the participants. There was no time limitation for the tasks. The participants were instructed to answer as accurately and as fast as possible. Once they found the solution, they had to confirm it by a mouse click to the correct position on screen (for Task 3, see Section 4.5) or by pressing a green-colored "YES" button or a red-colored "NO" button (for Tasks 1 and 2, see Section 4.5). The next stimulus was shown after the "NEXT" button had been pressed. Figure 4 shows a typical screenshot of the Java software employed for the user study.

4.5   Tasks

We tested three types of tasks in our study:

- **Task 1.** Is it possible to go from the node highlighted in green to the node highlighted in red by taking exactly one step, i.e., is there a directed edge starting at the green-colored node and pointing to the red-colored node?
- **Task 2.** Is it possible to go from the node highlighted in green to the node highlighted in red by taking exactly two steps, i.e., is there a path of length two starting at the green-colored node and ending at the red-colored node?

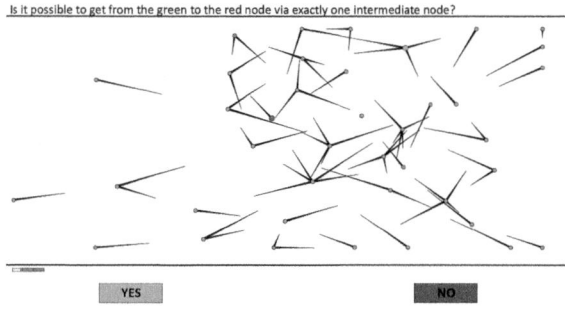

Fig. 4. Example screenshot from the user study. Here, a graph with 40 nodes and tapered links of 50% length is shown. The participant is asked for a path of length two starting at the green-colored node and ending at the red-colored node. (The figure is best viewed in the electronic color version of this paper.)

– **Task 3.** Which node has the highest number of outgoing edges?

Tasks 1 and 2 had to be answered by clicking on a button labeled with "YES" if the viewer agreed or "NO" if they disagreed. Task 3 had to be answered by clicking on the corresponding node on screen. We picked these tasks from the category of topology-based tasks for graphs according to Lee et al. [13]. Tasks 1 and 3 belong to the subcategory of adjacency-related tasks (direct connection between nodes); Task 2 is an accessibility-related task (here with an indirect connection).

5   Results

To evaluate the results, we first averaged the completion times and error rates over the 42 participants and 2 repetitions. This led to aggregated numbers for each of the 99 configurations. The scatterplot in Figure 5 and the line charts in Figure 6 show these averaged numbers.

In the scatterplot, the independent variables edge style, number of vertices, and length of links, as well as the task type are mapped to different visual attributes of glyphs: shape, border width, size, and color (shades of gray in black-and-white print). The scatterplot shows clusters for the different tasks: Task 1 (the direct edge search) has lowest completion times and error rates, with completion times ranging from 4.72 s (seconds) to 8.56 s, and error rates below 14%. The clusters for Task 2 and Task 3 spread more widely, but compared to Task 1, they show a clear tendency toward longer completion times and higher error rates with increasing number of vertices because thick-bordered glyphs lie right of and/or above thin-bordered glyphs.

For further analysis of the data, we turn to the line charts of Figure 6, which show completion times and error rates for all three tasks. Let us first focus on completion times (left column of Figure 6). Concerning the length of the links,

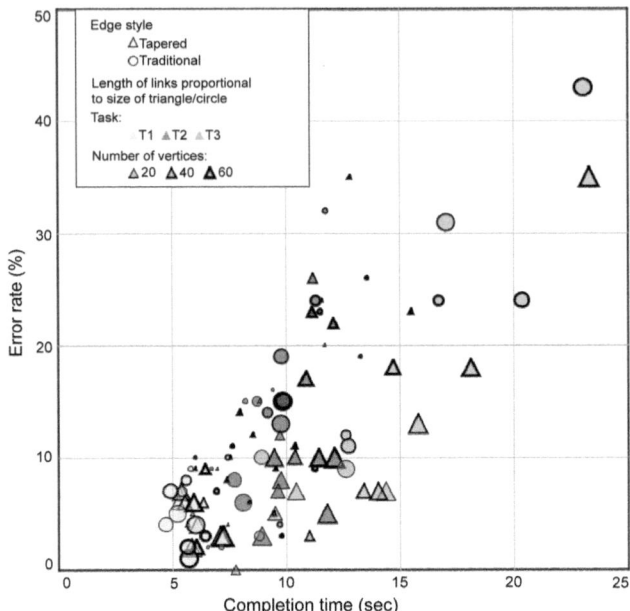

Fig. 5. Scatterplot of average error rates and completion times for all 99 configurations. The shape of the glyph denotes edge style, its size indicates length of the link, and its border width depends on the number of vertices. Different colors are used for the three different task types. (The figure is best viewed in the electronic color version of this paper; different colors correspond to shades of gray in black-and-white print.)

Tasks 1 and 2 show an interesting "dip" in the plots around 75% link length. This suggests that partially drawn links of 75% length provide the optimal balance between clutter reduction (supported by shorter links) and perception of node connections (supported by longer links) for these study parameters. Task 3 exhibits a different behavior: completion times become smaller and smaller with decreasing link length. This is reasonable because participants only had to find the "star" with the most jags to find the node with the highest number of outgoing edges. Therefore, we have indication that Research Question 1 (on link length) might be partially answered positively; however, there is a strong effect of type of task and, often, there might be an optimal length of medium size. Regarding edge style, the only visible difference appears for Task 3, where completion times are generally higher for traditional links. In contrast, Tasks 1 and 2 slightly tend toward lower completion times for traditional links. Therefore, the data to answer Research Question 3 (on edge style) is inconclusive; however, there might be an effect related to task type.

In general, completion times tend to increase with increasing number of vertices, independent of the type of task. The impact of graph size is most pronounced for Task 3, less so for Task 2, and even smaller for Task 1. Completion times are lowest for Task 1 (4.7 s – 8.5 s), medium for Task 2 (7.7 s – 15.5 s), and

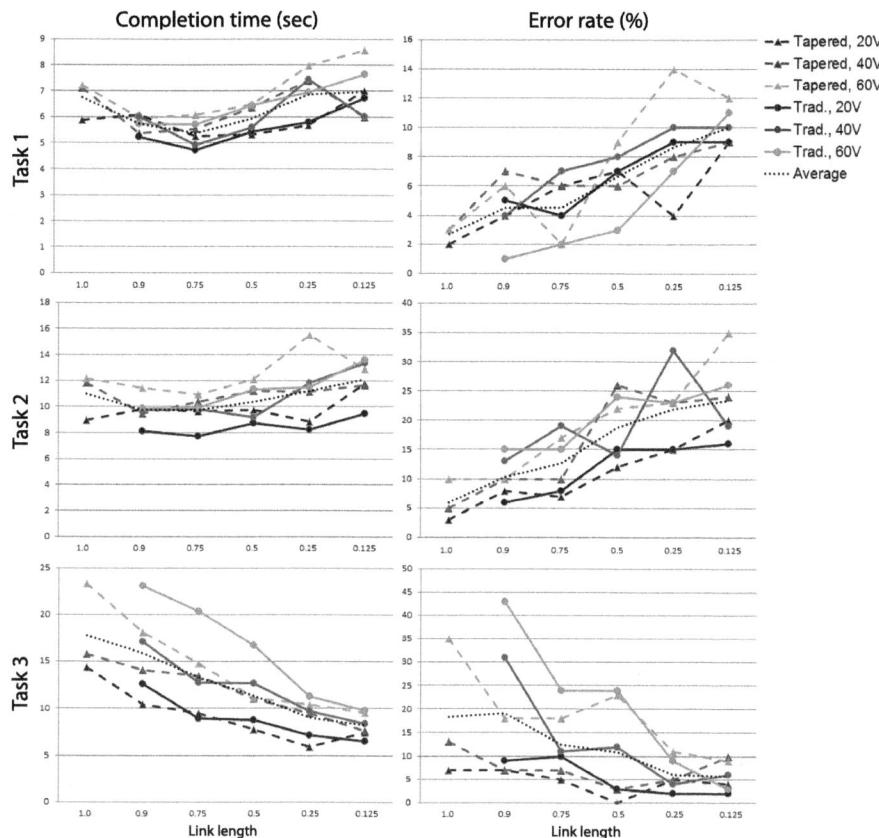

Fig. 6. The line charts plot the average completion times (left) and error rates (right) for all 99 configurations. Numbers were aggregated over 42 participants and 2 repetitions. The dotted line shows the average within the respective diagram. Link lengths are given as portions of the respective full length. Note that both completion times and error rates are in different scale for the three tasks.

longest for Task 3 (5.9 s – 23.3 s). We interpret this data as follows: Task 2 is more complex than Task 1, leading to longer completion times and larger spread thereof. This result is expected because Task 1 checks direct connections, whereas Task 2 indirect connections. Task 3 is even more complex, especially for large graphs, and thus leads to longer completion times and larger spread. However, overall we were not able to extract any structural impact of the number of vertices on how tasks were answered with different link length. Therefore, the data to answer Research Question 2 (on graph size) is inconclusive.

Let us now turn to error rates depicted in the right column of Figure 6. Error rates increase with decreasing link length for Tasks 1 and 2, whereas they improve for Task 3. Therefore, for Task 3, the error rates suggest a negative answer to Research Question 1. However, accuracy for Tasks 1 and 2 suggests a positive answer to Research Question 1. For Tasks 1 and 2, there is no clear difference in error rate for the two edge styles. In contrast, for Task 3, the traditional links lead to higher error rate for link lengths between 100% and 50%, but lower error rate for link lengths below 50%. Therefore, there is no clear answer to Research Question 3 in this case. Finally, the error rate depends on the size of the graph, as larger graphs lead to lower accuracy. Comparing the three tasks, the error rates are lowest for Task 1 (1%–14%) and much higher for Task 2 (3%–35%) and Task 3 (0%–43%), which is consistent with the task-specific completion times. However, the error-rate data does not lead to a clear answer to Research Question 2.

6   Conclusion and Future Work

We have conducted a user study with 42 participants to test whether node-link diagrams are still readable and interpretable when drawing links only partially. One result of the study is that the influence of the link length on completion time and error rate clearly depends on the type of task. The study suggests that partially drawn links can lead to shorter completion times. Depending on the task, the optimal link length varies between still rather long links of 75% length (in Tasks 1 and 2) and much shorter links (as low as 12.5% length for Task 3). In general, however, accuracy tends to suffer when links are drawn only partially— except for Task 3, for which error rate improves for shorter links. We have also tested tapered and traditional edge styles, but found no relevant general effect of those. In conclusion, the main message is that there is potential usefulness of partially drawn links, especially when completion time is more important than accuracy; however, there are substantial task-dependent effects.

Therefore, we plan to include other tasks, especially more complicated tasks that may focus on cliques or clusters. Also, statistical hypothesis testing should complement our current qualitative discussion of results to come up with statistically significant evidence. Other venues of future work could include further variation of independent variables. For example, a larger range of graph sizes could be considered, models apart from the Barabási-Albert model of scale-free graphs could be used, or the graph density could be varied. Finally, graph layouts different from the Fruchterman-Reingold layout could be employed, e.g., a circular layout may be of interest because the graph vertices would be equidistantly placed on a circle circumference and, hence, target vertex ambiguities could be reduced.

Acknowledgements. The project was in part funded by the German Research Foundation (DFG) grant DFG WE 2836/4-1.

References

1. Barabási, A.L., Albert, R.: Emergence of scaling in random networks. Science 286(5439), 509–512 (1999)
2. Becker, R.A., Eick, S.G., Wilks, A.R.: Visualizing network data. IEEE Transactions on Visualization and Computer Graphics 1(1), 16–28 (1995)

3. Di Battista, G., Eades, P., Tamassia, R., Tollis, I.G.: Graph Drawing: Algorithms for the Visualization of Graphs. Prentice Hall, Upper Saddle River (1999)
4. Eades, P.: A heuristic for graph drawing. Congressus Numerantium 42, 149–160 (1984)
5. Fruchterman, T.M.J., Reingold, E.M.: Graph drawing by force-directed placement. Software: Practice and Experience 21(11), 1129–1164 (1991)
6. Ghoniem, M., Fekete, J.D., Castagliola, P.: A comparison of the readability of graphs using node-link and matrix-based representations. In: Proc. IEEE Symposium on Information Visualization, pp. 17–24 (2004)
7. Holten, D., Isenberg, P., van Wijk, J.J., Fekete, J.D.: An extended evaluation of the readability of tapered, animated, and textured directed-edge representations in node-link graphs. In: Proc. IEEE Pacific Visualization Symposium, pp. 195–202 (2011)
8. Holten, D., van Wijk, J.J.: A user study on visualizing directed edges in graphs. In: Proc. SIGCHI Conference on Human Factors in Computing Systems, pp. 2299–2308 (2009)
9. Huang, W., Eades, P.: How people read graphs. In: Proc. Asia-Pacific Symposium on Information Visualisation, pp. 51–58 (2005)
10. Huang, W., Hong, S.-H., Eades, P.: Layout Effects on Sociogram Perception. In: Healy, P., Nikolov, N.S. (eds.) GD 2005. LNCS, vol. 3843, pp. 262–273. Springer, Heidelberg (2006)
11. Kamada, T., Kawai, S.: An algorithm for drawing general undirected graphs. Information Processing Letters 31(1), 7–15 (1989)
12. Koffka, K.: Principles of Gestalt Psychology. Harcourt, Brace (1935)
13. Lee, B., Plaisant, C., Parr, C.S., Fekete, J.-D., Henry, N.: Task taxonomy for graph visualization. In: Proc. AVI Workshop on BEyond time and errors: novel evaLuation methods for Information Visualization, BELIV 2006 (2006)
14. Purchase, H.C.: Which Aesthetic Has the Greatest Effect on Human Understanding? In: DiBattista, G. (ed.) GD 1997. LNCS, vol. 1353, pp. 248–261. Springer, Heidelberg (1997)
15. Purchase, H.C., Carrington, D., Allder, J.-A.: Empirical evaluation of aesthetics-based graph layout. Empirical Software Engineering 7(3), 233–255 (2002)
16. Purchase, H.C., Cohen, R.F., James, M.: Validating Graph Drawing Aesthetics. In: North, S.C. (ed.) GD 1996. LNCS, vol. 1190, pp. 435–446. Springer, Heidelberg (1997)
17. Rosenholtz, R., Li, Y., Mansfield, J., Jin, Z.: Feature congestion: a measure of display clutter. In: Proc. SIGCHI Conference on Human Factors in Computing Systems, pp. 761–770 (2005)
18. Rusu, A., Fabian, A.J., Jianu, R., Rusu, A.: Using the Gestalt principle of closure to alleviate the edge crossing problem in graph drawings. In: Proc. International Conference on Information Visualisation (IV 2011), pp. 488–493 (2011)
19. Ware, C., Purchase, H., Colpoys, L., McGill, M.: Cognitive measurements of graph aesthetics. Information Visualization 1(2), 103–110 (2002)

Realizing Planar Graphs as Convex Polytopes

Günter Rote

Institut für Informatik, Freie Universität Berlin,
Takustraße 9, 14195 Berlin, Germany
`rote@inf.fu-berlin.de`

Abstract. This is a survey on methods to construct a three-dimensional convex polytope with a given combinatorial structure, that is, with the edges forming a given 3-connected planar graph, focusing on efforts to achieve small integer coordinates.

Keywords: Convex polytope, spiderweb embedding.

1 Introduction

The graphs formed by the edges of three-dimensional polytopes are characterized by Steinitz' seminal theorem from 1916 [13]: they are exactly the planar 3-connected graphs. For such a graph G with n vertices, I will discuss different methods of actually constructing a polytope with this structure.

2 Inductive Methods

The original proof of Steinitz transforms G into simpler and simpler graphs by sequence of elementary operations, until eventually K_4, the graph of the tetrahedron, is obtained. By following this transformation in the reverse order, one can gradually turn the tetrahedron into a realization of G. The operations can be carried out with rational coordinates, and after clearing common denominators, one obtains integer coordinates. However, the required number of bits of accuracy for each vertex coordinate is exponential. In other words, the n vertices lie on an integer grid whose size is doubly exponential in n [9].

A *triangulated* (or simplicial) polytope, in which every face is a triangle, is easier to realize on the grid than a general polytope, since each vertex can be perturbed within some small neighborhood while maintaining the combinatorial structure of the polytope.

Das and Goodrich [5] showed that triangulated polytopes can be embedded with coordinates of size $O(2^{\mathrm{poly}(n)})$, by performing $O(\log n)$ stages of many independent Steinitz operations in parallel. (An explicit bound on the coordinates has not been worked out for this method.)

M. van Kreveld and B. Speckmann (Eds.): GD 2011, LNCS 7034, pp. 238–241, 2012.

3   Tutte Embeddings

The *Schlegel diagram* of a polytope P is obtained by a central projection from a point O that is outside P but sufficiently close to a face F of P such that F is the only face that F sees. In the Schlegel diagram, F will appear as the outer face, and the remaining faces will tile F without overlap. Thus, the Schlegel diagram is a plane drawing of the graph G with convex faces (including the outer face).

There are a number of methods that first construct such a plane drawing of G and then *lift* it to three dimensions. Convex faces are by no means sufficient to guarantee that a drawing is a Schlegel diagram. A characterization of Schlegel diagrams is provided by the so-called Maxwell–Cremona correspondence, observed by Maxwell in 1864 [8], which is described below. By a projective transformation, we can assume that the graph G is drawn in the xy-plane, and the projection center O is at infinity at the positive z-axis. In other words, the projection is vertical and consists in projecting away the z-coordinate.

An *equilibrium stress* assigns a force to every edge such that in every vertex, the forces cancel. The forces on an edge pull ("positive stress") or push ("negative stress") on both endpoints with the same magnitude, in the direction parallel to the edge.

Theorem 1 (Maxwell, Whiteley [18]). *Let G be a planar 3-connected graph drawn in the plane without crossings. The following are equivalent:*

- *G is the vertical projection of a convex polytope.*
- *There is an equilibrium stress on G which is positive on the interior edges and negative on the boundary edges.*

This theorem is constructive, in the sense that the lifting can be computed in a straightforward way from the equilibrium stress, and vice versa.

To construct a plane embedding that has an equilibrium stress, one can use the spider-web approach suggested by Tutte [15,16]: after fixing the positions of the vertices of the outer face in the shape of a convex polygon, we stipulate that the forces on the interior edges should be not just parallel to, but equal to the edge vectors. The equilibrium condition amounts now to requiring that every interior vertex should lie at the barycenter of its neighbors. This leads to a linear system of equations for the positions of the vertices. After solving this system, there is equilibrium at the interior vertices. However, equilibrium at the boundary vertices is only guaranteed when the outer face is a triangle. If this is not the case, one can realize the polar polytope P^*, whose graph G^* is the dual of G, instead: either G or G^* must contain a triangle. The calculations for the polarization operation increase the size of the coordinates, leading to bounds of $O(\text{const}^{n^2})$ [11]. A linear exponent of $O(188^n)$ has finally been achieved by Rote, Ribó and Schulz [10]: if the outer face is a quadrilateral or a pentagon, one can choose its shape in an appropriate way, in order to ensure that equilibrium also holds on the boundary, and polarization is not needed. This last paper establishes a connection between the size of the coordinates and the number of spanning trees of G. Due to improved upper bounds on the number of spanning trees of a planar graph [4], the best bound on the coordinates is currently $O(147.71^n)$.

3.1 Stacked Polytopes

A stacked polytope is obtained by starting with a tetrahedron and repeatedly gluing a new tetrahedron onto some face. Its graph is a *3-tree*: It is obtained from K_4 by repeatedly drawing a new vertex into a triangular face and connecting it to the three triangle vertices.

In a recent first breakthrough on the way towards providing polynomial grid embeddings for polytopes, Demaine and Schulz [6] (after some more specialized cases treated by Zickfeld [19]) showed that every stacked polytope with n vertices can be realized on a polynomial grid of size $O(n^4) \times O(n^4) \times O(n^{18})$.

Stacked polytopes are a special class of triangulated polytopes, and, due to their hierarchical structure, they are somewhat easier to handle. Sill, they are sufficiently varied so that one might hope to extend the techniques to, say, all triangulated polytopes.

4   Nonlinear Methods

For completeness, I will mention some other construction methods for polytopes, which, however, don't lend themselves to achieving integer realizations.

Midscribed Polytopes. An alternative proof of Steinitz' theorem applies the Koebe-Andreyev-Thurston Circle Packing Theorem (see for example [12]). This theorem can be used to produce a polytope whose *edges* are tangent to a sphere, that is, they are *mid-scribed* around the sphere (instead of circumscribed or inscribed). One can define a converging process that yields such a polytope. However, the exact mid-scribed realization (which is unique up to Möbius transformations) necessarily boils down to a nonlinear system of equations, and there are polytopes for which such a realization must have irrational coordinates. It is conceivable that an "approximately mid-scribed" polytope might be good enough, at least for triangulated graphs, but this has not been investigated.

The Colin de Verdière number. Lovász [7] showed that an $n \times n$ matrix of rank 3 that arises in the definition of the Colin de Verdière parameter $\mu(G)$ of a graph G (which equals 3 for graphs of polytopes), can be used to construct coordinates for a polytope realization. However, it is not easy to find this matrix.

5   Lower Bounds

The known lower bounds on a grid embedding of a 3-polytope as disappointingly weak. A convex n-gon with integral vertices needs an area of $\Omega(n^3)$ in the plane [1,2,14,17]. Therefore, realizing a 3-polytope with an $(n-1)$-gonal face requires at least one dimension of size $\Omega(n^{3/2})$. Given that only an exponential upper bound is known, this is very weak. If one is just interested in strictly convex faces, then a drawing on an $O(n^2) \times O(n^2)$ grid is possible [3]. The true bound is not known, but in this case the gap to the lower bound $\Omega(n^{3/2}) \times \Omega(n^{3/2})$ is not so big.

References

1. Acketa, D.M., Žunić, J.D.: On the maximal number of edges of convex digital polygons included into an $m \times m$-grid. J. Comb. Theory Ser. A 69(2), 358–368 (1995)
2. Andrews, G.E.: A lower bound for the volume of strictly convex bodies with many boundary lattice points. Trans. Amer. Math. Soc. 99, 272–277 (1961)
3. Bárány, I., Rote, G.: Strictly convex drawings of planar graphs. Documenta Math. 11, 369–391 (2006)
4. Buchin, K., Schulz, A.: On the Number of Spanning Trees a Planar Graph can have. In: de Berg, M., Meyer, U. (eds.) ESA 2010. LNCS, vol. 6346, pp. 110–121. Springer, Heidelberg (2010)
5. Das, G., Goodrich, M.T.: On the complexity of optimization problems for 3-dimensional convex polyhedra and decision trees. Comput. Geom. Theory Appl. 8(3), 123–137 (1997)
6. Demaine, E.D., Schulz, A.: Embedding stacked polytopes on a polynomial-size grid. In: Proceedings of the 22nd Annual ACM-SIAM Symposium on Discrete Algorithms (SODA), San Francisco, pp. 1177–1187 (2011)
7. Lovász, L.: Steinitz representations of polyhedra and the Colin de Verdière number. J. Comb. Theory, Ser. B 82, 223–236 (2000)
8. Maxwell, J.C.: On reciprocal figures and diagrams of forces. Phil. Mag. Ser. 27, 250–261 (1864)
9. Onn, S., Sturmfels, B.: A quantitative Steinitz' theorem. In: Beiträge zur Algebra und Geometrie, vol. 35, pp. 125–129 (1994)
10. Ribó Mor, A., Rote, G., Schulz, A.: Small grid embeddings of 3-polytopes. Discrete and Computational Geometry 45, 65–87 (2011), http://page.mi.fuberlin.de/rote/Papers/pdf/Small+grid+embeddings+of+3-polytopes.pdf
11. Richter-Gebert, J.: Realization Spaces of Polytopes. Lecture Notes in Mathematics, vol. 1643. Springer, Heidelberg (1996)
12. Schramm, O.: Existence and uniqueness of packings with specified combinatorics. Israel J. Math. 73, 321–341 (1991)
13. Steinitz, E.: Polyeder und Raumeinteilungen. In: Encyclopädie der mathematischen Wissenschaften, vol. III.1.2 (Geometrie), chap. IIIAB12, pp. 1–139. B. G. Teubner, Leipzig (1922)
14. Thiele, T.: Extremalprobleme für Punktmengen. Master's thesis, Freie Universität Berlin (1991)
15. Tutte, W.T.: Convex representations of graphs. Proceedings London Mathematical Society 10(38), 304–320 (1960)
16. Tutte, W.T.: How to draw a graph. Proceedings London Mathematical Society 13(52), 743–768 (1963)
17. Voss, K., Klette, R.: On the maximal number of edges of convex digital polygons included into a square. Počítače a umelá inteligencia 1(6), 549–558 (1982) (in Russian)
18. Whiteley, W.: Motion and stresses of projected polyhedra. Structural Topology 7, 13–38 (1982)
19. Zickfeld, F.: Geometric and Combinatorial Structures on Graphs. Ph.D. thesis, Technical University Berlin (December 2007)

Overloaded Orthogonal Drawings

Evgenios M. Kornaropoulos[1,2] and Ioannis G. Tollis[1,2]

[1] Department of Computer Science, University of Crete, Heraklion, Crete, Greece
[2] Institute of Computer Science, Foundation for Research and Technology-Hellas,
Vassilika Vouton, P.O. Box 1385, Heraklion, GR-71110 Greece
{kornarop,tollis}@ics.forth.gr

Abstract. Orthogonal drawings are widely used for graph visualization due to their high clarity of representation. In this paper we present a technique called Overloaded Orthogonal Drawing. We first place the vertices on grid points following a relaxed version of dominance drawing, called weak dominance condition. Edge routing is implied automatically by the vertex coordinates. In order to simplify these drawings we use an overloading technique. All algorithms are simple and easy to implement and can be applied to directed acyclic graphs, planar, non-planar and also undirected graphs. We also present bounds on the number of bends and the area. Overloaded Orthogonal drawings present several interesting properties such as efficient visual edge confirmation as well as simplicity and clarity of the drawing.

1 Introduction

An *orthogonal drawing* maps each edge into a chain of horizontal and vertical line segments. An *orthogonal grid drawing* is an orthogonal drawing such that vertices and bends along the edges have integer coordinates. Drawings in this style are useful in many applications due to the high clarity of the model. The problem of constructing an orthogonal drawing while minimizing several aesthetic criteria such as area, bends, maximum edge length and total edge length is an NP-hard problem [4]. Therefore most algorithms employ heuristics that try to layout the graph in a manner which is good for some set of aesthetics.

Various algorithms have been introduced to produce orthogonal drawings of planar graphs [18,2,20,19,4]. A necessary and sufficient condition for a plane graph with maximum degree three to have an orthogonal drawing without bends was presented in [17]. Another interesting result is that an outerplanar graph G with maximum degree at most three has an orthogonal drawing with no bends if and only if G contains no triangles [12]. Bertolazzi et al. presented [1] a branch and bound algorithm that computes an orthogonal representation with the minimum number of bends of a biconnected planar graph. For drawings of non-planar graphs [9,3,13], the required area can be as little as $0.76n^2$ [14], the total number of bends is no more than $2n + 2$ [2,14], and each edge has at most two bends. Experimental studies have been conducted where various proposed algorithms were tested on their performance on area, bends, crossings,

M. van Kreveld and B. Speckmann (Eds.): GD 2011, LNCS 7034, pp. 242–253, 2012.
© Springer-Verlag Berlin Heidelberg 2012

edge length, and time [21]. Dominance drawings are a widely used technique for visualizing planar st-graphs. These drawings have numerous useful features such as, small number of bends, small area, linear time complexity, detection and display of symmetries [4,5].

In this paper we introduce the overloaded orthogonal model which combines dominance and row/column reuse. We use a concept of relaxed dominance for vertex coordinate assignment, and orthogonal grid layout with overloaded use of rows/columns for edge routing. This type of routing has been used extensively in VLSI layout [11]. The concept of merging together groups of edges has been also used in the confluent drawing framework [6,7] to facilitate readability of the graph. This model can be applied to both planar and non-planar graphs. Also it can be efficiently applied to graphs with maximum degree four, and to graphs with degree higher than four. The presented algorithms produce drawings with at most $n - 1$ bends, $O(n^2)$ area, they run in linear time $O(n + m)$, and are easy to implement. Although a direct comparison with the bounds of traditional orthogonal drawings is a bit unfair (due to the reuse of rows and columns) our bounds on the number of bends and area are promising. Furthermore, every overloaded orthogonal drawing simplifies tremendously the visual confirmation of the existence of an edge and/or path between any two vertices.

This paper is organized as follows: in Section 2 we present an algorithm for constructing overloaded orthogonal drawings. In Section 3 we discuss some properties of the proposed model. In Section 4 we present properties and bounds of the overloaded orthogonal model in directed acyclic graphs. Section 5 gives an application of the proposed model to other graphs and finally Section 6 gives conclusions and open problems.

2   Overloaded Orthogonal Framework

In this framework, we propose to place the vertices in the grid so that edges flow from left-to-right and from bottom-to-top. Each vertex u is placed on a point in the grid with coordinates $X(u)$ and $Y(u)$. Dominance drawings achieve this vertex placement for st-planar graphs. A dominance drawing Γ of a graph $G = (V, E)$ has the following property: for any two vertices $u, v \in V$ there is a directed path from u to v in G, if and only if $X(u) \leq X(v)$ and $Y(u) \leq Y(v)$ in Γ. But, not every directed acyclic graph has a dominance drawing. Therefore we propose a relaxed condition, called *weak dominance condition*, that can be applied to any directed acyclic graph (dag):

Weak Dominance Condition: Let $G = (V, E)$ be a directed acyclic graph. For any two vertices $u, v \in V$ if there is a directed path from u to v in G, then $X(u) \leq X(v)$ and $Y(u) \leq Y(v)$.

Thus if v is in the upper-right quadrant of u, then v is not necessarily reachable from u. A path that is implied by the vertex coordinates but does not exist in G is called a *falsely implied path* (or *fip*). The problem of minimizing the number of falsely implied paths was introduced in [10], where it is shown that the corresponding decision problem is NP-complete.

Following the footsteps of the algorithm for dominance drawing for (reduced) planar st-graphs presented in [5], we formulate an algorithm for vertex placement that respects the weak dominance condition and is applicable to any dag. The main algorithm for planar st-graphs described in [5] consists of three phases. In the first phase, called 'Preprocessing Phase', a linked data structure is constructed in order to efficiently calculate coordinates. During the second phase called 'Preliminary Layout' distinct X, Y coordinates are given to each vertex. In the third and final phase, a compaction procedure is applied to reduce the area of the drawing.

We will construct a similar data structure as in 'Preprocessing step', but for general directed acyclic graphs. Let W be a representation of a dag G such that the incoming edges for each vertex u appear consecutively around u. Representation W will be called a *representation in consecutive form*. The representation in consecutive form is a method to force a left-to-right order in the incoming as well as outgoing edges of every vertex of G. Without loss of generality we assume that there is only one source, s. If not then we insert an artificial super-source s and connect it to all sources of G. The algorithm performs two topological sortings on the vertices of G. Successors of each vertex are scanned in clockwise order for the X coordinate assignment, and in counterclockwise order for the Y coordinate assignment. The order is imposed according to the representation in consecutive form that is given as an input. We will present the algorithm for clockwise scan, that computes the X-coordinate assignment.

Algorithm. TOPOLOGICAL-SORTING(Adj(G))
1. **for** each vertex $v \in V$
2. $X[v] \leftarrow \infty$
3. $X[s] \leftarrow 0$
4. time$\leftarrow 1$
5. VISIT-CLCK(s)
6. **return** X

Algorithm. VISIT-CLCK(u)
1. **for** each vertex $v \in \text{Adj}(u)$ such that (u, v)
             is the leftmost outgoing edge of u **do**
2.   **if** in-degree(v)=1
3. $X[v] \leftarrow$ time
4.   time$\leftarrow$time+1
5.   remove edge $e=(u, v)$
6.   VISIT-CLCK (v)
7.   **else**
8.   remove edge $e=(u, v)$

Algorithm TOPOLOGICAL-SORTING scans the outgoing edges of a vertex u in clockwise order (leftmost outgoing edge) and visits a direct successor v only if v has in-degree one. Otherwise, it removes edge (u, v) from the list. Analogously, we formulate an algorithm for the Y-coordinate assignment that performs a counterclockwise scan, by replacing VISIT-CLCK with VISIT-COCLCK. The difference between the two VISIT algorithms is Line 1, where instead of leftmost outgoing edge we now have rightmost outgoing edge. The two topological sortings are used by WDP algorithm for assigning X and Y coordinates to the vertices of G.

Algorithm. (WDP)WEAK DOMINANCE PLACEMENT (W)
1. X coordinates $\leftarrow$ TOPOLOGICAL SORTING(W) using VISIT-CLCK
2. Y coordinates $\leftarrow$ TOPOLOGICAL SORTING(W) using VISIT-COCLCK

We denote the number of vertices in G by n, and the number of edges in G by m. Since both topological sorting algorithms run in linear time $O(n + m)$, algorithm WDP also runs in linear time $O(n + m)$.

In the rest of this section we will see how the Algorithm WDP creates a natural separation between pq-components. A pq-component $G_{pq} = (V', E')$ of G is a maximally induced subgraph of G with a single source p and a single sink q that contains at least two edges and that is connected with the rest of G only through vertex p and vertex q. Thus, vertex p is a dominator of every vertex $v \in V'$ and q is a post-dominator of every vertex $v \in V'$. Due to space limitations, the proofs of the following results are omitted.

Lemma 1. *If dag $G=(V, E)$ includes a pq-component $G_{pq} = (V', E')$, then $X(q) = X(p) + |V'| - 1$ and $Y(q) = Y(p) + |V'| - 1$.*

Corollary 1. *If dag $G=(V,E)$ includes a pq-component $G' = (V', E')$, then for every vertex $u \in G'$, $X(p) \leq X(u) \leq X(p) + |V'| - 1$ and $Y(p) \leq Y(u) \leq Y(p) + |V'| - 1$.*

Let $X()$ and $Y()$ be the coordinates constructed by WDP algorithm. Also let $G' = (V', E')$ be a component where $V' \subseteq V$ and $E' \subseteq E$. A component G' is said to be *separated*, if the following property holds for $X()$ and $Y()$:

$$\forall u \in V', v \in V - V' \Rightarrow (X(u) \leq X(v) \wedge Y(u) \leq Y(v)) \vee (X(u) \geq X(v) \wedge Y(u) \geq Y(v))$$

This property is a guarantee that every vertex $v \in V - V'$ that is not a member of a component G' will not appear between the vertices of G'. We refer to this as the *separation property*.

Theorem 1. *Vertex placement $X()$ and $Y()$ constructed by algorithm WEAK DOMINANCE PLACEMENT respects the separation property for every pq-component.*

Proof. (Sketch) Let $G' = (V', E') \subseteq G$ be a pq-component. Then algorithm TOPO-LOGICAL - SORTING for G, returns a numbering of vertices of G' from $X(p)$ to $X(p) + |V'|$. Also holds for Y-coordinates, i.e., numbers vertices of G' from $Y(p)$ to $Y(p) + |V'|$. Thus, no vertex from $V - V'$ can be drawn inside a pq-component. $\square$

Lemma 2. *Let u and v be a pair of vertices of G such that $X(v) = X(u) + 1$. Then $Y(u) < Y(v)$ if and only if G has an edge (u,v).*

Lemma 3. *Let u and v be a pair of vertices of G such that $Y(v) = Y(u) + 1$. Then $X(u) < X(v)$ if and only if G has an edge (u,v).*

Our proposed framework contains the term 'overloaded' because all outgoing edges of a vertex use the same column in order to reach their corresponding destination vertex. We will first discuss how a single edge is routed, and then we will focus on unambiguously visualizing the edges of the drawing.

Edge routing is automatically implied by the coordinates of the vertices. Each edge (u, v) consists of a vertical edge segment from $(X(u),Y(u))$ to $(X(u),Y(v))$ and a horizontal segment from $(X(u),Y(v))$ to $(X(v),Y(v))$. Because various edges reuse segments of rows and columns we introduce e-points to resolve ambiguities, see Figure 1. Given an edge (u, v) an e-*point* is defined as an unlabeled point that is placed on point $(X(u), Y(v))$ to indicate a direct connection from u to v. A bend will appear in the final drawing instead of an e-point if: (a) vertex u does not have a successor w such that $Y(w) \geq Y(v)$ and (b) vertex v does not have a predecessor z such that $X(z) \leq X(v)$.

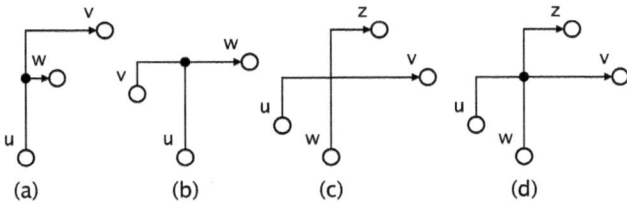

(a)   (b)   (c)   (d)

Fig. 1. (a) the vertical segment of (u, v) is overloaded by the vertical segment of (u, w). To visualize the edge from u to w, an e-point is placed at $(X(u), Y(w))$. (b) the horizontal segment of (v, w) is overloaded by the horizontal segment of (u, w). To visualize the edge from u to w, an e-point is placed at $(X(u), Y(w))$. If there is no e-point then $(w, v) \notin E$ (c), whereas if there is an e-point in $(X(w),Y(v))$ then $(w, v) \in E$ (d).

We will describe an algorithm that receives the vertex coordinates as an input, and outputs an overloaded orthogonal drawing. It routes the edges according to the given coordinates and places e-points where needed.

In order to construct an overloaded orthogonal drawing a linked data structure for G will be constructed. Each vertex $u \in V$ of G, points to the list of its direct successors sorted in decreasing order according to their Y-coordinate. This single linked list of u, can be traversed by means of pointer $next(u)$. It can also be accessed by pointer $getFirst(u)$, that is u's direct successor with the highest Y-coordinate (hence first in the list). In case of a tie, we can arbitrarily order vertices with the same coordinate without affecting the overall result.

Algorithm. (OOD) OVERLOADED ORTH. DRAWING(Adj(G) , X() ,Y())
1. **for** each vertex $u \in V$
2.   visited$[u] \leftarrow 0$
3. **for** each vertex $u \in V$ in *increasing* order of X-coordinate
4. $v \leftarrow \text{next}(u)$
5.   **while** $v \neq nil$
6.   Draw edge segment from (X(u),Y(u)) to (X(u),Y(v))
7.   Draw edge segment from (X(u),Y(v)) to (X(v),Y(v))
8.   **if** getFirst(u)$\neq v$ OR visited$[v]\neq0$
9.   New e-point $\leftarrow$ (X(u),Y(v))
10.   visited$[v] \leftarrow 1$
11. $v \leftarrow \text{next}(v)$
12. **end**

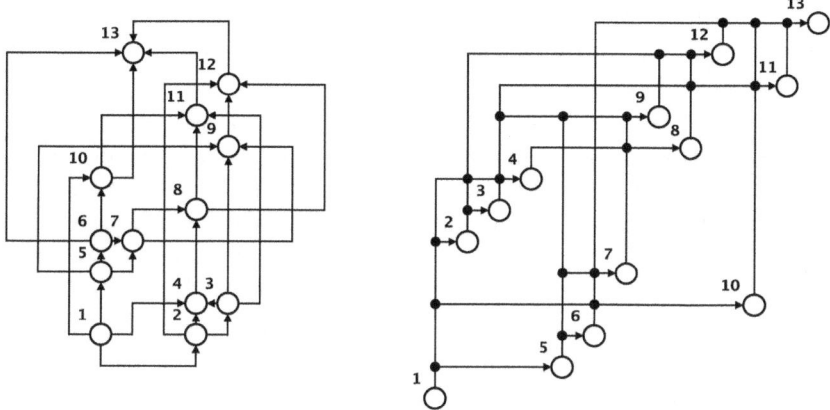

Fig. 2. Two different drawings of a regular degree four graph with 13 vertices and 26 edges. In the left picture an orthogonal grid drawing is depicted, the graph and the drawing are taken from [4]. While, in the right picture there is an overloaded orthogonal drawing of the same graph. No compaction was performed to the overloaded orthogonal drawing.

Theorem 2. *Algorithm OOD produces an overloaded orthogonal drawing Γ of G with vertex coordinates computed by algorithm WDP. Γ has at most $n-1$ bends, $O(n^2)$ area and is constructed in $O(n+m)$ time.*

2.1   Compaction

Compaction is applied as a post-processing step in an overloaded orthogonal drawing in order to reduce the X- and Y-coordinates. Our compaction follows the steps of the Algorithm in [4,5]. However since our graphs are not planar, and therefore we do not have planar embeddings, we need to be extra careful in order to produce a valid drawing. In this step we allow equality between vertex coordinates under the following conditions: (a) The compaction is performed between vertices $u, v \in V$ such that there is an edge $(u, v) \in E$. (b) Two distinct vertices cannot coincide in the same point. (c) Compaction on the X- or Y-coordinates will not be performed if an edge is forced to pass over u or any other vertex.

3   Clarity and Readability of the Model

In this section we outline some advantages of the overloaded orthogonal model.

• *Meaningful relation between vertex coordinates*: The weak dominance condition implies that: if there is a path from u to v then vertex v will appear in the upper right quadrant of vertex u.

• *Works for any pair of topological sortings as X, Y coordinates*: Since every pair of topological sortings respects the weak dominance condition, we can take any pair of topological sorting as X, Y coordinates.

• *Universality of the model*: The overloaded orthogonal model does not discriminate between graphs with maximum degree four, and graphs with higher degree. Furthermore, it can be efficiently applied to planar and to non-planar graphs. The overloaded orthogonal model can also be applied to undirected graphs, given that an *st*-numbering with various properties can be computed for any undirected graph [15,16] . An interesting example is presented in Section 5.

• *Efficient Visual Confirmation of an Edge*: We can visually confirm the existence of an edge (u, v) by checking if there is an e-point or a bend on point $(X(u), Y(v))$. If a compaction is performed u or v could replace the e-point at the location $(X(u), Y(v))$. In contrast, in the regular orthogonal model we would visually follow every outgoing edge of u successively, until we reach v. Consequently, the size of a graph does not affect the readability of an overloaded orthogonal drawing, as we can check if any two vertices are connected by inspecting only a single point i.e., in $O(1)$ time.

• *Efficient Visual Confirmation of Reachability*: An interesting extension of this graph drawing technique occurs when we use the transitive closure of a graph as input. In that case every possible path along the original directed acyclic graph $G = (V, E)$ will be represented by an edge in the transitive closure $G^* = (V, E^*)$. By applying the overloaded orthogonal model we can check if a vertex v is reachable from a vertex u by examining point $(X(u), Y(v))$ in the drawing. As shown in Figure 3, e-points of the corresponding transitive edges are colored grey. Notice that there is no e-point at $(X(4), Y(9))$, despite the fact that the coordinates of vertex 9 dominate the coordinates of vertex 4. In this context, crossings indicate the existence of falsely implied paths.

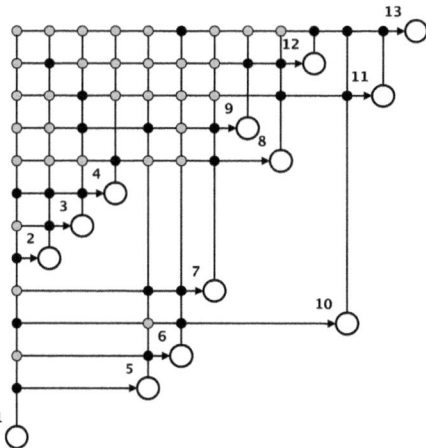

Fig. 3. An overloaded orthogonal drawing of the transitive closure. Reachability of any pair of vertices u-v can be confirmed by looking at point $(X(u), Y(v))$. By the the color of the e-point we can determine if there is an edge or a path between the vertices.

4   Directed Acyclic Graphs

In this section we present several properties and bounds of overloaded orthogonal drawings for directed acyclic graphs. If $X(u) \neq X(v)$ and $Y(u) \neq Y(v)$ for every pair of vertices $u, v \in V$, then every edge has a 'step'-like form and consequently produces either a bend or an e-point. Therefore we have:

Lemma 4. *Let Γ be an overloaded orthogonal drawing of dag G, where each vertex is placed in a distinct X, Y coordinate. Then $bends(\Gamma) + ePoints(\Gamma) = m$.*

If a compaction is performed on drawing Γ, then the sum $bends(\Gamma) + ePoints(\Gamma)$ would be less than the number of edges. Additionally, every vertex can have at most one bend on its row. That bend is produced from its direct predecessor with the lowest X-coordinate. Taking into consideration that sources do not have incoming edges, we have the following lemma:

Lemma 5. *Let Γ be any overloaded orthogonal drawing of a dag G. Let also n_s be the number of sources of G. Then $bends(\Gamma) \leq n - n_s$.*

The upper bound of the above lemma is tight as shown by the following theorem.

Theorem 3. *There exists a family of planar n-vertex graphs G_n, for $n \geq 3$, such that any overloaded orthogonal drawing Γ of G_n requires at least $n - 2$ bends, and $(n - 2) \times (n - 2)$ area.*

Proof. (Sketch) Consider the graph G_n shown in Figure 4. Each vertex u_i has two outgoing edges, (u_i, u_{i+2}) and (u_i, u_{i+1}). The transitive closure of this family of

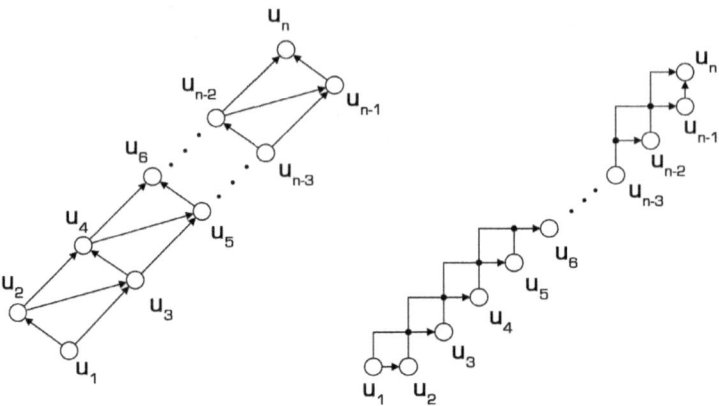

Fig. 4. An explanatory construction of Theorem 3

graphs is a complete directed acyclic graph, therefore the topological sorting for this graph is unique. Their drawings admit a single compaction in Y-coordinate between vertex u_1 and vertex u_2, and a single compaction in X-coordinate between vertex u_{n-1} and vertex u_n. Therefore an overloaded orthogonal drawing of this family of graphs has optimal area $(n-2) \times (n-2)$, and has at least $n-2$ bends. □

The dominance drawing technique was applied to reduced planar st-graphs in [5]. If we apply the edge routing technique using the vertex coordinates produced by the dominance drawing algorithm presented in [5], the drawing has zero bends.

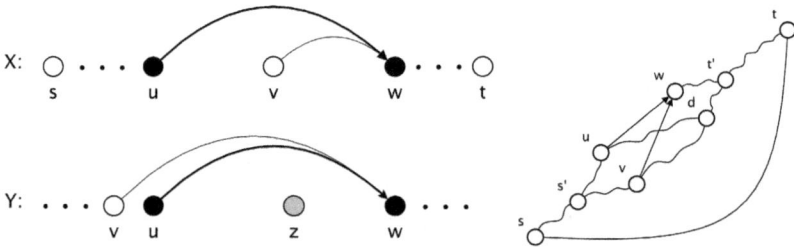

Fig. 5. Proof of Theorem 4. The left picture illustrates the difference between $(z1)$-case and $(z2)$-case. In the right picture there is a drawing of a $K_{3,3}$ that exists in $(z2)$-case.

Theorem 4. *Given a reduced planar st-graph $G = (V, E)$, an overloaded orthogonal drawing Γ with zero bends can be constructed in linear time, $O(n)$.*

Proof. (Sketch) Consider a reduced planar st-graph G with vertex coordinates obtained by the dominance drawing algorithm in [5]. Let an edge $(u, w) \in E$ such that it forms a bend that cannot be removed by a compaction. We construct such a scenario and prove that this edge cannot exist without contradicting the basic assumptions. Vertex u and vertex v cannot be consecutive in X-coordinate. Thus there must be a vertex v such that $X(u) < X(v) \leq X(w)$. Let also a vertex z such that $Y(u) < Y(z) < Y(w)$. Vertex z cannot be between u and w in X-coordinate due to the fact that G is reduced. Thus, we have two different cases: $(z1)$ where $X(s) < X(z) < X(u)$ and $(z2)$ where $X(w) < X(z) < X(t)$. Case $(z1)$ will conclude that edge (u, w) is transitive. Case $(z2)$ will conclude that there is a graph homeomorphic to $K_{3,3}$ and consequently G is not planar, a contradiction in both cases. $\qquad\square$

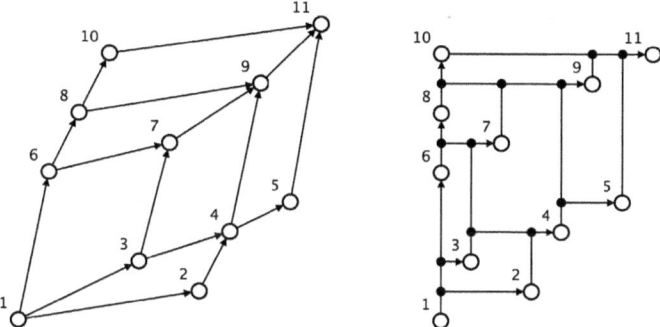

Fig. 6. In the left figure we have the straight-line dominance drawing of a reduced planar st-graph as described in [4]. In the right figure there is a compacted overloaded orthogonal drawing of the same graph with zero bends.

5   Other Graphs

In this section the overloaded orthogonal model is going to be extended to draw undirected graphs and directed graphs with cycles. Let G be an undirected graph and s, t be two distinct vertices of G. If the graph is planar we first construct a planar embedding and proceed, otherwise we ignore that step. An st-numbering for G is a numbering $v_1, v_2, \ldots, v_n$ of the vertices of G such that $s = v_1$, $t = v_n$, and every vertex v_j, other than s and t, is adjacent to at least two vertices v_i and v_k with $i < j < k$. Such a numbering can be constructed in linear time [8]. Given an st-numbering we orient the edges of E from the low-numbered vertex to the high numbered one. We name the resulting digraph D. The algorithm for st-orientation proposed in [15,16], parametrically controls the length of the longest path of the final st-oriented graph. As it was expected, different values of parameter p yield overloaded orthogonal drawings with different characteristics. We can apply the vertex placement algorithm to D, and then route the edges as described in Algorithm OOD. A compaction step can also be performed. As shown in Figure 7, the st-orientation with $p = 0$ results in an overloaded orthogonal drawing with area 19×19, while the st-orientation with $p = 1$ results in an overloaded orthogonal drawing with optimal area 2×19. We are conducting an experimental study in order to investigate the influence of an st-numbering of G, on the area of its overloaded orthogonal drawing Γ.

If G is a directed graph with cycles one could find a minimal feedback arc set F [4] and obtain an uncompacted overloaded orthogonal drawing of $G - F$. Complete the drawing by routing each edge $(u, v) \in F$ as follows: vertical segment from $(X(u), Y(u))$ to $(X(u), Y(v))$, horizontal segment from $(X(u), Y(v))$ to $(X(v), Y(v))$, placing e-points where necessary. Notice that rows and columns used for routing these edges, have not been used to route the edges of $G - F$.

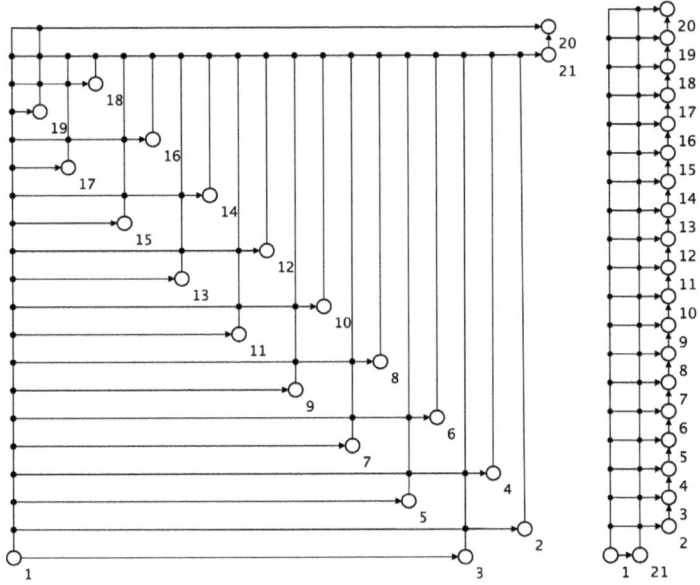

Fig. 7. Two overloaded orthogonal drawings of an originally undirected planar graph are shown. Left: the st-orientation was produced by algorithm [15] with parameter $p = 0$, right: same algorithm with parameter $p = 1$.

6   Conclusion and Open Problems

We presented algorithms that produce overloaded orthogonal drawings with at most $n - 1$ bends, $O(n^2)$ area, they run in linear time $O(n + m)$, and are easy to implement. An interesting open problem is to find algorithms for weak dominance placement that provide upper bounds on the number of crossings in an overloaded orthogonal drawing of the transitive closure.

References

1. Bertolazzi, P., Di Battista, G., Didimo, W.: Computing orthogonal drawings with the minimum number of bends. IEEE Transactions on Computers 49(8), 826–840 (2000)

2. Biedl, T., Kant, G.: A better heuristic for orthogonal graph drawings. Computational Geometry: Theory and Applications 9(3), 159–180 (1998)
3. Biedl, T.C., Madden, B.P., Tollis, I.G.: The Three-Phase Method: A Unified Approach to Orthogonal Graph Drawing. Int. J. Comput. Geometry Appl. 10(6), 553–580 (2000)
4. Di Battista, G., Eades, P., Tamassia, R., Tollis, I.G.: Graph Drawing: Algorithms for the Visualization of graphs. Prentice - Hall, New Jersey (1998)
5. Di Battista, G., Tamassia, R., Tollis, I.G.: Area Requirement and Symmetry Display of Planar Upward Drawings. Discrete and Comput. Geom. 7(4), 381–401 (1992)
6. Dickerson, M., Eppstein, D., Goodrich, M.T., Meng, J.Y.: Confluent Drawings: Vizualizing Non-planar Diagrams in a Planar Way. Journal of Graph Algorithms and Applications 9(1), 31–52 (2005)
7. Eppstein, D., Goodrich, M.T., Meng, J.Y.: Confluent Layered Drawings. Algorithmica 47(4), 439–452 (2007)
8. Even, S., Tarjan, R.: Computing an st-numbering. Theoretical Computer Science 2(3), 339–344 (1976)
9. Fößmeier, U., Kaufmann, M.: Algorithms and Area Bounds for Nonplanar Orthogonal Drawings. In: DiBattista, G. (ed.) GD 1997. LNCS, vol. 1353, pp. 134–145. Springer, Heidelberg (1997)
10. Kornaropoulos, E.M., Tollis, I.G.: Weak Dominance Drawings and Linear Extension Diameter, arXiv:1108.1439 (2011)
11. Lengauer, T.: Combinatorial algorithms for integrated circuit layout. John Wiley & Sons, Inc., New York (1990)
12. Nomura, K., Tayu, S., Ueno, S.: On the Orthogonal Drawing of Outerplanar Graphs. Journal IEICE Transactions on Fundamentals of Electronics, Communications and Computer Sciences E88-A(6), 1583–1588 (2005)
13. Papakostas, A., Tollis, I.G.: Efficient Orthogonal Drawings of High Degree Graphs. Algorithmica 26(1), 100–125 (2000)
14. Papakostas, A., Tollis, I.G.: Algorithms for Area-Efficient Orthogonal Drawings. Computational Geometry Theory and Applications 9(1-2), 83–110 (1998)
15. Papamanthou, C., Tollis, I.G.: Algorithms for computing a parameterized st-orientation. Theoretical Computer Science 408(2-3), 224–240 (2008)
16. Papamanthou, C., Tollis, I.G.: Applications of Parameterized st-Orientations. Journal of Graph Algorithms and Applications 14(2), 337–365 (2010)
17. Rahman, S., Nishizeki, T., Naznin, M.: Orthogonal Drawings of Plane Graphs Without Bends. Journal of Graph Algorithms and Applications 7(4), 335–362 (2003)
18. Storer, J.: On minimal node-cost planar embeddings. Networks 14(2), 181–212 (1984)
19. Tamassia, R.: On embedding a graph in the grid with the minimum number of bends. SIAM J. Computing 16(3), 421–444 (1987)
20. Tamassia, R., Tollis, I.G.: Planar Grid Embeddings in Linear Time. IEEE Transactions on Circuits and Systems 36(9), 1230–1234 (1989)
21. Vismara, L., Di Battista, G., Garg, A., Liotta, G., Tamassia, R., Vargiu, F.: Experimental studies on graph drawing algorithms. Software: Practice and Experience 30(11), 1235–1284 (2000)

Drawing Cubic Graphs
with the Four Basic Slopes

Padmini Mukkamala and Dömötör Pálvölgyi[*]

McDaniel College, Budapest and Eötvös University, Budapest

Abstract. We show that every cubic graph can be drawn in the plane with straight-line edges using only the four basic slopes, $\{0, \pi/4, \pi/2, 3\pi/4\}$. We also prove that four slopes have this property if and only if we can draw K_4 with them.

1 Introduction

A *straight-line drawing* of a graph represents the vertices by distinct points in the plane and represents the edges by the line-segments between the corresponding pairs of points, such that no edge passes through a vertex. If it leads to no confusion, in notation and terminology we make no distinction between a vertex and the corresponding point, and between an edge and the corresponding segment. The *slope* of an edge in a straight-line drawing is the slope of the corresponding segment. Wade and Chu [29] defined the *slope number*, $sl(G)$, of a graph G as the smallest number s with the property that G has a straight-line drawing with edges of at most s distinct slopes.

Obviously, if G has a vertex of degree d, then its slope number is at least $\lceil d/2 \rceil$. Dujmović et al. [12] asked if the slope number of a graph with bounded maximum degree d could be arbitrarily large. Pach and Pálvölgyi [28] and Barát, Matoušek, Wood [7] (independently) showed with a counting argument that the answer is yes for $d \geq 5$.

In [21], it was shown that cubic (3-regular) graphs could be drawn with five slopes. The major result from which this was concluded was that subcubic graphs[1] can be drawn with the four basic slopes, the slopes $\{0, \pi/4, \pi/2, 3\pi/4\}$, corresponding to the vertical, horizontal and the two diagonal directions.

This was improved in [26] to show that connected cubic graphs can be drawn with four slopes[2] while disconnected cubic graphs required five slopes.

---

[*] The second author was supported by the European Union and co-financed by the European Social Fund (grant agreement no. TAMOP 4.2.1/B-09/1/KMR-2010-0003) and EUROGIGA project GraDR 10-EuroGIGA-OP-003 (OTKA NN 102029). Part of this work was done in Lausanne and the authors gratefully acknowledge support from the Bernoulli Center at EPFL and from the Swiss National Science Foundation, Grant No. 200021-125287/1.

[1] A graph is subcubic if it is a proper subgraph of a cubic graph, i.e. the degree of every vertex is at most three and it is not cubic (not 3-regular).

[2] But not the four basic slopes.

M. van Kreveld and B. Speckmann (Eds.): GD 2011, LNCS 7034, pp. 254–265, 2012.
© Springer-Verlag Berlin Heidelberg 2012

It was shown by Max Engelstein [15] that 3-connected cubic graphs with a Hamiltonian cycle can be drawn with the four basic slopes.

We improve all these results by the following

Theorem 1. *Every cubic graph has a straight-line drawing with only the four basic slopes.*

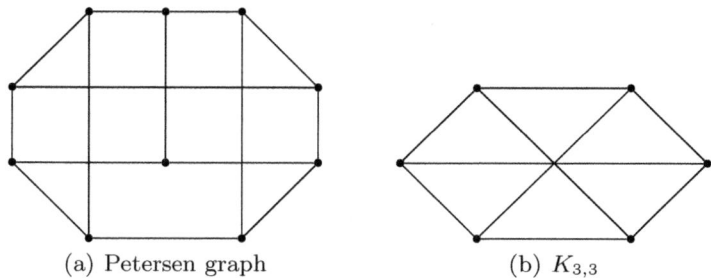

(a) Petersen graph   (b) $K_{3,3}$

Fig. 1. The Petersen graph and $K_{3,3}$ with the four basic slopes

This is the first result about cubic graphs that uses a nice, fixed set of slopes instead of an unpredictable set, possibly containing slopes that are not rational multiples of π. Also, since K_4 requires at least 4 slopes, this settles the question of determining the minimum number of slopes required for cubic graphs. In the last section we also prove

Theorem 2. *Call a set of slopes* good *if every cubic graph has a straight-line drawing with them. Then the following statements are equivalent for a set S of four slopes.*

1. *S is good.*
2. *S is an affine image of the four basic slopes.*
3. *We can draw K_4 with S.*

The problem whether the slope number of graphs with maximum degree four is unbounded or not remains an interesting open problem.

There are many other related graph parameters. The *thickness* of a graph G is defined as the smallest number of planar subgraphs it can be decomposed into [27]. It is one of the several widely known graph parameters that measures how far G is from being planar. The *geometric thickness* of G, defined as the smallest number of *crossing-free* subgraphs of a straight-line drawing of G whose union is G, is another similar notion [19]. It follows directly from the definitions that the thickness of any graph is at most as large as its geometric thickness, which, in turn, cannot exceed its slope number. For many interesting results about these parameters, consult [10], [12], [13], [14], [16], [17].

A variation of the problem arises if (a) two vertices in a drawing have an edge between them if and only if the slope between them belongs to a certain set S

and, (b) vertices may lie in the interior of a non-adjacent edge. This violates the condition stated before that an edge cannot pass through vertices other than its end points. For instance, K_n can be drawn with one slope. The smallest number of slopes that can be used to represent a graph in such a way is called the *slope parameter* of the graph. Under these set of conditions, Ambrus et al. [4] prove that the slope parameter of subcubic outerplanar graphs is at most 3. It was shown in Keszegh et al. [22] that the slope parameter of every cubic graph is at most seven. If only the four basic slopes are used, then the graphs drawn with the above conditions are called queens graphs and Ambrus and Barát [3] characterize certain graphs as queens graphs. Graph theoretic properties of some specific queens graphs can be found in Bell and Stevens [8].

Another variation for planar graphs is to demand a planar drawing. The *planar slope number* of a planar graph is the smallest number of distinct slopes with the property that the graph has a straight-line drawing with non-crossing edges using only these slopes. Dujmović, Eppstein, Suderman and Wood [11] raised the question whether there exists a function f with the property that the planar slope number of every planar graph with maximum degree d can be bounded from above by $f(d)$. Jelinek et al. [18] have shown that the answer is yes for *outerplanar* graphs, that is, for planar graphs that can be drawn so that all of their vertices lie on the outer face. Eventually the question was answered in [20] where it was proved that any bounded degree planar graph has a bounded planar slope number.

Finally we would mention a slightly related problem. Didimo et al. [9] studied drawings of graphs where edges can only cross each other in a right angle. Such a drawing is called an RAC (right angle crossing) drawing. They showed that every graph has an RAC drawing if every edge is a polygonal line with at most three bends (i.e. it consists of at most four segments). They also gave upper bounds for the maximum number of edges if less bends are allowed. Later Arikushi et al. [6] showed that such graphs can have at most $O(n)$ edges. Angelini et al. [5] proved that every cubic graph admits an RAC drawing with at most one bend. It remained an open problem whether every cubic graph has an RAC drawing with straight-line segments. If besides orthogonal crossings, we also allow two edges to cross at $45°$, then it is a straightforward corollary of Theorem 1 that every cubic graph admits such a drawing with straight-line segments.

In Section 2 we give the proof of the Theorem 1, while in Section 3 we prove Theorem 2 and discuss open problems.

2   Proof of Theorem 1

We start with some definitions we will use throughout the section. Then we prove in Corollary 1 that every cubic graph with many vertices contains a special cut. Finally in Lemma 4 we show how to use this and Theorem 3 to obtain a drawing with the four basic slopes.

2.1   Definitions and Subcubic Theorem

Throughout the paper log always denotes $\log_2$, the logarithm in base 2. We recall that the girth of a graph is the length of its shortest cycle.

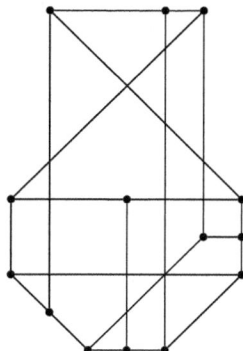

Fig. 2. The Heawood graph drawn with the four basic slopes

Definition 1. *Define a* supercycle *as a connected graph where every degree is at least two and not all are two. Note that a minimal supercycle will look like a "θ" or like a "dumbbell".*

We recall that a *cut* is a partition of the vertices into two sets. We say that an edge is in the cut if its ends are in different subsets of the partition. We also call the edges in the cut the *cut-edges*. The *size* of a cut is the number of cut-edges in it.

Definition 2. *We say that a cut is an* M*-cut if the cut-edges form a matching, in other words, if their ends are pairwise different vertices. We also say that an* M*-cut is* suitable *if after deleting the cut-edges, the graph has two components, both of which are supercycles.*

For any two points $p_1 = (x_1, y_1)$ and $p_2 = (x_2, y_2)$, we say that p_2 is *to the North* of p_1 if $x_2 = x_1$ and $y_2 > y_1$. Analogously, we say that p_2 is *to the Northwest* of p_1 if $x_2 + y_2 = x_1 + y_1$ and $y_2 > y_1$.

We will give the exact statement of the theorem of [21] about subcubic graphs here since it will be used in this proof.

Theorem 3 ([21]). *Let* G *be a connected graph that is not a cycle and whose every vertex has degree at most three. Suppose that* G *has at least one vertex of degree at most two and denote by* $v_1, \ldots, v_m$ *the vertices of degree at most two* $(m \geq 1)$.

Then, for any sequence $x_1, \ldots, x_m$ *of real numbers, linearly independent over the rationals,* G *has a straight-line drawing with the following properties:*

(1) Vertex v_i is mapped into a point with x-coordinate $x(v_i) = x_i$ $(1 \leq i \leq m)$
(2) The slope of every edge is $0, \pi/2, \pi/4$, or $-\pi/4$
(3) No vertex is to the North of any vertex of degree two.
(4) No vertex is to the North or to the Northwest of any vertex of degree one.

The proof of the theorem about subcubic graphs in [21] was incorrect. It used induction but during the proof the statement was also used for disconnected graphs. This can be a problem, since when drawing two components, it might happen that a degree three vertex of one component has to be above a degree two vertex of the other component. However, the proof can be easily fixed to hold for disconnected graphs as well and the theorem is true. For this, one can make the statement stronger, by saying that also for every graph one can select any sequence $x_{m+1}, \ldots, x_n$ of real numbers that satisfy that $x_1, \ldots, x_m, x_{m+1}, \ldots, x_n$ are linearly independent over the rationals, such that the x-coordinates of all the vertices are a linear combination with rational coefficients of $x_1, \ldots, x_n$. This way we can ensure that different components do not interfere. For details see the soon-to-appear errata or [24].

Note that Theorem 3 proves the result of Theorem 1 for subcubic graphs. Another minor observation is that we may assume that the graph is connected. Since we use the basic four slopes, if we can draw the components of a disconnected graph, then we just place them far apart in the plane so that no two drawings intersect. So we will assume for the rest of the section that the graph is cubic and connected.

2.2   Preliminaries

The results in this subsection are also interesting independent of the current problem we deal with. First we bound from above the girth of a cubic graph with its number of vertices. Our bound easily follows from the *Moore bound*, but as that bounds the inverse of our function, here we include a short proof for completeness.

Lemma 1. *Every connected cubic graph on n vertices contains a cycle of length at most $2\lceil \log(\frac{n}{3} + 1) \rceil$.*

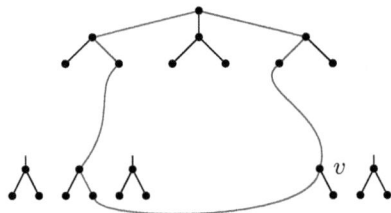

Fig. 3. Finding a cycle in the BFS tree using that the left child of v already occurred

Proof. Start at any vertex of G and conduct a breadth first search (BFS) of G until a vertex repeats in the BFS tree. We note here that by iterations we will (for the rest of the subsection) mean the number of levels of the BFS tree. Since G is cubic, after k iterations, the number of vertices visited will be $1 + 3 + 6 + 12 + \ldots + 3 \cdot 2^{k-2} = 1 + 3(2^{k-1} - 1)$. And since G has n vertices, some vertex must repeat after $k = \lceil \log(\frac{n}{3} + 1) \rceil + 1$ iterations. Tracing back along the two paths obtained for the vertex that reoccurs, we find a cycle of length at most $2\lceil \log(\frac{n}{3} + 1) \rceil$. $\qquad\square$

Lemma 2. *Every connected cubic graph on n vertices with girth g contains a supercycle with at most $2\lceil \log(\frac{n+1}{g}) \rceil + g - 1$ vertices.*

Proof. Contract the vertices of a length g cycle, obtaining a multigraph G' with $n - g + 1$ vertices, that is almost 3-regular, except for one vertex of degree g, from which we start a BFS. It is easy to see that the number of vertices visited after k iterations is at most $1 + g + 2g + 4g + \ldots + g \cdot 2^{k-2} = g(2^{k-1} - 1) + 1$. And since G' has $n - g + 1$ vertices, some vertex must repeat after $k = \lceil \log(\frac{n-g+1}{g} + 1) \rceil + 1 = \lceil \log(\frac{n+1}{g}) \rceil + 1$ iterations. Tracing back along the two paths obtained for the vertex that reoccurs, we find a cycle (or two vertices connected by two edges) of length at most $2\lceil \log(\frac{n+1}{g}) \rceil$ in G'. This implies that in G we have a supercycle with at most $2\lceil \log(\frac{n+1}{g}) \rceil + g - 1$ vertices. $\qquad\square$

Lemma 3. *Every connected cubic graph on $n > 2s - 2$ vertices with a supercycle with s vertices contains a suitable M-cut of size at most $s - 2$.*

Proof. The supercycle with s vertices, A, has at least two vertices of degree 3. The size of the $(A, G - A)$ cut is thus at most $s - 2$. This cut need not be an M-cut because the edges may have a common neighbor in $G - A$. To repair this, we will now add, iteratively, the common neighbors of edges in the cut to A, until no edges have a common neighbor in $G - A$. Note that in any iteration, if a vertex, v, adjacent to exactly two cut-edges was chosen, then the size of A increases by 1 and the size of the cut decreases by 1 (since, these two cut-edges will get added to A along with v, but since the graph is cubic, the third edge from v will become a part of the cut-edges). If a vertex adjacent to three cut-edges was chosen, then the size of A increases by 1 while the number of cut-edges decreases by 3. From this we can see that the maximum number of vertices that could have been added to A during this process is $s - 3$. Now there are three conditions to check.

The first condition is that this process returns a non-empty second component. This cannot occur if

$$(n - s) - (s - 3) > 0$$

or,

$$n > 2s - 3.$$

The second condition is that the second component should not be a collection of disjoint cycles. For this we note that it is enough to check that at every stage, the number of cut-edges is strictly smaller than the number of vertices in $G - A$. But since in the above iterations, the number of cut-edges decreases by a number greater than or equal to the decrease in the size of $G - A$, it is enough to check that before the iterations, the number of cut-edges is strictly smaller than the number of vertices in $G - A$. This is the condition

$$n - s > s - 2$$

or,

$$n > 2s - 2.$$

Note that if this inequality holds then the non-emptiness condition will also hold.

Finally, we need to check that both components are connected. A is always connected but $G - A$ need not be. Pick a component in $G - A$ that has more vertices than the number of cut-edges adjacent to it. Since the number of cut-edges is strictly smaller than number of vertices in $G - A$, there must be one such component, say B, in $G - A$. We add every other component of $G - A$ to A. Note that the size of the cut only decreases with this step. Since B is connected and has more vertices than the number of cut-edges, B cannot be a cycle.   □

Corollary 1. *Every connected cubic graph on $n \geq 18$ vertices contains a suitable M-cut.*

Proof. Using the first two lemmas, we have a supercycle with $s \leq 2\lceil \log(\frac{n+1}{g})\rceil + g - 1$ vertices where $3 \leq g \leq 2\lceil\log(\frac{n}{3} + 1)\rceil$. Then using the last lemma, we have an M-cut with both partitions being a supercycle if $n > 2s - 2$. So all we need to check is that n is indeed big enough. Note that

$$s \leq 2\log(\frac{n+1}{g}) + g + 1 = 2\log(n+1) + g - 2\log g + 1 \leq$$

$$\leq 2\log(n+1) + 2\log(\frac{n}{3} + 1) - 2\log(2\log(\frac{n}{3} + 1)) + 1$$

where the last inequality follows from the fact that $x - 2\log x$ is increasing for $x \geq 2/\log_e 2 \approx 2.88$. So we can bound the right hand side from above by $4\log(n+1) + 1$. Now we need that

$$n > 2(4\log(n+1) + 1) - 2 = 8\log(n+1)$$

which holds if $n \geq 44$.

The statement can be checked for $18 \leq n \leq 42$ with code that can be found in the Appendix of the full version [25]. It outputs for a given value of n, the g for which $2s - 2$ is maximum and this maximum value. Based on the output we can see that for $n \geq 18$, this value is smaller.   □

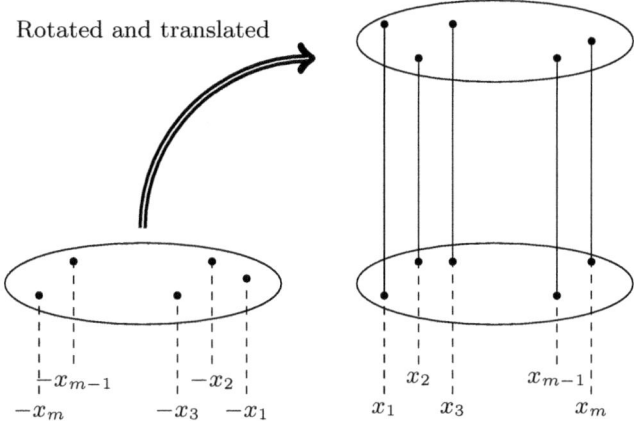

Fig. 4. The x-coordinates of the degree 2 vertices is suitably chosen and one component is rotated and translated to make the M-cut vertical

2.3   Proof

Lemma 4. *Let G be a connected cubic graph with a suitable M-cut. Then, G can be drawn with the four basic slopes.*

Proof. The proof follows rather straightforwardly from Theorem 3. Note that the two components are subcubic graphs and we can choose the x-coordinates of the vertices of the M-cut (since they are the vertices with degree two in the components). If we picked coordinates $x_1, x_2, \ldots, x_m$ in one component, then for the neighbors of these vertices in the other component we pick the x-coordinates $-x_1, -x_2, \ldots, -x_m$. We now rotate the second component by π and place it very high above the other component so that the drawings of the components do not intersect and align them so that the edges of the M-cut will be vertical (slope $\pi/2$). Also, since Theorem 3 guarantees that degree two vertices have no other vertices on the vertical line above them, hence the drawing we obtain above is a valid representation of G with the basic slopes.   □

By combining Lemma 1 and Lemma 4, we can see that Theorem 1 is true for all cubic graphs with $n \geq 18$. For smaller graphs, we give below some lemmas which help reduce the number of graphs we have to check. The lemmas below also occur in different papers and we give references where required.

Lemma 5. *A connected cubic graph with a cut vertex can be drawn with the four basic slopes.*

Proof. We observe that if the cubic graph has a cut vertex then it must also have a bridge. This bridge would be the suitable M-cut for using the previous Lemma 4, since neither of the components can be disconnected or cycles.   □

Lemma 6. *A connected cubic graph with a 2-vertex disconnecting set can be drawn with the four basic slopes.*

Proof. If a cubic graph has a 2-vertex disconnecting set, then it must have a cut of size two with non-adjacent edges. Again the two components we obtain must be connected (or the graph has a bridge) and cannot be cycles. Thus we can apply Lemma 4 again to get the required drawing.   □

The following theorem was proved by Engelstein [15].

Lemma 7. *Every 3-connected cubic graph with a Hamiltonian cycle can be drawn in the plane with the four basic slopes.*

Note that combining the last three lemmas, we even get

Corollary 2. *Every cubic graph with a Hamiltonian cycle can be drawn in the plane with the four basic slopes.*

The graphs which now need to be checked satisfy the following conditions:

1. the number of vertices is at most 16
2. the graph is 3-connected
3. the graph does not have a Hamiltonian cycle.

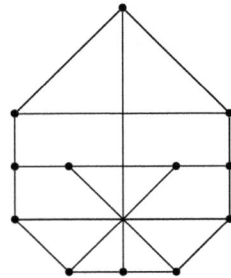

Fig. 5. The Tietze's graph drawn with the four basic slopes

Note that if the number of vertices is at most 16, then it follows from Lemma 1 that the girth is at most 6. Luckily there are several lists available of cubic graphs with a given number of vertices, n and a given girth, g.

If $g = 6$, then there are only two graphs with at most 16 vertices (see [1], [23]), both containing a Hamiltonian cycle.

If $g = 5$ and $n = 16$, then Lemma 2 gives a supercycle with at most 8 vertices, so using Lemma 3 we are done.

If $g = 5$ and $n = 14$, then there are only nine graphs (see [1], [23]), all containing a Hamiltonian cycle.

If $g \leq 4$ and $n = 16$, then Lemma 2 gives a supercycle with at most 8 vertices, so using Lemma 3 we are done.

If $g \leq 4$ and $n = 14$, then Lemma 2 gives a supercycle with at most 7 vertices, so using Lemma 3 we are done.

Finally, all graphs with at most 12 vertices are either not 3-connected or contain a Hamiltonian cycle, except for the Petersen graph and Tietze's Graph (see [2]). For the drawing of these two graphs, see the respective Figures.

3   Which Four Slopes? and Other Concluding Questions

After establishing Theorem 1 the question arises whether we could have used any other four slopes. Call a set of slopes *good* if every cubic graph has a straight-line drawing with them. In this section we prove Theorem 2 that claims that the following statements are equivalent for a set S of four slopes.

1. S is good.
2. S is an affine image of the four basic slopes.
3. We can draw K_4 with S.

Proof. Since affine transformation keeps incidences, any set that is the affine image of the four basic slopes is good.

On the other hand, if a set $S = \{s_1, s_2, s_3, s_4\}$ is good, then K_4 has a straight-line drawing with S. Since we do not allow a vertex to be in the interior of an edge, the four vertices must be in general position. This implies that two incident edges cannot have the same slope. Therefore there are two slopes, without loss of generality s_1 and s_2, such that we have two edges of each slope. These four edges must form a cycle of length four, which means that the vertices are the vertices of a parallelogram. But in this case there is an affine transformation that takes the parallelogram to a square. This transformation also takes S into the four basic slopes.   □

Note that a similar reasoning shows that no matter how many slopes we take, their set need not be good, because we cannot even draw K_4 with them unless they satisfy some correlation. The above proofs use the four basic slopes only in a few places (for rotation invariance and to start induction). Thus we make the following conjecture.

Conjecture 1. There is a (not necessarily connected, finite) graph such that a set of slopes is good if and only if this graph has a straight-line drawing with them.

This finite graph would be the disjoint union of K_4, maybe the Petersen graph and other small graphs. We could not even rule out the possibility that K_4 (or maybe another, connected graph) is alone sufficient. Note that we can define a partial order on the graphs this way. Let $G < H$ if any set of slopes that can be used to draw H can also be used to draw G. This way of course $G \subset H \Rightarrow G < H$ but what else can we say about this poset?

Is it possible to use this new method to prove that the slope parameter of cubic graphs is also four?

The main question remains to prove or disprove whether the slope number of graphs with maximum degree four is unbounded.

Acknowledgment. We would like to thank the anonymous referees their several useful remarks.

References

1. http://www.mathe2.uni-bayreuth.de/markus/reggraphs.html
2. http://en.wikipedia.org/wiki/Tableofsimplecubicgraphs
3. Ambrus, G., Barát, J.: A contribution to queens graphs: A substitution method. Discrete Mathematics 306(12), 1105–1114 (2006)
4. Ambrus, G., Barát, J., Hajnal, P.: The slope parameter of graphs. Acta Sci. Math (Szeged) 72, 875–889 (2006)
5. Angelini, P., Cittadini, L., Di Battista, G., Didimo, W., Frati, F., Kaufmann, M., Symvonis, A.: On the Perspectives Opened By Right Angle Crossing Drawings. In: Eppstein, D., Gansner, E.R. (eds.) GD 2009. LNCS, vol. 5849, pp. 21–32. Springer, Heidelberg (2010)
6. Arikushi, K., Fulek, R., Keszegh, B., Morić, F., Tóth, C.D.: Graphs that Admit Right Angle Crossing Drawings. In: Thilikos, D.M. (ed.) WG 2010. LNCS, vol. 6410, pp. 135–146. Springer, Heidelberg (2010)
7. Barát, J., Matousek, J., Wood, D.R.: Bounded-degree graphs have arbitrarily large geometric thickness. Electr. J. Comb. 13(1) (2006)
8. Bell, J., Stevens, B.: A survey of known results and research areas for n-queens. Discrete Mathematics 309, 1–31 (2009)
9. Didimo, W., Eades, P., Liotta, G.: Drawing Graphs With Right Angle Crossings. In: Dehne, F., Gavrilova, M., Sack, J.-R., Tóth, C.D. (eds.) WADS 2009. LNCS, vol. 5664, pp. 206–217. Springer, Heidelberg (2009)
10. Dillencourt, M.B., Eppstein, D., Hirschberg, D.S.: Geometric thickness of complete graphs. J. Graph Algorithms Appl. 4(3), 5–17 (2000)
11. Dujmović, V., Eppstein, D., Suderman, M., Wood, D.R.: Drawings of planar graphs with few slopes and segments. Comput. Geom. 38(3), 194–212 (2007)
12. Dujmović, V., Suderman, M., Wood, D.R.: Really Straight Graph Drawings. In: Pach, J. (ed.) GD 2004. LNCS, vol. 3383, pp. 122–132. Springer, Heidelberg (2005)
13. Dujmovic, V., Wood, D.R.: Graph treewidth and geometric thickness parameters. Discrete & Computational Geometry 37(4), 641–670 (2007)
14. Duncan, C.A., Eppstein, D., Kobourov, S.G.: The geometric thickness of low degree graphs. In: Snoeyink, J., Boissonnat, J.-D. (eds.) Symposium on Computational Geometry, pp. 340–346. ACM (2004)
15. Engelstein, M.: Drawing graphs with few slopes. Intel Competition for high school students (2005)
16. Eppstein, D.: Separating Thickness From Geometric Thickness. In: Goodrich, M.T., Kobourov, S.G. (eds.) GD 2002. LNCS, vol. 2528, pp. 150–161. Springer, Heidelberg (2002)
17. Hutchinson, J.P., Shermer, T.C., Vince, A.: On representations of some thickness-two graphs. Comput. Geom. 13(3), 161–171 (1999)
18. Jelínek, V., Jelínková, E., Kratochvíl, J., Lidický, B., Tesař, M., Vyskočil, T.: The Planar Slope Number Of Planar Partial 3-Trees Of Bounded Degree. In: Eppstein, D., Gansner, E.R. (eds.) GD 2009. LNCS, vol. 5849, pp. 304–315. Springer, Heidelberg (2010)
19. Kainen, P.C.: Thickness and coarseness of graphs. Abh. Math. Sem. Univ. Hamburg 39, 88–95 (1973)

20. Keszegh, B., Pach, J., Pálvölgyi, D.: Drawing Planar Graphs of Bounded Degree With Few Slopes. In: Brandes, U., Cornelsen, S. (eds.) GD 2010. LNCS, vol. 6502, pp. 293–304. Springer, Heidelberg (2011)
21. Keszegh, B., Pach, J., Pálvölgyi, D., Tóth, G.: Drawing cubic graphs with at most five slopes. Comput. Geom. 40(2), 138–147 (2008)
22. Keszegh, B., Pach, J., Pálvölgyi, D., Tóth, G.: Cubic graphs have bounded slope parameter. J. Graph Algorithms Appl. 14(1), 5–17 (2010)
23. Meringer, M.: Fast generation of regular graphs and construction of cages. J. Graph Theory 30, 137–146 (1999)
24. Mukkamala, P.: Obstacles, Slopes and Tic-Tac-Toe: An excursion in discrete geomety and combinatorial game theory. PhD thesis, Rutgers, The State University of New Jersey (2011), http://arxiv.org/abs/1106.1973
25. Mukkamala, P., Pálvölgyi, D.: Drawing cubic graphs with the four basic slopes. CoRR, abs/1106.1973 (2011)
26. Mukkamala, P., Szegedy, M.: Geometric representation of cubic graphs with four directions. Comput. Geom. 42(9), 842–851 (2009)
27. Mutzel, P., Odenthal, T., Scharbrodt, M.: The thickness of graphs: A survey. Graphs Combin. 14, 59–73 (1998)
28. Pach, J., Pálvölgyi, D.: Bounded-degree graphs can have arbitrarily large slope numbers. Electr. J. Comb. 13(1) (2006)
29. Wade, G.A., Chu, J.-H.: Drawability of complete graphs using a minimal slope set. Comput. J. 37(2), 139–142 (1994)

k-Quasi-Planar Graphs

Andrew Suk⋆

School of Basic Sciences, École Polytechnique Fédérale de Lausanne, Switzerland
suk@cims.nyu.edu

Abstract. A topological graph is *k-quasi-planar* if it does not contain k pairwise crossing edges. A topological graph is *simple* if every pair of its edges intersect at most once (either at a vertex or at their intersection). In 1996, Pach, Shahrokhi, and Szegedy [16] showed that every n-vertex simple k-quasi-planar graph contains at most $O\left(n(\log n)^{2k-4}\right)$ edges. This upper bound was recently improved (for large k) by Fox and Pach [8] to $n(\log n)^{O(\log k)}$. In this note, we show that all such graphs contain at most $(n\log^2 n)2^{\alpha^{c_k}(n)}$ edges, where $\alpha(n)$ denotes the inverse Ackermann function and c_k is a constant that depends only on k.

1 Introduction

A *topological graph* is a graph drawn in the plane such that its vertices are represented by points and its edges are represented by non-self-intersecting arcs connecting the corresponding points. The arcs are allowed to intersect, but they may not pass through vertices except for their endpoints. Furthermore, the edges are not allowed to have tangencies, i.e., if two edges share an interior point, then they must properly cross at that point in common. We only consider graphs without parallel edges or loops. A topological graph is *simple* if every pair of its edges intersect at most once. If the edges are drawn as straight-line segments, then the graph is *geometric*. Two edges of a topological graph *cross* if their interiors share a point.

Finding the maximum number of edges in a topological graph with a forbidden crossing pattern has been a classic problem in extremal topological graph theory (see [2,3,4,6,8,10,15,19,21]). It follows from Euler's Polyhedral Formula that every topological graph on n vertices and no crossing edges has at most $3n - 6$ edges. A topological graph is *k-quasi-planar*, if it does not contain k pairwise crossing edges. Hence 2-quasi-planar graphs are planar. An old conjecture (see Problem 1 in section 9.6 of [5]) states that for any fixed $k > 0$, every k-quasi-planar graph on n vertices has at most $c_k n$ edges, where c_k is a constant that depends only on k. Agarwal et al. [4] were the first to prove this conjecture for simple 3-quasi-planar graphs. Later, Pach, Radoičić, and Tóth [14] generalized the result for all (not simple) 3-quasi-planar graphs. Ackerman [1] proved the conjecture for $k = 4$.

For $k \geq 5$, Pach, Shahrokhi, and Szegedy [16] showed that every simple k-quasi-planar graph on n vertices has at most $c_k n(\log n)^{2k-4}$ edges. This bound can be improved to $c_k n(\log n)^{2k-8}$ by using a result of Ackerman [1]. Valtr [20] proved

⋆ The author gratefully acknowledges the support from the Swiss National Science Foundation Grant No. 200021-125287/1.

that every n-vertex k-quasi-planar geometric graph contains at most $O(n \log n)$ edges. Later, he extended this result to simple topological graphs with edges drawn as x-monotone curves [21]. Pach, Radoičić, and Tóth showed that every n-vertex (not simple) k-quasi-planar graph has at most $c_k n (\log n)^{4k-12}$ edges, which can also be improved to

$$c_k n (\log n)^{4k-16}$$

by a result of Ackerman [1].

Recently, Fox and Pach [8] improved (for large k) the exponent in the polylogarithmic factor for simple topological graphs. They showed that every simple k-quasi-planar graph on n vertices has at most

$$n(c \log n / \log k)^{c \log k}$$

edges, where c is an absolute constant. Our main result is the following.

Theorem 1. *Let $G = (V, E)$ be an n-vertex simple k-quasi-planar graph. Then*

$$|E(G)| \leq (n \log^2 n) 2^{\alpha^{c_k}(n)},$$

where $\alpha(n)$ denotes the inverse Ackermann function and c_k is a constant that depends only on k.

In the proof of Theorem 1, we apply results on generalized Davenport-Schinzel sequences. This method was used by Valtr [21], who showed that every n-vertex simple k-quasi-planar graph with edges drawn as x-monotone curves has at most $2^{2^{ck}} n \log n$ edges, where c is an absolute constant. Our next theorem extends his result to (not simple) topological graphs with edges drawn with x-monotone curves, and moreover we obtain a slightly better upper bound.

Theorem 2. *Let $G = (V, E)$ be an n-vertex (not simple) k-quasi-planar graph with edges drawn as x-monotone curves. Then $|E(G)| \leq 2^{ck^3} n \log n$, where c is an absolute constant.*

2 Generalized Davenport-Schinzel Sequences

The sequence $u = a_1, a_2, ..., a_m$ is called l-regular if any l consecutive terms are pairwise different. For integers $l, t \geq 2$, the sequence

$$S = s_1, s_2, ..., s_{lt}$$

of length $l \cdot t$ is said to be of type $up(l, t)$ if the first l terms are pairwise different and for $i = 1, 2, ..., l$

$$s_i = s_{i+l} = s_{i+2l} = \cdots = s_{i+(t-1)l}.$$

For example,

$$a, b, c, a, b, c, a, b, c, a, b, c,$$

would be an $up(3, 4)$ sequence. By applying a theorem of Klazar on generalized Davenport-Schinzel sequences, we have the following.

Theorem 3 ([11]). *For $l \geq 2$ and $t \geq 3$, the length of any l-regular sequence over an n-element alphabet that does not contain a subsequence of type $up(l, t)$ has length at most*

$$n \cdot l2^{(lt-3)} \cdot (10l)^{10\alpha^{lt}(n)}.$$

For $l \geq 2$, the sequence

$$S = s_1, s_2, ..., s_{3l-2}$$

of length $3l - 2$ is said to be of type *up-down-up*(l), if the first l terms are pairwise different, and for $i = 1, 2, ..., l$,

$$s_i = s_{2l-i} = s_{(2l-2)+i}.$$

For example,

$$a, b, c, d, c, b, a, b, c, d,$$

would be an *up-down-up*(4) sequence. Valtr and Klazar [12] showed that any l-regular sequence over an n-element alphabet containing no subsequence of type up-down-up(l) has length at most $2^{l^c} n$ for some constant c. Recently, Pettie made the following improvement.

Lemma 1 ([18]). *For $l \geq 2$, the length of any l-regular sequence over an n-element alphabet containing no subsequence of type up-down-up(l) has length at most $2^{O(l^2)} n$.*

For more results on generalized Davenport-Schinzel sequences, see [13,18,17].

3   Simple Topological Graphs

In this section, we will prove Theorem 1. For any partition of $V(G)$ into two disjoint parts, V_1 and V_2, let $E(V_1, V_2)$ denote the set of edges with one endpoint in V_1 and the other endpoint in V_2. The *bisection width* of a graph G, denoted by $b(G)$, is the smallest nonnegative integer such that there is a partition of the vertex set $V = V_1 \cup V_2$ with $\frac{1}{3} \cdot |V| \leq |V_i| \leq \frac{2}{3} \cdot |V|$ for $i = 1, 2$, and $|E(V_1, V_2)| = b(G)$. We will use the following result by Pach et al.

Lemma 2 ([16]). *If G is a graph with n vertices of degrees $d_1, ..., d_n$, then*

$$b(G) \leq 7cr(G)^{1/2} + 2\sqrt{\sum_{i=1}^{n} d_i^2},$$

where $cr(G)$ denotes the crossing number of G.

Since $\sum_{i=1}^{n} d_i^2 \leq 2n|E(G)|$ holds for every graph, we have

$$b(G) \leq 7cr(G)^{1/2} + 3\sqrt{|E(G)|n}. \tag{1}$$

Proof of Theorem 1. Let $k \geq 5$ and $f_k(n)$ denote the maximum number of edges in a simple k-quasi-planar graph on n vertices. We will prove that

$$f_k(n) \leq (n \log^2 n) 2^{\alpha^{c_k}(n)}$$

where $c_k = 10^5 \cdot 2^{k^2+2k}$. For sake of clarity, we do not make any attempts to optimize the value of c_k. We proceed by induction on n. The base case $n < 7$ is trivial. For the inductive step $n \geq 7$, let $G = (V, E)$ be a simple k-quasi-planar graph with n vertices and $m = f_k(n)$ edges, such that the vertices of G are labeled 1 to n. The proof splits into two cases.

Case 1. Suppose that $cr(G) \leq m^2/(10^4 \log^2 n)$. By (1), there is a partition $V(G) = V_1 \cup V_2$ with $|V_1|, |V_2| \leq 2n/3$ and the number of edges with one vertex in V_1 and one vertex in V_2 is at most

$$b(G) \leq 7cr(G)^{1/2} + 3\sqrt{mn} \leq 7\frac{m}{100 \log n} + 3\sqrt{mn}.$$

Let $n_1 = |V_1|$ and $n_2 = |V_2|$. Now if $7m/(100 \log n) \leq 3\sqrt{mn}$, then we have

$$m \leq 43n \log^2 n$$

and we are done since $\alpha(n) \geq 2$ and $k \geq 5$. Therefore, we can assume $7m/(100 \log n) > 3\sqrt{mn}$, which implies

$$b(G) \leq \frac{m}{7 \log n}. \tag{2}$$

By the induction hypothesis and equation (2), we have

$$m \leq f_k(n_1) + f_k(n_2) + b(G)$$

$$\leq \left(n_1 \log^2(2n/3)\right) 2^{\alpha^{c_k}(n)} + \left(n_2 \log^2(2n/3)\right) 2^{\alpha^{c_k}(n)} + b(G)$$

$$\leq \left(n \log^2(2n/3)\right) 2^{\alpha^{c_k}(n)} + \frac{m}{7 \log n}$$

$$\leq (n \log^2 n) 2^{\alpha^{c_k}(n)} - 2n2^{\alpha^{c_k}(n)} \log n \log(3/2) + n2^{\alpha^{c_k}(n)} \log^2(3/2) + \frac{m}{7 \log n}$$

which implies

$$m\left(1 - \frac{1}{7 \log n}\right) \leq (n \log^2 n) 2^{\alpha^{c_k}(n)} \left(1 - \frac{2\log(3/2)}{\log n} + \frac{\log^2(3/2)}{\log^2 n}\right).$$

Hence

$$m \leq (n \log^2 n) 2^{\alpha^{c_k}(n)} \frac{1 - 2\log(3/2)\log^{-1} n + \log^2(3/2)\log^{-2} n}{1 - 1/(7 \log n)}$$

$$\leq (n \log^2 n) 2^{\alpha^{c_k}(n)}.$$

Case 2. Now suppose that $cr(G) \geq m^2/(10^4 \log^2 n)$. By a simple averaging argument, there exists an edge $e = uv$ such that at least $2m/(10^4 \log^2 n)$ other edges cross e. Fix such an edge $e = uv$, and let E' denote the set of edges that cross e.

We order the edges in $E' = \{e_1, e_2, ..., e_{|E'|}\}$, in the order that they cross e from u to v. Now we create two sequences $S_1 = p_1, p_2, ..., p_{|E'|}$ and $S_2 = q_1, q_2, ..., q_{|E'|}$ as follows. For each $e_i \in E'$, as we move along edge e from u to v and arrive at edge e_i, we turn left and move along edge e_i until we reach its endpoint u_i. Then we set $p_i = u_i$. Likewise, as we move along edge e from u to v and arrive at edge e_i, we turn right and move along edge e_i until we reach its other endpoint v_i. Then set $q_i = v_i$. Thus S_1 and S_2 are sequences of length $|E'|$ over the alphabet $\{1, 2, ..., n\}$. See Figure 1 for a small example.

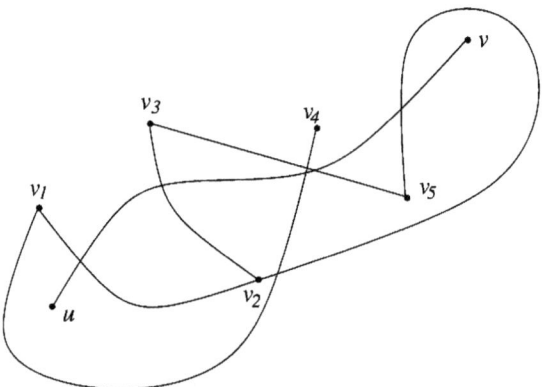

Fig. 1. In this example, $S_1 = v_1, v_3, v_4, v_3, v_2$ and $S_2 = v_2, v_2, v_1, v_5, v_5$

Now we need the following two lemmas. The first one is due to Valtr.

Lemma 3 ([21]). *For $l \geq 1$, at least one of the sequences S_1, S_2 defined above contains an l-regular subsequence of length at least $|E'|/(4l)$.*

Since each edge in E' crosses e exactly one, the proof of Lemma 3 can be copied almost verbatim from the proof of Lemma 4 in [21].

Lemma 4. *Neither of the sequences S_1 and S_2 contains a subsequence of type $up(2^{k^2+k}, 2^k)$.*

Proof. By symmetry, it suffices to show that S_1 does not contain a subsequence of type $up(2^{k^2+k}, 2^k)$. The argument is by contradiction. We will prove by induction on k, that such a sequence will produce k pairwise crossing edges in G. The base cases $k = 1, 2$ are trivial. Now assume the statement holds up to $k - 1$. Let

$$S = s_1, s_2, ..., s_{2^{k^2}+2k}$$

be our $up(2^{k^2+k}, 2^k)$ sequence of length 2^{k^2+2k} such that the first 2^{k^2+k} terms are pairwise different, and for $i = 1, 2, ..., 2^{k^2+k}$

$$s_i = s_{i+2^{k^2+k}} = s_{i+2 \cdot 2^{k^2+k}} = s_{i+3 \cdot 2^{k^2+k}} = \cdots = s_{i+(2^k-1)2^{k^2+k}}.$$

For each $i = 1, 2, ..., 2^{k^2+k}$, let $v_i \in V_1$ denote the label (vertex) of s_i. Moreover, let $a_{i,j}$ be the arc emanating from vertex v_i to the edge e corresponding to $s_{i+j2^{k^2+k}}$ for $j = 0, 1, 2, ..., 2^k - 1$. We will think of $s_{i+j2^{k^2+k}}$ as a point on $a_{i,j}$ very close but not on edge e. For simplicity, we will let $s_{2^{k^2+2k}+t} = s_t$ for all $t \in \mathbb{N}$ and $a_{i,j} = a_{i,j \bmod 2^k}$ for all $j \in \mathbb{Z}$. Hence there are 2^{k^2+k} distinct vertices $v_1, ..., v_{2^{k^2+k}}$, each vertex of which has 2^k arcs emanating from it to the edge e.

Consider the drawing of the 2^k arcs emanating from v_1 and the edge e. Since G is simple, this drawing partitions the plane into 2^k regions. By the Pigeonhole principle, there is a subset $V' \subset \{v_1, ..., v_{2^{k^2+k}}\}$ of size

$$\frac{2^{k^2+k} - 1}{2^k},$$

such that all of the vertices of V' lie in the same region. Let $j_0 \in \{0, 1, 2, ..., 2^k - 1\}$ be an integer such that V' lies in the region bounded by $a_{1,j_0}, a_{1,j_0+1}, e$. See Figure 2. In the case $j_0 = 2^k - 1$, V' lies in the unbounded region.

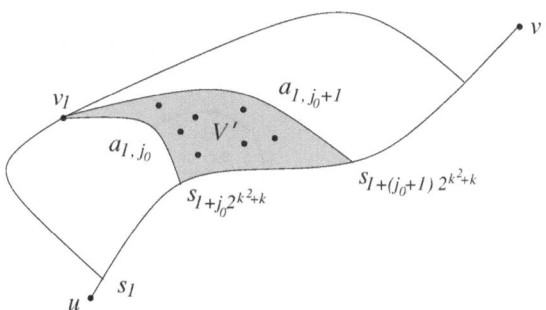

Fig. 2. Vertices of V' lie in the region enclosed by $a_{1,j_0}, a_{1,j_0+1}, e$.

Let $v_i \in V'$ and a_{i,j_0+j_1} be an arc emanating out of v_i for $j_1 \geq 1$. Notice that a_{i,j_0+j_1} cannot cross both a_{1,j_0} and a_{1,j_0+1} since G is simple. Suppose that a_{i,j_0+j_1} crosses a_{1,j_0+1}. Then the set of arcs (emanating out of v_i)

$$A = \{a_{i,j_0+1}, a_{i,j_0+2}, ..., a_{i,j_0+j_1-1}\}$$

must also cross a_{1,j_0+1}. Indeed, let γ be the simple closed curve created by the arrangement

$$a_{i,j_0+j_1} \cup a_{1,j_0+1} \cup e.$$

Since $a_{i,j_0+j_1}, a_{1,j_0+1}, e$ pairwise intersect at precisely one point, γ is well defined. We define points $x = a_{i,j_0+j_1} \cap a_{1,j_0+1}$ and $y = a_{1,j_0+1} \cap e$, and orient γ in the direction from x to y along γ.

Since a_{i,j_0+j_1} intersects a_{1,j_0+1}, v_i must lie to the right of γ. Moreover since the arc from x to y along a_{1,j_0+1} is a subset of γ, the points corresponding to the subsequence

$$S' = \{s_t \in S \mid 2 + (j_0 + 1)2^{k^2+k} \le t \le (i-1) + (j_0 + j_1)2^{k^2+k}\}$$

lie to the left of γ. Hence γ separates vertex v_i and the points of S'. Therefore each arc from A must cross a_{1,j_0+1} since G is simple (these arcs cannot cross a_{i,j_0+j_1}). See Figure 3.

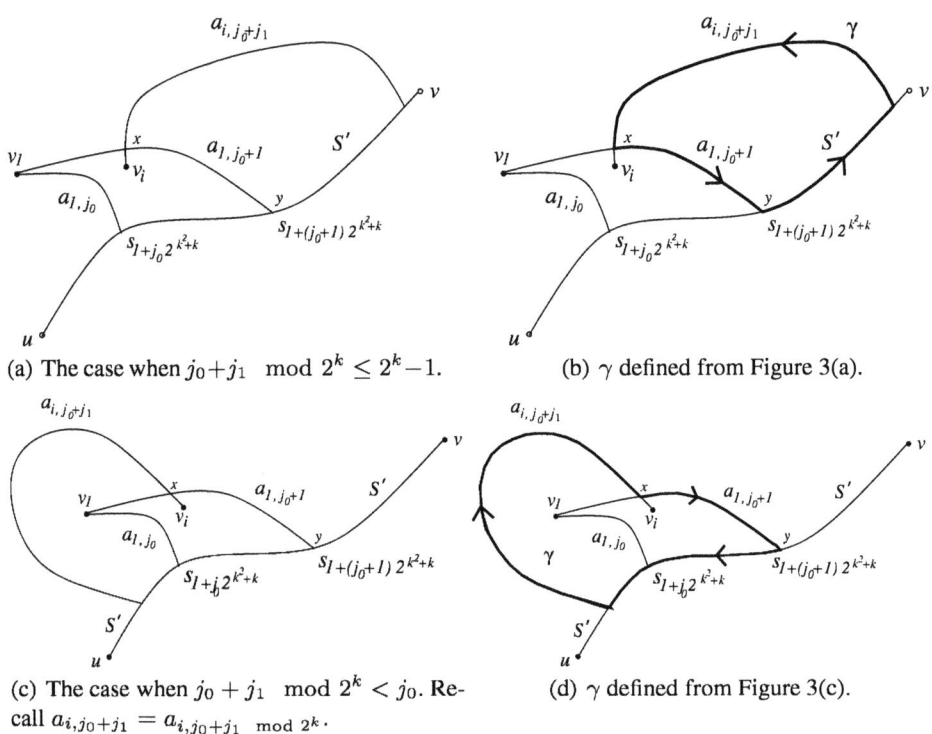

(a) The case when $j_0+j_1 \mod 2^k \le 2^k-1$.

(b) γ defined from Figure 3(a).

(c) The case when $j_0 + j_1 \mod 2^k < j_0$. Recall $a_{i,j_0+j_1} = a_{i,j_0+j_1 \mod 2^k}$.

(d) γ defined from Figure 3(c).

Fig. 3. Defining γ and its orientation

By the same argument, if the arc a_{i,j_0-j_1} crosses a_{1,j_0} for $j_1 \ge 1$, then the arcs (emanating out of v_i)

$$a_{i,j_0-1}, a_{i,j_0-2}, ..., a_{i,j_0-j_1+1}$$

must also cross a_{1,j_0}. Since $a_{i,j_0+2^k/2} = a_{i,j_0-2^k/2}$, we have the following observation.

Observation 4. *For half of the vertices $v_i \in V'$, the arcs emanating out of v_i satisfy*

1. $a_{i,j_0+1}, a_{i,j_0+2}, ..., a_{i,j_0+2^k}/2$ all cross a_{1,j_0+1}, or
2. $a_{i,j_0-1}, a_{i,j_0-2}, ..., a_{i,j_0-2^k}/2$ all cross a_{1,j_0}.

$\square$

Since

$$\frac{|V'|}{2} \geq \frac{2^{k^2+k} - 1}{2 \cdot 2^k} \geq 2^{(k-1)^2+(k-1)},$$

by Observation 4 we have a $(2^{(k-1)^2+(k-1)}, 2^{k-1})up$ sequence, whose corresponding arcs all cross either a_{1,j_0} or a_{1,j_0+1}. By the induction hypothesis, we have k pairwise crossing edges.

$\square$

Now we are ready to complete the proof of Theorem 1. By Lemma 3 we know that, say, S_1 contains a 2^{k^2+k}-regular subsequence of length $|E'|/(4 \cdot 2^{k^2+k})$. By Theorem 3 and Lemma 4, this subsequence has length at most

$$n2^{k^2+k}2^{2^{k^2+2k}-3}\left(10 \cdot 2^{k^2+k}\right)^{10\alpha^{2^{k^2+2k}}(n)}.$$

Therefore

$$\frac{2m}{10^4 \cdot 4 \cdot 2^{k^2+k}\log^2 n} \leq \frac{|E'|}{4 \cdot 2^{k^2+k}} \leq n2^{k^2+k}2^{2^{k^2+2k}-3}\left(10 \cdot 2^{k^2+k}\right)^{10\alpha^{2^{k^2+2k}}(n)}$$

which implies

$$m \leq 4 \cdot 10^4 \cdot 2^{2^{k^2+2k}}2^{2^{k^2+2k}-3}n\left(10 \cdot 2^{k^2+k}\right)^{10\alpha^{2^{k^2+2k}}(n)}\log^2 n.$$

Since $c_k = 10^5 \cdot 2^{k^2+2k}$, $\alpha(n) \geq 2$ and $k \geq 5$, we have

$$m \leq (n\log^2 n)2^{\alpha^{c_k}(n)}.$$

$\square$

4   *x*-Monotone

In this section we will prove Theorem 2.

Proof of Theorem 2. For $k \geq 2$, let $g_k(n)$ be the maximum number of edges in a (not simple) k-quasi-planar graph whose edges are drawn as x-monotone curves. We will prove by induction on n that

$$g_k(n) \leq 2^{ck^6}n\log n$$

where c is a sufficiently large absolute constant. The base case is trivial. For the inductive step, let $G = (V, E)$ be a k-quasi-planar topological graph whose edges are drawn as x-monotone curves, and let the vertices be labeled $1, 2, ..., n$. Then let L be the vertical line that partitions the vertices into two parts, V_1 and V_2, such that $|V_1| = \lfloor n/2 \rfloor$ vertices lie to the left of L, and $|V_2| = \lceil n/2 \rceil$ vertices lie to the right of L. Furthermore, let E_1 denote the set of edges induced by V_1, E_2 be the set of edges induced by V_2, and E' be the set of edges that intersect L. Clearly, we have

$$|E_1| \leq g_k(\lfloor n/2 \rfloor) \qquad \text{and} \qquad |E_2| \leq g_k(\lceil n/2 \rceil).$$

Hence it suffices that show that

$$|E'| \leq 2^{ck^6/2} n, \tag{3}$$

since this would imply

$$g_k(n) \leq g_k(\lfloor n/2 \rfloor) + g_k(\lceil n/2 \rceil) + 2^{ck^6/2} n \leq 2^{ck^6} n \log n.$$

For the rest of the proof, we will only consider the edges from E'. Now for each vertex $v_i \in V_1$, consider the graph G_i whose vertices are the edges with v_i as a left endpoint, and two vertices in G_i are adjacent if the corresponding edges cross at some point to the left of L. Since G_i is an *incomparability graph* (see [7], [9]) and does not contain a clique of size k, G_i contains an independent set of size $|E(G_i)|/(k-1)$. We keep all edges that correspond to the elements of this independent set, and discard all other edges incident to v_i. After repeating this process on all vertices in V_1, we are left with at least $|E'|/(k-1)$ edges.

Now we continue this process on the other side. For each vertex $v_j \in V_2$, consider the graph G_j whose vertices are the edges with v_j as a right endpoint, and two vertices in G_j are adjacent if the corresponding edges cross at some point to the right of L. Since G_j is an incomparability graph and does not contain a clique of size k, G_j contains an independent set of size $|E(G_j)|/(k-1)$. We keep all edges that corresponds to this independent set, and discard all other edges incident to v_j. After repeating this process on all vertices in V_2, we are left with at least $|E'|/(k-1)^2$ edges.

We order the remaining edges $e_1, e_2, ..., e_m$ in the order in which they intersect L from bottom to top. We define two sequences $S_1 = p_1, p_2, ..., p_m$ and $S_2 = q_1, q_2, ..., q_m$ such that p_i denotes the left endpoint of edge e_i and q_i denotes the right endpoint of e_i. Now we need the following lemma.

Lemma 5. *Neither of the sequences S_1 and S_2 contains a subsequence of type up-down-up$(k^3 + 2)$.*

Proof. By symmetry, it suffices to show that S_1 does not contain a subsequence of type *up-down-up*$(k^3 + 2)$. For the sake of contradiction, suppose S_1 did contain a subsequence of type *up-down-up*$(k^3 + 2)$. Then there is a sequence

$$S = s_1, s_2, ..., s_{3(k^3+2)-2}$$

such that the integers $s_1, ..., s_{k^3+2}$ are pairwise different and for $i = 1, 2, ..., k^3 + 2$ we have

$$s_i = s_{2(k^3+2)-i} = s_{2(k^3+2)-2+i}.$$

For each $i = 1, 2, ..., k^3 + 2$, let $v_i \in V_1$ denote the label (vertex) of s_i and let x_i denote the x-coordinate of vertex v_i. Moreover, let a_i be the arc emanating from vertex v_i to the point on L that corresponds to $s_{2(k^3+2)-i}$. Note that the set of arcs $A = \{a_2, a_3, ..., a_{k^3+1}\}$ are ordered downwards as they intersect L, and corresponds to the "middle" part of the up-down-up sequence. We define two partial orders on A as follows.

$$a_i \prec_1 a_j \text{ if } i < j, \ x_i < x_j \text{ and the arcs } a_i, a_j \text{ do not intersect,}$$

$$a_i \prec_2 a_j \text{ if } i < j, \ x_i > x_j \text{ and the arcs } a_i, a_j \text{ do not intersect.}$$

Clearly, $\prec_1$ and $\prec_2$ are partial orders. If two arcs are not comparable by either $\prec_1$ or $\prec_2$, then they must cross. Since G does not contain k pairwise crossing edges, by Dilworth's Theorem, there exist k arcs $\{a_{i_1}, a_{i_2}, ..., a_{i_k}\}$ such that they are pairwise comparable by either $\prec_1$ or $\prec_2$. Now the proof falls into two cases.

Case 1. Suppose that $a_{i_1} \prec_1 a_{i_2} \prec_1 \cdots \prec_1 a_{i_k}$. Then the arcs emanating from $v_{i_1}, v_{i_2}, ..., v_{i_k}$ to the points corresponding to

$$s_{2(k^3+2)-2+i_1}, s_{2(k^3+2)-2+i_2}, \cdots, s_{2(k^3+2)-2+i_k}$$

are pairwise crossing. See Figure 4.

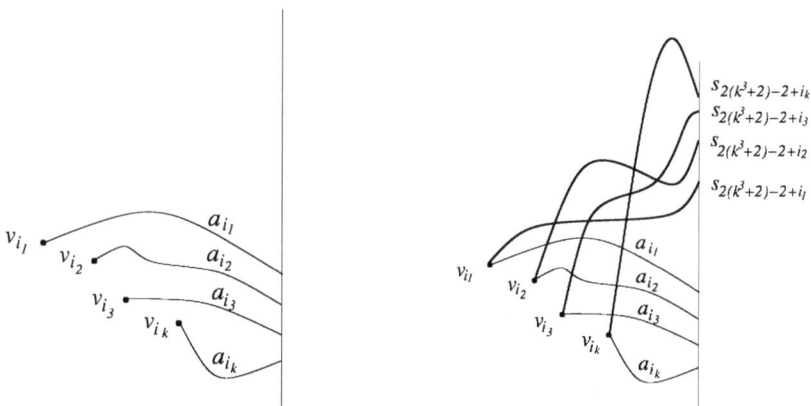

Fig. 4. Case 1

Case 2. Suppose that $a_{i_1} \prec_2 a_{i_2} \prec_2 \cdots \prec_2 a_{i_k}$. Then the arcs emanating from $v_{i_1}, v_{i_2}, ..., v_{i_k}$ to the points corresponding to $s_{i_1}, s_{i_2}, ..., s_{i_k}$ are pairwise crossing. See Figure 5.

□

We are now ready to complete the proof of Theorem 2. By Lemma 3, we know that, say, S_1 contains a $(k^3 + 2)$-regular subsequence of length

$$\frac{|E'|}{4(k^3 + 2)(k - 1)^2}.$$

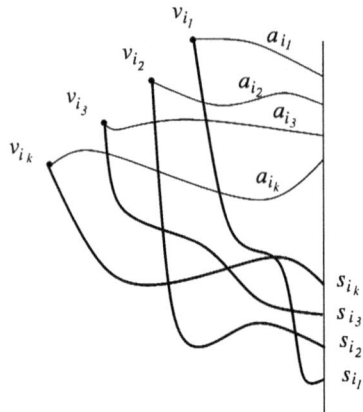

Fig. 5. Case 2

By lemma 1 and 5, this subsequence has length at most $2^{c'k^6}n$, where c' is an absolute constant. Hence

$$\frac{|E'|}{4(k^3+2)(k-1)^2} \le 2^{c'k^6}n$$

implies

$$|E'| \le 4k^5 2^{c'k^6}n \le 2^{ck^6/2}n$$

for a sufficiently large absolute constant c.   □

References

1. Ackerman, E.: On the Maximum Number of Edges in Topological Graphs with No Four Pairwise Crossing Edges. In: Proceedings of the Twenty-Second Annual Symposium on Computational Geometry, SCG 2006, pp. 259–263. ACM, New York (2006)
2. Ackerman, E., Fox, J., Pach, J., Suk, A.: On Grids in Topological Graphs. In: Proceedings of the 25th Annual Symposium on Computational Geometry, SCG 2009, pp. 403–412. ACM, New York (2009)
3. Ackerman, E., Tardos, G.: Note: On the Maximum Number of Edges in Quasi-Planar Graphs. J. Comb. Theory Ser. A 114(3), 563–571 (2007)
4. Agarwal, P.K., Aronov, B., Pach, J., Pollack, R., Sharir, M.: Quasi-Planar Graphs Have a Linear Number of Edges. In: Brandenburg, F.J. (ed.) GD 1995. LNCS, vol. 1027, pp. 1–7. Springer, Heidelberg (1995)
5. Brass, P., Moser, W., Pach, J.: Research Problems in Discrete Geometry. Springer, Berlin (2005)
6. Capoyleas, V., Pach, J.: A Turán-Type Theorem on Chords of a Convex Polygon. J. Combinatorial Theory, Series B 56, 9–15 (1992)

7. Dilworth, R.P.: A decomposition theorem for partially ordered sets. Annals of Math 51, 161–166 (1950)
8. Fox, J., Pach, J.: Coloring K_k-Free Intersection Graphs of Geometric Objects in the Plane. In: Proceedings of the Twenty-Fourth Annual Symposium on Computational Geometry, SCG 2008, pp. 346–354. ACM, New York (2008)
9. Fox, J., Pach, J., Tóth, C.: Intersection Patterns of Curves. Journal of the London Mathematical Society 83, 389–406 (2011)
10. Fulek, R., Suk, A.: Disjoint Crossing Families. In: EuroComb 2011 (2011, to appear)
11. Klazar, M.: A General Upper Bound in Extremal Theory of Sequences. Commentationes Mathematicae Universitatis Carolinae 33(4), 737–746 (1992)
12. Klazar, M., Valtr, P.: Generalized Davenport-Schinzel Sequences. Combinatorica 14, 463–476 (1994)
13. Nivasch, G.: Improved Bounds and New Techniques for Davenport–Schinzel Sequences and Their Generalizations. J. ACM 57(3), 3, Article 17 (2010)
14. Pach, J., Radoičić, R., Tóth, G.: Relaxing Planarity for Topological Graphs. In: Akiyama, J., Kano, M. (eds.) JCDCG 2002. LNCS, vol. 2866, pp. 221–232. Springer, Heidelberg (2003)
15. Pach, J., Pinchasi, R., Sharir, M., Tóth, G.: Topological Graphs with No Large Grids. Graph. Comb. 21(3), 355–364 (2005)
16. Pach, J., Shahrokhi, F., Szegedy, M.: Applications of the Crossing Number. J. Graph Theory 22, 239–243 (1996)
17. Pettie, S.: Generalized Davenport-Schinzel Sequences and Their 0-1 Matrix Counterparts. J. Comb. Theory Ser. A 118(6), 1863–1895 (2011)
18. Pettie, S.: On the Structure and Composition of Forbidden Sequences, with Geometric Applications. In: Proceedings of the 27th Annual ACM Symposium on Computational Geometry, SCG 2011, pp. 370–379. ACM, New York (2011)
19. Tardos, G., Tóth, G.: Crossing Stars in Topological Graphs. SIAM J. Discret. Math. 21(3), 737–749 (2007)
20. Valtr, P.: On Geometric Graphs with No k Pairwise Parallel Edges. Discrete Comput. Geom. 19(3), 461–469 (1997)
21. Valtr, P.: Graph Drawings with No k Pairwise Crossing Edges. In: Di Battista, G. (ed.) GD 1997. LNCS, vol. 1353, pp. 205–218. Springer, Heidelberg (1997)

Monotone Crossing Number

János Pach[*] and Géza Tóth

Rényi Institute, Budapest

Abstract. The *monotone crossing number* of G is defined as the smallest number of crossing points in a drawing of G in the plane, where every edge is represented by an x-monotone curve, that is, by a connected continuous arc with the property that every vertical line intersects it in at most one point. It is shown that this parameter can be strictly larger than the classical crossing number $\mathrm{CR}(G)$, but it is bounded from above by $2\mathrm{CR}^2(G)$. This is in sharp contrast with the behavior of the rectilinear crossing number, which cannot be bounded from above by any function of $\mathrm{CR}(G)$.

Keywords: crossing number, monotone drawing.

1 Introduction

Let $G = (V(G), E(G))$ be a graph with no loops and multiple edges, and let $V(G)$ and $E(G)$ denote its vertex set and edge set. A *drawing* of G is an embedding of G in the plane, where each vertex $v \in V(G)$ is mapped to a point and each edge $uv \in E(G)$ is mapped into a simple continuous arc connecting the images of its endpoints, but not passing through the image of any other vertex of G. The arcs representing the edges of G are allowed to cross, but we assume for simplicity that any two arcs have finitely many points in common and no three arcs pass through the same point. A common interior point p of two arcs is said to be a *crossing* if in a small neighborhood of p one arc passes through one side of the other arc to the other side. If it leads to no confusion, the vertices and their images, as well as the edges and the arcs representing them, will be denoted by the same symbols.

In the special case where G is a complete bipartite graph, the problem of minimizing the number of crossings in a drawing of G was first studied by Turán [17]. The question became known as the *brick factory problem*. It was generalized to all graphs by Erdős and Guy [3]. In two previous papers [10], [11], the authors of the present note pointed out some inconsistencies between various definitions of crossing numbers implicitly used in early publications on the subject. To distinguish between these notions, they introduced some new terminology and

---

[*] J. Pach is supported by NSF Grant CCF-08-32072, OTKA, Swiss National Science Foundation Grant 200021-125287/1, and EUROGIGA project GraDR 10-EuroGIGA-OP-003. G. Tóth is supported by OTKA-K-60427, OTKA-K-75016, Bernoulli Center at EPFL, and EUROGIGA project GraDR 10-EuroGIGA-OP-003.

M. van Kreveld and B. Speckmann (Eds.): GD 2011, LNCS 7034, pp. 278–289, 2012.

notation. The *crossing number* of G, denoted by $\mathrm{CR}(G)$, is the smallest number of crossings in a drawing of G in the plane. The *pairwise crossing number*, $\mathrm{PAIR\text{-}CR}(G)$, is the smallest number of crossing pairs of edges in a drawing of G. If two edges cross several times, they still count as a single crossing pair, so that we have $\mathrm{PAIR\text{-}CR}(G) \leq \mathrm{CR}(G)$ for every graph G. It is one of the most tantalizing open problems in this area to decide whether these two parameters coincide or at least $\mathrm{CR}(G) = O(\mathrm{PAIR\text{-}CR}(G))$ holds for all graphs G. It was shown in [10] that $\mathrm{CR}(G) = O(\mathrm{PAIR\text{-}CR}^2(G))$, which was successively improved in [19], [15], and [16] to $\mathrm{CR}(G) = O(\mathrm{PAIR\text{-}CR}^{7/4}(G)/\log^{3/2} \mathrm{PAIR\text{-}CR}(G))$. It is not easy to make any conjecture in this respect or even to experiment with concrete graphs. The computation of $\mathrm{CR}(G)$ and $\mathrm{PAIR\text{-}CR}(G)$ are both NP-hard problems [7], [6], [10].

On the other hand, there is another natural parameter that can be much larger than the above two crossing numbers. $\mathrm{LIN\text{-}CR}(G)$, the *rectilinear crossing number* of G, is the smallest number of crossings in a rectilinear drawing of G, that is, in a drawing where every edge is represented by a straight-line segment. We have $\mathrm{CR}(G) \leq \mathrm{LIN\text{-}CR}(G)$. Bienstock and Dean [1] constructed a series of graphs with crossing number 4, whose rectilinear crossing numbers are arbitrarily large.

An *x-monotone curve* is a connected, continuous arc with the property that every straight-line parallel to the y-axis intersects it in at most one point. A drawing of G is called *x-monotone* (or *monotone*, for short) if every edge of G is represented by an x-monotone curve. We define $\mathrm{MON\text{-}CR}(G)$, the *monotone crossing number* of G, as the smallest number of crossings in a monotone drawing of G. Obviously, every rectilinear drawing of G, in which no two vertices share the same x-coordinate, is a monotone drawing. Therefore, we have

$$\mathrm{CR}(G) \leq \mathrm{MON\text{-}CR}(G) \leq \mathrm{LIN\text{-}CR}(G),$$

for every graph G.

Monotone drawings and rectilinear drawings share many interesting properties. In particular, it was shown in [12] that every crossing-free monotone drawing of a (planar) graph G can be "stretched" without changing the x-coordinates of the vertices. In other words, there is a crossing-free rectilinear drawing of G, isomorphic to the original one, in which the vertices have the same x-coordinates. Another example, for drawings with many crossings, is related to Conway's famous *thrackle conjecture* [20], which says that if a graph can be drawn in the plane such that any two edges have *exactly one* common points (either a common endpoint, or a crossing) then the number of edges cannot exceed the number of vertices. (The conjecture has been verified for monotone drawings [9].) In sharp contrast to these analogies, there are no graphs with bounded crossing numbers that have arbitrarily large monotone crossing numbers. In the present note, we answer a question of Fulek, Pelsmajer, Schaefer, and Štefankovič [5] by establishing the following results.

Theorem 1. *Every graph G satisfies the inequality*

$$\mathrm{MON\text{-}CR}(G) < 2\mathrm{CR}^2(G).$$

Theorem 2. *There are infinitely many graphs G with arbitrarily large crossing numbers such that*

$$\text{MON-CR}(G) \geq \frac{7}{6}\text{CR}(G) - 6.$$

The proof of Theorem 1 is algorithmic. It is based on a recursive procedure to redraw a plane graph without changing its combinatorial structure so that in the resulting drawing any pair of vertices of the same cell can be connected by an x-monotone curve. See Theorem 2.2. One of the key ideas of the construction proving Theorem 2, the use of "weighted" edges or repeated paths, goes back to the paper of Bienstock and Dean [1] mentioned above. This idea was further developed and applied to related problems by Pelsmajer, Schaefer, and Štefankovič [13] and by Tóth [15].

2   Proof of Theorem 1

Two crossing-free (plane) drawings of a planar graph are said to be *isomorphic* if there is a homeomorphism of the plane which maps one to the other. In particular, it takes the unbounded cell of the first drawing to the unbounded cell of the second.

Definition 2.1. Let $\mathcal{D}$ be a crossing-free drawing of a planar graph G, and let $v \in V(G)$. We say that $\mathcal{D}$ is *v-spinal* if

1. $\mathcal{D}$ is a monotone drawing;
2. v is the leftmost vertex;
3. any two vertices belonging to the same (bounded or unbounded) cell C can be connected by an x-monotone curve that lies in the interior of C (with the exception of its endpoints);
4. every vertical ray starting at a boundary vertex of the unbounded cell C_0 and pointing downwards lies in the interior of C_0 (with the exception of its endpoint).

Theorem 1 is an easy corollary of the following result.

Theorem 2.2. *For any crossing-free drawing $\mathcal{D}$ of a planar graph and for any vertex v of the unbounded cell, there is a v-spinal drawing isomorphic to $\mathcal{D}$.*

It follows from the result of [12] mentioned in the introduction that every v-spinal drawing can be "stretched" without changing the x-coordinates of the vertices. That is, we can assume without loss of generality that the drawing whose existence is guaranteed by Theorem 2.2 is rectilinear. However, in the recursive argument proving Theorem 2.2, we will not need this fact. It will be sufficient to assume that the edges are represented by x-monotone polygonal paths, so that in a small neighborhood of their endpoints it will make sense to talk about the *slopes* of these paths.

Fig. 1. A plane drawing and a v-spinal drawing

Before turning to the proof of Theorem 2.2, we show how Theorem 2.2 implies Theorem 1.

Proof of Theorem 1 (using Theorem 2.2). Let G be any graph, and let $\mathcal{D}$ be a drawing of G with $\mathrm{CR}(G)$ crossings. Let $G' \subseteq G$ denote the subgraph consisting of all vertices of G and all edges not crossed by any other edge in this drawing. Clearly, G' is a planar graph. Let $\mathcal{D}'$ stand for the corresponding crossing-free subdrawing of $\mathcal{D}$.

Let v be a vertex of the unbounded cell. By Theorem 2.2, there is a v-spinal drawing $\mathcal{D}''$ of G', isomorphic to $\mathcal{D}'$. Consider now an edge $v_1 v_2 \in E(G) \setminus E(G')$. In $\mathcal{D}$, this edge was represented by a curve that, with the exception of its endpoints, lied in the interior of a single cell C' in the subdrawing $\mathcal{D}'$. Let C'' denote the cell in $\mathcal{D}''$, which corresponds to C'. In view of condition 3 in Definition 2.1, the points representing v_1 and v_2 can be connected by an x-monotone curve within the cell C''. Let us choose such an x-monotone connecting curve for each edge in $E(G) \setminus E(G')$, so that the total number of crossings between them is as small as possible. Observe that any two such curves can cross at most once, otherwise by swapping their sections between two consecutive crossing points and slightly separating them, we could reduce the total number of crossings by 2. During this transformation, both curves remain x-monotone.

Therefore, in the resulting x-monotone drawing of G, the total number of crossings is at most $\binom{|E(G)| - |E(G')|}{2}$. This yields that

$$\mathrm{MON\text{-}CR}(G) \leq \binom{|E(G)| - |E(G')|}{2}.$$

On the other hand, taking into account that every edge in $E(G) \setminus E(G')$ participates in at least one crossing in $\mathcal{D}$, we have

$$|E(G)| - |E(G')| \leq 2\mathrm{CR}(G).$$

Comparing the last two inequalities, the theorem follows. □

Proof of Theorem 2.2. We proceed by induction on the number of vertices of $\mathcal{D}$. The theorem is obviously true for graphs with one or two vertices. Suppose now that $\mathcal{D}$ has n vertices and that the theorem has already been proved for all drawings of graphs with fewer than n vertices. Let v be a vertex of the unbounded cell in $\mathcal{D}$.

CASE 1: $\mathcal{D}$ is not connected. Suppose for simplicity that it has two connected components, $\mathcal{D}_1$ and $\mathcal{D}_2$; the other cases can be treated analogously. Assume without loss of generality that $v \in \mathcal{D}_1$.

Subcase 1.1: $\mathcal{D}_2$ has a vertex v' that belongs to the unbounded cell in $\mathcal{D}$. Take a v-spinal drawing isomorphic to $\mathcal{D}_1$, and place a v'-spinal drawing isomorphic $\mathcal{D}_2$ completely to the right of it, so that every vertex of the latter has a larger x-coordinate than any vertex of the former. The resulting drawing meets the requirements.

Subcase 1.2: $\mathcal{D}_2$ does not have a vertex that belongs to the boundary of the unbounded cell in $\mathcal{D}$. Let C denote the cell in $\mathcal{D}_1$ that contains $\mathcal{D}_2$, and fix a vertex w of C. Let v' be a vertex of the unbounded cell in $\mathcal{D}_2$. Take a v-spinal drawing isomorphic to $\mathcal{D}_1$, and place a very small copy of a v'-spinal drawing isomorphic to $\mathcal{D}_2$ in the cell C' of $\mathcal{D}_1$ that corresponds to C, in a small neighborhood of the vertex that corresponds to w.

The resulting drawing $\mathcal{D}$ obviously satisfies conditions 1, 2, and 4 in Definition 2.1. As for condition 3, we have to verify only that any two vertices, v_1 and v_2, that belong to the union of the boundary of C' and the outer boundary of the small v'-spinal drawing isomorphic to $\mathcal{D}_2$ can be connected by an x-monotone curve that does not cross $\mathcal{D}$. This readily follows by the induction hypothesis, unless v_1 belongs to the boundary of C' and v_2 belongs to the outer boundary of the small drawing isomorphic to $\mathcal{D}_2$. In the latter case, move slightly downward from v_2 and then closely follow the x-monotone curve connecting w to v_1.

CASE 2: $\mathcal{D}$ has a cut vertex v'. Suppose that $\mathcal{D} = \mathcal{D}_1 \cup \mathcal{D}_2$, where the only point that $\mathcal{D}_1$ and $\mathcal{D}_2$ have in common is v'. Assume without loss of generality that v is a vertex of $\mathcal{D}_1$. Note that v and v' may be identical.

Let C denote the cell in $\mathcal{D}_1$ that contains $\mathcal{D}_2$. In particular, v' is a vertex of C. In $\mathcal{D}_2$, the vertex v' belongs to the unbounded cell.

Take a v-spinal drawing isomorphic to $\mathcal{D}_1$, and fix a very short non-vertical segment s, which is incident to the point $p(v')$ representing v' and which lies in the cell C' that corresponds to C. In the special case where $v' = v$ and C' is the unbounded cell, make sure that the x-coordinates of the points of s are larger than the x-coordinate of $p(v')$. In addition, take a very small v'-spinal drawing isomorphic to $\mathcal{D}_2$ such that the point representing v' coincides with $p(v')$. Applying a suitable orientation preserving linear transformation to this second drawing, it can be achieved that it becomes very "flat" and small, and lies in a very small neighborhood of the segment s, within C'. Putting these two drawings together, the resulting drawing meets the requirements.

Note that, if the x-coordinates of the points of s are smaller than the x-coordinate of $p(v')$, then the above linear transformation reverses the order of the x-coordinates in the v'-spinal drawing isomorphic to $\mathcal{D}_2$. In order to preserve the combinatorial structure of the cell decomposition, we have to make sure that we use a linear transformation that preserves the orientation of the plane.

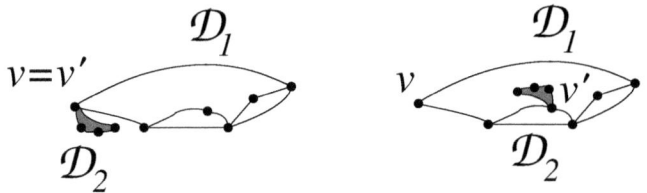

Fig. 2. Case 2. $\mathcal{D}$ has a cut vertex v'

CASE 3: $\mathcal{D}$ is 2-connected. We need the following well known result.

Lemma 2.3. ([2], Proposition 3.1.2) *For every 2-connected graph other than a cycle, there exists a path whose internal vertices have degree two, such that removing all edges and all internal vertices of this path, the remaining graph is still 2-connected.*

Let $\mathcal{D}$ be a drawing of a cycle with vertices $v = v_1, v_2, \ldots, v_n$, in counterclockwise order. Then the rectilinear drawing induced by the points $p(v_i) = (i, i^2)$ is v-spinal and isomorphic to $\mathcal{D}$.

If $\mathcal{D}$ is not a cycle, then, according to the lemma, it can be obtained from a 2-connected drawing $\mathcal{D}_0$, by adding a path P between two vertices, u and w, of $\mathcal{D}_0$), which, with the exception of its endpoints, lies in the interior of a cell C. We distinguish two subcases.

Subcase 3.1: v is a vertex of $\mathcal{D}_0$. Take a v-spinal drawing isomorphic to $\mathcal{D}_0$. Let C' denote the cell that corresponds to C in this drawing. The vertices u and w belong to the boundary of this cell. Therefore, by condition 3 in Definition 2.1, u and w can be connected by an x-monotone curve within C'. Put all internal vertices of P along this curve, very close to u. The resulting drawing meets the requirements.

Subcase 3.2: v is an internal vertex of P. Since v is a vertex of the unbounded cell in $\mathcal{D}$, the cell C in $\mathcal{D}_0$ that contains P, must be the unbounded cell.

Let $P = uu_1 \cdots u_m vw_1 w_2 \cdots w_k w$. Assume without loss of generality that in $\mathcal{D}$ the unbounded cell lies on the *left-hand side* of P, as we traverse it from u to w. Take a u-spinal drawing $\mathcal{D}_1$ isomorphic to $\mathcal{D}_0$. Place v to the left and w_1 to the right of all vertices of $\mathcal{D}_0$.

Connect u and v by an x-monotone curve in $\mathcal{D}_1$, and place the vertices $u_1, \ldots, u_m$ on this curve, in this order. Then connect v to w_1 by an x-monotone curve running above all previously drawn vertices and edges. Finally, connect w_1 to w by an x-monotone curve which does not cross any previously drawn edges, and place the vertices $w_2, \ldots, w_k$ on this curve, in this order, very close to w_1. Adding these three curves that represent P to $\mathcal{D}_1$, we obtain a v-spinal drawing isomorphic to $\mathcal{D}$, as required.   □

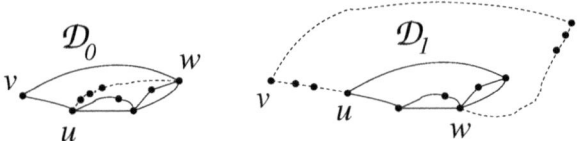

Fig. 3. Case 3. $\mathcal{D}$ is two-connected

3   Proof of Theorem 2

Throughout this section, let k be a fixed positive integer. We construct a graph G_k with $\mathrm{CR}(G_k) = 6k + 6$ and $\mathrm{MON\text{-}CR}(G) = 7k + 6$, as follows.

First, we define an auxiliary graph on the vertex set $V(H) = \{u, w, v_1, \ldots, v_9\}$ such that each of its edges is red, blue, or black. Let w be connected to every element of $v_1, \ldots, v_9$ by a red edge. Let $v_1, \ldots, v_9$ form a red cycle, in this order. Finally, let H have three blue edges, uv_2, uv_5, and uv_8, and three black edges, v_1v_6, v_7v_3, v_4v_9. See Figure 4. Let H' be a colored graph isomorphic to H with $V(H') = \{u', w', v_1', \ldots, v_9'\}$ and $V(H') \cap V(H) = \emptyset$.

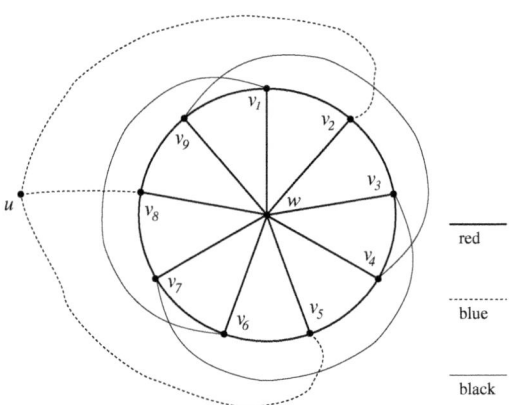

Fig. 4. Graph H

Let H_k denote the graph obtained from H by substituting each of its red edges by $10k$ paths of length two and each of its blue edges by k paths of length two such that the middle vertices of these paths are disjoint from one another and from all previously listed vertices. We will refer to these paths as *red paths* and *blue paths*, respectively. Let H_k' denote the graph with $V(H_k') \cap V(H_k) = \emptyset$ which can be obtained from H' in exactly the same way as H_k was constructed from H.

Finally, connect u to u' by a red edge, and replace this edge by $10k$ vertex disjoint red paths of length two, as above. Denote the resulting graph by G_k.

We start with the following simple observation.

Claim 3.1. $\mathrm{CR}(G_k) \leq 6k + 6$ *and* $\mathrm{MON\text{-}CR}(G_k) \leq 7k + 6$.

Proof. A drawing of G_k with $6k + 6$ crossings and a monotone drawing with $7k + 6$ crossings are depicted on Figure 5, and Figure 6, respectively. The thick edges and the dotted edges represent bundles consisting of $10k$ red paths and k blue paths, respectively. The paths representing the same colored edge run very close to one another and do not cross. The only difference between the two drawings is that in the first one v_4v_9 crosses uv_2, while in the second it crosses uv_5 and uv_8. $\square$

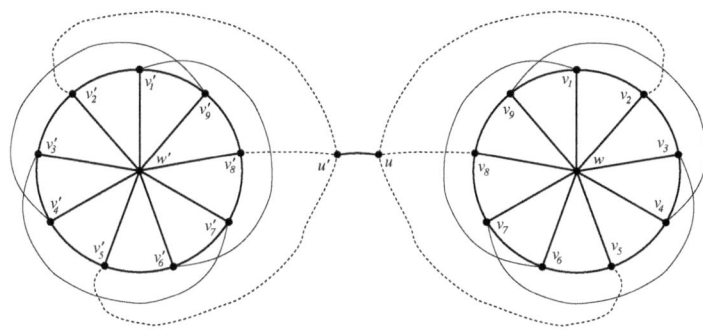

Fig. 5. A CR-optimal drawing of G

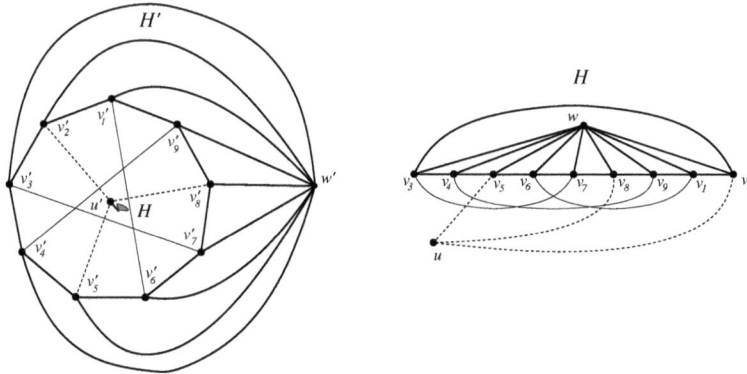

Fig. 6. Left: a $\mathrm{MON\text{-}CR}$-optimal drawing of G. Right: the drawing of H, from the left.

A drawing of a graph G is called CR-*optimal* if the number of crossings in it is $\mathrm{CR}(G)$. Analogously, a $\mathrm{MON\text{-}CR}$-optimal drawing is a monotone drawing in which the number of crossings is $\mathrm{MON\text{-}CR}(G)$.

Claim 3.2. *Each of the graphs G_k, H_k, and H'_k has a CR-optimal drawing and a MON-CR-optimal drawing satisfying the the following conditions. (i) The red paths substituting the same red edge run very close to one another and do not cross any edge. (ii) The blue paths substituting the same blue edge run very close to one another, do not cross one another, and cross exactly the same edges.*

Proof. Let G stand for one of the graphs G_k, H_k, or H'_k. Let $P_1, \ldots, P_m$ ($m = 10k$ or k) denote the paths substituting the same red or blue edge. Consider a CR-optimal or a MON-CR-optimal drawing of G. Suppose without loss of generality that among all P_is the path P_1 participates in the smallest number of crossings. Redraw $P_2, \ldots, P_m$ so that they run "parallel" to P_1 and very close to it. Clearly, this transformation does not increase the total number of crossings, so that the resulting drawing remains optimal.

Suppose that $P_1, \ldots, P_m$ ($m = 10k$) are *red paths* that substitute the same red edge and run parallel to one another. If any of them crosses an edge, then all of them do. This alone creates a total of at least $10k$ crossings, which contradicts the assumption the drawing was optimal. □

Claim 3.3. $CR(H_k) = MON\text{-}CR(H_k) = 3k + 3$. *Consequently, we have* $CR(G_k) = 6k + 6$.

Proof. The right part of Figure 6 shows a monotone drawing of H. From this one can easily construct a monotone drawing of H'_k with $3k + 3$ crossings. Therefore, we have $CR(H_k) \leq MON\text{-}CR(H_k) = MON\text{-}CR(H'_k) \leq 3k + 3$. As before, the thick and the dotted edges represent bundles of $10k$ parallel red paths and bundles of k parallel blue paths.

Consider a CR-optimal drawing of H_k which satisfies the conditions in Claim 3.2. Replace now the red paths substituting the same red edge by a single red edge running along any one of those paths. The red cycle $C = v_1 v_2 \cdots v_9$ divides the rest of the plane into a bounded and an unbounded region. All points that belong to the bounded (unbounded) region are said to be *inside (outside)* of C. Assume without loss of generality that the vertex w lies *inside* of C. Since no red edge is allowed to cross any other edge, the edges $v_3 v_7$, $v_1 v_6$, and $v_4 v_9$, as well as the vertex u with all edges incident to it, must lie *outside* of C. Thus, the edges $v_3 v_7$, $v_1 v_6$, and $v_4 v_9$ are pairwise crossing. Moreover, the path $v_2 u v_5$ must cross the edges $v_3 v_7$ and $v_4 v_9$, and the path $v_2 u v_8$ must cross the edge $v_1 v_6$. This already guarantees the existence of $3k + 3$ crossings, so that we have $CR(H_k) = MON\text{-}CR(H_k) = 3k + 3$. □

To complete the proof of Theorem 2, it remains to verify the following.

Claim 3.4. $MON\text{-}CR(G_k) \geq 7k + 6$.

Proof. Fix a MON-CR-optimal drawing of G_k, satisfying the conditions in Claim 3.2. As in the proof of Claim 3.3, replace every bundle of red paths substituting the same red edge by a single red edge. Let C and C' denote the red cycles induced by the vertices $v_1, v_2, \ldots, v_9$ and $v'_1, v'_2, \ldots, v'_9$. Both of them divide the plane into a bounded and an unbounded region, so that it makes sense to say that a point is inside or outside of C or C'.

By Claim 3.2, in the original drawing of G_k, the red edges cannot cross any other edge. Suppose that a blue edge belonging to $H_k \subset G_k$ crosses an edge belonging to $H'_k \subset G_k$. Then the number of crossings is at least $k + \text{MON-CR}(H_k) + \text{MON-CR}(H'_k) = 7k + 6$, and we are done. Thus, we can assume that in the drawing of G_k, the blue edges of H_k do not cross any edge of H'_k, and analogously, the blue edges of H'_k do not cross any edge of H_k.

Let v be the vertex of G_k with the smallest x-coordinate, and suppose without loss of generality that $v \in V(H'_k)$. Consider now separately the drawing of H_k and the induced cell decomposition. By definition, v lies in the unbounded cell. Observe, that if we remove edges $v'_1 v'_6$, $v'_7 v'_3$, $v'_4 v'_9$ from H'_k, that is, if we keep only the red and blue edges, we still have a connected graph. The red and blue edges of H'_k cannot cross any edge of H_k. Hence, all vertices of H'_k must lie in the unbounded cell of the cell decomposition induced by H_k.

The vertices u and u' are connected by a red edge in G_k. Hence, u must lie on the boundary of the unbounded cell of the cell decomposition induced by H_k. In particular, u is *outside* of the cycle C. Since w is connected to each edge of C by a red edge, u and w lie on different sides of C. Thus, w must be *inside* of C. Therefore, the edges $v_3 v_7$, $v_1 v_6$, $v_4 v_9$, as well as the vertex u together with all edges incident to it, must lie *outside* of C. Consequently, the edges $v_3 v_7$, $v_1 v_6$, $v_4 v_9$ must be pairwise crossing. The edges $v_3 v_7$, $v_1 v_6$, $v_4 v_9$ together with C divide the plane into *eight* cells, one of which is unbounded, and u must belong to this cell Γ.

Let v_i be the vertex of C with the smallest x-coordinate. Since $v_3 v_7$, $v_1 v_6$, $v_4 v_9$ are represented by monotone curves, v_i has to lie on the boundary of the unbounded cell Γ. We can assume without loss of generality that $1 \leq i \leq 3$. (If this is not the case, we can add 3 or 6 to all indices modulo 9.) So, v_i is on the boundary of the unbounded cell, and u is in the unbounded cell. Using the fact that the edges $v_1 v_2$ and $v_2 v_3$ do not cross any other edge, we can conclude that v_1, v_2, and v_3 all lie on the boundary of the unbounded cell Γ. See Figure 6. Since we started with a MON-CR-optimal drawing, the edge $u v_2$ does not cross $v_4 v_9$. The path $v_2 u v_5$ crosses $v_4 v_9$, so that $u v_5$ must cross $v_4 v_9$. Analogously, $v_2 u v_8$ crosses $v_4 v_9$, so that $u v_8$ crosses $v_4 v_9$. Moreover, the path $v_2 u v_5$ crosses $v_3 v_7$, and $v_2 u v_8$ crosses $v_1 v_6$. Recall from the previous paragraph that the edges $v_3 v_7$, $v_1 v_6$, and $v_4 v_9$ are pairwise crossing. Summarizing, there are at least $4k + 3$ crossings between edges of H_k. By Claim 3.3, $\text{MON-CR}(H'_k) \geq 3k + 3$, so that altogether $\text{MON-CR}(G_k) \geq (4k + 3) + (3k + 3) \geq 7k + 6$, as required. $\square$

4   Concluding Remarks

1. Another important parameter of a graph, the *odd-crossing number*, was introduced implicitly by Tutte [18]. It is defined as the minimum number $\text{ODD-CR}(G)$ of all pairs of edges that cross an odd number of times, over all drawings of G. Clearly, for any graph G, we have $\text{ODD-CR}(G) \leq \text{PAIR-CR}(G) \leq \text{CR}(G) \leq \text{MON-CR}(G) \leq \text{LIN-CR}(G)$. Theorem 1 can be strengthened as follows.

Corollary 4.1. *Every graph G satisfies the inequality*

$$\text{MON-CR}(G) < 2\text{ODD-CR}^2(G).$$

Proof. Let $\mathcal{D}$ be a drawing of G, in which the number of pairs of edges that cross an odd number of times is $\text{ODD-CR}(G)$. Let $G' \subseteq G$ denote the subgraph consisting of all vertices of G and all edges that do not cross any other edge an odd number of times. It was shown in [10] that G has another drawing, $\mathcal{D}'$, in which the edges belonging to G' do not participate in any crossing, and hence they form a plane graph. Every edge in $E(G) \setminus E(G')$ is represented by a curve that lies entirely in a cell of this plane graph. According to our Theorem 2.2, this plane graph admits a v-spinal (monotone) drawing for some $v \in V(G)$. By definition, we can add to this drawing all edges in $E(G) \setminus E(G')$, so that all of them are represented by monotone curves, and they do not cross any edge of G'. Among all such monotone drawings of G, consider one that minimizes the total number of crossings. In this drawing, any two edges cross at most once. Thus, we have

$$\text{MON-CR}(G) \leq \binom{|E(G)| - |E(G')|}{2}.$$

On the other hand, taking into account that every edge in $E(G) \setminus E(G')$ participates in at least one pair of edges in $\mathcal{D}$ which cross an odd number of times, we obtain that

$$|E(G)| - |E(G')| \leq 2\text{ODD-CR}(G).$$

Comparing the last two inequalities, the corollary follows.   □

In [11], we introduced the following variant of the odd-crossing number. Two edges of a graph G are called *independent* if they do not share a vertex. Let $\text{ODD-CR}_-(G)$ denote the smallest number of pairs of independent edges that cross an odd number of times, over all drawings of G. That is, we do not count those pairs of edges that are incident to the same vertex, even if they cross an odd number of times. Pelsmajer, Schaefer, and Štefankovič [14] managed to strengthen the result of [10], used in the proof of Corollary 4.1. They established the following result. Consider a drawing of G in the plane. An edge $e \in E(G)$ is called *independently even* if it crosses every other edge of G which is independent of e an even number of times. Then G has another drawing in which no independently even edge crosses any edge. Plugging this result into the above proof, we obtain the following strengthening of Corollary 4.1.

Corollary 4.1'. *Every graph G satisfies the inequality*

$$\text{MON-CR}(G) \leq 2\text{ODD-CR}_-^2(G).$$

2. As mentioned in the Introduction, Tóth [16] proved that every graph G satisfies the inequality

$$\text{CR}(G) = O(\text{PAIR-CR}^{7/4}(G)/\log^{3/2} \text{PAIR-CR}(G)).$$

Restricting the notion of pair-crossing number to monotone drawings, we obtain another closely related graph parameter. The *monotone pair-crossing number* of G, MON-PAIR-CR(G), is defined as the smallest number of crossing pairs of edges over all *monotone* drawings of G. Obviously, we have that ODD-CR$(G) \leq$ PAIR-CR$(G) \leq$ MON-PAIR-CR(G), for any graph G. Valtr [19] proved that every graph G satisfies the inequality MON-CR$(G) = O($MON-PAIR-CR$^{4/3}(G))$.

References

1. Bienstock, D., Dean, N.: Bounds for rectilinear crossing numbers. J. Graph Theory 17, 333–348 (1993)
2. Diestel, R.: Graph Theory, 3rd edn. Graduate Texts in Mathematics, vol. 173. Springer, Berlin (2005)
3. Erdős, P., Guy, R.K.: Crossing number problems, Amer. Math. Monthly 80, 52–58 (1973)
4. Fáry, I.: On straight line representation of planar graphs. Acta Univ. Szeged. Sect. Sci. Math. 11, 229–233 (1948)
5. Fulek, R., Pelsmajer, M.J., Schaefer, M., Štefankovič, D.: Hanani-Tutte, monotone drawings, and level-planarity (to appear)
6. Garey, M.R., Johnson, D.S.: Crossing number is NP-complete, SIAM J. Alg. Disc. Meth. 4, 312–316 (1983)
7. Garey, M.R., Johnson, D.S., Stockmeyer, L.J.: Some simplified NP-complete graph problems. Theoretical Computer Science 1, 237–267 (1976)
8. Guy, R.K.: The decline and fall of Zarankiewicz's theorem. In: Guy, R.K. (ed.) Proof Techniques in Graph Theory, pp. 63–69. Academic Press, New York (1969)
9. Pach, J., Sterling, E.: Conways conjecture for monotone thrackles. Amer. Math. Monthly 118, 544–548 (2011)
10. Pach, J., Tóth, G.: Which crossing number is it, anyway? J. Combin. Theory Ser. B 80, 225–246 (2000)
11. Pach, J., Tóth, G.: Thirteen problems on crossing numbers. Geombinatorics 9, 194–207 (2000)
12. Pach, J., Tóth, G.: Monotone drawings of planar graphs. J. Graph Theory 46, 39–47 (2004)
13. Pelsmajer, M.J., Schaefer, M., Štefankovič, D.: Odd crossing number and crossing number are not the same. Discrete Comput. Geom. 39, 442–454 (2008)
14. Pelsmajer, M.J., Schaefer, M., Štefankovič, D.: Removin independently even crossings. SIAM J. Discrete Math. 24, 379–393 (2010)
15. Tóth, G.: Note on the pair-crossing number and the odd-crossing number. Discrete Comput. Geom. 39, 791–799 (2008)
16. Tóth, G.: A better bound for the pair-crossing number (manuscript)
17. Turán, P.: A note of welcome. J. Graph Theory 1, 7–9 (1977)
18. Tutte, W.T.: Toward a theory of crossing numbers. J. Combinatorial Theory 8, 45–53 (1970)
19. Valtr, P.: On the pair-crossing number. In: Combinatorial and Computational Geometry. Math. Sci. Res. Inst. Publ., vol. 52, pp. 569–575. Cambridge Univ. Press, Cambridge (2005)
20. Woodall, D.R.: Thrackles and deadlock. In: Welsh, D.J.A. (ed.) Combinatorial Mathematics and Its Applications, pp. 335–348. Academic Press (1969)

Upper Bound Constructions
for Untangling Planar Geometric Graphs

Javier Cano[1], Csaba D. Tóth[2], and Jorge Urrutia[3]

[1] Posgrado en Ciencia e Ingeniería de la Computación,
Universidad Nacional Autónoma de México, D.F. México
j_cano@uxmcc2.iimas.unam.mx
[2] Department of Math., University of Calgary, Canada
cdtoth@ucalgary.ca
[3] Instituto de Matemáticas,
Universidad Nacional Autónoma de México, D.F. México
urrutia@matem.unam.mx

Abstract. For every $n \in \mathbb{N}$, there is a straight-line drawing D_n of a planar graph on n vertices such that in any *crossing-free* straight-line drawing of the graph, at most $O(n^{.4982})$ vertices lie at the same position as in D_n. This improves on an earlier bound of $O(\sqrt{n})$ by Goaoc *et al.* [6].

1 Introduction

A *straight-line drawing* of a graph G is a representation of G in the plane where the vertices are mapped to distinct points in the plane, and each edge is represented by a line segment joining pairs of points representing adjacent vertices. A drawing is *crossing-free* if no two edges intersect, except perhaps at a common endpoint. A *geometric graph* is a graph given with a straight-line drawing. Every planar graph has a crossing-free straight-line drawing by Fary's Theorem [5], however, not all straight-line drawings are crossing-free. Suppose that we are given a *planar geometric graph* G. Since G is planar, it can be redrawn (by relocating some of its vertices) such that no two edges cross anymore. The process of redrawing G to obtain a crossing-free straight-line drawing, is called an *untangling* of G.

In this paper we study the following problem: For an integer $n \in \mathbb{N}$, what is the maximum number $f(n)$ such that every planar geometric graph with n vertices can be untangled such that at least $f(n)$ vertices remain in their original position.

The first question on untangling planar geometric graphs was posed by Mamoru Watanabe in 1998: Is it true that every polygon P with n vertices can be untangled in at most ϵn steps, for some absolute constant $\epsilon < 1$, where in each step, we move a vertex of G to a new location. Watanabe's question was proved to be false by Pach and Tardos [9]: they showed that every n-gon can be untangled in at most $n - \sqrt{n}$ moves, but there are n-gons where no more than $O((n \log n)^{2/3})$ vertices can be fixed. Recently, Cibulka [3] proved that every n-gon can be untangled while keeping $\Omega(n^{2/3})$ vertices fixed.

M. van Kreveld and B. Speckmann (Eds.): GD 2011, LNCS 7034, pp. 290–295, 2012.
© Springer-Verlag Berlin Heidelberg 2012

The problem of untangling planar geometric graphs was studied by Goaoc *et al.* [6]. They constructed planar geometric graphs showing that $f(n) \leq \sqrt{n} + 2$. Kang *et al.* [8] explored several families of graphs in which no more than $O(\sqrt{n})$ of n vertices can be fixed. Bose *et al.* [2] devised an untangling algorithm that fixes at least $(n/3)^{1/4}$ of n vertices, which proves $f(n) \geq (n/3)^{1/4}$.

In this note, we improve the upper bound for $f(n)$ to $O(n^{1/(3-\log_{38} 37)}) \subset O(n^{.4982})$. We construct planar geometric graphs such that any untangling of them fixes $O(n^{1/(3-\log_{38} 37)})$ of n vertices. The framework of our construction leads to new problems in graph drawing, which we discuss in Section 5. Any improvement in these problems would immediately improve the upper bound for $f(n)$.

2   Preliminaries

Monotone Subsequences. Erdős and Szekeres showed that every permutation of $[n] = \{0, 1, \ldots, n-1\}$ contains a monotonically increasing or degreasing subsequence of length at least $\lceil \sqrt{n} \rceil$, and this bound is the best possible. The lower bound is attained on many different permutations. The best known construction consists of $\lceil \sqrt{n} \rceil$ monotonically increasing subsequences of consecutive elements, where the minimum element of each subsequence is larger than the maximum element of the next. We will use permutations in which monotone subsequences "spread out" more evenly. In a permutation $(\sigma_1, \sigma_2, \ldots, \sigma_n)$, we define the *spread* of a subsequence $(\sigma_{j_1}, \sigma_{j_2}, \ldots, \sigma_{j_k})$, $1 \leq j_1 < j_2 < \ldots < j_k \leq n$, to be $j_k - j_1$.

Lemma 1. *For every $m \in \mathbb{N}$, there is a permutation π_n of $[n] = [4^m]$ such that*

- *the length of every monotone subsequence is at most $2^m = \sqrt{n}$; and*
- *the spread of every monotone subsequence of length $k \geq 2$ is at least $\frac{k^2+2}{6}$.*

Proof. We construct the permutation π_n by induction on m. For $m = 1$, let $\pi_4 = (2, 3, 0, 1)$ and observe that it has the desired properties. Assume that $\pi_n = (\sigma_1, \ldots, \sigma_n)$ is a permutation of $[n]$ with the desired properties. We construct a permutation π_{4n} of $[4n]$ by replacing each σ_i with the 4-tuple

$$(4\sigma_i + 2, 4\sigma_i + 3, 4\sigma_i + 0, 4\sigma_i + 1).$$

Let L be a monotone subsequence of length k in π_{4n}. Note that L has at most two elements from each 4-tuple. The sequence of these 4-tuples corresponds to a monotone subsequence of π_n, which we denote by L'. The length of L' is at least $k/2$, with equality iff L contains exactly two elements from each of the 4-tuples involved. By induction, the length of L' is $k/2 \leq 2^m$. Hence, we have $k \leq 2^{m+1}$, as required. If the length of L' is exactly $k/2$, then its spread is at least $\frac{(k/2)^2+2}{6}$ in π_n, and so the spread of L is at least $4(\frac{(k/2)^2+2}{6}) - 1 = \frac{k^2+2}{6}$. If the length of L' is more than $k/2$, then its spread is at least $\frac{(k/2+1)^2+2}{6}$, and the spread of L is at least $4(\frac{(k/2+1)^2+2}{6}) - 1 \geq \frac{k^2+2}{6}$, as required.   □

A Recursive Construction. We say that a planar straight-line graph T is an (a, b, c)-*triangulation* for integers $a \geq b > c > 0$ if T is a 3-connected triangulation such that it has a total of a faces, b of which are marked, and any line intersects at most c marked faces in any plane straight-line drawing of T.

Note that, by Steiniz's theorem, a 3-connected triangulation is the 1-skeleton of a combinatorially unique 3-dimensional polytope. Hence an (a, b, c)-triangulation has a unique embedding in the plane up to homeomorphisms and the choice of the outer face. In the following lemma, we recursively construct a larger triangulation from an (a, b, c)-triangulation.

Lemma 2. *If there exists an (a, b, c)-triangulation for constants $a \geq b > c > 0$, then for every $n \in \mathbb{N}$, there is an (a', b', c')-triangulation with $a' = \Theta(n)$, $b' = \Theta(n)$, and $c' = \Theta(n^{\log_b c})$.*

Proof. Let $T_{a,b,c}$ be an (a, b, c)-triangulation. Plug in $T_{a,b,c}$ in all marked faces of $T_{a,b,c}$ recursively k times, where k is specified shortly. We obtain a 3-connected triangulation $T_{a,b,c}^k$ (that is, $T_{a,b,c} = T_{a,b,c}^0$), which has $b' = b^{k+1}$ marked faces, a line intersects at most $c' = c^{k+1}$ marked faces in any plane straight-line drawing, and the total number of faces is $a' = b^{k+1} + (a - b)(b^{k+2} - 1)/(b - 1)$. If we denote by v the number of vertices of $T_{a,b,c}^k$, then it has $2v - 4$ faces, $\Theta(v)$ of which are marked, and a line intersects at most $\Theta(v^{\log_b c})$ marked faces in any plane straight-line drawing of $T_{a,b,c}^k$. Choose k such that $a' = \Theta(v)$.   □

3   Upper Bound Constructions

Theorem 1. *If there exists an (a, b, c)-triangulation for constants $a \geq b > c > 0$, then $f(n) \in O(n^\kappa)$ for $\kappa = 1/(3 - \log_b c)$.*

Note that $b > c$, and so we have $0 < \log_b c < 1$ and $0 < \kappa < 1/2$. That is, the existence of *any* (a, b, c)-triangulation implies an upper bound $f(n) \in O(n^{\frac{1}{2} - \varepsilon})$ for some $\varepsilon > 0$. We discuss (a, b, c)-triangulations in Section 4.

Proof. For every $n \in \mathbb{N}$, we construct a drawing of a planar graph G_n with $\Theta(n)$ vertices such that in any untangling of G_n, at most $O(n^\kappa)$ vertices remain fixed.

Fig. 1. Triangulation $S = P_2 * P_5$.

Construction. We first construct the planar graph G_n. By Lemma 2, there is a 3-connected triangulation T with $\Theta(n^\kappa)$ vertices and $\Theta(n^\kappa)$ marked faces such that any line intersects at most $\Theta(n^{\kappa \log_b c})$ marked faces in any plane straight-line drawing of T. Let S be the join $P_2 * P_{s+1}$ of two paths with 2 and $s + 1$

vertices, respectively, where $s = \Theta(n^{1-\kappa})$ and s is a power of 4 (see Fig. 1). Note that S has exactly s interior vertices, which have a natural order along an interior path. We construct G_n by plugging in a copy of S into each marked face of T. Denote the copies of S by S_i, for $i = 1, 2, \ldots, \Theta(n^{\kappa})$. The total number of vertices of G_n is $\Theta(n^{\kappa} + n^{\kappa} \cdot n^{1-\kappa}) = \Theta(n)$.

Next, we describe a straight-line drawing of G_n. Embed the vertices of the triangulation T arbitrarily in general position above the x-axis. Embed the interior vertices of S_1 into integer points $\{0, 1, \ldots, s-1\} \times \{0\}$ on the x-axis such that their natural order is permuted by π_s from Lemma 1. The interior vertices of S_i, for each $i > 1$, are embedded into a translated copy of this permutation, translated along the x-axis by δi for some small $0 < \delta \ll n^{-\kappa}$.

Bounding the Number of Fixed Vertices. Consider a crossing-free straight-line drawing of G_n. The $\Theta(n^{\kappa})$ vertices of T may be fixed. It is sufficient to consider the interior vertices of S_i, $i = 1, 2, \ldots, \Theta(n^{\kappa})$. Suppose that ℓ_i interior vertices of S_i are fixed, for $i = 1, 2, \ldots, \Theta(n^{\kappa})$. Since the x-axis intersects at most $O(n^{\kappa \log_b c})$ triangles of T, all but at most $O(n^{\kappa \log_b c})$ values of ℓ_i are zero.

Consider now a triangulation S_i where $\ell_i > 0$. Note that S_i contains a sequence of $s+1$ nested triangles that share a common edge (the horizontal edge in Fig. 1). In *any* straight-line drawing of S_i (independent of the choice of the outer face), at least $(s+1)/2$ of these triangles form a nested sequence. Hence, at least $\ell_i/2$ fixed interior vertices of S_i are vertices in a sequence of nested triangles in the crossing-free straight-line drawing of G_n. The intersection of the x-axis with a sequence of nested triangles is a line segment. It can be partitioned into two directed segments, with opposite directions, such that each of them is directed towards the deepest point in the arrangement of nested triangles. At least $\ell_i/4$ fixed points of S_i lie on the same directed segment, and these points must form a monotone sequence along the x-axis. Furthermore, the elements of this monotone subsequence are all contained in the largest triangle from the nested sequence of triangles in S_i, therefore, their convex hull is disjoint from the convex hulls of similar sequences in any other S_j, $j \neq i$.

By Lemma 1, the spread of the monotone subsequence of length at least $\ell_i/4$ is at least $(\ell_i^2 + 32)/96$. Hence these fixed points "occupy" an interval of length $(\ell_i^2 + 32)/96$ on the x-axis. As noted above, the convex hulls of monotone sequences from distinct copies of S are disjoint, and so we have

$$\sum_{i=1}^{\Theta(n^{\kappa})} \frac{\ell_i^2 + 32}{96} \leq 2s. \tag{1}$$

Recall that at most $O(n^{\kappa \log_b c})$ values of ℓ_i are nonzero. By Jensen's inequality, the sum $\sum_{i=1}^{\Theta(n^{\kappa})} \ell_i$ is maximized if all nonzero values of ℓ_i are equal. Suppose, by relabeling the copies of S if necessary, that $\ell_i = \ell$ for $i = 1, 2, \ldots, \Theta(n^{\kappa \log_b c})$;

and $\ell_i = 0$ for all other i. In this case, Inequality (1) becomes $\Theta(n^{\kappa \log_b c}) \cdot \ell^2 \leq \Theta(n^{1-\kappa})$, or $\ell \in O(n^{(1-\kappa(1+\log_b c))/2})$. Therefore, the number of fixed vertices is at most

$$\sum_{i=1}^{\Theta(n^{\kappa})} \ell_i \leq \Theta(n^{\kappa \log_b c}) \cdot \ell = \Theta(n^{(1+\kappa(\log_b c-1))/2}) = \Theta(n^{\kappa}),$$

as required.   □

4 (a, b, c)-Triangulations

Non-hamiltonian Triangulations. By Steinitz's theorem, every 3-connected cubic planar graph G is the 1-skeleton of a convex polytope. The dual graph G^*, corresponding to the dual polytope, is a 3-connected triangulation. Tait [10] conjectured in 1884 that every 3-connected cubic planar graph is Hamiltonian. Tutte [11] found a counterexample with 44 vertices in 1946. The smallest known counterexample, due to Bernette, Bosák, and Lenderberg, has 38 vertices, and it is known that there is no counterexample with 36 or fewer vertices [7].

A Hamiltonian cycle of G corresponds to a simple closed curve visiting every face exactly once in any plane drawing of G^*. In a straight-line drawing, every face of a triangulation is convex and thus it is visited by a line at most once. Therefore, if G is not Hamiltonian, then G^* has no plane straight-line drawing in which a line visits every face (including the outer face). The smallest known counterexample to Tait's conjecture implies that there is a $(38, 38, 37)$-triangulation. Combined with Theorem 1, we obtain a new upper bound for $f(n)$.

Corollary 1. $f(n) \in O(n^{1/(3-\log_{38} 37)}) \subset O(n^{.4982})$.

5 Conclusion

Our upper bounds for $f(n)$ depend on the value $\log_b c$ of an (a, b, c)-triangulation. The (a, b, c)-triangulations we considered are all derived from counterexamples for Tait's conjecture. Since these are counterexamples for Hamiltonicity, they all have $a = b > c > 0$. It is conceivable, though, that there are better constructions for (a, b, c)-triangulations in which $a > b$.

The best possible upper bound for $f(n)$ achievable with our framework would come from the minimum value of $\log_b c$, leading to the following problems.

Problem 1. What is the minimum value of $\log_b c$ over all (a, b, c)-triangulations?

Problem 2. What is the minimum value of $\log_b c$ over all 3-connected cubic planar graphs G, where G has b has marked vertices and any simple cycle visits at most c marked vertices?

The latter problem is purely graph theoretical. But the two problems are, in fact, equivalent. The dual of Problem 2 asks for the minimum value of $\log_b c$ over all 3-connected plane triangulations T with a faces, b of which are marked, such that any closed Jordan curve γ that visits every face at most once can visit at most c marked faces. One can show that every such Jordan curve γ is "stretchable." That is, T has a plane straight-line drawing T' in which a line L visits the exact same faces as γ visited in T (in the same cyclic order). See Fig. 2. Details are omitted, and will be given in the full version of this paper.

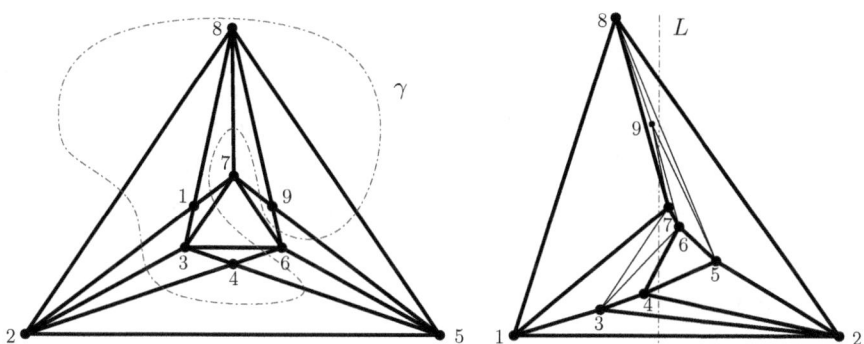

Fig. 2. Left: a plane 3-connected triangulation T, where curve γ visits every face exactly once. Right: a plane straight-line drawing T' of T, where line L stabs every face.

References

1. Arkin, E.M., Held, M., Mitchell, J.S.B., Skiena, S.: Hamiltonian triangulations for fast rendering. The Visual Computer 12(9), 429–444 (1996)
2. Bose, P., Dujmovic, V., Hurtado, F., Langerman, S., Morin, P., Wood, D.R.: A polynomial bound for untangling geometric planar graphs. Discrete Comput. Geom. 42(4), 570–585 (2009)
3. Cibulka, J.: Untangling polygons and graphs. Discrete Comput. Geom. 43, 402–411 (2010)
4. Erdős, P., Szekeres, G.: A combinatorial problem in geometry. Compositio Mathematica 2, 463–470 (1935)
5. Fáry, I.: On straight line representation of planar graphs. Acta Univ. Szeged, Acta Sci. Math. 11, 229–233 (1948)
6. Goaoc, X., Kratochvíl, J., Okamoto, Y., Shin, C.S., Spillner, A., Wolff, A.: Untangling a planar graph. Discrete Comput. Geom. 42(4), 542–569 (2009)
7. Holton, D.A., McKay, B.D.: The smallest non-Hamiltonian 3-connected cubic planar graphs have 38 vertices. J. Combin. Theory Ser. B 45(3), 305–319 (1988)
8. Kang, M., Pikhurko, O., Ravsky, A., Schacht, M., Verbitsky, O.: Untangling planar graphs from a specified vertex position—Hard cases. Discrete Appl. Math. 159(8), 789–799 (2011)
9. Pach, J., Tardos, G.: Untangling a polygon. Discrete Comput. Geom. 28(4), 585–592 (2002)
10. Tait, P.G.: Listing's Topologie. Philosophical Magazine 17, 30–46 (1884)
11. Tutte, W.T.: On Hamiltonian circuits. J. LMS 21(2), 98–101 (1946)

Triangulations with Circular Arcs⋆

Oswin Aichholzer[1], Wolfgang Aigner[2], Franz Aurenhammer[2],
Kateřina Čech Dobiášová[3], Bert Jüttler[3], and Günter Rote[4]

[1] Institute for Software Technology, Graz University of Technology, Austria
[2] Institute for Theoretical Computer Science, Graz University of Technology, Austria
[3] Institute of Applied Geometry, Johannes Kepler University Linz, Austria
[4] Institut für Informatik, Freie Universität Berlin, Germany

Abstract. An important objective in the choice of a triangulation is that the smallest angle becomes as large as possible. In the straight-line case, it is known that the Delaunay triangulation is optimal in this respect. We propose and study the concept of a circular arc triangulation—a simple and effective alternative that offers flexibility for additionally enlarging small angles—and discuss its applications in graph drawing.

1 Introduction

Geometric graphs and especially triangular meshes (often called triangulations) are an ubiquitous tool in geometric data processing [4,17,26]. The quality of a given triangular mesh naturally depends on the size and shape of its composing triangles. In particular, the angles arising in the mesh are among the critical issues in main application areas like modeling, drawing, and finite element methods [26].

For practical purposes, quite often the Delaunay triangulation (see, e.g., [17]) is the mesh of choice, because it maximizes the smallest angle over all possible triangulations of a given finite set of points in the plane. Still, the occurrence of badly shaped triangles cannot be avoided sometimes, especially near the boundary of the input domain, or due to the presence of mesh vertices of high degree.

The situation becomes different (and interesting again) if the requirement that triangulation edges be straight is dropped. Indeed, certain applications are not confined to straight-line triangular meshes, or even are not really suited for it. In applications from graph drawing, for example, staying with straight edges might mean a hindrance to the readability of the drawing. Moreover, in finite element methods, the respective bivariate functions may be defined, in a natural way and with certain advantages, over 'triangles' with nonlinear boundaries. In these and other applications, the calculational and aesthetical benefits of a graph that potentially grants nice angles can be exploited fully only if curved edges are permitted.

In this paper, we want to encourage the use of so-called *arc triangulations*, which simply are triangulations whose edges are circular arcs. Maximizing the smallest angle in a combinatorially fixed arc triangulation of a point set can be

---

⋆ Supported by FWF NRN 'Industrial Geometry' S92. A preliminary version of this work appeared as [1].

M. van Kreveld and B. Speckmann (Eds.): GD 2011, LNCS 7034, pp. 296–307, 2012.

formulated as a linear program (Section 2), which for most settings can even be transformed to a simple graph-theoretic problem (Section 3). This guarantees a fast solution of this (and of related) optimization problems for arc triangulations in practice and in theory. Moreover, the linear program will tell us whether a given domain admits an arc triangulation of a pre-specified combinatorial type, by checking whether its feasible region is nonempty. In particular, flips for arcs can be defined (Section 4), by optimizing the triangulation that is obtained after applying the flip combinatorially. Preliminary inspection shows that small angles tend to enlarge significantly under such heuristics.

We believe that arc triangulations constitute a useful tool especially in two important application areas—graph drawing and finite element methods. In particular, so-called π-triangulations (Section 5) can be used with advantage, based on the fact that arc triangles whose angles sum to π are images of straight triangles under a Möbius transformation. In view of graph drawing applications [10,15,23], it is desirable to extend our approach to optimizing angles in general plane graphs (Section 6). This cannot be done dirctly, but by completing the graph to a suitable triangulation (for example, its constrained Delaunay triangulation [9]), and treating the sums of triangulation angles between the graph arcs as single entities to be maximized. A simple and efficient method for optimally redrawing a straight-line graph with circular arcs is obtained. Applications to finite element methods will be discussed in the full version of this paper.

2   Angle Optimization

Consider a straight-line triangulation, $\mathcal{T}$, in a given domain D of the plane. No restrictions on D are required but, for the ease of presentation, let D be simply connected and have a piecewise circular (or linear) boundary. In general, $\mathcal{T}$ will use vertices. in the interior of D. Throughout the paper, we assume general position of the vertex set. We are interested in the following optimization problem: Replace each interior (i.e., non-boundary) edge of $\mathcal{T}$ by some circular arc, in a way such that the smallest angle in the resulting arc triangulation is maximized.

To see that this problem is well defined, notice that the optimal solution, call it $\mathcal{T}^*$, cannot contain negative angles: The smallest angle between arcs has to be at least as large as the smallest angle that arises in $\mathcal{T}$. As a consequence, for each vertex in S, the order of its incident arcs in $\mathcal{T}^*$ coincides with the order of its incident edges in the input triangulation $\mathcal{T}$. In other words, each arc triangle in $\mathcal{T}^*$ is *well-oriented*, i.e., it has the same orientation as its straight-line equivalent. Therefore, no overlap of arcs or arc triangles in $\mathcal{T}^*$ can occur. Interestingly, this is a specialty of triangulations; the last conclusion remains no longer true if faces with more than three arcs are present. An arc quadrangle, for instance, may have self-overlaps in spite of being well-oriented, whereas this is not possible for an arc triangle; see Figures 1 and 2. We postulate for the rest of this paper that arc triangles be well-oriented.

We now formulate the angle optimization problem as a linear program. For each straight-line edge $e = pq$ in the triangulation $\mathcal{T}$, we introduce two variables

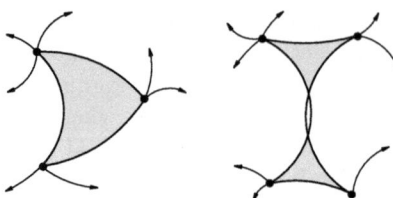

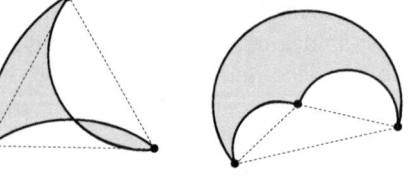

Fig. 1. Well-oriented arc triangle and quadrangle

Fig. 2. These arc triangles are not well-oriented

ϕ_{pq} and ϕ_{qp}. The variable ϕ_{pq} describes the (signed) angle at which the circular arc $\widehat{pq}$ deviates to the left from the straight connection, when seen from p, and ϕ_{qp} describes this deviation angle, when seen from q. We have

$$\phi_{pq} = -\phi_{qp} \tag{1}$$

for all edges pq. For each edge e' of $\mathcal{T}$ on the input boundary ∂D, we fix the two deviation variables to the values $d_{e'}$ and $-d_{e'}$ given by ∂D. Thus, for a boundary edge $e' = pq$, we have

$$\phi_{pq} = -\phi_{qp} = d_{e'}. \tag{2}$$

We have $d_{e'} = 0$ if e' is supposed to stay a line segment. Alternatively, and preferably in certain applications, we could keep $\phi_{pq} = -\phi_{qp}$ variable and bound it by some threshold. The inequalities for the linear program now stem from the angles α_{qpr} arising in $\mathcal{T}$. The two edges pq and pr that define α_{qpr} are adjacent around p in the drawing, such that pr is the next edge counterclockwise from pq. We are interested in the angle between the corresponding two circular arcs, which is $\beta_{qpr} = -\phi_{pq} + \alpha_{qpr} + \phi_{pr}$, and we put

$$\delta \leq \beta_{qpr} . \tag{3}$$

The linear objective function L, which is to be maximized, is just $L = \delta$.

Clearly, maximizing δ will maximize the smallest angle $\beta_{\min}$ in the arc triangulation. Note that we may have $\beta_{\min} > \frac{\pi}{3}$ in $\mathcal{T}^*$ because, due to its piecewise circular shape, the sum of inner angles for ∂D may be larger than $\pi(h-2)$, for h being the number of vertices on ∂D. There are $O(n)$ (in)equalities and $O(n)$ variables, if n is the total number of vertices.

Sometimes the objective is to optimize not only the smallest angle, but rather to maximize lexicographically the sorted list of all arising angles, as is guaranteed by the Delaunay triangulation in the straight-line case. This can be achieved by repeatedly solving the linear program above, keeping angles that have been optimized already as constants. Care has to be taken however, because, depending on the solver, minimum angles do typically occur at several places, and the optimal ones among them have to be singled out. This type of problems has been called *lexicographic bottleneck optimization* in [6], in the context of combinatorial optimization problems. In [22] a general solution procedure in the context of linear optimization is given, which amounts to repeatedly solving some slightly modified linear programs.

Angles larger than π may arise in the optimal triangulation. If this is undesirable in a particular application, constraints like

$$-\phi_{pq} + \alpha_{qpr} + \phi_{pr} \leq \gamma$$

for $\gamma < \pi$ may be added. In particular, choosing $\gamma = \pi - \delta$ will simultaneously decrease large angles, and thus will lead to arc triangles 'as equilateral as possible'. However, the demand of maximizing the smallest angle over the space of all possible arc triangulations (with the same combinatorics as $\mathcal{T}$) is then lost. Various other linear restrictions on angles can be added to the linear program, like fixing the angle sum in each arc triangle to π, or keeping each arc triangle inside the circumcircle of its three vertices. The relevance of these and other conditions will be substantiated in Sections 5 and 6. We consider the flexibility of our simple approach as an important feature in practice.

3   Graph-Theoretic Approach

The special setting of our linear program allows us to apply a purely graph-theoretic approach for its resolution.

Theorem 1. *The linear-programming problem of maximizing δ under restrictions (1–3) can be solved by a combinatorial (graph-theoretic) algorithm in $O(n^2)$ time.*

The remainder of this section gives a proof of Theorem 1. We have two variables ϕ_{pq} and ϕ_{qp} for each edge pq in the given straight-line triangulation, and the variable δ. Since a triangulation is a planar graph, there are $O(n)$ variables, $O(n)$ inequalities of type (3) induced by the angles between adjacent edges, and $O(n)$ equations of types (1) and (2).

First we consider a fixed value of δ and ask whether the system (1–3) is feasible. By using a method in [27] (see also [13,25]), we can transform the system into an equivalent system, in which every constraint has one of the following forms

$$X \leq Y + c, \tag{4}$$

$$X \leq 0 + c, \tag{5}$$

$$0 \leq Y + c, \tag{6}$$

where X and Y are two variables and c is a constant.

By substituting β_{qpr} we can easily rewrite (3) in this form, namely

$$\phi_{pq} \leq \phi_{pr} + (\alpha_{qpr} - \delta). \tag{7}$$

If we have bounds on the variables, $a \leq X \leq b$, we can also bring them into the desired form, and hence each equation (2) can be also handled, by first converting it into two inequalities.

We still have to deal with the equations (1) between 'opposite' variables. To this end, let us consider a system of inequalities of the form (4–6) in $2m$ variables $\mathcal{V} = \{x_1, \ldots, x_m, x'_1, \ldots, x'_m\}$ that come in 'opposite pairs'

$$x_i = -x'_i, \text{ for } i = 1, \ldots, m. \tag{8}$$

For a variable X, we will denote by $\bar{X}$ its opposite partner, $\bar{x}_i = x'_i$, $\bar{x}'_i = x_i$, $\bar{\bar{X}} = X$. The system we have at hands is of this form, with $\bar{\phi}_{pq} = \phi_{qp}$. Now, for each inequality of the form (4–6), we can form an equivalent *opposite inequality*, in which each variable is replaced by the opposite variable on the other side. For example,

$$X \leq Y + c \tag{4}$$

is turned into $\bar{Y} \leq \bar{X} + c$. In view of (8), the opposite inequality is equivalent to the original one. Thus, when we add all opposite inequalities, we will create some redundancy but we will not change the solution. It is easy to prove the following:

Lemma 1. *Consider a system of the equations (8) together with inequalities of the form (4–6), that also contains with each inequality its opposite inequality. Then this system has a solution if and only if the system without the equations (8) has a solution.*

This means that we can ignore the equations (1), at the expense of doubling the number of inequalities. All inequalities have the form (4–6). By introducing a new variable Z_0 representing zero, the inequalities (5–6) that contain only one variable can also be brought into the standard form (4). This new system is equivalent to the original one: Since all inequalities now have the form (4), one can add an arbitrary constant to all variables without invalidating the inequalities, and thus one can assume, without loss of generality, that $Z_0 = 0$.

It is well known that a system of inequalities of the form (4) can be tested by checking whether an associated graph G has a negative cycle [7,27], and a solution can be found by a shortest path calculation. The graph G has a node for each variable, and for each inequality of the form (4) it contains an arc of weight c from X to Y. Moreover, consider an augmented graph G^+, that has an additional start node S and an edge of weight 0 from S to every node of G.

Lemma 2. *A system of inequalities of the form (4) has a solution iff the associated graph G (or equivalently, G^+) has no negative cycle. If a solution exists, it can be found by computing shortest distances from S to all nodes in G^+.*

The running time of this test, with the Bellman–Ford algorithm, is given by the number of nodes or variables ($2m = O(n)$ in our case), times the number of arcs or inequalities ($O(n)$ as well). Thus, finding a solution of the angle drawing problem for a given value of δ takes $O(n^2)$ time.

Now we will consider δ as a variable and come back to the problem of maximizing δ. This amounts to checking for a negative cycle in a graph whose weights are of the form $c - \delta$, for constants c and a parameter δ. This problem is known as the *minimum cycle mean problem*: For a cycle with k edges the weight has the form $w - k\delta$, where w is the sum of all positive edge constants c along the cycle. The weight is negative for $\delta > w/k$. So w/k, the *mean weight* of the cycle, is the largest value for δ which does not result in a negative cycle. For the entire graph, this means that the largest possible value of δ for which the graph is free of negative cycles is determined by the minimum cycle mean. The minimum cycle mean problem has been solved in [19], and the algorithm takes the same running time as the Bellman–Ford algorithm, that is, $O(n^2)$ time, but it takes $O(n^2)$ space.

4   Flipping in Arc Triangles

The fact that every simple polygon can be triangulated with straight line segments is folklore. However, a domain D with piecewise circular boundary need not admit *any* triangulation, even if circular arcs may be used. It is known that a linear number of Steiner points is required in the worst case to ensure an arc triangulation [3].

One of the arising questions is: Given the domain D and a (combinatorial) triangulation $\mathcal{T}_c$ in D, possibly with (fixed) interior points, can $\mathcal{T}_c$ be realized by circular arcs? Clearly, if only straight-line edges are to be used, then this is merely a segment intersection problem. For deciding the general case, we can now utilize the linear program formulated in Section 2. A realizing arc triangulation exists if and only if the feasible region of the linear program is nonempty.[1] As a particularly nice feature, this enables us to define flip operations in arc triangulations, as is described below.

Consider some arc triangulation $\mathcal{A}$ in the domain D. Each interior arc $\overset{\frown}{pq}$ of $\mathcal{A}$ lies on the boundary of two arc triangles. Let r and s be the two vertices of these arc triangles different from p and q. Flipping $\overset{\frown}{pq}$ by definition means removing $\overset{\frown}{pq}$ from $\mathcal{A}$, establishing an arc between r and s combinatorially, and optimizing over the resulting triangulation. Note that 'well-oriented' in this case has to refer to the *combinatorial* order of the edges around a vertex of a triangulation.

For the linear program that describes this optimization problem, we have to know the angles α of the corresponding straight-line embedding; see Section 2. Note that after a flip, the straight-line realization of the graph is not necessarily a valid geometric triangulation. In such a case, the combinatorial order around a vertex is different from the geometric one. As a consequence, some angles α have to take negative values to obtain a valid setting for the linear program that optimizes δ. See Figure 4 for an example with the combinatorial order being 1 to 5, while the geometrical order is $1, 4, 2, 3, 5$.

Unlike for the original setting in Section 2, here a positive solution for δ is not guaranteed. In fact, the sign of the optimized value δ indicates whether or not the combinatorial triangulation (after a flip) is realizable as an arc triangulation.

---

[1] Note that the following related problem is NP-complete [20]: Given a point set S and *some* set E of straight-line edges on S, decide whether E contains a triangulation of S.

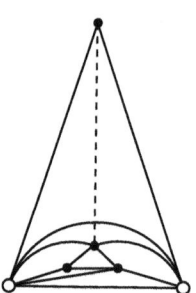

Fig. 3. A double edge connecting bottom vertices

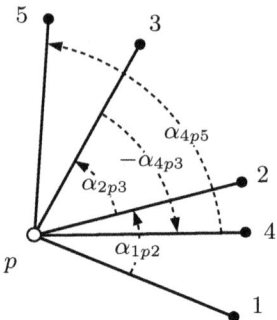

Fig. 4. Combinatorial order at p

If $\delta > 0$ after the optimization, then the new arc triangulation exists and contains a circular arc between r and s that satisfies the criterion of being geometrically well-oriented. In case of nonexistence (if $\delta \leq 0$), the combinatorial triangulation is not realizable as an arc triangulation, and we declare the arc $\widehat{pq}$ as nonflippable. Observe that an arc flip may change various circular arcs geometrically, as we optimize over their curvature afterwards.

Sometimes we may not want to perform an arc flip even if it exists. For example, flipping an arc a can lead to an inner vertex of degree 2, a property of arc triangulations which is possibly unwanted in the application. Arc a can easily be declared as not flippable, by putting the restriction that angles in triangles be less than π. Note that this does not necessarily prevent the occurrence of double-edges between two vertices of an arc triangulation. For example, see Figure 3, where all angles are smaller than π. However, a check if an edge already exists can be done before the optimization step, and thus does not have to be incorporated into the linear program.

Optimizing angles with arc flips is a powerful (though maybe costly) tool. We demonstrate the positive effect of sequences of such flips with Figures 5 and 6. A significant improvement over the Delaunay triangulation becomes possible (in fact, the smallest angle is doubled in this example) by reducing the degree of a particular vertex, v. Note that this configuration is quite 'robust' in the sense that v retains its high degree in the Delaunay triangulation even if the placement of the other vertices is changed moderately. Repeated appearance of patterns as in Figure 5 may lead to an overall poor quality of a given triangular mesh.

In general, we observe that small angles in a straight-line triangulation stem from one of two reasons: (1) The geometry of the underlying domain D (plus its vertex set) forces slim triangles in the vicinity of ∂D. These 'boundary effects' can usually be mildened by mere geometric optimization of the corresponding arc triangulation. (2) Vertices of degree k naturally impose an upper bound of $\frac{2\pi}{k}$ on the smallest arising angle. This situation can be remedied only with combinatorial changes, and in contrast to the straight edge case, this is indeed possible for arc triangulations. For straight edges, the combinatorics of the Delaunay triangulation is already optimal.

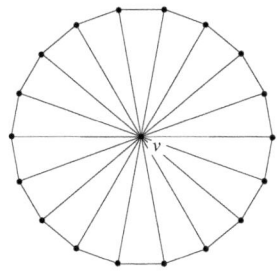

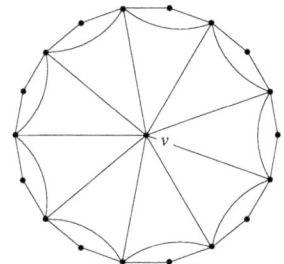

Fig. 5. Delaunay triangulation   **Fig. 6.** Optimized arc triangulation

A challenging open question is whether repeated application of angle-improving arc flips always leads to the global optimum, that is, to the combinatorial type of arc triangulation which admits the largest possible minimum angle for the given domain. A more basic question is whether the set of combinatorial triangulations that are realizable as arc triangulations is connected by flips. We leave these problems as a subject for future research.

5 Special Arc Triangles

Before discussing the relevance of arc triangulations to the area of graph drawing, we have a look at special types of arc triangles. Recall from Section 2 the convention that arc triangles are geometrically well-oriented.

An arc triangle ∇ is termed a π-*triangle* if the sum of its interior angles is π. These triangles are interesting because they are images of a straight-line triangle under a unique Möbius transformation [24]. Moreover, any π-triangle is contained in the circumcircle of its vertices, a possibly useful regularity condition. We study arc triangulations that are composed of π-triangles. Such π-*triangulations* will not always exist, but they do, of course, if the domain D is a simple polygon, because every straight-line triangulation is a π-triangulation. If ∂D is composed of circular arcs, a necessary (though not sufficient) existence condition is that the sum of interior angles at the h boundary vertices of D is $\pi(h-2)$.

For the remainder of this section, let D be a simple polygon, and $\mathcal{T}$ be some straight-line triangulation in D. The geometry of any arc triangulation $\mathcal{A}$ in D that is combinatorially equivalent to $\mathcal{T}$ is determined by the vector $\Phi(\mathcal{A})$ of deviation angles ϕ_{pq}, for the interior arcs $\widehat{pq}$ of $\mathcal{A}$. (The opposite value, ϕ_{qp}, is fixed by ϕ_{pq}; see Section 2). Interpreting $\Phi(\mathcal{A})$ as a point in high dimensions, we can talk of the space of arc triangulations for $\mathcal{T}$. The next lemma is important in view of optimizing a given π-triangulation. Let us assume that there exists an arc triangulation for D where all interior angles are positive.

Lemma 3. *Let $\mathcal{T}$ have n vertices, h of which lie on the boundary of D. The dimension of the space of π-triangulations for $\mathcal{T}$ is $n-h$.*

The proof is omitted due to space constraints. Lemma 3 remains true if $\mathcal{T}$ is replaced by any π-triangulation of D. For applications, the input is most likely a

Table 1. Angle improvement in arc triangulations

angle sum	smallest angle	improvement over Delaunay
Delaunay (180°)	18.03°	0
180°	22.52°	25 %
179°–181°	22.92°	26 %
175°–185°	24.88°	38 %
170°–190°	27.53°	50 %
160°–200°	31.77°	72 %

straight-line triangulation, which is to be optimized into a π-triangulation with maximum smallest angle. The boundary of D might be given as a spline curve, approximated smoothly by circular arcs. The inner angle sum for D is $\pi \cdot h$ in this case (rather than $\pi(h-2)$), such that a π-triangulation does not exist. Still, the approximating circular arcs will be close to line segments for most practical data, such that an 'almost straight' π-triangulation is likely to exist. Also, one could start with some combinatorial triangulation suitable for D, to be able to treat a larger class of domains.

Table 1 shows experimental data for Delaunay meshes optimized into (almost) π-triangulations, for 500 random points, postprocessed to keep a certain interpoint distance as in realistic meshes. The gain is quite significant, especially if the condition on the angle sum is relaxed from π to a small interval around that value. For several applications, there is sometimes a certain threshold (typically around 25°) beyond which a mesh is considered as poor-quality [5].

Note that, by Lemma 3, optimization is only possible in subdomains of D where interior points are present. Thus, the diagonals of D defined by T (if any) separate optimizable subdomains from each other. Again, such diagonals are unlikely to appear in the dense meshes used in practical applications. In any case, extraneous points can be inserted into the π-triangulation while keeping all angle sums in arc triangles to π. In particular, we can put such points on arcs, in order to split obstructive diagonals of D.

6   Graph Drawing

Literature on drawing graphs nicely in the plane is large; see e.g. [10,23,28]. Most algorithms take as input an abstract graph G and produce a layout of the vertices of G such that the resulting straight-line (or orthogonal) drawing is aesthetically pleasing, and preferably is even optimal with respect to certain application criteria. On the theoretical side, bounds on the achievable angular resolution are known for various classes of graphs [16,21]. A characterization of all planar drawings of a triangular graph through a system of equations and inequalities relating its angles is given in [11].

Results for curvilinear drawings of graphs are comparatively sparse. See, for example, [8,18] and references therein, who give lower bounds and algorithms for drawing graphs on a grid with curved edges (including circular multiarcs), and [15] where a method based on physical simulation is proposed. In [14], crossing-free drawings of graphs with circular arcs as edges are considered from an algorithmic viewpoint. The vertices are fixed and each edge has to be chosen from a given number of arcs. Recently, circular arc graphs with equiangular edges around each vertex have been studied in [12].

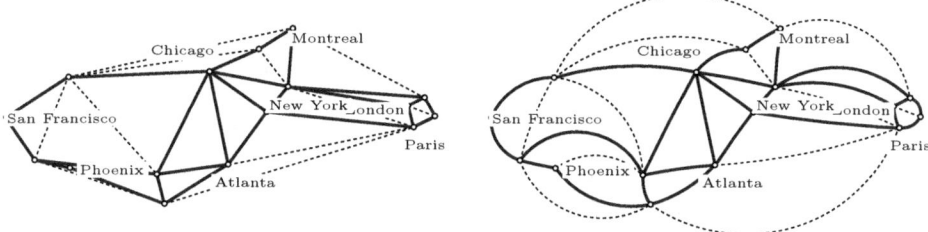

Fig. 7. IP backbone graph   **Fig. 8.** Backbone optimally redrawn

Here we actually consider a simpler setting, namely, for a given planar straight-line embedding of a graph G, the problem of *redrawing* G with curved edges in an optimal way. In a redrawing, the positions of the vertices are kept fixed. This may be a natural demand, for instance, in certain geographical applications. Recently it has been shown [2] that redrawings of G with tangent-continuous *biarcs* or quadratic Bézier curves (parabolic arcs) always exist such that every vertex is pointed, i.e., has an incident angle of at least π. Potential applications concern labeling the graph vertices with high readability. Redrawing a plane graph G with circular arcs in a pointed way is not always possible.

Let us describe how maximizing the smallest angle in a circular arc redrawing of G can be achieved. It is tempting to apply the linear optimization method from Section 2 to G directly. This, however, bears the risk of arc overlaps getting out of control. (Recall that overlap-free optimization is guaranteed only for full triangulations. This is possibly the reason why this simple approach has not been used in practice yet.) One way out is to embed G in some triangulation $\mathcal{T}$ first, and treat respective sums of angles as single entities to be optimized. That is, for each angle ϱ in G, given by the concatenation of angles $\alpha_1, \ldots, \alpha_k$ in $\mathcal{T}$, we use the constraint $\delta \leq \beta_1, \ldots, \beta_k$, with each β_i expressed by the corresponding straight-line triangulation angle α_i and its two assigned deviation variables $\beta_i = -\phi_1 + \alpha_i + \phi_2$ as in Section 2.

The quality of optimization depends on the chosen triangulation, which will be subject of future research; cf. Section 4. Note that, however, even if we try out all possible triangulations, this may not lead to the optimal solution, as there are arc polygons that cannot be triangulated without additional vertices. If the optimal drawing contains such a face, then no triangulation will yield the optimum drawing.

If we wish to optimize the entire angle vector $\varrho_1, \ldots, \varrho_m$ for G, this can be achieved too, in an iterative way as before. Additional restrictions may be posed, like $\varrho_j < \pi$ or $\varrho_j < \frac{\pi}{2}$, in order to preserve obtuse or sharp angles in G.

The adjacency graph in Figures 7 and 8, and the layer graph in Figures 9 and 10 exemplify the effect of our circular arc redrawing method. The results

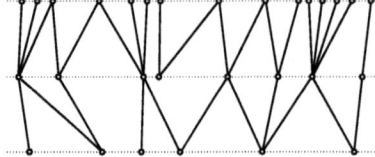
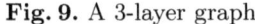
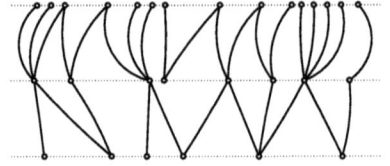

Fig. 9. A 3-layer graph   **Fig. 10.** Arc redrawing

seem satisfactory, in spite of the fact that vertices are required not to move. For geographic structures as in Figure 7, or certain graph structures arising in physics, this is quite often a desired property. Our results compare well to, e.g. [15], who use for optimization the additional freedom of placing vertices, though at a price of high computation cost. For our method, the number of vertices of the input graph is no limitation, as far as applications from graph drawing are concerned.

7   Open Questions

For non-triangulated regions in the input graph (compare the quadrangle in Figure 1), the requirement that arcs do not intersect induces a nonlinear constraint between the corresponding angles. It would be interesting to know if this constraint has some structure (for example, convexity), which would allow it to be accommodated in the optimization process. Further open questions raised here are the convergence of the angle-increasing arc flipping process in Section 4, and an extension of the presented results to three dimensions.

References

1. Aichholzer, O., Aigner, W., Aurenhammer, F., Čech Dobiášová, K., Jüttler, B.: Arc triangulations. In: Proc. 26th European Workshop Comput. Geometry, pp. 17–20 (2010)
2. Aichholzer, O., Rote, G., Schulz, A., Vogtenhuber, B.: Pointed drawings of planar graphs. In: Proc. 19th Ann. Canadian Conf. Comput. Geometry, pp. 237–240 (2007)
3. Aichholzer, O., Aurenhammer, F., Hackl, T., Juettler, B., Oberneder, M., Sir, Z.: Computational and structural advantages of circular boundary representation. Int'l J. Computational Geometry & Applications 21, 47–69 (2011)
4. Bern, M., Eppstein, D.: Mesh generation and optimal triangulation. Computing in Euclidean Geometry. LN Series on Computing, vol. 4, pp. 47–123. World Scientific (1995)
5. Boivin, C., Ollivier-Gooch, C.: Guaranteed-quality triangular mesh generation for domains with curved boundaries. International Journal for Numerical Methods in Engineering 55, 1185–1213 (2002)
6. Burkard, R.E., Rendl, F.: Lexicographic bottleneck problems. Operations Research Letters 10, 303–308 (1991)
7. Carré, B.: Graphs and networks. Oxford University Press (1979)

8. Cheng, C.C., Duncan, C.A., Goodrich, M.T., Kobourov, S.G.: Drawing Planar Graphs With Circular Arcs. In: Kratochvíl, J. (ed.) GD 1999. LNCS, vol. 1731, pp. 117–126. Springer, Heidelberg (1999)

9. Chew, L.P.: Constrained Delaunay triangulations. Algorithmica 4, 97–108 (1989)

10. Di Battista, G., Eades, P., Tamassia, R., Tollis, I.G.: Graph Drawing—Algorithms for the Visualization of Graphs. Prentice-Hall (1999)

11. Di Battista, G.D., Vismara, L.: Angles of planar triangular graphs. SIAM J. Discrete Mathematics 9, 349–359 (1996)

12. Duncan, C.A., Eppstein, D., Goodrich, M.T., Kobourov, S.G., Nöllenburg, M.: Lombardi Drawings of Graphs. In: Brandes, U., Cornelsen, S. (eds.) GD 2010. LNCS, vol. 6502, pp. 195–207. Springer, Heidelberg (2011)

13. Edelsbrunner, H., Rote, G., Welzl, E.: Testing the necklace condition for shortest tours and optimal factors in the plane. Theor. Comput. Sci. 66, 157–180 (1989)

14. Efrat, A., Erten, C., Kobourov, S.G.: Fixed-Location Circular-Arc Drawing of Planar Graphs. Journal of Graph Algorithms and Applications 11, 145–164 (2007)

15. Finkel, B., Tamassia, R.: Curvilinar Graph Drawing Using The Force-Directed Method. In: Pach, J. (ed.) GD 2004. LNCS, vol. 3383, pp. 448–453. Springer, Heidelberg (2005)

16. Formann, M., Hagerup, T., Haralambides, J., Kaufmann, M., Leighton, F.T., Symvonis, A., Welzl, E., Wöginger, G.: Drawing graphs in the plane with high resolution. SIAM J. Computing 22, 1035–1052 (1993)

17. Fortune, S.: Voronoi diagrams and Delaunay triangulations. Computing in Euclidean Geometry. LN Series on Computing, vol. 4, pp. 225–265. World Scientific (1995)

18. Goodrich, M.I., Wagner, C.G.: A framework for drawing planar graphs with curves and polylines. J. Algorithms 37, 399–421 (2000)

19. Karp, R.M.: A characterization of the minimum cycle mean in a digraph. Discrete Mathematics 23, 309–311 (1978)

20. Lloyd, E.L.: On triangulations of a set of points in the plane. In: Proc. 18th IEEE Symp. on Foundations of Computer Science, pp. 228–240 (1977)

21. Malitz, S., Papakostas, A.: On the angular resolution of planar graphs. In: Proc. 24th Ann., pp. 527–538 (1992)

22. Marchi, E., Oviedo, J.A.: Lexicographic optimality in the multiple objective linear programming: The nucleolar solution. European Journal of Operational Research 57, 355–359 (1992)

23. Nishizeki, T., Rahman, M.S.: Planar graph drawing. World Scientific (2004)

24. Pedoe, D.: A course of geometry for colleges and universities. Cambridge University Press (1970)

25. Rote, G.: Two solvable cases of the traveling salesman problem. PhD Thesis, TU Graz, Institute for Mathematics (1988)

26. Shewchuk, J.: What is a good linear element? Interpolation, conditioning, and quality measures. In: Proc. 11th International Meshing Roundtable, pp. 115–126 (2002)

27. Shostak, R.: Deciding linear inequalities by computing loop residues. Journal of the ACM 28, 769–779 (1981)

28. Sugiyama, K.: Graph Drawing and Applications for Software and Knowledge Engineers. World Scientific (2002)

Planar and Poly-arc Lombardi Drawings

Christian A. Duncan[1], David Eppstein[2], Michael T. Goodrich[2],
Stephen G. Kobourov[3], and Maarten Löffler[2]

[1] Department of Computer Science, Louisiana Tech Univ., Ruston, Louisiana, USA
[2] Department of Computer Science, University of California, Irvine, California, USA
[3] Department of Computer Science, University of Arizona, Tucson, Arizona, USA

Abstract. In Lombardi drawings of graphs, edges are represented as circular arcs, and the edges incident on vertices have perfect angular resolution. However, not every graph has a Lombardi drawing, and not every planar graph has a planar Lombardi drawing. We introduce k-Lombardi drawings, in which each edge may be drawn with k circular arcs, noting that every graph has a smooth 2-Lombardi drawing. We show that every planar graph has a smooth planar 3-Lombardi drawing and further investigate topics connecting planarity and Lombardi drawings.

1   Introduction

Motivated by the work of the American abstract artist Mark Lombardi [21], who specialized in drawings that illustrate financial and political networks, Duncan et al. [9,10] proposed a graph visualization called *Lombardi drawings*. These types of drawings attempt to capture some of the visual aesthetics used by Mark Lombardi, including his use of circular-arc edges and well-distributed edges around each vertex.

A vertex with circular arc edges extending from it has ***perfect angular resolution*** if the angles between consecutive edges, as measured by the tangents to the circular arcs at the vertex, all have the same degree. A ***Lombardi drawing*** of a graph $G = (V, E)$ is a drawing of a graph where every vertex is represented as a point, the edges incident on each vertex have perfect angular resolution, and every edge is represented as a line segment or circular arc between the points associated with adjacent vertices.

One drawback of previous work on Lombardi drawings is that (as we prove here) not every graph has a Lombardi drawing. In this paper we attempt to remedy this by considering drawings in which edges are represented by multiple circular arcs. This added generality allows us to draw any graph.

k-Lombardi Drawings. We define a k-***Lombardi drawing*** to be a drawing with at most k circular arcs per edge, with a 1-Lombardi drawing being equivalent to the earlier definition of a Lombardi drawing. We say that a k-Lombardi drawing is ***smooth*** if every edge is continuously differentiable, i.e., no edge in the drawing has a sharp bend. If a k-Lombardi drawing is not smooth, we say it is ***pointed***. Fortunately, we do not need large values of k to be able to draw all graphs: as we show, every graph has a smooth 2-Lombardi drawing. Interestingly, this result is hinted at in the work of Lombardi himself—Figure 1 shows a portion of a drawing by Lombardi that uses smooth edges consisting of two near-circular arcs.

M. van Kreveld and B. Speckmann (Eds.): GD 2011, LNCS 7034, pp. 308–319, 2012.
© Springer-Verlag Berlin Heidelberg 2012

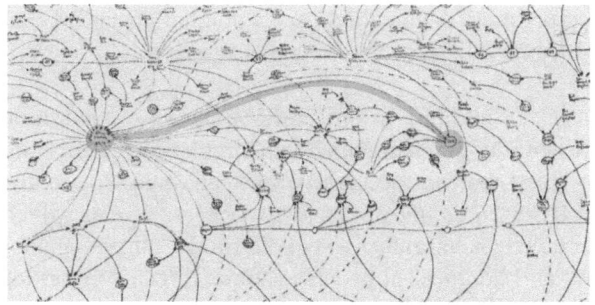

Fig. 1. A portion of Mark Lombardi, *Chicago Outfit and Satellite Regimes, ca. 1931–83*, 1998, 48.125 × 96.6225 inches (cat. no. 11) [21]. Note the highlighted smooth two-arc edge.

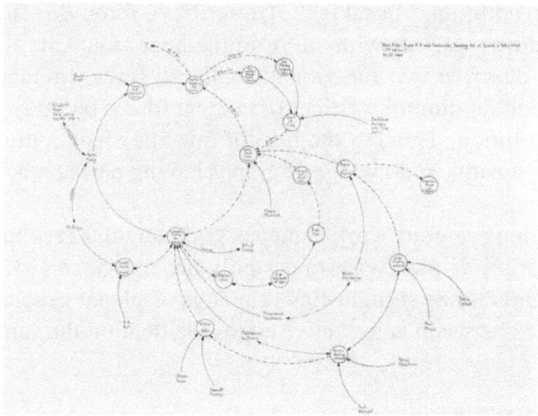

Fig. 2. Mark Lombardi, *Hans Kopp, Trans K-B and Shakarchi Trading AG of Zurich, ca. 1981–89* (3rd Version), 1999, 20.25 × 30.75 inches (cat. no. 22) [21]

Planar Lombardi Drawings. Drawing planar graphs without crossings is a natural goal for graph drawing algorithms and is easily achieved when angular resolution is ignored. Lombardi himself avoided crossings in many of his drawings, as shown in Fig. 2. In previous work on Lombardi drawings, Duncan et al. [10] showed that there exist embedded planar graphs that have Lombardi drawings but do not have *planar Lombardi drawings*. Here we continue this investigation of planar Lombardi drawings and extend it to planar k-Lombardi drawings.

New Results. In this paper we provide the following results: (1) We find examples of graphs that do not have a Lombardi drawing, regardless of the ordering of edges around each vertex, thus strengthening an example from [10] of graphs for which a specific edge ordering cannot be drawn. (2) We find examples of planar 3-trees with no planar Lombardi drawing, strengthening an example from [10] of a planar graph with treewidth greater than three that is not planar Lombardi. (3) We show how to construct a smooth 2-Lombardi drawing for any graph, a smooth planar 2-Lombardi drawing of planar graph with maximum degree three, and a pointed planar 2-Lombardi drawing or a smooth planar 3-Lombardi drawing of any planar graph.

Other Related Work. In addition to the earlier work on Lombardi drawings, there is considerable prior work on graph drawing with circular-arc or curvilinear edges for the sake of achieving good, but not necessarily perfect, angular resolution [4,16]. There is also significant work on *confluent drawings* [7,11,12,18,19], which use curvilinear edges not to separate edges but rather to bundle similar edges together and avoid edge crossings. Brandes and Wagner [3] provide a force-directed algorithm for visualizing train schedules using Bézier curves for edges and fixed positions for vertices. Finkel and Tamassia [14] extend this work by giving a force-directed method for drawing graphs with curvilinear edges where vertex positions are not fixed. Aichholzer et al. [1] show, for a given embedded planar triangulation with fixed vertex positions, it is possible to find a circular-arc drawing that maximizes the minimum angular resolution by solving a linear program. In addition, Matsakis [23] describes a force-directed approach to producing Lombardi drawings, but without an implementation. Chernobelskiy et al. [5], on the other hand, describe two functional Lombardi force-directed schemes that are based on the use of either dummy vertices or tangent forces but may not always achieve perfect angular resolution. Thus, to the best of our knowledge, none of this other related work correctly results in drawings of graphs having perfect angular resolution and curvilinear edges.

Alternatively, some previous work achieves good angular resolution using straight-line drawings [6,15,22] or piecewise-linear poly-arc drawings [13,17,20]. Di Battista and Vismara [6] characterize straight-line drawings of planar graphs with a prescribed assignment of angles between consecutive edges incident on the same vertex.

2   *k*-Lombardi Drawings

2.1   Non-Lombardi Graphs

Before investigating *k*-Lombardi drawings, we first establish the need for using poly-arc edges to be able to draw any graph. Although Duncan et al. [10] show a graph, Fig. 3(a), for which no Lombardi drawing is possible *while preserving the given ordering of edges around each vertex*, as Fig. 3(b) shows, if the ordering is not fixed, it is possible to create a valid Lombardi drawing for the graph. In this section, we provide a graph that has no Lombardi drawing *irrespective of the edge ordering*.

There are some complications in proofs of non-Lombardi counterexamples that differ from counterexamples in straight-line planar drawings. For example, if graph G is non-Lombardi, this does not imply that all graphs $H \supset G$ are non-Lombardi because the addition of edges changes the angular resolution and can therefore dramatically change the subsequent placement of vertices. In addition, since the edge ordering is not fixed by the input, we must argue that any ordering forces a conflict.

Additional complications concern the density and symmetry of any possible counterexample. A *k-degenerate graph* is a graph that can be reduced to the empty graph by iteratively removing vertices of degree at most k. The graph in Fig. 3 is 3-degenerate,

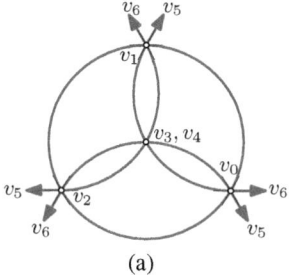
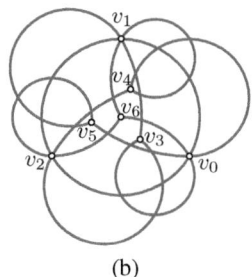

(a)   (b)

Fig. 3. A 7-vertex 3-degenerate graph that has no Lombardi drawing with the given edge ordering. (a) A Möbius transformation makes triangle $v_0v_1v_2$ equilateral, forcing both vertices v_3 and v_4 to be placed at the centroid and vertices v_5 and v_6 at the point at infinity; (b) A different ordering that does provide a Lombardi drawing.

and 3-degenerate graphs can be drawn Lombardi-style if we are willing to ignore vertex-vertex and vertex-edge overlaps.[1] Consequently, if a 3-degenerate graph is to be a counterexample, we must show that all vertex orderings force two vertices to overlap. Intuitively, 4-degenerate graphs should be more restrictive, but the simplest 4-degenerate graph, K_5, nevertheless has a circular Lombardi drawing. One reason is the fact that K_5 is extremely symmetrical. Therefore, we shall modify this graph to break its symmetry. We define our counterexample graph G_8 to be K_5 with the addition of three degree-one vertices causing one of the vertices of the original K_5 to have degree 5 and another to have degree 6, while the other three remain with degree 4; see Fig. 4(a).

Before we can establish our main theorem, we need to present a few geometric properties related to Lombardi drawings.

Property 1 ([10]). Let A be a circular arc or line segment connecting two points p and q that both lie on circle O. Then A makes the same angle to O at p that it makes at q. Moreover, for any p and q on O and any angle $0 \leq \theta \leq \pi$, there exist either two arcs or a line segment and pair of collinear rays connecting p and q, making angle θ with O, one lying inside and one outside of O.

We defer the proof of the next property, partially established in [10], to the full version of this paper [8].

Property 2. Suppose we are given two points $p = (p_x, p_y)$ and $q = (q_x, q_y)$ and associated angles θ_{ph} and θ_{qh} and an angle θ_{pq}. Consider all pairs of circular arcs that leave p and q with angles θ_{ph} and θ_{qh} respectively (measured with respect to the positive horizontal axis) and meet at an angle θ_{pq}. The locus of meeting points for these pairs of arcs is a circle. Moreover, the circle has radius $r_c = d_{pq} \csc \alpha / 2$ and center $(p_x + r_c \sin(\alpha + \beta), p_y - r_c \cos(\alpha + \beta))$, where $\alpha = (\theta_{ph} - \theta_{qh} - \theta_{pq})/2$, β is the angle formed by the ray from p through q with respect to the positive horizontal axis, and d_{pq} is the distance between the points p and q.

---

[1] Note that a drawing with vertex-vertex overlaps still must obey the perfect angular resolution constraints on the (possibly zero-length) edges, assuming such edge lengths are even allowed.

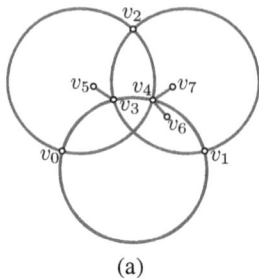
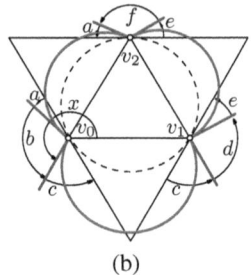

(a)   (b)

Fig. 4. (a) G_8 with the K_5 subgraph drawn Lombardi-style and additional edges shown. (b) Computing the twist for the three vertices 0, 1, and 2. The twist for vertex 0 is x.

Theorem 1. *The graph G_8 is non-Lombardi.*

Proof. Let v_0, v_1, v_2 be the three vertices of G_8 with degree four. Let v_3 and v_4 be the vertices with degree five and six respectively. We do not care about the final placement of the degree-one vertices, whose main purpose is to alter the angular resolution of v_3 and v_4. Using a Möbius transformation we can assume that the first three vertices v_0, v_1, and v_2 are placed on the corners of a unit equilateral triangle such that v_0 and v_1 have positions $(0,0)$ and $(1,0)$ respectively. We shall show that for every edge ordering, the two vertices v_3 and v_4 cannot both be placed to maintain correctly their angular resolution and be connected to each other. We do this by establishing the algebraic equations for their positions based on the edge orderings of all vertices. We then show that such a set of equations has no solution for any valid assignment of orderings.

We first establish a notation for representing a specific edge ordering. For every vertex v_i with neighbor v_j, let k_{ij} represent the counterclockwise cyclic ordering of edge (v_i, v_j) about v_i with $k_{01} = 0$ and $k_{i0} = 0$ for $i > 0$. For example, in Fig. 4(a), the edge ordering around v_4 has $k_{41} = 2$, $k_{42} = 4$, $k_{43} = 5$, $k_{46} = 1$, and $k_{47} = 3$. The *twist* t_i of a vertex v_i is the angle made by the arc extending from v_i to the neighbor v_j with $k_{ij} = 0$. From the initial placement of v_0, v_1, and v_2 on an equilateral triangle and their respective edge orderings, we can uniquely determine the twists for each of these vertices; see Fig. 4(b). Since the three vertices lie on an equilateral triangle, the tangents to the circle defined by the three points also form an equilateral triangle. From Property 1, the angles formed by the arcs connecting each pair of vertices to the tangents at the circle yield matching (but undetermined) angles, labeled a, c, and e. The angles b, d, and f are determined uniquely by the edge orderings as follows:

$$b = 2\pi - k_{02}\pi/2, \qquad d = k_{12}\pi/2, \qquad f = 2\pi - k_{21}\pi/2 \qquad (1)$$

Noting that certain triplets of angles yield a value of π, we have the following three equations on three unknowns: $a + b + c = \pi + 2i_0\pi$, $c + d + e = \pi + 2i_1\pi$, and $e + f + a = \pi + 2i_2\pi$. Solving for a yields: $2a = \pi - f - b + d + 2(i_0 - i_1 + i_2)\pi$. For the twist for v_0, we wish to know the value of x, the angle for the arc from v_0 to v_1. Noting that $x = a + b + 2\pi/3 - 2i_0\pi$ and substituting in (1) yields $t_0 = x = 7\pi/6 + \pi(k_{12} + k_{21} - $

$k_{02})/4 + (i_2 - i_0 - i_1)\pi$. Noting that $t_0 + c + \pi/3 = 2\pi$ yields $t_1 = \pi - t_0$. Similarly, $t_2 = \pi - a = 5\pi/3 - t_0 - k_{02}\pi/2 + 2\pi(1 - i_0)$.

The positions and orienting twists of the first three vertices also yield a unique position and twist for vertices v_3 and v_4. After determining these values, we shall show that in all orderings it is not possible to connect v_3 to v_4 with a single circular arc while still maintaining the proper angular resolution.

From Property 2, v_3 must lie on a circle C_{01} defined by the neighbors v_0 and v_1 and their corresponding arc tangents. Similarly, it must lie on circles C_{02} and C_{12}. The intersection of these three circles determines the position and orientation of v_3. Let us proceed to determine C_{01}. Letting $p = v_0$ and $q = v_1$, we have $\theta_{ph} = t_0 + \pi k_{03}/4$ and $\theta_{qh} = t_1 + \pi k_{13}/4$ and $\theta_{pq} = \pi(k_{31} - k_{30})/5 = \pi k_{31}/5$. From Property 2 and the fact that $d_{pq} = 1$, we can determine that C_{01} has radius $r_{01} = \csc \alpha_{01}/2$ and center $c_{01} = (r_{01} \sin \alpha_{01}, -r_{01} \cos \alpha_{01}) = (1/2, -\cot \alpha_{01}/2)$ with $\alpha_{01} = (\theta_{ph} - \theta_{qh} - \theta_{pq})/2 = t_0 - \pi/2 + \pi(5k_{03} - 5k_{13} - 4k_{31})/40$. Similarly, C_{02} has radius $r_{02} = \csc \alpha_{02}/2$ and center $c_{02} = (r_{02} \sin(\alpha_{02} + \pi/3), -r_{02} \cos(\alpha_{02} + \pi/3))$ with $\alpha_{02} = t_0 - 5\pi/6 + \pi(5k_{03} + 10k_{02} - 5k_{23} - 4k_{32})/40 + (i_0 - 1)\pi$.

Given the circles and the position of v_0 at the origin, it is easy to determine the intersection of the two circles, one of which is v_0 and the other, if it even exists, must be v_3. Since v_0 must lie on the intersection, the line from v_0 to v_3 is perpendicular to the line, ℓ, through the two centers. Moreover, v_3 is the reflection of p about ℓ. Thus, letting $v = (v_x, v_y) = c_{02} - c_{01}$, $c = v_0 - c_{01} = -c_{01}$, and $v^\perp = (-v_y, v_x)$ yields $v_3 = \frac{-2c \cdot v^\perp}{v \cdot v} v^\perp$. To establish the twist t_3 at v_3 we observe from Property 1 that the angle α formed by the line ℓ_{03} from v_0 to v_3 and the tangent of the curve from v_0 to v_3 is the same as the tangent of the curve from v_3 to v_0 and the line ℓ_{03}. Moreover, $\theta_{03} = t_0 + k_{03}\pi/4 = \alpha + \beta_{03}$ and $t_3 = \theta_{30} = \pi - \alpha + \beta_{03}$ where $\beta_{03} = \arctan(v_3(y)/v_3(x))$ is the slope of ℓ_{03}. From this, we can deduce that $t_3 = \pi - t_0 - k_{03}\pi/4 + 2\beta_{03}$. The exact same calculations can be used to compute v_4 and t_4.

As with the twists for t_3 and t_4, we can use Property 1 to determine the angles formed by the arc from v_3 to v_4 given their positions and twists. We know that the angles of the tangents to the arc at v_3 and v_4 are $\theta_{34} = t_3 + k_{34}\pi/5$ and $\theta_{43} = t_4 + k_{43}\pi/6$ respectively. Letting $\beta_{34} = \arctan((v_4(y) - v_3(y))/(v_4(x) - v_3(x)))$ be the slope of the line from v_3 to v_4, we have that $\theta_{34} - \beta_{34} = \alpha$ and $\pi - \alpha = \theta_{43} - \beta_{34}$. Consequently, we have

$$\theta_{34} + \theta_{43} = \pi + 2\beta_{34}. \tag{2}$$

Each specific edge ordering therefore yields a unique set of positions and twists for v_3 and v_4 as outlined above. To show that no Lombardi drawing is possible one must simply show that (2) does not hold for *any* edge ordering. Though there are a finite number of possible orderings and though symmetries could be used to reduce that number, the individual case analysis for such a proof appears to be quite unwieldy. Instead, we simply iterate over every possible edge ordering, applying these equations to a numerical algorithm that searches for a valid non-contradictory assignment. The full version of this paper [8] contains the Python code for this program. By running this program, one can see that no valid assignments are possible, concluding our proof.   □

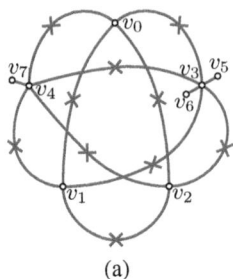 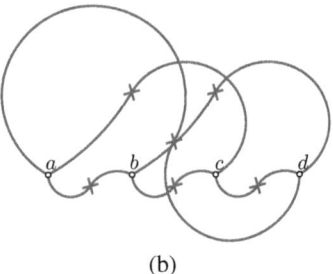

(a)   (b)

Fig. 5. (a) An example 2-Lombardi drawing of G_8. The bend points (not all of which are necessary) are shown with crossed marks. (b) An example 2-Lombardi drawing of K_4 with the vertices placed on a line and tangents oriented to force numerous inflection points.

Observing that we can take any Lombardi graph, subdivide an edge, and split the resulting new vertex into two degree-one vertices to produce a new Lombardi graph, we can get the following corollary, whose complete proof is provided in [8].

Corollary 1. *There are an infinite amount of connected non-Lombardi graphs.*

2.2   Smooth 2-Lombardi Drawings

If we want to draw Lombardi-style drawings for any given graph we have to relax one of the two requirements that specify Lombardi drawings. Here, we would like to avoid relaxing the requirement that edges have perfect angular resolution. Fortunately, we can achieve a Lombardi methodology for drawing any graph if we allow two circular arcs per edge; for example, see Fig. 5(a).

Since every 2-degenerate graph has a Lombardi drawing [10, Thm. 3] and since subdividing every edge in a graph results in a 2-degenerate graph, we readily obtain the following corollary, whose complete proof is found in the full version of the paper [8].

Corollary 2. *Every graph has a smooth 2-Lombardi drawing. Furthermore, the vertices can be chosen to be in any fixed position.*

As Fig. 5(b) illustrates, although we can place the vertices in any position with any initial orientation, an arc's smooth bend point might be an inflection point.

3   Planar k-Lombardi Drawings

3.1   A Planar 3-Tree with No Planar Lombardi Drawing

It is known that planar graphs do not necessarily have planar (non-crossing) Lombardi drawings. For example, Duncan et al. [10] show that the nested triangles graph must have edge crossings whenever there are 4 or more levels of nesting. While this graph

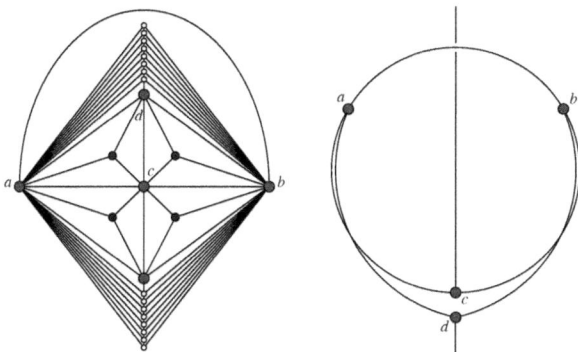

Fig. 6. Left: A planar 3-tree that has no planar Lombardi drawing. Right: For the K_4 subgraph defined by the four vertices a, b, c, and d, a drawing with the correct angles at each vertex will necessarily have crossings.

is 4-degenerate, even more constrained classes of planar graphs have no planar Lombardi drawings. Specifically, we can show that there exists a planar 3-tree that has no planar Lombardi realization. The planar 3-trees, also known as Apollonian networks, are the planar graphs that can be formed, starting from a triangle, by repeatedly adding a vertex within a triangular face, connected to the three triangle vertices, subdividing the face into three smaller triangles. These graphs have attracted much attention within the physics research community both as models of porous media with heterogeneous particle sizes and as models of social networks [2]. In addition, 3-trees are relevant for Lombardi drawings because they are examples of 3-degenerate graphs, which have nonplanar Lombardi drawings if vertex-vertex and vertex-edge overlaps are allowed.

Theorem 2. *There exists a planar 3-tree that has no planar Lombardi drawing.*

Proof. An example of a planar 3-tree that has no planar Lombardi drawing is given in Fig. 6; in the figure, sixteen small white vertices are shown, but our construction requires a sufficient number (which we do not specify precisely) in order to force the angle between arcs ad and ab to be arbitrarily close to π. The numbers of white vertices on the top and bottom of the figure should be equal. Because of this equality, the three arcs ab, bc, and ca split the graph into two isomorphic subgraphs, and due to this symmetry they must each meet at angle π, necessarily forming a circle in any Lombardi drawing. By performing a Möbius transformation on the drawing, we may assume without loss of generality that these three points form the vertices of an equilateral triangle inscribed within the circle, as shown in the right of the figure. Then, according to our previous analysis of 3-degenerate Lombardi graph drawing, there is a unique point in the plane at which vertex d may be located so that the arcs ad, bd, and cd form the correct angle of $2\pi/3$ to each other and the correct angles to the three previous arcs ab, bc, and ca. However, as shown on the right of the figure, that unique point lies outside circle abc and causes multiple edge crossings in the drawing.   □

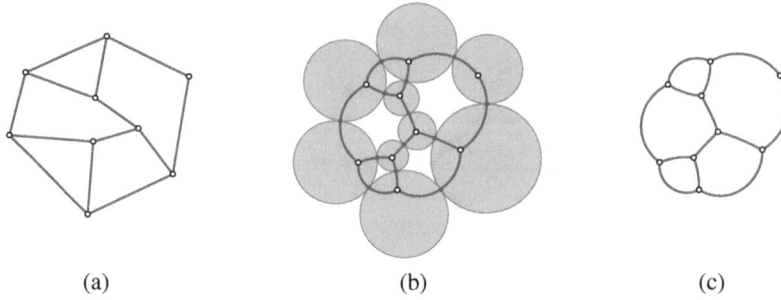

Fig. 7. (a) A planar graph of maximum degree 3. (b) A representation of the graph as tangent circles according to the Koebe–Andreev–Thurston theorem, together with arcs connecting each vertex perpendicularly to the disk tangency points. Layout generated using Ken Stephenson's CirclePack software. (c) The final smooth 2-Lombardi drawing.

3.2 Smooth Planar 2-Lombardi Drawings of Planar Max-Degree-3 Graphs

Lemma 1. *Given a circle C and three points a, b, and c on it, there exists a point p inside C such that we can draw three edges from p to a, b, and c as circular arcs that are all perpendicular to C and meet inside p at angle $2\pi/3$.*

Proof. We can find a Möbius transform τ that maps the circle to itself, mapping a, b, and c to three points a', b' and c' that are $2\pi/3$ radians apart on the circle. For these three points, the three edges can be drawn as radii of the circle meeting at the center point p'. The inverse transformation to τ maps p' to p and maps these three radii to circular arcs with the desired property.   □

Theorem 3. *Every planar graph G of maximum degree 3 has a smooth planar 2-Lombardi drawing.*

Proof. We apply the Koebe–Andreev–Thurston theorem to create a representation of G as the intersection graph of tangent circles; see Fig. 7(b). Each circle has three contact points that will be the bend points of its incident edges. Applying Lemma 1 to the circles yields a vertex and half-edge drawing inside each disk. At each contact point two half-edges meet at angle π, resulting in a smooth planar 2-Lombardi drawing of G.   □

3.3 Pointed Planar 2-Lombardi Drawings of Planar Graphs

We now show that every planar graph allows a planar 2-Lombardi drawing with pointed joints. The approach is similar to the previous section, but the drawing method inside the disks is different. We need the following lemmas, which are illustrated in Fig. 8. We defer the proof of the first lemma to the full version of the paper [8].

Lemma 2. *Let C be a circle and P be a set of n points on C. Additionally suppose that the four integers n_1, n_2, n_3, n_4 sum up to n and satisfy the inequalities $\lfloor n/4 \rfloor \leq$*

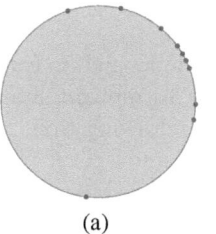 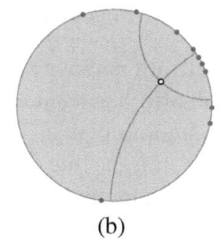 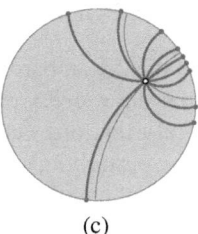

(a)   (b)   (c)

Fig. 8. (a) A disk with a set of connection points on its boundary. (b) A placement for the vertex in the disk that divides the connection points into four quadrants. (c) The actual connections are not fixed and guaranteed not to intersect.

$n_i \leq \lceil n/4 \rceil$ and $\lfloor n/2 \rfloor \leq n_i + n_{(i+1) \bmod 4} \leq \lceil n/2 \rceil$. Then there exist two circles A and B disjoint from P such that A, B, and C are pairwise perpendicular and such that A and B subdivide P into four sets of cardinalities n_1, n_2, n_3 and n_4.

Lemma 3. *Given a circle C and a set P of n points on C, there exists a point p in C such that we can draw n edges from p to the points in P as circular arcs that lie completely inside C, do not cross each other, and meet in p at angle $2\pi/n$.*

Proof. Draw n ports around a point with equal angles, and draw two perpendicular lines through the point (not coinciding with any ports), and count the number of points in each quadrant. Let these numbers be $n_1, \ldots, n_4$ and find two circles A and B as in Lemma 2. Then we place p at their intersection point inside C. Now orient the ports at p such that each quadrant has the correct number of ports.

Within any quadrant, there is a circular arc tangent to C at the point where it is crossed by B and tangent to A at point p; this can be seen by using a Möbius transformation to transform A and B into a pair of perpendicular lines, after which the desired arc has half the radius of C. By the intermediate value theorem, there are two circular arcs from p to any point q on the boundary arc of the quadrant that remain entirely within the quadrant and are tangent to A and B respectively. By a second application of the intermediate value theorem, there is a unique circular arc that connects p to each connection point on the boundary of C such that the outgoing direction at p matches the port and such that the arc remains entirely within its quadrant.

Any two arcs lying in the same quadrant belong to two circles that cross at p and at one more point. Whether that second crossing point is inside or outside of the quadrant can be determined by the relative ordering of the two arcs at p and on the boundary of the quadrant. However, since the ordering of the ports and of the connection points is the same, none of the crossings of these circles are within C, so no two arcs cross.   □

Theorem 4. *Every planar graph has a pointed planar 2-Lombardi drawing.*

Proof. As before, we first obtain a touching-circles representation of graph G using the Koebe–Andreev–Thurston theorem. Each vertex v in G is represented by a circle C; place v together with arcs connecting it to the set of contact points on C using Lemma 3. The arcs meet up at the contact points to form (non-smooth) 2-Lombardi edges.   □

3.4   Smooth Planar 3-Lombardi Drawings of Planar Graphs

Although the 2-Lombardi planar realization above has non-smooth bends in each edge, as we now show, every planar graph also has a smooth planar 3-Lombardi drawing. It seems likely that one can obtain a smooth planar 3-Lombardi drawing from a planar graph G by perturbing each edge of a straight-line drawing of G into a curve formed by two very small circular arcs near each endpoint of the edge, connected to each other by a straight segment. However, the details of this construction are messy. An alternative construction is much simpler, once Theorem 4 is available.

Theorem 5. *Every planar graph has a smooth planar 3-Lombardi drawing.*

Proof. Find a pointed planar 2-Lombardi drawing by Theorem 4. For each pointed bend of the drawing formed by two circular arcs a_1 and a_2, replace the bend by a third circular arc tangent to both a_1 and a_2, with the two points of tangency close enough to the bend to avoid crossing any other edge.   □

4   Conclusions

We have proven several new results about the planarity of Lombardi drawings and about the classes of graphs that can be drawn as k-Lombardi drawings rather than as Lombardi drawings. However, several problems remain open, including the following:

1. Characterize the subclasses of planar graphs having Lombardi planar realizations and those having smooth 2-Lombardi planar realizations.
2. Bound the (change in) curvature of edge segments in k-Lombardi drawings.
3. Address area and resolution requirements for Lombardi drawings of graphs.

Acknowledgments. This research was supported in part by the National Science Foundation under grants CCF-0830403, CCF-0545743, and CCF-1115971, by the Office of Naval Research under MURI grant N00014-08-1-1015, and by the Louisiana Board of Regents through PKSFI Grant LEQSF (2007-12)-ENH-PKSFI-PRS-03.

References

1. Aichholzer, O., Aigner, W., Aurenhammer, F., Dobiášová, K.Č., Jüttler, B.: Arc triangulations. In: Proc. 26th Eur. Worksh. Comp. Geometry (EuroCG 2010), Dortmund, Germany, pp. 17–20 (2010)
2. Andrade Jr., J.S., Herrmann, H.J., Andrade, R.F.S., da Silva, L.R.: Apollonian networks: Simultaneously scale-free, small world, Euclidean, space filling, and with matching graphs. Physics Review Letters 94, 018702 (2005); arXiv:cond-mat/0406295
3. Brandes, U., Wagner, D.: Using graph layout to visualize train interconnection data. J. Graph Algorithms Appl. 4(3), 135–155 (2000),
   http://jgaa.info/accepted/00/BrandesWagner00.4.3.pdf
4. Cheng, C.C., Duncan, C.A., Goodrich, M.T., Kobourov, S.G.: Drawing planar graphs with circular arcs. Discrete Comput. Geom. 25(3), 405 (2001), doi:10.1007/s004540010080

5. Chernobelskiy, R., Cunningham, K., Goodrich, M.T., Kobourov, S.G., Trott, L.: Force-directed Lombardi-style graph drawing. In: van Kreveld, M., Speckmann, B. (eds.) GD 2011. LNCS, vol. 7034, pp. 320–331. Springer, Heidelberg (2011)
6. Di Battista, G., Vismara, L.: Angles of planar triangular graphs. SIAM J. Discrete Math. 9(3), 349–359 (1996), doi:10.1137/S0895480194264010
7. Dickerson, M., Eppstein, D., Goodrich, M.T., Meng, J.Y.: Confluent drawings: Visualizing non-planar diagrams in a planar way. J. Graph Algorithms Appl. 9(1), 31–52 (2005), http://jgaa.info/accepted/2005/Dickerson+2005.9.1.pdf
8. Duncan, C.A., Eppstein, D., Goodrich, M.T., Kobourov, S.G., Löffler, M.: Planar and poly-arc Lombardi drawings. ArXiv e-prints abs/1109.0345, arXiv:1109.0345 (September 2011)
9. Duncan, C.A., Eppstein, D., Goodrich, M.T., Kobourov, S.G., Nöllenburg, M.: Drawing Trees With Perfect Angular Resolution and Polynomial Area. In: Brandes, U., Cornelsen, S. (eds.) GD 2010. LNCS, vol. 6502, pp. 183–194. Springer, Heidelberg (2011), doi:10.1007/978-3-642-18469-7_17; arXiv:1009.0581
10. Duncan, C.A., Eppstein, D., Goodrich, M.T., Kobourov, S.G., Nöllenburg, M.: Lombardi Drawings of Graphs. In: Brandes, U., Cornelsen, S. (eds.) GD 2010. LNCS, vol. 6502, pp. 195–207. Springer, Heidelberg (2011), doi:10.1007/978-3-642-18469-7_18; arXiv:1009.0579
11. Eppstein, D., Goodrich, M.T., Meng, J.Y.: Delta-Confluent Drawings. In: Healy, P., Nikolov, N.S. (eds.) GD 2005. LNCS, vol. 3843, pp. 165–176. Springer, Heidelberg (2006), doi:10.1007/11618058_16; arXiv:cs/0510024v1
12. Eppstein, D., Goodrich, M.T., Meng, J.Y.: Confluent layered drawings. Algorithmica 47(4), 439–452 (2007), doi:10.1007/s00453-006-0159-8
13. Eppstein, D., Löffler, M., Mumford, E., Nöllenburg, M.: Optimal 3D Angular Resolution for Low-Degree Graphs. In: Brandes, U., Cornelsen, S. (eds.) GD 2010. LNCS, vol. 6502, pp. 208–219. Springer, Heidelberg (2011), doi:10.1007/978-3-642-18469-7_19; arXiv:1009.0045
14. Finkel, B., Tamassia, R.: Curvilinear Graph Drawing Using the Force-Directed Method. In: Pach, J. (ed.) GD 2004. LNCS, vol. 3383, pp. 448–453. Springer, Heidelberg (2005), doi:10.1007/978-3-540-31843-9_46
15. Garg, A., Tamassia, R.: Planar Drawings and Angular Resolution: Algorithms and Bounds. In: van Leeuwen, J. (ed.) ESA 1994. LNCS, vol. 855, pp. 12–23. Springer, Heidelberg (1994), doi:10.1007/BFb0049393
16. Goodrich, M.T., Wagner, C.G.: A framework for drawing planar graphs with curves and polylines. Journal of Algorithms 37(2), 399–421 (2000), doi:10.1006/jagm.2000.1115
17. Gutwenger, C., Mutzel, P.: Planar Polyline Drawings With Good Angular Resolution. In: Whitesides, S.H. (ed.) GD 1998. LNCS, vol. 1547, pp. 167–182. Springer, Heidelberg (1999), doi:10.1007/3-540-37623-2_13
18. Hirsch, M., Meijer, H., Rappaport, D.: Biclique Edge Cover Graphs and Confluent Drawings. In: Kaufmann, M., Wagner, D. (eds.) GD 2006. LNCS, vol. 4372, pp. 405–416. Springer, Heidelberg (2007), doi:10.1007/978-3-540-70904-6_39
19. Holten, D., van Wijk, J.J.: Force-directed edge bundling for graph visualization. Computer Graphics Forum 28, 983–990 (2009), doi:10.1111/j.1467-8659.2009.01450.x
20. Kant, G.: Drawing planar graphs using the canonical ordering. Algorithmica 16, 4–32 (1996), doi:10.1007/BF02086606
21. Lombardi, M., Hobbs, R.: Mark Lombardi: Global Networks. Independent Curators (2003)
22. Malitz, S., Papakostas, A.: On the angular resolution of planar graphs. SIAM J. Discrete Math. 7(2), 172–183 (1994), doi:10.1137/S0895480193242931
23. Matsakis, N.: Transforming a random graph drawing into a Lombardi drawing. ArXiv ePrints abs/1012.2202, arXiv:1012.2202 (December 2010)

Force-Directed Lombardi-Style Graph Drawing⋆

Roman Chernobelskiy[1], Kathryn I. Cunningham[1], Michael T. Goodrich[2],
Stephen G. Kobourov[1], and Lowell Trott[2]

[1] Department of Computer Science, University of Arizona, Tucson, AZ, USA
[2] Department of Computer Science, University of California, Irvine, CA, USA

Abstract. A *Lombardi drawing* of a graph is one in which vertices are represented as points, edges are represented as circular arcs between their endpoints, and every vertex has perfect angular resolution (equal angles between consecutive edges, as measured by the tangents to the circular arcs at the vertex). We describe two algorithms that create "Lombardi-style" drawings (which we also call *near-Lombardi* drawings), in which all edges are still circular arcs, but some vertices may not have perfect angular resolution. Both of these algorithms take a force-directed, spring-embedding approach, with one using forces at edge tangents to produce curved edges and the other using dummy vertices on edges for this purpose. As we show, these approaches produce near-Lombardi drawings, with one being slightly better at achieving near-perfect angular resolution and the other being slightly better at balancing edge placements.

1 Introduction

The American artist, Mark Lombardi, was known for his drawings of social networks of conspiracy theories, which use circular arcs for edges and have a nice aesthetic placement for both vertices and edges (e.g., see Figure 1).

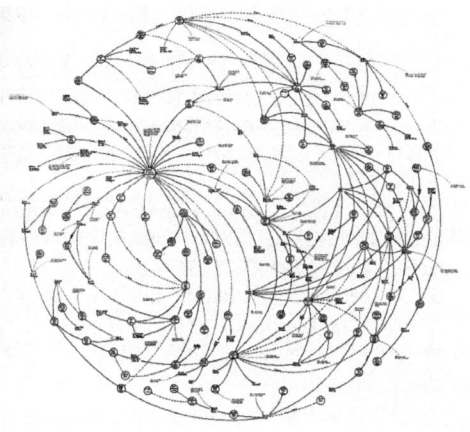

Fig. 1. Mark Lombardi's WFC 1970-84 [24]

⋆ Research funded in part by NSF grants CCF-0545743 and CCF-1115971.

M. van Kreveld and B. Speckmann (Eds.): GD 2011, LNCS 7034, pp. 320–331, 2012.

Inspired by Lombardi's work, Duncan *et al.* [11, 12] introduce the concept of a *Lombardi drawing*, which is a drawing that uses circular arcs for edges and achieves the maximum (i.e., *perfect*) amount of angular resolution possible at each vertex. Their methods are deterministic and not force-directed, but, as they show, there exist graphs that do not have perfect Lombardi drawings. These negative results motivate a relaxation of the requirement that drawings achieve perfect angular resolution at every vertex.

At the same time, experimental studies have shown that angular resolution has a significant impact on the readability of a graph [27, 28]. Thus, our goal in this paper is to study the degree to which one can achieve good angular resolution at vertices by using the Lombardi-inspired approach of embedding edges as circular arcs.

Force-directed layout algorithms, also known as "spring embedders," are well-known for the "organic" type of drawings they produce, in terms of vertex and edge placement, using straight-line edges (e.g., see [7, 16–18]). Still, straight-line segments rarely occur in nature; hence, it is not clear that humans prefer straight-line segments for the sake of graph readability. With this in mind, we consider force-directed graph-drawing algorithms that allow for circular-arc edges and include forces that tend to spread those edges more evenly around vertices. We feel this approach can result in drawings that appear more "alive" than can be achieved using straight-line edges; see Fig. 2.

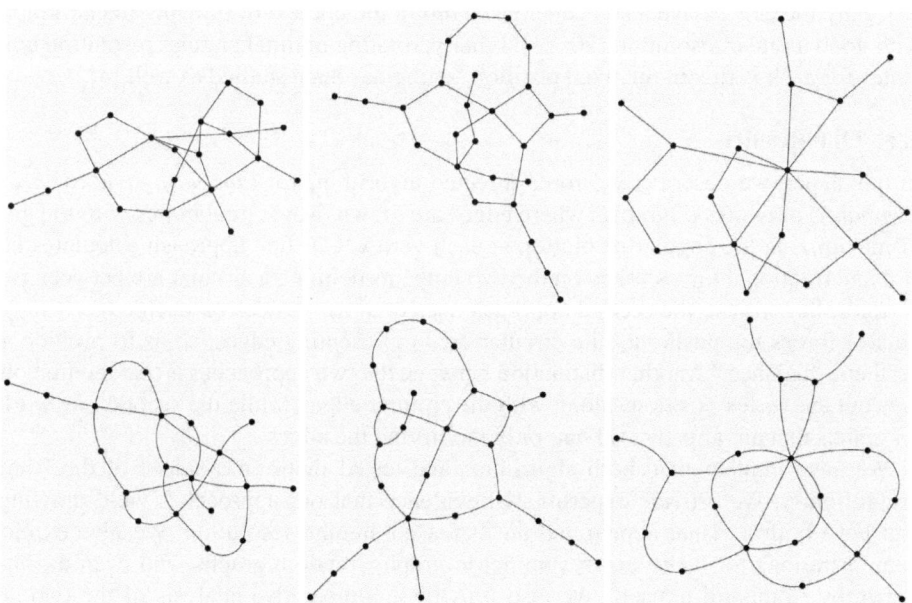

Fig. 2. Examples of standard straight-line and Lombardi-style drawings

1.1   Related Work

There are several graph drawing methods that use circular-arc edges or curvilinear poly-edges. For example, Goodrich and Wagner [20] give algorithms for drawing planar graphs using Bézier splines for edges, and Cheng *et al.* [6] describe a scheme for drawing graphs using circular arc poly-edges. *Confluent drawings* [9, 14, 22], bundle edges together in smooth curves so as to reduce crossings.

There is also a great deal of prior work on force-directed graph drawing and we refer the reader to some excellent surveys (e.g., see [1, 2, 7]). Holten and van Wijk [23] give a force-directed method for producing an edge-bundled drawing that is similar to a confluent drawing. Brandes and Wagner [5] describe a force-directed method for drawing train connections, where the vertex positions are fixed but transitive edges are drawn as Bézier curves (see also [3]). Finkel and Tamassia [15], on the other hand, describe a force-directed method for drawing graphs using curvilinear edges where vertex positions are free to move. Their method is based on adding dummy vertices, as one of our methods does, but their dummy vertices serve as control points for Bézier curves, rather than circular arcs, and their drawings do not achieve locally-optimal edge resolution at the vertices. Matsakis [26] describes a force-directed approach to producing a Lombardi-style drawing, which is based on iteratively visiting each vertex v and making adjustments locally with respect to v. Unfortunately, he does not evaluate his method experimentally and it is not clear that it always converges.

Angular resolution in the straight-line setting is also a well-studied problem [8, 19, 25]. Polyline edges have also been considered in the context of drawing planar graphs with good angular resolution [20, 21]. Finally, rotating optimal angular resolution templates for each vertex in the fixed position setting has been studied as well [4].

1.2   Our Results

In this paper, we describe two force-directed algorithms for *Lombardi-style* (or *near-Lombardi*) drawings of graphs, where edges are drawn using circular arcs with the goal of maximizing the angular resolution at each vertex. Our first approach calculates lateral and rotational forces based on the two tangents defining a circular arc between two vertices. In contrast, the second approach uses dummy vertices on each edge with repulsive forces to "push out" the circular arcs representing edges, so as to provide an aesthetic "balance." Another distinction between the two approaches is that the first one lays out the vertex positions along with the circular edges, while the second one works on graphs that are already laid out, only modifying the edges.

We have implemented both algorithms and tested them on a subset of the Rome graph library. We provide experimental evidence that our approaches yield drawings that have both a visual appeal and an increased angular resolution. We give explicit demonstrations for well-known symmetric graphs, random graphs, and even a graph drawn by Lombardi himself. We also provide a comparative analysis of the two approaches, which suggests that the tangent-based method is in general better at achieving the highest angular resolution possible, while the dummy-vertex approach is better in general at balancing the placement of edges.

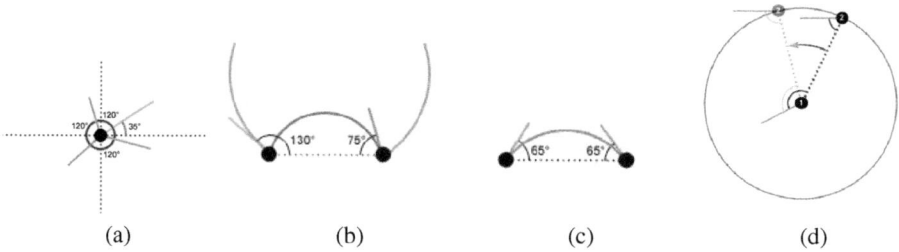

Fig. 3. Lombardi forces move and rotate vertices so that all corresponding tangents have matching angles, allowing for feasible circular arcs. (a) An illustration of pre-assigned tangents for a given degree-3 vertex. The angles between the tangents are equal and remain fixed, while the vertex itself can be rotated by changing its orientation with respect to the origin (currently 35° as indicated by the green line); (b) If the angles differ, then there cannot be a circular arc tangential to both tangents; (c) If the angles are equal, there is a unique circular arc; (d) The tangential force for vertex 2 with respect to vertex 1 moves vertex 2 to make the tangent angles equal.

2   A Tangent-Based Lombardi Spring Embedder Formulation

Force-directed algorithms treat a given graph as a physical system, where the vertices represent points in an N-body mechanics problem. In the system set up by Eades [13], vertices are treated as steel rings and the edges are springs that obey Hooke's Law. Fruchterman and Reingold describe a model in which a strong nuclear force attracts two protons within the atomic nucleus at close range, while an electrical force repels them at farther range [16]. Although inspired by physics, most force-directed algorithms do not attempt to mimic physical laws precisely.

Similar to most force-directed layout algorithms, our tangent-based Lombardi spring embedder assigns a force to each vertex and aims to minimize the overall energy of the system. There are three forces which affect vertex position, and one force which affects the radius of the circular arcs between a pair of vertices connected by an edge.

The attractive force, F_a, pulls vertices connected by edges closer together. It is applied to every pair of vertices connected by an edge as follows: $F_a = (d - k)/d$, where d is the current distance between the two vertices and k is a constant representing the ideal distance between them. The repulsive force, F_r, pushes vertices apart. It is applied to every pair of vertices using the following formula: $F_r = k^2/d^3$.

The tangential and rotational forces make it possible for circular arcs to be drawn between vertices, while maintaining a perfect (or near-perfect) angular resolution. To help compute the two forces, we augment each vertex with an orientation and fixed tangents, which dictate how to draw the arcs. The angles between the tangents are equal and remain fixed, while the vertex itself can be rotated by changing its orientation with respect to the origin. Note that the angle of a tangent at one vertex must equal the angle of a tangent at the other vertex for an arc to be possible between them. Here, angles are measured with respect to the segment connecting two vertices; see Fig. 3.

The tangential force, F_t, attempts to move vertices so as to make a circular arc possible between any pair of vertices connected by an edge. To compute this force we need to find the optimal position of a vertex with respect to its neighbor. The magnitude of

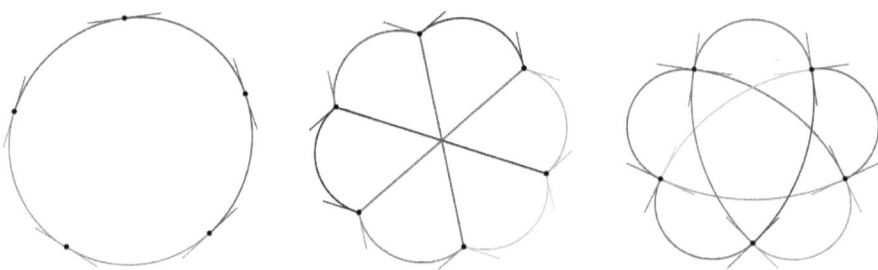

Fig. 4. Perfect Lombardi drawings of C_5, $K_{3,3}$ and K_5, with shown tangents

the force is proportional to the distance between this optimal position and the current position, and the direction is straight towards the optimal position. It is applied to every pair of vertices that share an edge as follows: $F_t = A \times d$, where d is the distance between current and optimal positions, and A is the tangential force constant.

The rotational force, F_ρ, does not attempt to move a vertex, but to rotate a vertex and its tangent template so as to make the tangent angles match, thereby making the arc between two vertices possible. To compute the rotational force we find the optimal angle of a tangent and subtract the current angle as follows:$F_\rho = B \times \Delta angle$, where $\Delta angle$ is the rotation required and B is a small constant. For each vertex v, the three appropriately scaled movement forces are added together to the rotational force in order to determine the overall force acting on the vertex: $F(v) = F_a + F_r + F_t + F_\rho$.

The following cooling function is used to determine the magnitude of the force in terms of the number of iterations: $T(i) = (T_0 * (M - i))/M$, where i is the iteration number and M is the maximum number of iterations (adapted from graphviz). T_0 is calculated as follows: $T_0 = K * \sqrt{n}/5$, where K is the ideal spring length and n is the number of vertices. Experimentally determined values for the constants we use are $K = 0.3$, $M = 600$, $A = 0.9$, $B = 0.5$. For many small graphs the algorithm succeeds in computing perfect Lombardi drawings; see Fig. 4.

Note that as described, the algorithm operates on a fixed ordering of the tangents around each vertex. This strict order hinders our algorithm as it attempts to find a Lombardi drawing. To give our algorithm more flexibility, we add a "shuffling" method that can modify the relative order of tangents around a vertex. This rearrangement occurs at the beginning of each iteration, and tangents are reordered only if a lower energy state is found within a reasonable number of calculations. In this case, the energy we seek to minimize is the total amount of rotation generated by all tangents to match the angle of their counterpart across their edges. This is the sum of the absolute value of the rotation generated by each tangent, rather than the net sum of the rotations, as we calculated when finding the rotational force. If the degree of a vertex is small, we calculate the total amount of rotation for every permutation of tangent orderings to find the one with the least energy. If the number of tangents is large, we test a subset of tangent orderings created by all possible pairwise swaps.

2.1   A Tangent-Based Near-Lombardi Spring Embedder

As not all graphs are Lombardi graphs [10], and our algorithm cannot guarantee that it will find a Lombardi drawing even if one does exist, when needed we relax the perfect

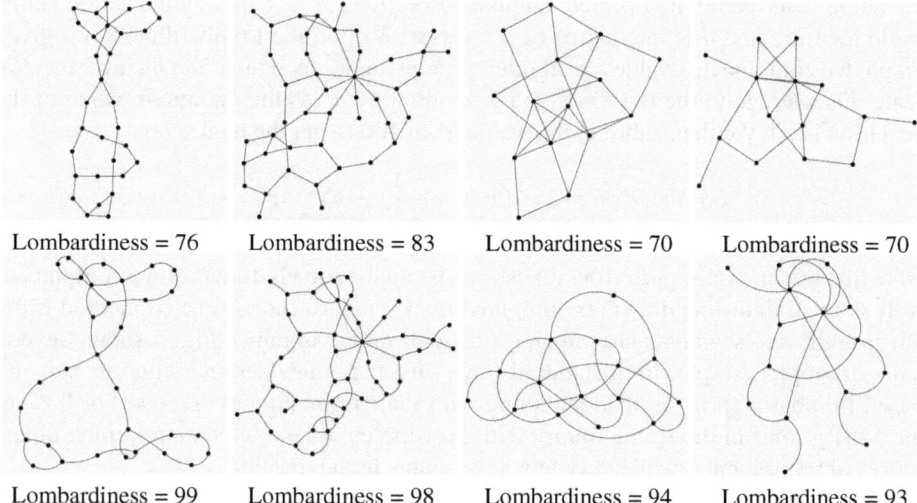

Fig. 5. Standard force-directed drawings (above) and near-Lombardi drawings (below)

angular resolution constraint. If the above tangent-based Lombardi Spring Embedder has failed to find optimal positions for every vertex, we modify the tangents of vertices which have infeasible edge constraints.

For tangents of adjacent vertices that have unequal angles, we move each tangent to the average of both tangents' positions. Now we can draw circular arcs between all connected vertices with minimal loss of angular resolution. As this change in tangent angles may drastically affect the angular resolution, we improve the resolution with another round of force-directed simulation. Rather than rotate or change the position of vertices as before, here we only modify angles between adjacent tangents (which were fixed before). This new force is based on the observation that, ideally, a tangent should bisect the angle formed by the two tangents immediately clockwise and counterclockwise from itself. We find this midpoint and compute the force proportional to the required rotation to move the tangent to this location: $F_{NL} = C \times \Delta angle$, where $\Delta angle$ is the difference between the current angle of the tangent and the angle that bisects the neighboring tangents, and C is a constant. In order to maintain equal tangent angles for both tangents across an edge, which is necessary to draw a perfectly circular arc, we compute this force for both of the tangents incident to the edge, average it, and then apply it equally to both.

2.2 Lombardi Metric

For near-Lombardi drawings we need a measure of quality. As edges produced by the algorithm are always perfect circular arcs, the only violations of the Lombardi criteria are at vertices where perfect angular resolution could not be achieved. With this in mind, we define the *Lombardiness* of a drawing to be a number in the range 0 to 100, based on the average deviation from perfect angular resolution across all inter-tangent angles. This deviation is the difference between the actual angle measure and the measure of the angle if its vertex had perfect angular resolution: $|a - \frac{2\pi}{d}|$, where a is the actual angle measure and d is the degree of a's vertex. To find the Lombardiness of a given graph, we compute this value for all inter-tangent angles, and then find the average. We scale this average to the 0-100 range by dividing by π (as the maximum value of the deviation is π). We then subtract this value from 100 to get the final score:

$$Lombardiness = 100 - \frac{1}{\pi \times 2|E|} \sum_{a \in A} |a - \frac{2\pi}{d}|.$$

Note that this measure of Lombardiness can be applied to all drawings we compute, as well as to straight-line drawings computed by a standard force-directed method (after all, straight-line segments are circular arcs with radius infinity). Fig. 5 shows several pairs of graphs drawn with a standard force directed embedder and with our tangent-based Lombardi spring embedder, along with their Lombardiness scores. For 80% of the 5451 graphs in the Rome library with 50 vertices or less, we obtain Lombardiness scores of 98 or higher, while very few have scores in the low 90s.

A web-enabled demo, as well as complete python source code, image libraries, and several movies illustrating this tangent-based algorithm at work can be found at `http://lombardi.cs.arizona.edu`.

3   A Dummy-Vertex Approach to Lombardi-Style Drawings

Brandenburg *et al.* have experimentally shown [1] that different force-directed methods can produce results with various trade-offs for aesthetic criteria. Thus, to allow for freedom with respect to these criteria, we built a second Lombardi-style force-directed method that allows for choice in the underlying force-directed algorithm. This method relies on a simple two-step process so as to allow an augmentation that can be applied to existing force-directed approaches. The first step involves using an existing straight-line force-directed method to place vertices and fix the order of edges around them, and the second step applies a force-directed approach based on the use of dummy vertices to maximize angular resolution at the vertices through the use of circular-arc edges.

Once the nodes have been placed with a user-selected force-directed method we begin our second phase. First, we assign to each edge an additional "dummy" vertex that is placed at the midpoint of that edge. Note that once the endpoints of an edge have been placed, only one more point is required to uniquely determine a circular arc

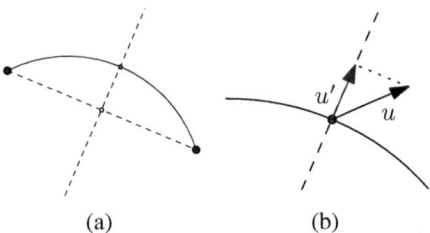

(a)   (b)

Fig. 6. (a) Points along the perpendicular bisector will determine an arc. (b) The update vector u' used will be the projection of the sum force vector u.

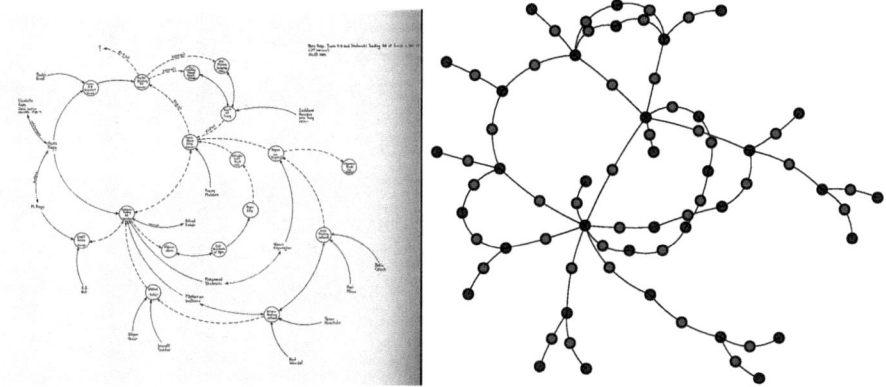

Fig. 7. Lombardi's Hans Kopp, Trans K-B and Shakarchi Trading [24], shown as rendered by Lombardi and as rendered by our dummy-vertex force-directed method

between these points. Thus, we can describe all possible arcs between nodes by the set of points along the perpendicular bisector of their straight-line connection; see Fig. 6(a).

The responsibility for moving an edge will be given to the additional node we have added to that edge. We then proceed to use the force-directed method to place these edge-nodes. Each (dummy) edge-node will consider the nodes that it connects as neighbors, and the partial edges as springs with a fractional resting length. Moreover, each edge node will repulse from all other nodes, both the original graph nodes and other edge-nodes. The sum force vector is calculated as before, but will be used to move the node in a modified way. If u is the sum force update vector we consider only its motion along the perpendicular bisector. This projection will determine a new update vector u' that we will use to move the edge node; see Fig. 6(b). Using u' we can determine the movement of the edge-node, while maintaining a circular arc edge. The edge-node positions are updated iteratively until an equilibrium is reached.

In Figure 7, we provide a scan of a drawing of Lombardi and the result of our method applied to the same graph. In Figure 8, we show the evolution of our algorithm through various substeps of the two phases.

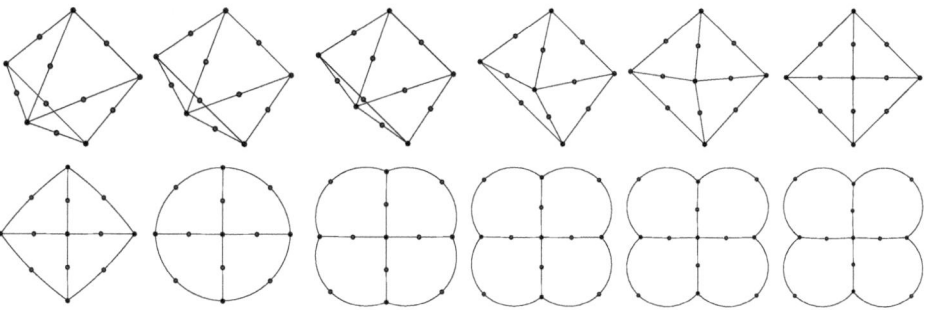

Fig. 8. A 5-node graph with center node initially displaced. Selected stages of the force-directed placement are shown. The top row shows phase 1 and the bottom phase 2.

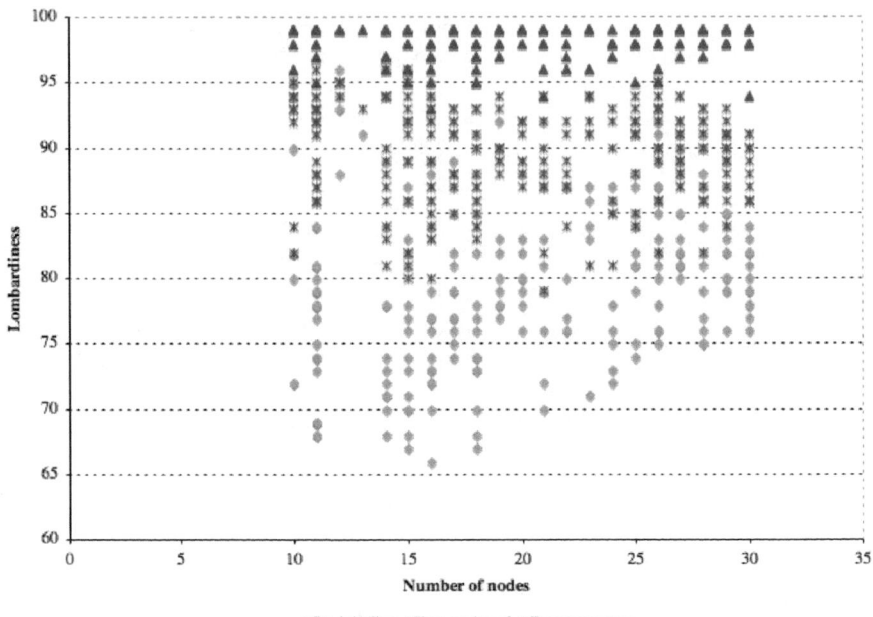

Number of nodes

◆ Straight-line ▲ Tangent-based ✳ Dummy-vertex

Fig. 9. A scatter plot of the Lombardiness of a collection of 250 graphs with straight-line, tangent-based, and dummy-vertex embeddings (with many data points overlapping)

4   A Comparative Analysis

In this section, we provide a small comparative analysis of our two methods. From the visual examples of drawings generated by the two methods it can already be seen that although both use circular arcs and aim to provide near-perfect angular resolution, the drawings seem to optimize different aesthetic qualities; see Fig. 10. For instance, in the tangent-based approach, the tangents of a node's edges have direct control over the angular resolution of that node, and this method does not take node positions as fixed. Thus, the tangent-based approach is able to achieve near-perfect angular resolution on all nodes. The dummy-vertex approach, on the other hand, starts from node positions determined by a straight-line force-directed method and moves edges into open space using dummy vertices. Since it does not directly consider the angle of other outgoing edges incident on the same vertex, it is not as successful in approaching perfect angular resolution. Nevertheless, it does improve angular resolution over straight-line drawings. To verify these observations, we performed an experimental analysis involving 250 graphs in the Rome library, and visualized their Lombardiness scores against their size; see Fig. 9. The data confirms this, showing a near-perfect separation between the three approaches (the third one given by the straight-line drawing).

The primary difficulty in drawing comparisons between these two methods is that they are inherently different, as evidenced by a visual comparison; see Fig. 10. While the tangent-based method arranges both vertices and edges, the dummy-vertex method focuses on edges alone. And since the underlying force-based layout algorithm used

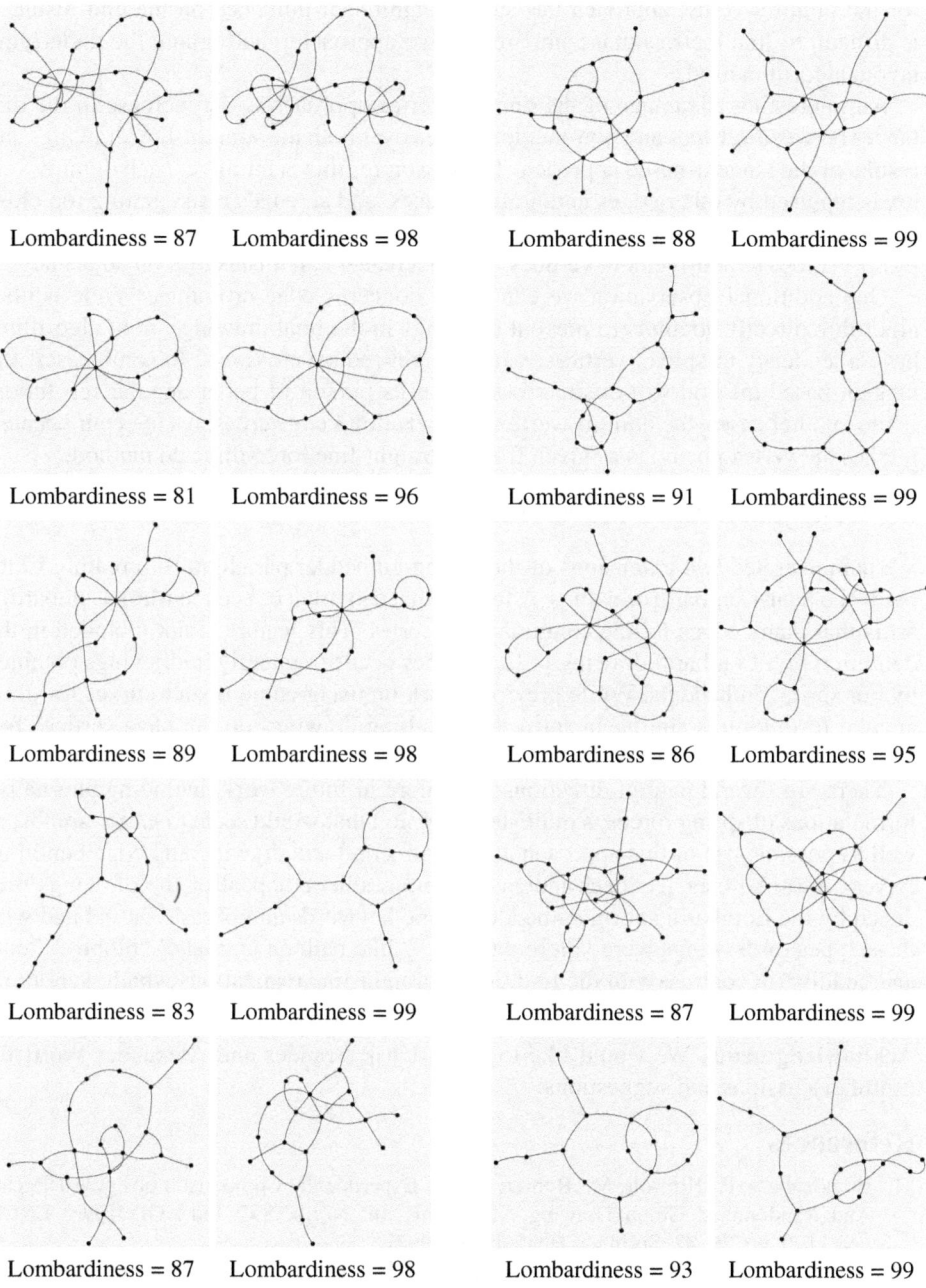

Lombardiness = 87   Lombardiness = 98   Lombardiness = 88   Lombardiness = 99

Lombardiness = 81   Lombardiness = 96   Lombardiness = 91   Lombardiness = 99

Lombardiness = 89   Lombardiness = 98   Lombardiness = 86   Lombardiness = 95

Lombardiness = 83   Lombardiness = 99   Lombardiness = 87   Lombardiness = 99

Lombardiness = 87   Lombardiness = 98   Lombardiness = 93   Lombardiness = 99

Fig. 10. Some example Lombardi-style drawings using the two force-directed approaches. For each pair, the drawing on the left was done using the dummy-vertex approach and the drawing on the right was done using the tangent-based approach. The Lombardiness score for each is given below.

for the dummy-vertex approach has such a significant influence on the end result, it is difficult to find metrics that compare the two approaches and ignore the underlying layout algorithm used.

An interesting advantage of the dummy-vertex approach is the increase in the distances between vertices and non-incident edges over both the straight-line drawings and results of the tangent-based approach. The reason for this is intuitive: each dummy vertex is repulsed by other edges and graph vertices, and so edges resist getting too close to other edges or non-incident vertices. This helps alleviate a distraction when edges pass too close to non-incident vertices, which a reader can mistake for an adjacency.

One additional observation we can make concerns edge crossings. While neither algorithm directly attempts to prevent crossings in the final drawings, both algorithms have a tendency to spread vertices, which might reduce crossings. In some cases, the tangent-based method will create crossings in its pursuit of better angular resolution, while in other cases, the dummy-vertex method allows edge crossings to occur because it takes the vertex positions as given from a straight-line force-directed method.

5   Conclusion and Future Work

We demonstrated two extensions of the spring-embedder paradigm for creating Lombardi and near-Lombardi drawings. A feature that can often be seen in Mark Lombardi's art is that many edges follow common trajectories. This feature is not included in the definition of a Lombardi drawing [12], but does occur frequently in drawings obtained by our spring embedders. While previous work on using cubic Bézier curves for good angular resolution is similar in spirit, the resulting drawings do not have vertices following common trajectories.

There are several natural directions to explore in future work, including alternative formulations of spring forces, a multi-level version that would scale to larger graphs, as well as possible use of this approach along with confluent drawing and edge bundling. A very informal user feedback indicates some aesthetic appeal of the drawings produced by the Lombardi spring embedder. Some keywords and phrases associated with these types of drawings were "more natural, " "like balloon animals," "blobby," "cute and cuddly," in contrast with the traditional straight-line realizations which were more "jagged" and "angular."

Acknowledgments. We would like to thank Ulrik Brandes and Alexander Wolff for useful discussions and suggestions.

References

1. Brandenburg, F., Himsolt, M., Rohrer, C.: An Experimental Comparison of Force-Directed And Randomized Graph Drawing Algorithms. In: North, S.C. (ed.) GD 1996. LNCS, vol. 1190, pp. 76–87. Springer, Heidelberg (1997)
2. Brandes, U.: Drawing on Physical Analogies. In: Kaufmann, M., Wagner, D. (eds.) Drawing Graphs. LNCS, vol. 2025, pp. 71–86. Springer, Heidelberg (2001)
3. Brandes, U., Schlieper, B.: Angle and Distance Constraints On Tree Drawings. In: Kaufmann, M., Wagner, D. (eds.) GD 2006. LNCS, vol. 4372, pp. 54–65. Springer, Heidelberg (2007)
4. Brandes, U., Shubina, G., Tamassia, R.: Improving angular resolution in visualizations of geographic networks. In: 2nd TCVG Symp. Visualization, pp. 23–32 (2000)

5. Brandes, U., Wagner, D.: Using Graph Layout to Visualize Train Interconnection Data. J. Graph Algorithms Appl. 4(3), 135–155 (2000)
6. Cheng, C.C., Duncan, C.A., Goodrich, M.T., Kobourov, S.G.: Drawing planar graphs with circular arcs. Discrete Comput. Geom. 25(3), 405–418 (2001)
7. Di Battista, G., Eades, P., Tamassia, R., Tollis, I.G.: Graph Drawing: Algorithms for the Visualization of Graphs. Prentice Hall PTR, Upper Saddle River (1998)
8. Di Battista, G., Vismara, L.: Angles of planar triangular graphs. SIAM J. Discrete Math. 9(3), 349–359 (1996)
9. Dickerson, M., Eppstein, D., Goodrich, M.T., Meng, J.Y.: Confluent drawings: Visualizing non-planar diagrams in a planar way. J. Graph Algorithms Appl. 9(1), 31–52 (2005)
10. Duncan, C.A., Eppstein, D., Goodrich, M.T., Kobourov, S.G., Löffler, M.: Planar and Poly-Arc Lombardi Drawings. In: van Kreveld, M., Speckmann, B. (eds.) GD 2011. LNCS, vol. 7034, pp. 308–319. Springer, Heidelberg (2011)
11. Duncan, C.A., Eppstein, D., Goodrich, M.T., Kobourov, S.G., Nöllenburg, M.: Drawing Trees with Perfect Angular Resolution and Polynomial Area. In: Brandes, U., Cornelsen, S. (eds.) GD 2010. LNCS, vol. 6502, pp. 183–194. Springer, Heidelberg (2011)
12. Duncan, C.A., Eppstein, D., Goodrich, M.T., Kobourov, S.G., Nöllenburg, M.: Lombardi Drawings of Graphs. In: Brandes, U., Cornelsen, S. (eds.) GD 2010. LNCS, vol. 6502, pp. 195–207. Springer, Heidelberg (2011)
13. Eades, P.: A heuristic for graph drawing. Congressus Numerantium 42, 149–160 (1984)
14. Eppstein, D., Goodrich, M.T., Meng, J.Y.: Confluent layered drawings. Algorithmica 47(4), 439–452 (2007)
15. Finkel, B., Tamassia, R.: Curvilinear Graph Drawing Using the Force-Directed Method. In: Pach, J. (ed.) GD 2004. LNCS, vol. 3383, pp. 448–453. Springer, Heidelberg (2005)
16. Fruchterman, T., Reingold, E.: Graph drawing by force-directed placement. Softw. – Pract. Exp. 21(11), 1129–1164 (1991)
17. Gajer, P., Goodrich, M.T., Kobourov, S.G.: A multi-dimensional approach to force-directed layouts of large graphs. Comp. Geometry: Theory and Applications 29(1), 3–18 (2004)
18. Gajer, P., Kobourov, S.G.: GRIP: Graph dRawing with Intelligent Placement. Journal of Graph Algorithms and Applications 6(3), 203–224 (2002)
19. Garg, A., Tamassia, R.: Planar drawings and angular resolution: algorithms and bounds. In: 2nd European Symposium on Algorithms, London, UK, pp. 12–23 (1994)
20. Goodrich, M.T., Wagner, C.G.: A framework for drawing planar graphs with curves and polylines. J. Algorithms 37(2), 399–421 (2000)
21. Gutwenger, C., Mutzel, P.: Planar Polyline Drawings With Good Angular Resolution. In: Whitesides, S.H. (ed.) GD 1998. LNCS, vol. 1547, pp. 167–182. Springer, Heidelberg (1999)
22. Hirsch, M., Meijer, H., Rappaport, D.: Biclique Edge Cover Graphs and Confluent Drawings. In: Kaufmann, M., Wagner, D. (eds.) GD 2006. LNCS, vol. 4372, pp. 405–416. Springer, Heidelberg (2007)
23. Holten, D., van Wijk, J.J.: Force-directed edge bundling for graph visualization. Computer Graphics Forum 28, 983–990 (2009)
24. Lombardi, M., Hobbs, R.: Mark Lombardi: Global Networks. Independent Curators (2003)
25. Malitz, S., Papakostas, A.: On the angular resolution of planar graphs. SIAM J. Discrete Math. 7(2), 172–183 (1994)
26. Matsakis, N.: Transforming a random graph drawing into a Lombardi drawing. arXiv ePrints, abs/1012.2202 (2010)
27. Purchase, H.: Which Aesthetic Has The Greatest Effect On Human Understanding? In: DiBattista, G. (ed.) GD 1997. LNCS, vol. 1353, pp. 248–261. Springer, Heidelberg (1997)
28. Purchase, H.C., Cohen, R.F., James, M.: Validating Graph Drawing Aesthetics. In: North, S.C. (ed.) GD 1996. LNCS, vol. 1190, pp. 435–446. Springer, Heidelberg (1997)

Every Graph Admits
an Unambiguous Bold Drawing

János Pach[*]

EPFL, Lausanne and Rényi Institute, Budapest
pach@cims.nyu.edu

Abstract. Let r and w be a fixed positive numbers, $w < r$. In a *bold drawing* of a graph, every vertex is represented by a disk of radius r, and every edge by a narrow rectangle of width w. We solve a problem of van Kreveld [K09] by showing that every graph admits a bold drawing in which the region occupied by the union of the disks and rectangles representing the vertices and edges does not contain any disk of radius r other than the ones representing the vertices.

1   Introduction

In this note, we adopt a "realistic" view of graph drawing, proposed by Marc van Kreveld [K09]. Let G be a graph with vertices $v_1, \ldots, v_n$, represented by points in the plane, and let the edges be drawn as possibly crossing straight-line segments. Now fix two positive numbers r and w, $w < 2r$, and replace each vertex by a disk of radius r centered at v_i, and each edge $v_i v_j$ by a rectangle such that its midsegment is $v_i v_j$ and its width, the length of its side perpendicular to $v_i v_j$, is w. We call the union D of these disks and rectangles a *bold drawing* of G. A bold drawing is said to be *unambiguous* if it satisfies the following two conditions.

1. No two disks representing vertices of G intersect.
2. The set D contains no disk of radius r other than the disks representing its vertices.

The first condition is equivalent to saying that $2r$ is smaller than the minimum distance between two points v_i and v_j. It follows from the second condition that a bold drawing of a graph which has at least one edge can be unambiguous only if $w < 2r$. It was shown in [K09] that if $w > r$, then the maximum degree of the vertices of all graphs that admit an unambiguous bold drawing is bounded from above by a constant depending only on w and r. On the other hand, van Kreveld proved that for $w < r$, any *star* consisting of a central vertex connected to an arbitrary number of other vertices admits an unambiguous bold drawing.

[*] Supported by NSF Grant CCF-08-30272, by NSA, by OTKA under EUROGIGA project GraDR 10-EuroGIGA-OP-003, and by Swiss National Science Foundation Grant 200021-125287/1.

M. van Kreveld and B. Speckmann (Eds.): GD 2011, LNCS 7034, pp. 332–342, 2012.

He also raised the question whether there exists a fixed pair of values w, r with $w < r$ such that with these parameters every finite graph admits an unambiguous bold drawing. The aim of this note is to answer this question in the affirmative in the following strong sense.

Theorem 1. *Let w and r be any positive constants with $w < r$. Then, for every positive integer n, the complete graph K_n admits an unambiguous bold drawing, in which the vertices are represented by disks of radius r and the edges by rectangles of width w.*

In the next statement, we describe our construction in full detail.

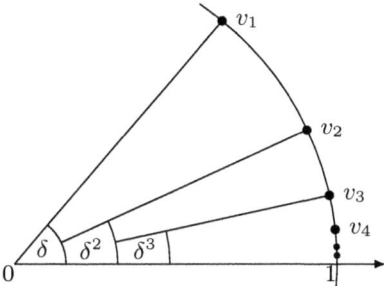

Fig. 1. Construction for Theorem 2

Theorem 2. *Let w and r be any positive constants with $w < r$. Let C be a circle of radius 1 around the origin, and let v_i $(1 \leq i \leq n)$ denote the intersection point of C and the ray obtained from the positive x-axis by a counterclockwise rotation through angle δ^i, where $\delta = \min(\frac{1}{2}, 1 - \frac{w}{r})$.*

For every n, there exists a sufficiently small $\varepsilon = \varepsilon(n) > 0$ such that replacing each v_i by a disk of radius εr centered at v_i and each edge $v_i v_j$ by a rectangle of width εw with midsegment $v_i v_j$, the union of these disks and rectangles contains no disk of radius εr other than the ones representing the vertices.

Theorem 2 immediately implies Theorem 1. Indeed, if we choose $\varepsilon(n) > 0$ so small that in addition to the property in Theorem 2, it satisfies the inequality $2\varepsilon r < \min_{1 \leq i < j \leq n} |v_i v_j| = |v_{n-1} v_n|$, and we blow up the drawing described in Theorem 2 by a factor of $1/\varepsilon$, then we obtain a bold drawing of K_n that meets both requirements for unambiguity stated above.

In [K09], van Kreveld listed seven properties that a "good" bold drawing of a graph G must satisfy. These include the two conditions for unambiguous drawings stated above, so that every good bold drawing of G is also unambiguous. It is easy to see that if we choose the constant $\varepsilon(n)$ small enough, then our drawing of K_n will also meet the five additional properties formulated in [K09].

Before turning to the proof, we would like to argue that in some sense we are "forced" to consider constructions of the type described in Theorem 2. We say that a set of points in the plane is in *general position* if no three of them are collinear. According to the Erdős-Szekeres theorem [ES35], for any integer K, every sufficiently large set of points in general position in the plane contains K elements that form the vertex set of a convex K-gon. This readily implies, that for any K there exists $N(K)$ such that any set of $N(K)$ points in general position has K elements that lie on a convex curve whose total turning angle is small. By rotating the coordinate axes if necessary, the coordinates of these points can be written as $(x_i, f(x_i))$, where $x_1 < x_2 < \ldots < x_K$ and $f(x)$ is a smooth convex function whose derivative is bounded by a small constant. Let $\gamma = \frac{\sqrt{5}+1}{2} \approx 1.618$, the golden ratio. Color the triples (i, j, k), $1 \le i < j < k \le K$, with *red, blue,* or *green*, according to whether $\frac{x_k - x_j}{x_j - x_i}$ is at most γ^{-1}, belongs to the interval (γ^{-1}, γ), or is at least γ, respectively. According to Ramsey's theorem [R30, GRS90], for every $n \ge 4$ we can choose $K = K(n)$ so large that there is a sequence $1 \le i_1 < i_2 < \ldots < i_n \le K$ with the property that all triples determined by its members are of the same color. It is easy to check that there exists no sequence of length 4 such that all of its triples are blue. Therefore, we can assume that all triples determined by the sequence $1 \le i_1 < i_2 < \ldots < i_n \le K$ are red or all of them are green. In the first case the distances $x_{i+1} - x_i$ decrease, in the second one increase at least exponentially fast, as i grows ($1 \le i \le n$). Summarizing: for every $n \ge 4$, there is an integer N with the property that from any set of N points in general position in the plane we can select a sequence of length n which lies on an arc of a convex curve with small total turning angle and the distances between its consecutive elements decrease at least exponentially. (We can reverse the numbering of the elements, if necessary.) Suppose now that K_N admits an unambiguous bold drawing. Applying the last statement to the centers of the disks representing the vertices, we obtain an unambiguous bold subdrawing of a complete graph K_n such that the centers of the disks representing its vertices lie on a convex curve and the distances between them are fast decreasing. Our construction in Theorem 2 is motivated by this observation.

The proof of Theorem 2 is somewhat subtle. In Sect. 2, we introduce some definitions that simplify the presentation and we state two easy but useful lemmas that can be proved by direct computation. The heart of the proof lies in Lemma 5, stated and established in Sect. 3. After this preparation, the proof of Theorem 2 presented in Sect. 4 is rather straightforward.

Several graph drawing programs for straight-line drawing offer the option to draw the vertices and the edges bold (see, for example, NEATO [N04]). Some algorithmic aspects of bold drawing were addressed in [K09]. In particular, given a drawing of a graph G with possibly crossing straight-line edges, van Kreveld applied a line segment intersection algorithm [CE92], [CS89], [M88] to find the smallest w for which, if we draw the edges as closed rectangles of width w, we find three edges, not all incident to the same vertex, such that the corresponding rectangles have a point in common. Duncan, Efrat, Kobourov, and Wenk [DEKW06] presented an efficient algorithm to determine the largest w, for a given planar embedding of a graph G, such that G admits an equivalent drawing in which the edges are represented by nonoverlapping, not necessarily straight bold curves of width w.

2   Terminology and Two Preliminary Lemmas

In the rest of this note, w and r are fixed positive numbers with $w < r$. Throughout the next two sections, we also fix the parameter $\varepsilon > 0$, which will be varied only in Sect. 4, in the proof of Theorem 2.

First, we introduce some notation and terminology. Let v be a point of the plane, and let $R_1, \ldots, R_s$ be a set of infinite rays (half-lines) emanating from v, listed in clockwise order. Assume that all rays R_i point into the same half-plane bounded by a line passing through v. Replace v by a closed disk of radius εr centered at v, and replace each R_i by a closed one-way infinite half-strip of width εw with R_i as its mid-ray. The union of the disk and these half-strips is called a *palm* and is denoted by $P = P(v, R_1, \ldots, R_s)$. The point v is said to be the *apex* of the palm, the half-strips are said to be its *fingers*, and the largest angle between the rays defining two (not necessarily consecutive) fingers is the *angle* of the palm.

If we go far enough from v, the fingers start to bifurcate. For any two consecutive fingers corresponding to the rays R_i and R_{i+1}, we define the *distance* from v at which they *bifurcate*, as the maximum radius of a disk centered at v with the property that its intersection with the complement of the union of the fingers (half-strips) is connected. Analogously, for any two (two-way infinite) *strips S* and S' such that their midlines cross at a point v, we define the distance from v at which they bifurcate as the maximum radius of a disk centered at v with the property that its intersection with the complement of $S \cup S'$ has at most *two* connected components.

The following two simple statements can be established by straightforward trigonometric calculations.

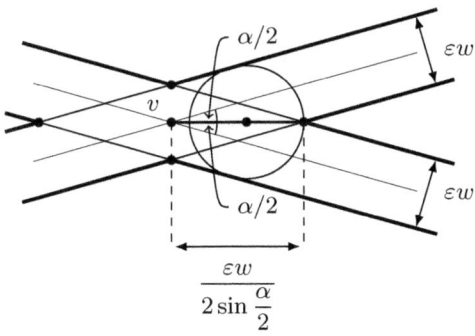

Fig. 2. For Lemma 3

Lemma 3. *Let S and S' be two strips of width εw such that their midlines cross at a point v and the angle between them is $\alpha \leq \frac{\pi}{2}$. Then*

1. *$S \cup S'$ contains no disk of radius εw;*
2. *S and S' bifurcate at distance $\frac{\varepsilon w}{2 \sin \frac{\alpha}{2}}$ from v;*
3. *any two consecutive fingers of a palm such that the angle between the rays defining them is $\alpha \leq \pi/2$ bifurcate at distance $\frac{\varepsilon w}{2 \sin \frac{\alpha}{2}}$ from the apex.*

Lemma 4. *Let $P = P(v, R_1, \ldots, R_s)$ be a palm as above, and assume that its angle is smaller than $2 \arcsin \frac{1}{4} < \frac{\pi}{6}$. Let $\overline{P} \supset P$ denote the union of the disk of radius εr centered at v and the convex hull of the union of the first and last fingers, corresponding to R_1 and R_s.*

Then $\overline{P}$ contains no disk of radius εr that intersects the disk of radius εr centered at its apex v. Hence, the same is true for P.

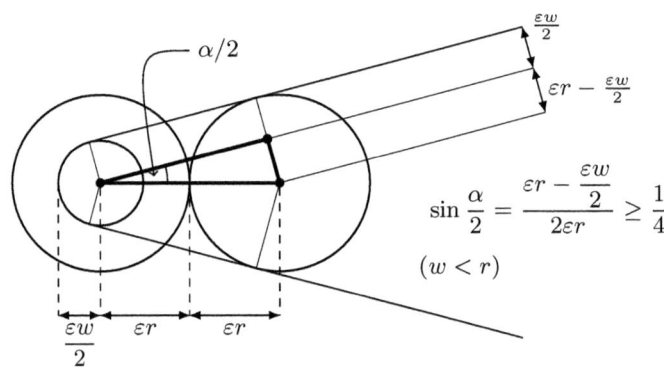

Fig. 3. For Lemma 4

3   The Main Lemma

As in the previous section, w, r, and ε are fixed positive constants, $w < r$. The main component of the proof of Theorem 2 is the following lemma, which guarantees that if the angles between the consecutive fingers of a palm P decrease sufficiently fast, then P cannot contain a disk of radius εr. The proof of this fact requires some detailed calculations, but heuristically it is clear that in this case only the first two fingers play an important role, and the situation is similar to the setting of Lemma 3, part 1.

Lemma 5. *Let $\delta = \min(\frac{1}{2}, 1 - \frac{w}{r})$, and let $P = P(v, R_1, \ldots, R_s)$ be a palm of angle $\alpha < \delta^{1/2}$. Let α_i denote the angle between R_i and R_{i+1}, and assume that for every i $(1 \leq i < s)$ we have $\frac{\alpha_{i+1}}{\alpha_i} \leq \delta$.*

Then P contains no disk of radius εr.

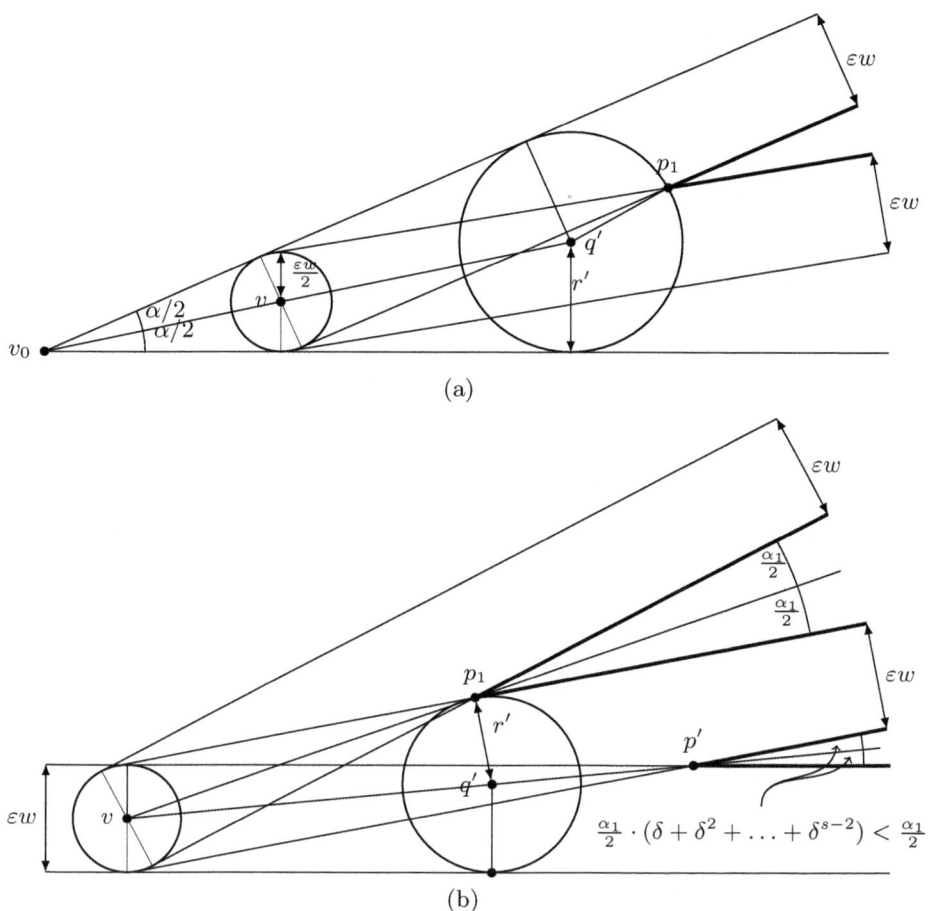

Fig. 4. For Lemma 5

Proof. If the fingers corresponding to R_i and R_{i+1} bifurcate at distance d_i from v, then they share a boundary point p_i with $|vp_i| = d_i$ ($1 \leq i < s$). These points are called *points of bifurcation*. It follows from the condition in Lemma 5 about the ratios α_{i+1}/α_i that $d_1 < d_2 < d_3 < \ldots$ is a fast increasing sequence. If P has at most 2 fingers, then Lemma 5 is true by Lemma 3, part 1. Therefore, we can assume that s, the number of fingers, is at least 3 and that we have already proved the lemma for all palms with fewer than s fingers.

Suppose that $|vp_1| = d_1 = \min_{1 \leq i < s} d_i \leq \varepsilon r$. Then P is the union of two palms $P(v, R_1)$ and $P(v, R_2, \ldots, R_s)$, each having fewer than s fingers, so that any disk of radius εr other than the one centered at v must belong to one of them. Thus, in this case we are done, by induction. From now on assume that p_1 and hence all other points p_i lie outside of the disk of radius εr centered at v. Note that the part of the ray vp_i beyond the point p_i does not belong to P.

In fact, it lies in an infinite *open* cone C_i, symmetric about vp_i, which belongs to the complement of P. By rotating the coordinate system if necessary, we can assume without loss of generality that R_s is parallel to the positive x-axis, so that all other rays $R_1, \ldots, R_{s-1}$ point into the positive quadrant $x, y \geq 0$. Then it makes sense to talk about the *lower* and the *upper boundary* of a finger. The cone C_i is bounded by two half-lines: one belongs to the lower boundary of the finger corresponding to R_i and the other to the upper boundary of the finger corresponding to R_{i+1}.

Suppose for contradiction that P contains a disk D of radius at least εr, other than the disk of radius εr centered at v. It follows from Lemma 4 that D cannot intersect the disk of radius εr centered at v. We also know that D must have a point that belongs only to the first finger, but not to the second one, otherwise we can remove the first finger and obtain a contradiction using the induction hypothesis.

Let $\overline{P}$ be the same as in Lemma 4, and let $P' \supset P$ denote the region obtained from $\overline{P}$ by deleting all points that belong to the infinite cone C_1 with apex p_1. Let D' be a disk of *maximum radius* in P' with the property that it has a point that belongs to the first finger of P, but not to the interior of the second one. Let q' and r' denote the center and the radius of D'. By our assumption, we have that $r' \geq \varepsilon r$, and it follows from Lemma 4 that D' does not intersect the disk of radius εr centered at v.

It is easy to verify that

1. p_1 lies on the boundary of D';
2. D' is tangent to the lower (horizontal) boundary half-line of $\overline{P}$;
3. D' is tangent either to the upper boundary half-line of the second finger or to the upper boundary half-line of $\overline{P}$.

Indeed, it follows from the maximality of D' that D' is "fixed" by the boundary of P'. One point cannot fix a disk. The same is true for two points, one lying on the lower, one on the upper boundary half-line of $\overline{P}$. In other words, if D' is tangent to the lower and to the upper boundary half-lines of $\overline{P}$, by maximality, it must also touch the boundary of the cone C_i. Suppose first that D' is tangent to the upper boundary half-line of $\overline{P}$ and to the upper boundary half-line of C_1. If condition 1 is not satisfied, that is, D' touches a point of the upper boundary half-line of C_1 other than p_1, then D' must lie entirely in the first finger, and its radius cannot exceed $\varepsilon w/2 < \varepsilon r$, which is impossible. Therefore, condition 1 is satisfied and, unless D' also satisfies condition 2, D' can be enlarged without violating the requirements.

Suppose next that D' is not tangent to the upper boundary half-line of $\overline{P}$. Then D' must be tangent to the lower boundary half-line of $\overline{P}$ and to the lower boundary half-line of C_1. Moreover, the point at which D' touches the lower boundary half-line of C_1 must be p_1, otherwise D' cannot have a point that belongs to the first finger of P, but not to the interior of the second one. If D' has such a point *strictly* above the upper boundary of the second finger then it could be slightly enlarged without violating the conditions. Indeed, q' belongs to the locus of all points equidistant from p_1 and the (horizontal) supporting line of the lower boundary half-line of $\overline{P}$, which is a parabola Π with a vertical axis of symmetry. If q' is on the left side of this parabola, then we can enlarge the radius of D' by moving q' along Π slightly to the left, if it is on the right side of Π, then by moving it slightly to the right. Therefore, we can conclude that D' must be tangent to the upper boundary of the second finger at point p_1, and condition 3 holds.

Now we can easily complete the proof of Lemma 5.

If conditions 1, 2, and the first option in condition 3 hold, then consider the triangle vp_1q'. Using that $\delta \leq 1/2$, we obtain

$$
\begin{aligned}
\angle vp_1q' = \frac{\pi}{2} - \frac{\alpha_1}{2} &\leq \frac{\pi}{2} - \frac{\delta\alpha_1}{2(1-\delta)} \\
&< \frac{\pi}{2} - \frac{\alpha_1}{2}(\delta + \delta^2 + \ldots + \delta^{s-2}) \\
&\leq \frac{\pi}{2} - \frac{\alpha_2 + \alpha_3 + \ldots + \alpha_{s-1}}{2} = \angle vq'p_1 \ .
\end{aligned}
$$

This yields that $|vq'| < |vp_1|$. As was used above, the angle α_1 between R_1 and R_2 is larger than $\alpha_2 + \ldots + \alpha_{s-1}$, the angle between R_2 and R_s. Therefore, the fingers corresponding to R_2 and R_s bifurcate at a point p' which is farther away from v than p_1 is. This implies that $|vq'| < |vp_1| < |vp'|$. The points v, q', and p' are collinear, so that it follows from the last inequality that q' lies in the interior of the second finger. Since $r' = |q'p_1|$ is equal to the distance of q' from the upper boundary half-line of the second finger, we obtain that $r' < \varepsilon w < \varepsilon r$, which is a contradiction.

In the other case, when conditions 1, 2, and the second option in condition 3 hold, just like in the first case, we have $|vq'| < |vp_1|$. (In fact, it is easy to argue that the part of the parabola Π which lies below the line vp_1 and to the left of the line through p_1 perpendicular to R_s is entirely contained in the interior of the circle through p_1 centered at v. The point q' belongs to this arc.)

Let v_0 denote the intersection point of the supporting lines of the upper boundary ray of the first finger (that corresponds to R_1) and the lower boundary ray of the last finger (that corresponds to R_s). The points v_0, v, and q' are collinear. Using the notation $\alpha = \alpha_1 + \ldots + \alpha_{s-1}$, we have

$$
\begin{aligned}
r' = |v_0q'| \sin\frac{\alpha}{2} = (|v_0v| + |vq'|) \sin\frac{\alpha}{2} &< (|v_0v| + |vp_1|) \sin\frac{\alpha}{2} \\
&\leq \left(\frac{\varepsilon w}{2\sin\frac{\alpha}{2}} + \frac{\varepsilon w}{2\sin\frac{\alpha_1}{2}} \right) \sin\frac{\alpha}{2} = \frac{\varepsilon w}{2}\left(1 + \frac{\sin\frac{\alpha}{2}}{\sin\frac{\alpha_1}{2}} \right) \ .
\end{aligned}
$$

Here we used Lemma 3, part 2 to estimate $|vp_1|$.

In view of the assumption on the angles between consecutive fingers, we have that

$$
\alpha = \alpha_1 + \alpha_2 + \ldots + \alpha_{s-1} = \alpha_1(1 + \delta + \ldots + \delta^{s-2}) < \frac{\alpha_1}{1-\delta} \ .
$$

Hence, the above upper bound on r' can be rewritten as

$$r' < \frac{\varepsilon w}{2}\left(1 + \frac{\sin\frac{\alpha_1}{2(1-\delta)}}{\sin\frac{\alpha_1}{2}}\right) < \frac{\varepsilon w}{2}\left(1 + \frac{\frac{\alpha_1}{2(1-\delta)}}{\sin\frac{\alpha_1}{2}}\right) .$$

Using the Taylor series of the $\sin x$ function, it is easy to verify that, given any δ, $0 < \delta < 1$, the inequality $\sin\frac{\alpha_1}{2} > \frac{\alpha_1}{2(1+\delta)}$ holds for all $\alpha_1 \leq \delta^{1/2}$. By the assumptions in the lemma, this condition is satisfied, so that we have

$$r' < \frac{\varepsilon w}{2}\left(1 + \frac{1+\delta}{1-\delta}\right) = \frac{\varepsilon w}{1-\delta} .$$

By our choice of δ, we have $\delta \leq 1 - \frac{w}{r}$. That is,

$$r' < \frac{\varepsilon w}{1-\delta} \leq \varepsilon r ,$$

the desired contradiction. $\square$

4   The Proof of Theorem 2

In the previous two sections, apart from n, w, and r, we also fixed the constant $\varepsilon > 0$. In the proof of Theorem 2 presented in this section, we keep n, w, and r fixed, but we will vary ε.

Let $S(\varepsilon)$ denote the union of the disks of radius εr representing the vertices v_i ($1 \leq i \leq n$) and the rectangles of width εw representing the edges $v_i v_j$ ($1 \leq i < j \leq n$).

For a given v_i, consider the rectangles representing the edges incident to v_i and extend them to one-way infinite half-strips pointing away from v_i. More precisely, for any $j \neq i$, let $R_{i,j}$ denote the ray $\overrightarrow{v_i v_j}$ emanating from v_i and pointing to the direction of v_j. Let $F_{i,j}(\varepsilon)$ be the half-strip of width εw, the mid-ray of which is $R_{i,j}$. The union of the disk of radius εr centered at v_i and the sets $F_{i,j}(\varepsilon)$ for all $j \neq i$ is denoted by $P_i(\varepsilon)$. Any two distinct half-strips $F_{i,j}(\varepsilon)$ and $F_{i,j'}(\varepsilon)$ bifurcate at a certain distance from v_i. Let $\varrho_i(\varepsilon)$ denote the maximum of these $\binom{n-1}{2}$ distances plus εr.

Let us fix a small $\varepsilon > 0$ such that the following three conditions are satisfied.

1. No three rectangles representing distinct edges, not all of which are incident to the same vertex, have a point in common.
2. Any rectangle representing an edge $v_j v_k$ is disjoint from any disk of radius $\varrho_i(\varepsilon)$ centered at v_i, for all $i \neq j, k$.
3. For every pair $i \neq j$, the disk of radius $\varrho_i(\varepsilon)$ centered at v_i is disjoint from the disk of radius $\varrho_j(\varepsilon)$ centered at v_j.

It follows from the second condition that no rectangle representing an edge $v_j v_k$ can intersect any disk representing a vertex v_i with $i \neq j, k$. The last condition implies that the disk of radius $\varrho_i(\varepsilon)$ centered at v_i cannot contain any disk of

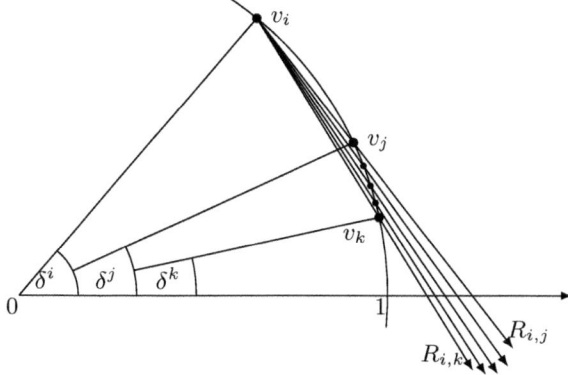

Fig. 5. For Proof of Theorem 2

radius εr representing a vertex v_j with $j \neq i$. If three edges share an interior point, then the first condition cannot be satisfied. However, it is easy to argue that in our case this cannot occur.

From now on ε will be fixed, so that in notation we can drop the parameter ε. In particular, instead of $S(\varepsilon), \varrho_i(\varepsilon)$, and $P_i(\varepsilon)$, we will write S, ϱ_i, and P_i.

Suppose for contradiction that the set S contains a disk D of radius εr which is not one of the disks representing the vertices. Where can such a disk D lie? The only possibility is that for some i $(1 \leq i \leq n)$, it lies in the part of S contained in the disk of radius ϱ_i centered at v_i. Otherwise, by the conditions listed above, D would be contained in the union of two strips of width w, contradicting part 1 of Lemma 3. Observe that the part of S contained in the disk of radius ϱ_i centered at v_i is exactly the same as the part of P_i contained in the disk of radius ϱ_i centered at v_i. Therefore, to finish the proof of Theorem 2, it is sufficient to show that no set P_i contains a disk of radius εr $(1 \leq i \leq n)$.

To see this, notice that for every i $1 \leq i \leq n$, the set P_i can be written as the union of at most two palms of angle smaller than δ (see the beginning of Sect. 2). We have $P_1 = P(v_1, R_{1,2}, R_{1,3}, \ldots, R_{1,n})$, $P_n = P(v_n, R_{n,1}, R_{n,2}, \ldots, R_{n,n-1})$, and

$$P_i = P(v_i, R_{i,1}, R_{i,2}, \ldots, R_{i,i-1}) \cup P(v_i, R_{i,i+1}, R_{i,i+2}, \ldots, R_{i,n}) \ ,$$

for every i, $1 < i < j$,. If $i \neq 1, n$, then the smallest angle between a finger of $P(v_i, R_{i,1}, R_{i,2}, \ldots, R_{i,i-1})$ and a finger of $P(v_i, R_{i,i+1}, R_{i,i+2}, \ldots, R_{i,n})$ is the angle between $R_{i,1}$ and $R_{i,n}$, which is equal to $\pi - \frac{\delta - \delta^n}{2} > \pi - \frac{\delta}{2}$. It follows from here that the fingers corresponding to $R_{i,1}$ and $R_{i,n}$ bifurcate within the disk of radius εr centered at v_i. This, in turn, implies that any disk D of radius εr which lies in P_i and is different from the disk representing v_i is entirely contained in one of the two palms comprising P_i. Applying Lemma 5 to this palm, we obtain the desired contradiction. The only thing that remains to be checked is that the conditions of the lemma about the angles α and α_i are satisfied.

The maximum angle of the palms of the form $P(v_i, R_{i,i+1}, R_{i,i+2}, \ldots, R_{i,n})$ and $P(v_i, R_{i,1}, R_{i,2}, \ldots, R_{i,i-1})$, for $1 \leq i \leq n$, is the angle of $P(v_n, R_{n,1}, R_{n,2}, \ldots, R_{n,n-1})$, which is equal to

$$\angle v_1 v_n v_{n-1} = \frac{\angle v_1 0 v_{n-1}}{2} = \frac{\delta - \delta^{n-1}}{2} < \frac{\delta}{2} ,$$

so that the condition on the angle of the palm is satisfied. (Here 0 denotes the origin, the center of the circle containing all points v_i.) As for the condition on the angles α_i, we have that the angle between two consecutive rays $R_{i,t}$ and $R_{i,t+1}$ is equal to

$$\frac{\angle v_t 0 v_{t+1}}{2} = \frac{\delta^t - \delta^{t+1}}{2} = \frac{1 - \delta}{2} \delta^t .$$

Analogously, the angle between $R_{i,t+1}$ and $R_{i,t+2}$ is equal to $\frac{1-\delta}{2}\delta^{t+1}$. Hence, all ratios $\frac{\alpha_s}{\alpha_{s+1}}$ are equal to δ, and the conditions of Lemma 5 are satisfied.

This completes the proof of Theorem 2. $\qquad\qquad\square$

Acknowledgement. The author is grateful to Mark van Kreveld for calling his attention to the problem addressed in this paper, and to Radoslav Fulek, Fabrizio Frati, and Deniz Sarıöz for valuable discussions, and Deniz Sarıöz also for coding the figures in TikZ.

References

[BGR04] Barequet, G., Goodrich, M.T., Riley, C.: Drawing planar graphs with large vertices and thick edges. J. Graph Algorithms Appl. 8, 3–20 (2004)

[CE92] Chazelle, B., Edelsbrunner, H.: An optimal algorithm for intersecting line segments in the plane. J. ACM 39, 1–54 (1992)

[CS89] Clarkson, K.L., Shor, P.W.: Application of random sampling in computational geometry, II. Discrete & Computational Geometry 4, 387–421 (1989)

[DEKW06] Duncan, C.A., Efrat, A., Kobourov, S.G., Wenk, C.: Drawing with fat edges. Int. J. Found. Comput. Sci. 17, 1143–1164 (2006)

[ES35] Erdös, P., Szekeres, G.: A combinatorial problem in geometry. Compositio Mathematica 2, 463–470 (1935)

[GRS90] Graham, R., Rothschild, B., Spencer, J.H.: Ramsey Theory. John Wiley and Sons, New York (1990)

[K09] van Kreveld, M.: Bold graph drawings. In: Proc. Canadian Conference on Computational Geometry, CCCG 2009 (2009), http://cccg.ca/proceedings/2009/cccg09_31.pdf; Also Computational Geometry: Theory & Applications (to appear)

[M88] Mulmuley, K.: A fast planar partition algorithm, I. In: Proc. 29th FOCS, pp. 580–589 (1988)

[N04] North, S.C.: Drawing Graphs with Neato (2004), http://www.graphviz.org/Documentation/neatoguide.pdf

[R30] Ramsey, F.P.: On a problem of formal logic. Proc. London Math. Soc. Series 30(2), 264–286 (1930)

Adjacent Crossings Do Matter

Radoslav Fulek[1,*], Michael J. Pelsmajer[2,**],
Marcus Schaefer[3], and Daniel Štefankovič[4]

[1] Ecole Polytechnique Fédérale de Lausanne, Lausanne, Switzerland
radoslav.fulek@epfl.ch
[2] Illinois Institute of Technology, Chicago, IL 60616, USA
pelsmajer@iit.edu
[3] DePaul University, Chicago, IL 60604, USA
mschaefer@cs.depaul.edu
[4] University of Rochester, Rochester, NY 14627, USA
stefanko@cs.rochester.edu

Abstract. In a drawing of a graph, two edges form an *odd pair* if they cross each other an odd number of times. A pair of edges is *independent* if they share no endpoint. For a graph G, let $\mathrm{ocr}(G)$ be the smallest number of odd pairs in a drawing of G and let $\mathrm{iocr}(G)$ be the smallest number of independent odd pairs in a drawing of G. We construct a graph G with $\mathrm{iocr}(G) < \mathrm{ocr}(G)$, answering a question by Székely, and—for the first time—giving evidence that crossings of adjacent edges may not always be trivial to eliminate.

The graph G is based on a separation of iocr and ocr for monotone drawings of ordered graphs. A drawing of a graph is *x-monotone* if every edge intersects every vertical line at most once and every vertical line contains at most one vertex. A graph is *ordered* if each of its vertices is assigned a distinct x-coordinate. We construct a family of ordered graphs such that for x-monotone drawings, the monotone variants of ocr and iocr satisfy $\mathrm{mon\text{-}iocr}(G) < O(\mathrm{mon\text{-}ocr}(G)^{1/2})$.

1 Introduction

When drawing a graph some assumptions are natural: there are only finitely many crossings, no more than two edges cross in a point, edges do not pass through vertices, and edges do not touch.[1] Sometimes these assumptions are relaxed (*degenerate drawings* allow more than two edges to cross in a point), and sometimes more restrictions are added, for example adjacent edges may not be allowed to cross.

The *crossing number* $\mathrm{cr}(G)$ of a graph G is the smallest number of crossings in a drawing of G. It is easy to see that in an optimal drawing, adjacent edges

---

* The first author gratefully acknowledges support from the Swiss National Science Foundation Grant No. 200021-125287/1.
** The second author gratefully acknowledges the support from NSA Grant H98230-08-1-0043 and the Swiss National Science Foundation Grant No. 200021-125287/1.
[1] For a detailed discussion see [14].

M. van Kreveld and B. Speckmann (Eds.): GD 2011, LNCS 7034, pp. 343–354, 2012.

of G do not cross (such crossings can always be removed). This may have led researchers on crossing numbers to think that adjacent crossings are irrelevant or even to prohibit them in drawings.[2] Another source for ignoring adjacent crossings may be the fact that graph drawings are often straight-line drawings in which adjacent edges naturally cannot cross.

Pach and Tóth point out in "Which Crossing Number is It Anyway?" that there have been many different ideas on how to define a notion of crossing number, including the following (see [6,14]):

pair crossing number: $\mathrm{pcr}(G)$, the smallest number of pairs of edges crossing in a drawing of G,

odd crossing number: $\mathrm{ocr}(G)$, the smallest number of pairs of edges crossing oddly (*odd pairs*) in a drawing of G.

Tutte introduced another type of crossing number by orienting edges arbitrarily, then letting $\lambda(e, f)$ be the difference in the number of crossings where e is pointed to the left of f and the number of crossings where e is pointed to the right of f. Changing the orientation of e or f will only change the sign of $\lambda(e, f)$, so one can define:

algebraic crossing number: $\mathrm{acr}(G)$, the minimum of $\sum |\lambda(e, f)|$ in a drawing of G, where the sum is taken over pairs of edges e, f.

By definition we have $\mathrm{ocr}(G) \leq \mathrm{pcr}(G) \leq \mathrm{cr}(G)$ and $\mathrm{ocr}(G) \leq \mathrm{acr}(G) \leq \mathrm{cr}(G)$.

For each of these notions, one can ask whether adjacent crossings matter. In [5], Pach and Tóth suggest a systematic study of this issue (see also [1, Section 9.4]) by introducing two rules: "Rule $+$" restricts the drawings to drawings in which adjacent edges are not allowed to cross. "Rule $-$" allows crossings of adjacent edges, but does not count them towards the crossing number. Each parameter ocr, pcr, acr, and cr can be modified by either rule, but since $\mathrm{cr}_+ = \mathrm{cr}$ (implied by the discussion at the beginning of the section), this yields up to eleven possible distinct variants.

The tables below are based on a figure from [1]. The notion of ocr_- was introduced as the *independent odd crossing number*, iocr, by Székely [14].[3]

Rule $+$	ocr_+	pcr_+	cr		ocr_+	acr_+	cr
	ocr	pcr			ocr	acr	
Rule $-$	$\mathrm{iocr} = \mathrm{ocr}_-$	pcr_-	cr_-		$\mathrm{iocr} = \mathrm{ocr}_-$	acr_-	cr_-

It immediately follows from the definitions that the values in each table increase monotonically as one moves from the left to the right and from the bottom to the top. Not much more is known about the relationships between these crossing number variants. In [5], Pach and Tóth write, "We cannot prove anything

---

[2] Székely discusses this issue in [14].

[3] Székely credits Tutte [18] with the (implicit) definition of iocr, but Tutte is really concerned with the algebraic crossing numbers only, acr and acr_-; he does not consider parity.

else about $\mathrm{iocr}(G)$, $\mathrm{pcr}_-(G)$, and $\mathrm{cr}_-(G)$. We conjecture that these values are very close to $\mathrm{cr}(G)$, if not the same. That is, we believe that by letting pairs of incident edges cross an arbitrary number of times, we cannot effectively reduce the total number of crossings between independent pairs of edges."[4] Tutte [18] seems to have had a similar opinion, when he explained his choice to study acr_-, writing, "We are taking the view that crossings of adjacent edges are trivial, and easily got rid of." Székely [14] later commented "We interpret this sentence as a philosophical view and not a mathematical claim." West [20] and Székely [15] mention the specific question of whether there are graphs with $\mathrm{iocr}(G) < \mathrm{ocr}(G)$.

There are situations when the entire system of crossing numbers collapses. The classic Hanani-Tutte theorem states that if a graph can be drawn in the plane so that no pair of independent edges crosses an odd number of times, then it is planar [3,18]. In other words, $\mathrm{iocr}(G) = 0$ implies that $\mathrm{cr}(G) = 0$ and, thus, that all of the eleven variants are equal (to zero). This was extended to show that all eleven variants are equal as long as $\mathrm{iocr}(G) \leq 2$ [12]. Székely gave an explicit criterion for when all variants are equal [16]. It is also known that all eleven variants are within a square of each other, since $\mathrm{cr}(G) \leq \binom{2\,\mathrm{iocr}(G)}{2}$ [12]. For drawings of G on the projective plane N_1, we know that $\mathrm{iocr}_{N_1}(G) = 0$ implies that $\mathrm{cr}_{N_1}(G) = 0$, so again all variants are equal (to zero) in this case [8].

Setting aside the Rule $-$ variants, there are some strong results for the remaining seven variants, ocr, ocr_+, acr, acr_+, pcr, pcr_+ and cr. If $\mathrm{ocr}(G) \leq 3$ then all these seven variants are equal [9]. For drawings on any surface S, if $\mathrm{ocr}_S(G) = 0$ then all seven variants are equal (to zero) [11]. Valtr [19] showed that $\mathrm{cr}(G) = O(\mathrm{pcr}^2(G)/\log \mathrm{pcr}(G))$, which Tóth [17] improved to $\mathrm{cr}(G) = O(\mathrm{pcr}^2(G)/\log^2 \mathrm{pcr}(G))$.

On the other hand, we know that ocr and pcr differ: there is an infinite family of graphs with $\mathrm{ocr}(G) < 0.867 \cdot \mathrm{pcr}(G)$ [10]. Tóth improved this by giving a family of graphs with $\mathrm{acr}(G) < 0.855 \cdot \mathrm{pcr}(G)$ [17] (so $\mathrm{ocr}(G) < 0.855 \cdot \mathrm{pcr}(G)$ as well). For such G it immediately follows that $\mathrm{ocr}(G) < \mathrm{cr}(G)$ and $\mathrm{acr}_-(G) < \mathrm{cr}(G)$, answering questions of Pach and Tóth [6] and Tutte [18]; additional consequences can be deduced from the tables above. However, none of these results address the intuitions expressed by Tutte and by Pach and Tóth about how Rule $-$ may or may not affect cr, pcr, ocr, or acr.

We can finally give a result of this nature.

Theorem 1. *For every n, there is a graph G with $\mathrm{iocr}(G) < \mathrm{ocr}(G) - n$.*

In short, adjacent crossings matter.[5]

To prove Theorem 1, we will first prove a separation for monotone drawings of ordered graphs. An *ordered graph* is a graph with a total ordering of its vertices. For our purposes, we will assume that the vertex set of an ordered graph is a subset of the integers, and we will only consider drawings where each vertex

---

[4] Some authors write *incident edges* to mean two edges that share an endpoint, but we will only use *adjacent edges*. Non-adjacent edges are also called *independent edges*.

[5] Among other things, Theorem 1 justifies the rather baroque **NP**-completeness proof for iocr in [13]. **NP**-completeness of ocr is simpler in comparison [7].

n has x-coordinate equal to n. A drawing of a graph is x-*monotone* if every edge intersects every vertical line at most once and every vertical line contains at most one vertex. We can generalize each crossing number variant to x-monotone drawings of ordered graphs G, which we denote mon-cr(G), mon-ocr(G), mon-iocr(G), etc.

Pach and Tóth proved that mon-ocr$(G) = 0$ implies mon-cr$(G) = 0$ [7]. We strengthened this by showing that mon-iocr$(G) = 0$ implies mon-cr$(G) = 0$ [2], which had been left as an open problem in [7]. On the other hand, in the same paper we showed that for every n there is a graph G such that mon-cr$(G) \geq n$ and mon-ocr$(G) = 1$. In this paper, we will show that there can also be an arbitrary gap between mon-ocr and mon-iocr.

Theorem 2. *For every $n \geq 3$ there is an ordered graph G with* mon-iocr$(G) = 3n < n^2 + n = $ mon-ocr(G).

Note that for such G, we have mon-iocr$(G) = O($mon-ocr$(G)^{1/2})$. We will use Theorem 2 to prove Theorem 1.

2   Separating Monotone Crossing Numbers

We generalize the crossing number definitions for graphs with weighted edges. Suppose that G is a graph and each edge e has weight $w(e)$. A crossing between edges e and f is assigned *crossing weight* equal to the product $w(e)w(f)$. Let D be an arbitrary drawing of G, and define

cr$(D) = $ the sum of crossing weights, taken over all crossing in D,
ocr$(D) = $ the sum of $w(e)w(f)$, taken over all odd pairs e, f in D,
iocr$(D) = $ the sum of $w(e)w(f)$, taken over all independent odd pairs e, f in D.

Let cr$(G) = \min_D$ cr(D), ocr$(G) = \min_D$ ocr(D), and iocr$(G) = \min_D$ iocr(D), with each minimum taken over all drawings D of G. If we assign every edge weight equal to 1, then these definitions revert back to their original, unweighted versions.

Consider an ordered graph $G = ([7], \{13, 16, 23, 24, 25, 27, 35, 37, 46, 47, 56\})$ with edge weights $w(16) = w(23) = w(25) = w(27) = w(46) = w(47) = 2x^2$, $w(24) = w(37) = x$, $w(13) = w(35) = w(56) = 1$ (see the left of Figure 1 for a drawing of G).

Theorem 3. *For the weighted ordered graph G in Figure 1 with $x \geq 3$, we have*

$$\text{mon-iocr}(G) = 3x < x^2 + x = \text{mon-ocr}(G). \tag{1}$$

Proof. In the drawing on the left side of Figure 1, the only independent pairs of edges that cross oddly are $(13, 24)$, $(24, 35)$, $(56, 37)$, showing

$$\text{mon-iocr}(G) \leq 3x. \tag{2}$$

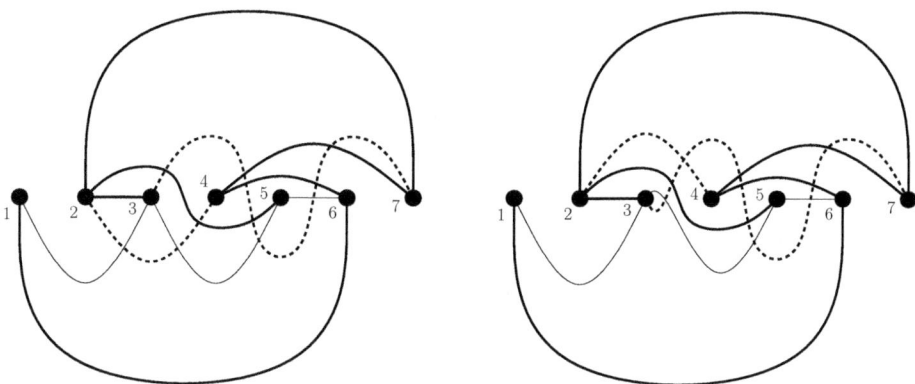

Fig. 1. Two drawings of a weighted ordered graph G with mon-iocr$(G) <$ mon-ocr(G); thick solid edges have weight $2x^2$, the thick dashed edges have weight x, and the thin solid edges have weight 1. The left drawing shows that mon-iocr$(G) \leq 3x$; the right drawing that mon-ocr$(G) \leq x^2 + x$.

The drawing also shows mon-ocr$(G) \leq 4x + 2x^3$, since for mon-ocr the odd pairs $(24, 25)$ and $(35, 37)$ count. If we reroute edge 24 to go above 25, it crosses 37 (instead of 13, 35, and 25). Close to 3, we can twist 35 and 37 so they cross evenly. This yields the drawing on the right in Figure 1. It shows that

$$\text{mon-ocr}(G) \leq x^2 + x. \tag{3}$$

Suppose that we have a drawing D of G with mon-iocr$(D) < x^2 + x$. Since $x^2 + x < 2x^2$ for $x \geq 3$, no thick edge (that is an edge of weight $2x^2$) is crossed oddly by an independent edge. We claim that this forces most of the drawing to be as depicted in Figure 1: Without loss of generality assume that 46 passes above 5. Then 35 must pass below 4 (to avoid crossing 46 oddly) and 47 must pass above 5 and 6 (to avoid crossing 35 and 56 oddly). Now 16 has to pass below 4 (to avoid 47) and hence below 2, 3, 5 (to avoid 24, 23, 35). Since 16 goes below 2 we have that 27 is above 6 (to avoid 16) and also above 3, 4, 5 (to avoid 56, 35, 46). Then 13 has to be below 2 (to avoid 27) and 25 has to be above 3 (to avoid 13) and below 4 (to avoid 46). The edge 37 has to go below 5 (to avoid 25) above 6 (to avoid 16) and hence above 4 (to avoid crossing 46 oddly). Note that we have determined the above-below relationship for all relevant edge-vertex pairs (when the vertex lies between the endpoints of the edge) except for those with the edge 24, using only the fact that thick edges cannot be crossed oddly by independent edges. Note that thus far, $(37, 56)$ is the only independent odd pair of edges.

Consider how the edge 24 can be drawn. If we draw it above 3 then it will cross 37 oddly bringing the total number of odd crossings between independent pairs of edges to $x^2 + x$. Thus 24 has to go below 3. To summarize: we have shown that any drawing D of G with mon-iocr$(D) < x^2 + x$ must have the same (or mirrored) above-below relationships as in the drawing on Figure 1. Note that

24 crosses 13 and 35 oddly, bringing the total number of odd crossings between independent pairs of edges to $3x$. This proves the left equality in (1).

We next prove the right equality in (1). For this we need only show mon-ocr$(G) \geq x^2 + x$, due to (3). Suppose that we have a drawing D of G such that mon-ocr$(D) < x^2 + x$. This implies mon-iocr$(D) < x^2 + x$ so by the earlier argument we may assume that every relevant edge-vertex pair has the same above-below relationship in D as in the drawing of Figure 1.

If 24 leaves 2 above 23 then 23 and 24 cross oddly (since 24 goes below 3) showing mon-ocr$(D) \geq 2x^3 + 3x$, a contradiction. Thus 24 leaves 2 below 23. If 24 leaves 2 below 25 then 24 and 25 cross oddly (since 25 goes below 4) showing mon-ocr$(D) \geq 2x^3 + 3x$, a contradiction. Thus 24 leaves 2 above 25. Now, using transitivity, 23 leaves 2 above 25 but that means that 23 and 25 cross oddly (since 25 goes above 3) showing mon-ocr$(D) \geq 2x^2 + 3x$, a contradiction. Hence there is no drawing D of G with mon-ocr$(D) < x^2 + x$, finishing the proof of (1).

2.1   From Weighted Edges to Unweighted Edges

Suppose that G is a graph or ordered graph with edges of positive integer weight. Let G' be the graph obtained by replacing each edge of weight w with w edges of weight 1, equivalently, with w unweighted edges. Choose any of the eleven crossing variants mentioned in Section 1, and consider a drawing of G' (which is x-monotone if G is an ordered graph) that optimizes that crossing variant. Suppose that e_1 and e_2 are copies of the same edge e of G. Without loss of generality, we may assume that e_1 contributes less than or equal to what e_2 contributes to the chosen crossing parameter. We can redraw e_2 along the side of e_1 so that they do not cross; then e_2 will contribute the same to the crossing parameter as e_1, so the new drawing is still optimal. Hence, we may assume that in an optimal drawing of G', multiple edges are drawn in a bundle, all with essentially the same behavior.[6] It follows that all crossing parameters are the same for G and G'.

Lemma 1. *Subdividing an edge of a graph does not change* ocr *or* iocr. *Subdividing an edge of an ordered graph near one of its endpoints does not change* mon-ocr *or* mon-iocr. *These results hold for graphs with multiple edges as well.*

Proof. Let G be a graph or ordered graph, possibly with multiple edges. If G is an ordered graph, we will restrict all drawings to be x-monotone drawings.

Fix an ocr-optimal (iocr-optimal) drawing of G, and choose any edge uv. Subdivide uv with a vertex z, which is added to the drawing of uv near the endpoint u. Then for each edge $e \neq uv$, e will cross zv oddly if and only if e crossed uv oddly, and e does not cross uz at all. Hence ocr is unchanged; iocr is also unchanged unless e shares an endpoint with uv but not with zv, which means that e is incident to u but not v. In this case, we can deform a small section of e until it passes over z (while maintaining its monotonicity, if G is

---

[6] This argument was probably first made in Kainen [4] for the standard crossing number.

ordered); do this for all such e. This yields a drawing with iocr no bigger than in the initial drawing.

Now consider any drawing of the new graph. We can erase z from that drawing to obtain a drawing of the original, unsubdivided graph. If G is ordered, then we erase z from an x-monotone drawing where z lies strictly between u and v, so we obtain an x-monotone drawing of G. Erasing z moves all odd pairs of edges with uz or zv to become odd pairs with uv (and if an edge crosses both uz and zv oddly, then these cancel and it crosses uv evenly). Hence the number of odd pairs and independent odd pairs does not increase. □

Consider any integer $x \geq 3$. Replace the weighted edges of the graph in Figure 1 by multiple edges, and then apply Lemma 1 to every edge. We obtain an unweighted ordered graph H with mon-iocr$(H) = 3x < x^2 + x = $ mon-ocr(H). Thus, Theorem 2 is proved.

Before moving on, note that for any drawing of a graph G, we can remove self-intersections of edges without adding any crossing or odd pair, by redrawing locally near the crossing as shown in Figure 2 (originally from [10]).

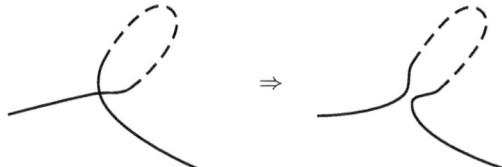

Fig. 2. Removing a self-intersection

3   Adjacent Crossings Are Not Trivial

Given an ordered graph $G = (V, E)$ with $V = \{v_1 < v_2 \cdots < v_n\}$ let G' be obtained from G by adding the following framework: start with a cycle C_{2n+2} formed from two paths $s, u_1, \ldots, u_n, t$ and $s, w_1, \ldots, w_n, t$; call this the *outer framework*. Add paths $Q_i = u_i v_i w_i$ for $1 \leq i \leq n$; call this the *inner framework*. Assign a weight of $w_I = n^4 + 1$ to the edges in the inner framework and a weight of $w_O = n^4 + n^3 w_I + 1$ to the edges in the outer framework. Edges originally in G remain at weight 1 (unweighted). From the weighted graph G' we will obtain the unweighted graph G'' by replacing each edge of weight $w > 1$ in G' by w copies of P_3.

Lemma 2. *With G' as defined above we have $\psi(G'') = $ mon-$\psi(G) + c$ for any connected graph G, where ψ is one of the crossing numbers $\{\text{iocr}, \text{ocr}, \text{cr}\}$ and $c = w_I \sum_{v_i v_j \in E(G), i < j} (j - i - 1)$.*

Lemma 2 and Theorem 2 immediately yield Theorem 1. In [2] we showed that for every n there is an ordered graph G such that mon-cr$(G) \geq n$ and mon-ocr$(G) = 1$. Together with Lemma 2, this yields a new graph G' with ocr$(G') < $ cr(G'), joining the earlier examples from [10] and [17]. In the journal version of this paper, we show that Lemma 2 can be made to work for other crossing numbers as well, so that it has the potential to lead to further separations.

For the proof of Lemma 2, we need the following lemma. An *even* edge is an edge that crosses every other edge an even number of times (possibly zero times).

Lemma 3 (Pelsmajer, Schaefer, and Štefankovič [9]). *If D is a drawing of G in the plane and E is the set of even edges in D, then G has a redrawing in which all edges in E are crossing-free, there are no new pairs of edges that cross an odd number of times, and the cyclic order of edges at each vertex does not change.*

Proof (of Lemma 2). First note that $\psi(G'') \leq$ mon-$\psi(G) + c$ is immediate: take a monotone drawing realizing mon-$\psi(G)$ and overlay it with a planar drawing of the framework, call the resulting drawing D' (see Figure 3 for an example). Then $\psi(D') =$ mon-$\psi(G)+c$ since the only crossings are single crossings between pairs of non-adjacent edges that count the same whatever ψ is. From D' we can obtain a drawing D'' of G'' by replacing the weighted edges in the drawing by parallel P_3s; then $\psi(D'') = \psi(D')$ (since the framework edges are not involved in any adjacent crossings), so $\psi(G'') \leq \psi(D'') =$ mon-$\psi(G) + c$.

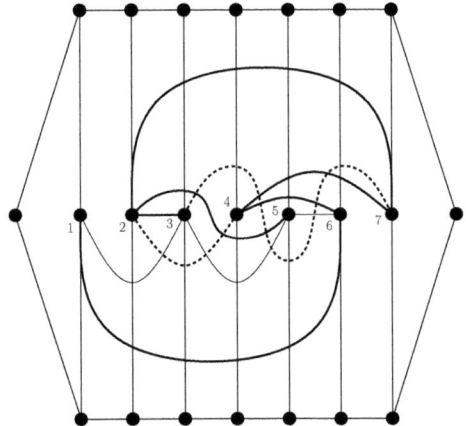

Fig. 3. Overlay of G from Figure 1 with framework (note that in the construction, G will be an unweighted graph)

It remains to prove $\psi(G'') \geq$ mon-$\psi(G) + c$ for $\psi \in \{\mathrm{cr}, \mathrm{ocr}, \mathrm{iocr}\}$. It is easy to see that $\psi(G'') \geq \psi(G')$: fix an ψ-optimal drawing of G''. Consider w parallel paths P_3 that were used to replace an edge of weight w in G'. Pick one of these paths P that contributes the smallest amount to $\psi(G'')$. Now redraw the remaining $w - 1$ paths to run very close to P and without crossing each other. This redrawing cannot increase the value of ψ of the drawing. But now we can bundle the parallel paths into a single weighted edge to obtain a drawing D' of G' with $\psi(D') \leq \psi(G'')$. So $\psi(G') \leq \psi(G'')$.

Hence, to establish the lemma it is sufficient to show that $\psi(G') \geq \text{mon-}\psi(G) + c$. We proceed in three steps; we first show that there is a ψ-minimal drawing of G' in which the edges of the outer framework are crossing-free. In the second step we show that we can assume that the edges of the inner framework do not cross each other. In the third step we show that from such a drawing of G', we can construct a monotone drawing of G with at most $\psi(G') - c$ crossings. It follows that $\text{mon-}\psi(G) \leq \psi(G') - c$.

For the first step, fix an ψ-minimal drawing of G'. For $\psi = \text{cr}$ the claim is immediate: any edge crossing an edge of the outer framework contributes at least w_O to $\psi(G')$. However, we already proved that $\psi(G') \leq \text{mon-}\psi(G) + c \leq n^4 + n^3 w_I < w_O$, so all edges of the outer framework must be crossing-free. If $\psi = \text{ocr}$ then edges of the outer framework cannot be involved in any odd pairs, since any such odd pair would contribute w_O to ocr and, as above, $\psi(G') \leq \text{mon-}\psi(G) + c \leq n^4 + n^3 w_I < w_O$. So all the edges in the outer framework are even. We can then apply Lemma 3 to make all edges in the outer framework crossing-free without introducing any new pair of edges crossing oddly (in particular, ψ does not increase). This leaves the case $\psi = \text{iocr}$. The argument here is similar to ocr. In any iocr-minimal drawing, edges of the outer framework cannot be involved in any independent odd pairs, so all odd pairs involving these edges must have adjacent edges. However, all vertices in the outer framework have degree 2 or 3, so we can modify the drawing near each of these vertices to ensure that all the edges in the outer framework are actually even. We then proceed as in the case of ocr.

This completes the first step: we know that we can assume that the outer framework is entirely free of crossings. Since we assumed that G is a connected graph, all vertices of G must lie in the same face of C_{2n+2}, without loss of generality, the inner face. Since every edge not in the outer framework is incident to a vertex of G this also implies that all edges lie in the inner face and the outer face is therefore empty.

In the second step we show that we can assume that edges of the inner framework do not cross each other. Recall that $Q_i = u_i v_i w_i$ is the inner framework path passing through v_i with endpoints u_i and w_i on C_{2n+2}, for $1 \leq i \leq n$.

For $\psi = \text{cr}$ the claim is immediate again, since any such crossing would contribute $w_I^2 = w_I(n^4 + 1) = n^4 w_I + w_I > n^3 w_I + n^4 + 1 = w_O$ to $\psi(G')$, but we already know that $\psi(G') \leq w_O$.

For $\psi = \text{ocr}$, we can similarly conclude that any two edges of the inner framework cross evenly, and for $\psi = \text{iocr}$, we know that any independent pair of edges in the inner framework crosses evenly. Suppose that $\psi = \text{iocr}$ and two adjacent edges of the inner framework, $u_i v_i$ and $v_i w_i$, cross oddly. In that case, we perform a $(u_i v_i, v_i)$-move (that is, we deform a small section of $u_i v_i$, bring it close to v_i and then make it pass over v_i); this does not affect iocr and ensures that $u_i v_i$ and $v_i w_i$ cross evenly. We conclude that for $\psi \in \{\text{ocr, iocr}\}$ any two edges of the inner framework cross an even number of times. We next show how to remove crossings between edges of the inner framework.

To this end, let us consider $Q_1 = u_1 v_1 w_1$. Let e be an edge of the inner framework that crosses $u_1 v_1$ (we allow the case $e = v_1 w_1$). Deform e near each such crossing so that it follows along $u_1 v_1$ toward v_1 and then over v_1. Since e must have crossed $u_1 v_1$ an even number of times, this procedure will not change the value of ψ for the drawing. Performing this for all such edges e of the inner framework leaves $u_1 v_1$ free of crossings with edges of the inner framework. This redrawing process may have introduced self-crossings of $v_1 w_1$ which can be removed without affecting ψ, as described at the end of Section 2. So $u_1 v_1$ crosses no edge of the inner framework and $v_1 w_1$ crosses every other edge of the inner framework evenly. Without loss of generality, we can assume that t is in the exterior of $s u_1 v_1 w_1 s$. Then the interior of $s u_1 v_1 w_1 s$ does not contain any vertices: every vertex (other than t) has a path consisting of edges of weight at least w_I to t, contributing at least w_I^2 to ψ, which we know to be impossible. Now cut each edge e of the inner framework where it crosses $v_1 w_1$. We can partition the crossings of e and $v_1 w_1$ into pairs since they cross evenly, and then for each pair we add curves that run along each side of $v_1 w_1$ that connect the severed ends of e. Thus, e is replaced by a curve that may have more than one component, all but one of which are closed curves with no vertex, and none of the components intersect $v_1 w_1$. Because of the way the connecting curves are added in pairs, the value of ψ is unchanged. The components lying within $s u_1 v_1 w_1 s$ are all closed curves without vertices. Moreover, since there is no vertex within that region, they can be deleted without affecting ψ. Any two of the curves on the other side of Q_1 can be merged by erasing a tiny bit of each curve and adding two parallel curves within the region that join the erased bits of opposite curves, giving a wide berth to all vertices, which ensures that ψ is unchanged. Repeating this process merges all curve components in that region into a single curve, and after removing self-intersections we obtain a valid drawing of e within that region. We can now repeat this argument with Q_2 and $s u_1 u_2 v_2 w_2 w_1 s$, and so on, to establish that none of the Q_i, $1 \leq i \leq n$ have crossings with any edges of the inner framework. This completes the second step.

Hence, for the third step, we can assume that every crossing is between two edges of G or between an edge of G and an edge of the inner framework.

At this point, let us deform the whole drawing so that $C_{2n+2} \cup \{Q_1, Q_n\} - \{s, t\}$ is a rectangle and all the Q_i are parallel straight-line segments orthogonal to the outer framework.

For $\psi = \mathrm{cr}$ we are nearly done: a G-edge e connecting v_i to v_j must cross all Q_k with $i < k < j$, forcing at least c crossings. This leaves $\psi(G') - c \leq$ mon-$\psi(G) \leq n^4 < w_I$ crossings counting towards $\psi(G')$. Since a crossing with an edge of the inner framework contributes at least w_I to $\psi(G')$ this accounts for all crossings with edges of the inner framework. So an edge $e = v_i v_j$ crosses all Q_k with $i < k < j$ and no other Q_ks. The actual behavior of e between two neighboring Q_ks is irrelevant and within each such region we can replace e by a straight-line segment connecting its crossings between neighboring Q_ks.

This does not affect ψ and results in a monotone drawing of G with $\psi(G') - c$ crossings, proving that mon-$\psi(G) \leq \psi(G') - c$ which is what we had to prove.

For $\psi \in \{\text{ocr}, \text{iocr}\}$ we need to do a bit more work. A G-edge e connecting v_i to v_j must cross all Q_k with $i < k < j$ oddly. So the crossings of G-edges with the inner framework contribute at least c to the value of ψ. This leaves at most $\psi(G') - c \leq$ mon-$\psi(G) < w_I$ in $\psi(G')$ unaccounted for. So there are no non-adjacent odd pairs with edges of the inner framework except those absolutely necessary to connect the endpoints of every edge in G. The only case in which odd pairs with inner framework edges can still occur is in the iocr case (where such crossings do not count) if an edge $v_i v_j$, $i < j$ crosses an adjacent inner framework edge ($u_i v_i$, $v_i w_i$, $u_j v_j$, or $v_j w_j$) oddly. In this case we redraw $v_i v_j$ near each endpoint (if necessary) so that the ends of $v_i v_j$ at v_i and v_j lie between Q_i and Q_j; this does not affect iocr and results in $v_i v_j$ crossing both Q_i and Q_j an even number of times. It is possible at this point that $v_i v_j$ crosses both $u_k v_k$ and $v_k w_k$ oddly, where $k \in \{i, j\}$. In that case we perform a $(v_i v_j, v_k)$-move; this does not affect iocr and ensures that $v_i v_j$ crosses both $u_k v_k$ and $v_k w_k$ evenly.

Thus for $\psi \in \{\text{ocr}, \text{iocr}\}$ we can now assume that if an edge $e = v_i v_j$ crosses $u_k v_k$ or $v_k w_k$ with $k \leq i$ or $k \geq j$ it must do so evenly. As we did above for the inner framework edges, we push all crossings of e with $u_k v_k$ along $u_k v_k$ and over v_k to $v_k w_k$ so that $u_k v_k$ does not cross e at all; pushing e off $u_k v_k$ does not affect ψ, since e crossed $u_k v_k$ evenly. For all $k \leq i$ and $k \geq j$ cut e at $v_k w_k$; pair up crossings of e with $v_k w_k$ and reconnect severed ends of e on both side of $v_k w_k$ for all $k \leq i$, $k \geq j$. Closed components of e between Q_i and Q_j can be reconnected to the arc-component of e without affecting ψ. Every other closed component of e is entirely contained in a region which does not contain a vertex, so all such components are even and can be dropped without affecting ψ. In the end, all of e lies in the region formed by C_{2n+2} and Q_i and Q_j.

Now for any $i < k < j$ we have either $\text{ocr}(e, u_k v_k) = 0$ and $\text{ocr}(e, v_k w_k) = w_I$ or $\text{ocr}(e, u_k v_k) = w_I$ and $\text{ocr}(e, v_k w_k) = 0$ (since we have already accounted for all crossings with edges of weight at least w_I). For every k push all crossings of e with Q_k from the edge with ocr $= 0$ to the other edge (not affecting the value of ψ); that is, e avoids one of the edges of Q_k for every $i < k < j$. Let e' be any other curve in the region in C_{2n+2} bounded by Q_i, Q_j that shares ends with e (here, an end is an endpoint together with a small, crossing-free part of the edge incident to the endpoint); furthermore, suppose that e' avoids the same edge in each Q_k as does e. Then $\text{ocr}(e, g) = \text{ocr}(e', g)$ for every edge g (other than e), since e can be continuously deformed to e' without passing over any vertex. In particular, we can replace e with a monotone polygonal arc without changing the value of ψ. Repeating this for all edges of G gives us a monotone drawing of G with mon-ψ crossings. This completes the argument for $\psi \in \{\text{ocr}, \text{iocr}\}$. $\square$

References

1. Brass, P., Moser, W., Pach, J.: Research Problems in Discrete Geometry. Springer, New York (2005)
2. Fulek, R., Pelsmajer, M.J., Schaefer, M., Štefankovič, D.: Hanani-Tutte, monotone drawings, and level-planarity. Accepted for WG (2011)

3. Chojnacki, C., Hanani, H.: Über wesentlich unplättbare Kurven im drei-dimensionalen Raume. Fundamenta Mathematicae 23, 135–142 (1934)
4. Kainen, P.C.: A lower bound for crossing numbers of graphs with applications to K_n, $K_{p,q}$, and $Q(d)$. J. Combinatorial Theory Ser. B 12, 287–298 (1972)
5. Pach, J., Tóth, G.: Thirteen problems on crossing numbers. Geombinatorics 9(4), 194–207 (2000)
6. Pach, J., Tóth, G.: Which crossing number is it anyway? J. Combin. Theory Ser. B 80(2), 225–246 (2000)
7. Pach, J., Tóth, G.: Monotone drawings of planar graphs. J. Graph Theory 46(1), 39–47 (2004)
8. Pelsmajer, M.J., Schaefer, M., Stasi, D.: Strong Hanani–Tutte on the projective plane. SIAM Journal on Discrete Mathematics 23(3), 1317–1323 (2009)
9. Pelsmajer, M.J., Schaefer, M., Štefankovič, D.: Removing even crossings. J. Combin. Theory Ser. B 97(4), 489–500 (2007)
10. Pelsmajer, M.J., Schaefer, M., Štefankovič, D.: Odd crossing number and crossing number are not the same. Discrete Comput. Geom. 39(1), 442–454 (2008)
11. Pelsmajer, M.J., Schaefer, M., Štefankovič, D.: Removing even crossings on surfaces. European Journal of Combinatorics 30(7), 1704–1717 (2009)
12. Pelsmajer, M.J., Schaefer, M., Štefankovič, D.: Removing independently even crossings. SIAM Journal on Discrete Mathematics 24(2), 379–393 (2010)
13. Pelsmajer, M.J., Schaefer, M., Štefankovič, D.: Crossing numbers of graphs with rotation systems. Algorithmica 60, 679–702 (2011), doi:10.1007/s00453-009-9343-y
14. Székely, L.A.: A successful concept for measuring non-planarity of graphs: the crossing number. Discrete Math. 276(1-3), 331–352 (2004)
15. Székely, L.A.: Progress on Crossing Number Problems. In: Vojtáš, P., Bieliková, M., Charron-Bost, B., Sýkora, O. (eds.) SOFSEM 2005. LNCS, vol. 3381, pp. 53–61. Springer, Heidelberg (2005)
16. Székely, L.A.: An optimality criterion for the crossing number. Ars Math. Contemp. 1(1), 32–37 (2008)
17. Tóth, G.: Note on the pair-crossing number and the odd-crossing number. Discrete Comput. Geom. 39(4), 791–799 (2008)
18. Tutte, W.T.: Toward a theory of crossing numbers. J. Combinatorial Theory 8, 45–53 (1970)
19. Valtr, P.: On the pair-crossing number. In: Combinatorial and Computational Geometry. Math. Sci. Res. Inst. Publ., vol. 52, pp. 569–575. Cambridge University Press, Cambridge (2005)
20. West, D.: Open problems - graph theory and combinatorics, http://www.math.uiuc.edu/~west/openp/ (accessed April 7, 2005)

Low Distortion Delaunay Embedding
of Trees in Hyperbolic Plane

Rik Sarkar

Institut Für Informatik,
Freie Universität Berlin, Germany
sarkar@inf.fu-berlin.de

Abstract. This paper considers the problem of embedding trees into
the hyperbolic plane. We show that any tree can be realized as the Delaunay graph of its embedded vertices. Particularly, a weighted tree can
be embedded such that the weight on each edge is realized as the hyperbolic distance between its embedded vertices. Thus the embedding
preserves the metric information of the tree along with its topology. The
distance distortion between non adjacent vertices can be made arbitrar-
ily small – less than a $(1 + \varepsilon)$ factor for any given ε. Existing results
on low distortion of embedding discrete metrics into trees carry over to
hyperbolic metric through this result. The Delaunay character implies
useful properties such as guaranteed greedy routing and realization as
minimum spanning trees.

1   Introduction

Embedding given data into a standard space lets us use properties of the target
space as additional structure in the original dataset, and brings to front information that is hard to detect in the raw input. If the target space is Euclidean,
that allows us to visualize and treat the data geometrically. In general, if it is a
metric space, the properties of the metric can aid in understanding the original
data and answering queries. This approach has been found relevant to a variety
of subjects such as data visualization, network analysis, routing, localization,
machine learning, statistics, biology and many others.

Trees are an important class of data structures. They occur commonly in
natural scenarios, therefore associating trees with geometric spaces can be of
benefit in many domains. Realizing trees as Delaunay graphs lets us combine
the structural properties of trees with those of Delaunay graphs as well as those
of the ambient space.

In this paper, we show that this can always be achieved in the hyperbolic
spaces. In particular, we tackle the question for weighted trees and show that
any given edge weights can be realized exactly (upto a choice of unit) in the
delaunay embedding, while keeping the overall distortion arbitrarily low.

M. van Kreveld and B. Speckmann (Eds.): GD 2011, LNCS 7034, pp. 355–366, 2012.

1.1   Related Work

Trees in Euclidean Plane. Monma and Suri [11] address embedding minimum spanning trees in the Euclidean space. They analyze questions of perturbations to the embedding and their effect on the topology of the MST. Relevant to us, they consider the problem of which trees can be realized as minimum spanning tree in the Euclidean plane, and show that any tree with maximum vertex degree of 5 or less admits an embedding as a minimum spanning tree. The topic of distortion of a tree embedded in euclidean metric is analyzed in [10].

Low Distortion Metric Embeddings. This is an extensively studied subject that we do not have the space to discuss. We just note that metrics can have nice probabilistic embeddings into trees. N-point metrics have distributions over embeddings in trees, and weighted graphs have distributions over spanning trees with small expected distortions. See [6,3,4].

The hyperbolic metric behaves like a tree in many respects. One way to model this is the concept of $\delta-$hyperbolic metric[8]. Chepoi and co-authors[1] show that a graph that has an n-node $\delta-$hyperbolic graph admits an approximating tree of additive error $O(\delta \log n)$.

Hyperbolic Embeddings Embedding graphs into hyperbolic spaces can provide advantages that we do not get in Euclidean spaces or trees. Kleinberg [9] considers the problem of *greedy routing* in wireless networks. This is the method, where a node routes a message by forwarding it to the neighbor that is nearest to the destination. The idea in [9] is to embed a spanning tree of the network into the hyperbolic plane such that this routing always works successfully on the tree, and thereby on the original network. Eppstein and Goodrich [5] show a related method that uses small sized coordinates. Cvetkovski et al. [2] extend [9] to incorporate dynamic insertion of edges. Papadopoulos et al. [12] show that hyperbolic embedding can naturally give rise to scale free networks and as before, can support greedy routing. Zeng et al. [15] embed the universal covering space of a network into the hyperbolic plane as a Riemann surface and show that this can be used to easily find paths of different homotopy types in the network.

1.2   Our Contributions

The choice of the target space determines the properties we can expect to obtain from the embedding. In this paper, we use the hyperbolic plane as the standard target space and show that this works remarkably well for all kinds of trees – weighted as well as unweighted, and induces many desirable features. We present the important ideas and theorems in the main body of the paper. Details of proofs can be found in the full version of the paper online [13].

Delaunay Embedding. Delaunay graphs are known to have many useful properties. If a graph can be treated as the Delaunay graph in some space, we can expect to leverage some of these properties for our purposes. With this motivation, we take up the question of realizing trees as Delaunay graphs, and show that the vertices of any tree can be embedded in the hyperbolic plane such that their Delaunay graph is the original tree. This leads to two immediate consequences: Realization of any tree as the minimum spanning tree of its vertices and guaranteed delivery by greedy routing. The first overcomes the constant degree bound of Euclidean case, the second implies the result of [9]. The embedding in [9] is a specific Delaunay embedding based on tiling of $\mathbb{H}$ by ideal regular d-gons, where d is the max degree. We generalize this to not depend on a tiling, which gives greater flexibility needed to create the low distortion embeddings in sections 4 and 5.

Weighted Trees. Suppose the given tree has weights or lengths defined for its edges. This induces a metric and we refer to the input as a *metric tree*. We show that the vertices of any such metric tree has a Delaunay embedding in the hyperbolic plane, such that each Delaunay edge has a length that is the multiple of the edge's prescribed length by a constant. As before, the properties implied by Delaunay embedding (MST and greedy routing) are retained.

Thus, the embedding not only preserves the topology of the tree, it also preserves the geometry – the metric information from the tree. The effect of the constant is negligible: since it is same globally, once it has been computed we can scale our unit of length to eliminate the factor. Our definition of distortion is oblivious to such global scaling.

Distortion of Metric Trees. While we preserve edge lengths precisely, the overall metric may not be preserved. The hyperbolic distance between v_i and v_j will be less than or equal to the path length between the two vertices in the tree. We show that this distortion can be kept arbitrarily low. Given any ε, we show it is possible to do the Delaunay embedding such that the distortion is less than $1 + \varepsilon$. Thus the tree can be seen as a hyperbolic spanner of the embedded vertices.

This implies distortion bounds for embedding of metrics in general. The results in [6,3,4] show low distortion bounds (probabilistic and average) for embedding arbitrary n point metrics into trees. It is therefore possible to embed arbitrary metrics into hyperbolic metric with the same bounds, using the tree embedding as an intermediate step.

2   Basics and Notations

This section introduces some basic facts and notations about the geometry of hyperbolic plane that will be used later. We denote the hyperbolic plane by $\mathbb{H}$. For a more detailed exposition of these ideas, see [7]. We present the ideas in terms of the plane $\mathbb{H}$, but they generalize directly to higher dimensions. We use $|.|_{\mathbb{H}}$ to denote distances in hyperbolic metric. Correspondingly, we use $|.|_T$ to denote distances in the metric of the input tree.

In hyperbolic geometry, the axioms of euclidean geometry are all true, except the parallel axiom, which is replaced by the *Hyperbolic Axiom: There is a line l and point P not on l, such that there are at least 2 different lines through P parallel to l.*

This leads to the general property that given a line, there are an infinite number of different lines parallel to it through an outside point. Parallel lines can be of different types. A pair of parallel lines are said to be *limiting parallel* if they approach each-other asymptotically without intersecting. Such a pair does not admit a common line perpendicular to both, and the distance between the two does not have a minimum. A pair of parallel lines are called *divergent parallel* if they move away from each-other in both directions. They have a common segment perpendicular to both. This segment achieves the minimum distance between the two lines.

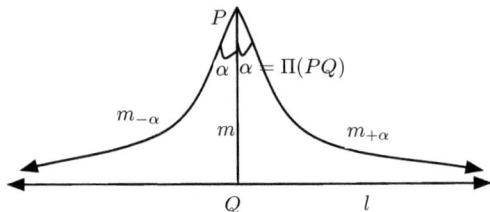

Fig. 1. Line $m \perp l$ and rays $m_{-\alpha}$ and $m_{+\alpha}$ are limiting parallel to l. The angle α depends on the length of the segment PQ. Hyperbolic straight lines $m_{-\alpha}$ and $m_{+\alpha}$ look curved because our figure is in euclidean space.

Given a line l and a point P outside, there is always a point Q on l such that $PQ \perp l$. Through P, there are always rays $m_{+\alpha}$ and $m_{-\alpha}$ that are limiting parallel to l in the two directions. The angle between PQ and $m_{+\alpha}$ (or symmetrically, the angle between PQ and $m_{-\alpha}$) is called the angle of parallelism and represented by $\Pi(PQ)$. The Bolyai and Lobachevsky formula gives its value in terms of the length $|PQ|_{\mathbb{H}}$:

$$\tan \frac{\Pi(PQ)}{2} = e^{-|PQ|_{\mathbb{H}}/k}, \tag{1}$$

where k is a constant for the hyperbolic plane in consideration. Note that $\Pi(PQ)$ is always less than $\pi/2$ radians, since PQ cannot be perpendicular to a ray that is limiting parallel to l. Given l and P, the limiting parallel rays and the angle $\Pi(PQ)$ are unique. A ray that creates a larger angle with PQ will be divergent parallel, while a ray with smaller angle will intersect l.

The region bounded by rays $m_{-\alpha}$ and $m_{+\alpha}$ and containing the ray $m = \overrightarrow{PQ}$ will be important to our discussion. Let us refer to it as a closed cone $\bar{C}(\overrightarrow{PQ}, \alpha)$. We are particularly interested in the set $\bar{C}(\overrightarrow{PQ}, \alpha) \setminus \{P\}$, we call it a cone of angle α at P (or rooted at P) and denote it by $C(\overrightarrow{PQ}, \alpha)$.

The usual Euclidean axioms of betweenness, incidence and angles hold in the hyperbolic case. Therefore two cones $C(m, \alpha)$ and $C(n, \beta)$ at P do not intersect if and only if the angle between m and n is greater than $\alpha + \beta$. We say such pairs of cones are *disjoint*.

Observation 1. *Given any finite integer d we can always construct d mutually disjoint cones at P by taking d different rays and cones of suitably small angles around them.*

Our goal is to compute an embedding function $\Phi : V \to \mathbb{H}$, where $V = v_0, v_1, v_2, \ldots$ is the set of vertices of the tree. To abbreviate notations, we write $\varphi_i = \Phi(v_i)$. We sometimes abbreviate our notations for cones as $C_{ij}^\alpha = C(\overrightarrow{\varphi_i \varphi_j}, \alpha)$.

We consider distortion over some set W of pairs of distinct vertices in question. For example, W can be the set of edges in the tree, or it can be the set of all pairs of vertices.

Distortion. We define the *contraction factor* over W as $\delta_c = \max_{(i,j) \in W} \frac{|v_i v_j|_T}{|\varphi_i \varphi_j|_\mathbb{H}}$, and similarly the *expansion factor* as $\delta_e = \max_{(i,j) \in W} \frac{|\varphi_i \varphi_j|_\mathbb{H}}{|v_i v_j|_T}$. The *distortion* is defined as $\delta = \delta_c \cdot \delta_e$.

Observe that if the embedding globally scales all distances for pairs in W by the same factor, then $\delta = 1$.

We consider Voronoi diagrams in $\mathbb{H}$. Given a finite set of vertices $v_0, v_1, \cdots \subset \mathbb{H}$, the *Voronoi cell* of v_i, denoted $\mathcal{V}(v_i)$ is the set of points whose distance to v_i is not larger than the distance to v_j for any $j \neq i$. The **Delaunay Graph** is its dual: given a set of vertices in $\mathbb{H}$ their Delaunay graph is one where a pair of vertices are neighbors if their Voronoi cells intersect. As with the euclidean case, this delaunay graph contains the MST.

Delaunay Embedding of Graphs: Given a graph G, its Delaunay embedding in $\mathbb{H}$ is an embedding of the vertices such that their Delaunay graph is G.

3   Delaunay Embedding of Trees

In this section, we describe the basic construct of embedding a tree as a Delaunay graph. The Delaunay graph automatically has the minimum spanning tree embedding and greedy embedding property. The basic idea is not new, and has been used in [9,2,14]. But we wish to make it more general and write in terms of cones rooted at the embedded vertices. This makes it easier to handle weighted trees and low distortion embedding in following sections.

Reorganizing equation 1, we have that given rays $m_{+\alpha}$ and $m_{-\alpha}$ through P, the distance $|PQ|_\mathbb{H}$ to a point Q on m so that the line $l \perp PQ$ at Q is limiting parallel to $m_{-\alpha}$ and $m_{+\alpha}$ is given by:

$$|PQ|_\mathbb{H} = -k \ln \left(\tan \frac{\alpha}{2} \right). \tag{2}$$

Since α is always less than $\pi/2$ radians, we have $\tan \frac{\alpha}{2} < 1$. Therefore $|PQ|_\mathbb{H}$ is positive and monotone decreasing in α. This means in particular, that if $|PQ|_\mathbb{H}$ is larger, then l is limiting parallel to the bounding lines of a smaller cone, and therefore divergent parallel to $m_{-\alpha}$ and $m_{+\alpha}$, and fully contained in $C(m, \alpha)$.

We construct a function Φ that embeds the vertices of a tree T into $\mathbb{H}$. Function Φ is designed such that T is the Delaunay graph of the embdded vertices.

Lemma 1. *Given two cones $C(\overrightarrow{PR}, \alpha)$ and $C(\overrightarrow{RP}, \beta)$, and $\gamma = \min(\alpha, \beta)$, if $|PR|_{\mathbb{H}} \geq -2k \ln\left(\tan\frac{\gamma}{2}\right)$ the perpendicular bisector of the segment PR lies in the intersection $C(\overrightarrow{PR}, \gamma) \cap C(\overrightarrow{RP}, \gamma)$ which is contained in the intersection $C(\overrightarrow{PR}, \alpha) \cap C(\overrightarrow{RP}, \beta)$.*

Proof. Suppose without loss of generality that $\gamma = \alpha$. Let us say that the perpendicular bisector l intersects PR at its midpoint Q.

Then $|PQ|_{\mathbb{H}} \geq -k \ln\left(\tan\frac{\alpha}{2}\right)$ and therefore $l \subset C(\overrightarrow{PR}, \alpha)$. Symmetrically, $|RQ|_{\mathbb{H}} \geq -k \ln\left(\tan\frac{\alpha}{2}\right) \implies l \subset C(\overrightarrow{RP}, \alpha)$. Since $\alpha \leq \beta$, we have $C(\overrightarrow{RP}, \alpha) \subseteq C(\overrightarrow{RP}, \beta)$. Therefore $l \subset C(\overrightarrow{PR}, \alpha) \cap C(\overrightarrow{RP}, \alpha)$ and $l \subset C(\overrightarrow{PR}, \alpha) \cap C(\overrightarrow{RP}, \beta)$. □

This means in general we can consider the smaller of the two angles and consider the two cones to be of this same angle. See Figure 2. We say a segment PR is Delaunay for angle γ if it satisfies the conditions of the lemma for $\gamma = \alpha = \beta$.

Now we describe embedding an edge into a given cone. The goal is to embed in a way such that the Voronoi cell of one embedded vertex is completely contained inside a cone at the other vertex, and the edge is realized as the Delaunay edge between the embedded vertices, and the two cones can have the same angle. Again, see Figure 2.

Edge embedding function Φ. Given an edge $v_i v_j$ and an angle α, we select a cone $C(m, \alpha)$ at point P, and embed as follows. Vertex v_i is embedded at $\varphi_i = P$. We select a point R on the ray m such that $|PR|_{\mathbb{H}} \geq -2k \ln\left(\tan\frac{\alpha}{2}\right)$, and embed v_j as $\varphi_j = R$. We call this a Delaunay embedding of the edge for angle α.

This construction satisfies the conditions of lemma 1 with $\gamma = \alpha = \beta$, and therefore cones of angle α at φ_i and φ_j contain the perpendicular bisector of $\varphi_i \varphi_j$.

Now we extend the embedding to the entire tree. The idea is to embed such that each edge is Delaunay for a suitable angle, and the corresponding cones rooted at a vertex are disjoint.

Definition 1. *Tree Embedding Function Φ. The vertices of a tree are embedded in a way that allows an assignment of an angle $\theta(e)$ to each edge e such that:*

1. *Cones determined by θ over all edges incident on any vertex are disjoint.*
2. *For any edge $v_i v_j$, its embedding $\varphi_i \varphi_j$ is Delaunay for angle $\theta(v_i v_j)$.*

The following algorithm describes the construction of such a Φ. Without loss of generality, we treat the input as a rooted tree.

Algorithm: *Construction of Φ for T.* Embed the root v_0 to an arbitrary point φ_0. If v_0 has d children, they are Delaunay embedded individually into d disjoint cones at φ_0. We embed all other vertices inductively as follows. Suppose

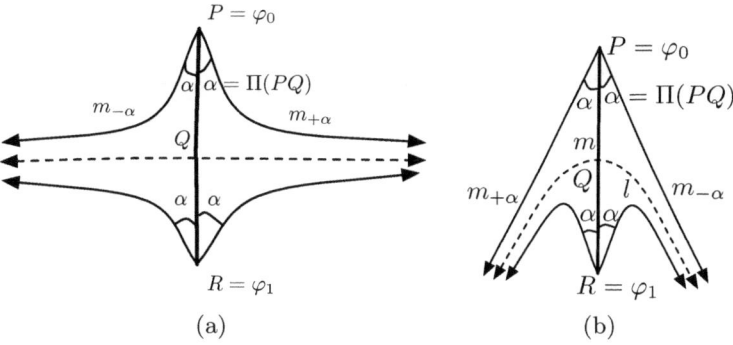

Fig. 2. Embedding an edge v_0v_1. (a) Shows the symmetry in the embedding – it is the same from the point of view of both end points. The Voronoi edge is the dashed line. (b) shows a different view, which is more akin to the view from point P. This view gives the intuition that we can embed any number of cones of small enough angles at P, and by symmetry, same is true at point R.

v_j is a child of v_i, and has been embedded in a cone of angle α at φ_i. The children of v_j are Delaunay embedded in mutually disjoint cones that are also disjoint from the cone $C(\overrightarrow{\varphi_j\varphi_i}, \alpha)$.

By construction, this algorithm produces an embedding that satisfies definition 1. The construction works for infinite trees as well.

Lemma 2. *If v_i is the parent of v_{i+1}, and Φ embeds the edge v_iv_{i+1} in the cone $C(\overrightarrow{\varphi_i, \varphi_{i+1}}, \alpha)$ then*

1. *The Voronoi cells of all nodes in the subtree rooted at φ_{i+1} are contained in $C(\overrightarrow{\varphi_i, \varphi_{i+1}}, \alpha)$.*
2. *The Voronoi cell of any node not in the subtree rooted at φ_{i+1} are contained in the cone $C(\overrightarrow{\varphi_{i+1}, \varphi_i}, \alpha)$.*

The hyperbolic plane contains many mutually parallel lines. Voronoi edges can be aligned to such lines, therefore, the Voronoi diagram consists of disjoint lines, carving out many disjoint half planes. The consequence is that Voronoi cells of nodes from different subtrees do not intersect. The only pairs of Voronoi cells that can intersect correspond to pairs of nodes that form edges of the tree. Therefore:

Theorem 2. *The function Φ embeds T as a Delaunay graph in $\mathbb{H}$.*

This directly implies that T is embedded as the minimum spanning tree of its vertices, since the Delaunay graph contains the MST.

Greedy Embedding. Greedy routing is a well studied problem in wireless networks. Delaunay graphs are significant to this problem as well:

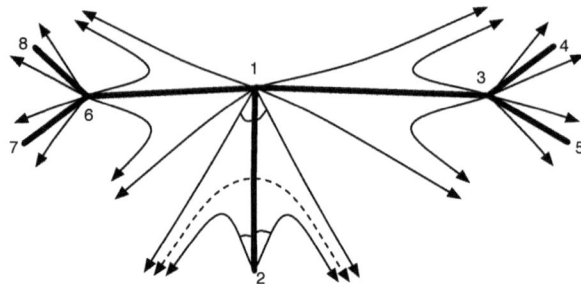

Fig. 3. Embedding a tree, by using cones of the type shown in Fig. 2 for each edge. 1, 2, 3 . . . are the vertices of the tree. The rays form the cones. The thick edges are the edges of the tree.

Theorem 3. *A Delaunay embedding guarantees delivery by greedy routing.*

This implies the result of [9]. In fact the embedding of [9] is a specific instance of Delaunay embedding.

Edge Insertion. It is easy to adapt the algorithm to allow dynamic insertion into trees. At every vertex, we select the cones such that there is always enough angle left over to create more cones. This can always be done, since we can make the new cone small enough to not cover the entire available space. A new edge can be inserted into a new cone created in this space. This implies the edge insertion result of [2].

4   Delaunay Embedding of Metric Trees

Now we show a stronger embedding. Suppose $T = (V, E, w)$ is a weighted tree, where $w : E \to \mathbb{R}$ is the weight or length function on the edges. The goal is to realize the weight $w(v_i v_j)$ on each edge $v_i v_j$ as the length $|\varphi_i \varphi_j|_{\mathbb{H}}$ of the edge in the Delaunay embedding of the tree. For now, we are interested only in the distortion of individual edges of the tree, and show that in that sense there is an embedding with no distortion. The general distortion case is handled in the next section.

In the previous section we saw that each edge has to be embedded to a minimum length depending on the cone in which it is embedded. Based on this idea, we proceed as follows.

For each edge e, we compute a minimum length $L(e)$ needed to embed it as a Delaunay edge along with its neighboring edges. This gives a minimum scaling factor $\eta(e) = L(e)/w(e)$ for each edge. Next we compute $\eta_{\max} = \max\limits_{e \in E} \eta(e)$ as the worst scaling needed at any edge. Then the tree is embedded such that each edge has length $\eta_{\max} \cdot w(e)$, that is, each edge is scaled by the same amount. Therefore, for unit length $\eta_{\max}$ the Delaunay graph realizes the metric tree.

Algorithm: *Computation of $\eta_{\max}$*. Let us write $d(v)$ for the degree of a vertex v. The algorithm executes 5 passes over the tree:

1. Select for each vertex a maximum cone angle $\mu(v_i) < 2\pi/d(v_i)$.
2. Select for each edge $v_i v_j$ the maximum cone angle as $\alpha_{ij} = \min(\mu(v_i), \mu(v_j))$.
3. Compute for each edge the minimum required length $L(v_i v_j) = -2k \ln\left(\tan \frac{\alpha_{ij}}{2}\right)$.
4. Compute for each edge the minimum scaling factor $\eta(v_i v_j) = L(v_i v_j)/w(v_i v_j)$.
5. Compute the max value of η over all edges : $\eta_{\max} = \max\limits_{v_i v_j \in E} \eta(v_i v_j)$.

Algorithm: *Embedding function Φ_w of metric tree T*. The function is a special case of Φ, and proceeds the same way. The cone construction and embedding steps are made more specific as follows:

1. At vertex φ_i with parent φ_h, we create $d(v_i) - 1$ disjoint cones of angle $\mu(v_i)$, disjoint from cone $C(\overrightarrow{\varphi_i \varphi_h}, \mu(v_i))$.
2. A child v_j of v_i is embdded in the cone $C(\overrightarrow{\varphi_i \varphi_j}, \mu(v_i))$ such that $|\varphi_i \varphi_j|_{\mathbb{H}} = \eta_{\max} \cdot w(v_i v_j)$.

For any edge $v_i v_j$ in E, the function Φ_w embeds it in cones $C(\overrightarrow{\varphi_i \varphi_j}, \mu(v_i))$ and $C(\overrightarrow{\varphi_j \varphi_i}, \mu(v_j))$ respectively.

The need to compute $\eta_{\max}$ implies that algorithm works only on finite trees. But if constant bounds are known for the maximum vertex degree and the minimum edge length, then $\eta_{\max}$ can be computed beforehand and the algorithm can be applied to infinite trees.

Lemma 3. $|\varphi_i \varphi_j|_{\mathbb{H}} \geq -2k \ln\left(\tan \frac{\alpha_{ij}}{2}\right)$.

Therefore, the embedding satisfies definition 1 and:

Theorem 4. *The embedding Φ_w is a Delaunay embedding of a metric tree with distortion 1 over the set of edges of the tree.*

5   Delaunay Embedding with $(1 + \varepsilon)$ Distortion: Hyperbolic Spanner

In this section, we address the question of reducing the overall distortion to arbitrarily close to 1. Thus, in such an embedding, the tree acts as hyperbolic spanner of the embedded points. This embedding also implies that known results that show existence of *(tree, embedding)* pairs translate to existence of embeddings in the hyperbolic plane.

Definition 2. β separated cones. *Suppose cones $C(\overrightarrow{\varphi_j \varphi_i}, \alpha)$ and $C(\overrightarrow{\varphi_j \varphi_x}, \gamma)$ are adjacent with the same root φ_j. Then the cones are β separated if the two cones are an angle 2β apart. That is, for arbitrary points $p \in C(\overrightarrow{\varphi_j \varphi_i}, \alpha)$ and $q \in C(\overrightarrow{\varphi_j \varphi_x}, \gamma)$, the angle $\angle p \varphi_j q > 2\beta$.*

An embedding is globally β separated if every pair of adjacent cones used for embedding are β separated.

Lemma 4. *If cones* $C(\overrightarrow{\varphi_j\varphi_i}, \alpha)$ *and* $C(\overrightarrow{\varphi_j\varphi_x}, \gamma)$ *are* β *separated and* $\varphi_r \in C(\overrightarrow{\varphi_j\varphi_i}, \alpha)$ *and* $\varphi_s \in C(\overrightarrow{\varphi_j\varphi_x}, \gamma)$ *then there is a constant* ν *depending only on* β *such that* $|\varphi_r\varphi_j|_{\mathbb{H}} + |\varphi_s\varphi_j|_{\mathbb{H}} > |\varphi_r\varphi_s|_{\mathbb{H}} > |\varphi_r\varphi_j|_{\mathbb{H}} + |\varphi_s\varphi_j|_{\mathbb{H}} - \nu$.

Proof. Suppose l is the line limiting parallel to both the rays $\overrightarrow{\varphi_j\varphi_r}$ and $\overrightarrow{\varphi_j\varphi_s}$. Let us refer to the point φ_j as P, and the perpendicular on l as PQ. Then we know that $|PQ|_{\mathbb{H}} \leq -k \ln\left(\tan \frac{\beta}{2}\right)$.

Now, $\varphi_j, \varphi_r, \varphi_s$ are on the same side of l. Therefore, $\varphi_r\varphi_s$ intersects PQ, say at W. This implies that $PW < PQ$.

By triangle inequalities : $|\varphi_r\varphi_j|_{\mathbb{H}} + |\varphi_s\varphi_j|_{\mathbb{H}} > |\varphi_r\varphi_s|_{\mathbb{H}} > |\varphi_r\varphi_j|_{\mathbb{H}} + |\varphi_s\varphi_j|_{\mathbb{H}} - 2|PW|_{\mathbb{H}} > |\varphi_r\varphi_j|_{\mathbb{H}} + |\varphi_s\varphi_j|_{\mathbb{H}} - 2|PQ|_{\mathbb{H}}$. Substituting $\nu = 2|PQ|_{\mathbb{H}} \leq -2k \ln\left(\tan \frac{\beta}{2}\right)$, we get the result. $\qquad\square$

If we can embed a tree such that neighboring edges, that is, edges incident on the same vertex, are always Delaunay in β separated cones for some suitably small beta, then for such embeddings, we have the following theorem:

Theorem 5. *If all edges of T are scaled by a constant factor $\tau \geq \eta_{\max}$ such that each edge is longer than $\nu \frac{(1+\varepsilon)}{\varepsilon}$ and the Delaunay embedding of T is β separated, then the distortion over all vertex pairs is bounded by $1 + \varepsilon$.*

Proof. Let $v_i, v_{i+1}, v_{i+2}, \ldots, v_{i+p}$ be the path in the tree between the two end points. For any vertex v_{i+j} on the path, observe that φ_i is contained in a cone $C(\overrightarrow{\varphi_{i+j}\varphi_{i+j-1}}, \alpha)$ – a consequence of lemma 2. This cone, by construction, is β separated from cone $C(\overrightarrow{\varphi_{i+j}\varphi_{i+j+1}}, \gamma)$ containing φ_{i+j+1}.

Thus we have : $|\varphi_i\varphi_{i+2}|_{\mathbb{H}} \geq |\varphi_i\varphi_{i+1}|_{\mathbb{H}} + |\varphi_{i+1}\varphi_{i+2}|_{\mathbb{H}} - \nu \geq |\varphi_i\varphi_{i+1}|_{\mathbb{H}} + |\varphi_{i+1}\varphi_{i+2}|_{\mathbb{H}} - \frac{\varepsilon}{1+\varepsilon}|\varphi_{i+1}\varphi_{i+2}|_{\mathbb{H}}$, since each edge is longer than $\nu \frac{1+\varepsilon}{\varepsilon}$. Repeating, we have

$$|\varphi_i\varphi_{i+2}|_{\mathbb{H}} \geq |\varphi_i\varphi_{i+1}|_{\mathbb{H}} + |\varphi_{i+1}\varphi_{i+2}|_{\mathbb{H}} - \frac{\varepsilon}{1+\varepsilon} \cdot |\varphi_{i+1}\varphi_{i+2}|_{\mathbb{H}}$$

$$|\varphi_i\varphi_{i+3}|_{\mathbb{H}} \geq |\varphi_i\varphi_{i+2}|_{\mathbb{H}} + |\varphi_{i+2}\varphi_{i+3}|_{\mathbb{H}} - \frac{\varepsilon}{1+\varepsilon} \cdot |\varphi_{i+2}\varphi_{i+3}|_{\mathbb{H}}$$

$$|\varphi_i\varphi_{i+4}|_{\mathbb{H}} \geq |\varphi_i\varphi_{i+3}|_{\mathbb{H}} + |\varphi_{i+3}\varphi_{i+4}|_{\mathbb{H}} - \frac{\varepsilon}{1+\varepsilon} \cdot |\varphi_{i+3}\varphi_{i+4}|_{\mathbb{H}}$$

$$\cdots$$

$$|\varphi_i\varphi_{i+p}|_{\mathbb{H}} \geq |\varphi_i\varphi_{i+p-1}|_{\mathbb{H}} + |\varphi_{i+p-1}\varphi_{i+p}|_{\mathbb{H}} - \frac{\varepsilon}{1+\varepsilon} \cdot |\varphi_{i+p-1}\varphi_{i+p}|_{\mathbb{H}}$$

Adding:

$$|\varphi_i\varphi_{i+p}|_{\mathbb{H}} \geq \sum_{x=i}^{i+p-1} |\varphi_x\varphi_{x+1}|_{\mathbb{H}} - \sum_{x=i}^{i+p-1} \frac{\varepsilon}{1+\varepsilon}|\varphi_x\varphi_{x+1}|_{\mathbb{H}}$$

Therefore, $|\varphi_i\varphi_{i+p}|_{\mathbb{H}} \geq \dfrac{1}{1+\varepsilon} \cdot \displaystyle\sum_{x=i}^{i+p-1} |\varphi_x\varphi_{x+1}|_{\mathbb{H}}.$

Since τ is fixed for all edges, $|\varphi_i\varphi_{i+p}|_{\mathbb{H}} \geq \frac{1}{1+\varepsilon} \cdot \tau |v_i v_{i+p}|_T$.

All edges are assumed to be scaled by a factor $\tau \geq \eta_{\max}$ such that they are longer than $\nu \frac{(1+\varepsilon)}{\varepsilon}$. We can therefore assume without loss of generality that $\tau > 1$. Now, for vertices v_i and v_j, $|v_i v_j|_{\mathbb{H}} \leq \tau |v_i v_j|_T$. That is, hyperbolic distance is at most τ times the tree distance.

Thus the expansion factor δ_e is dominated by the edges of the tree satisfying equality, and $\delta_e = \tau$. The contraction factor is dominated by the pair with maximum distortion: $\delta_c \leq (1+\varepsilon)/\tau$. Therefore, distortion $\delta = \delta_c \cdot \delta_e \leq (1+\varepsilon)$.

$\square$

Based on the theorem, the algorithm for $1 + \varepsilon$ distortion embedding becomes simple:

Algorithm: $(1+\varepsilon)$ ***Distortion embedding.*** Suppose d is the maximum degree of any node. Then a low distortion embedding algorithm follows in these steps:

1. Compute a cone separation angle $\beta < \pi/d$, and an angle for cones $\alpha = 2\pi/d - 2\beta$. Set $\nu = -2k \ln\left(\tan\dfrac{\beta}{2}\right)$.
2. Compute $\eta_{\max}$ as in the previous section.
3. Select $\tau > \eta_{\max}$ such that all edges are longer than $\nu\frac{1+\varepsilon}{\varepsilon}$.
4. Embed edges as before, but into β separated cones, and edges scaled by a factor τ.

The bounds of embedding arbitrary metrics into trees with low distortion extend to the hyperbolic plane via this result. From [6] we have that for any metric, there is a distribution over embeddings in the hyperbolic plane with expected distortion $O(\log n)$. Similarly [4] implies that for every graph there is an embedding into hyperbolic plane such that the average distortion of edges is bounded by $O(\log^2 n \log\log n)$. As before, the algorithm applies to infinite trees with known degree and length bounds.

6   Conclusion

We presented a method to embed trees into hyperbolic plane that simultaneously has several desirable properties. It is a Delaunay realization of the tree, preserves edge lengths exactly, and distance between non-neighbors is distorted by at most $1 + \varepsilon$. This suggests hyperbolic plane as a useful general target to investigate embedding questions. It remains to be seen what additional properties can be obtained in $\mathbb{H}$, for the more general embedding questions.

Acknowledgement. Thanks to Jie Gao and the anonymous reviewers for many helpful comments on the original draft of the article. This work is funded by the German Research Foundation (DFG) through the research training group Methods for Discrete Structures (GRK 1408).

References

1. Chepoi, V., Dragan, F., Estellon, B., Habib, M., Vaxès, Y., Xiang, Y.: Additive spanners and distance and routing labeling schemes for hyperbolic graphs. Algorithmica, 1–20 (2010), doi:10.1007/s00453-010-9478-x
2. Cvetkovski, A., Crovella, M.: Hyperbolic embedding and routing for dynamic graphs. In: Proceedings of Infocom 2009 (April 2009)
3. Dhamdhere, K., Gupta, A., Räcke, H.: Improved embeddings of graph metrics into random trees. In: Proceedings of the Seventeenth Annual ACM-SIAM Symposium on Discrete Algorithm, SODA 2006, pp. 61–69 (2006)
4. Elkin, M., Emek, Y., Spielman, D.A., Teng, S.-H.: Lower-stretch spanning trees. In: Proceedings of the Thirty-Seventh Annual ACM Symposium on Theory of Computing, STOC 2005, pp. 494–503 (2005)
5. Eppstein, D., Goodrich, M.T.: Succinct Greedy Graph Drawing in the Hyperbolic Plane. In: Tollis, I.G., Patrignani, M. (eds.) GD 2008. LNCS, vol. 5417, pp. 14–25. Springer, Heidelberg (2009)
6. Fakcharoenphol, J., Rao, S., Talwar, K.: A tight bound on approximating arbitrary metrics by tree metrics. In: Proceedings of the Thirty-fifth Annual ACM Symposium on Theory of Computing, pp. 448–455 (2003)
7. Greenberg, M.J.: Euclidean and Non-Euclidean Geometries. W.H. Freeman (1993)
8. Gromov, M.: Hyperbolic groups. In: Essays In Group Theory, pp. 75–263. Springer, New York (1987)
9. Kleinberg, R.: Geographic routing using hyperbolic space. In: Proceedings of the 26th Conference of the IEEE Communications Society (INFOCOM 2007), pp. 1902–1909 (2007)
10. Lee, J., Naor, A., Peres, Y.: Trees and markov convexity. Geometric and Functional Analysis 18, 1609–1659 (2009), doi:10.1007/s00039-008-0689-0
11. Monma, C., Suri, S.: Transitions in geometric minimum spanning trees (extended abstract). In: Proceedings of the Seventh Annual Symposium on Computational Geometry, pp. 239–249 (1991)
12. Papadopoulos, F., Krioukov, D., Boguñá, M., Vahdat, A.: Greedy forwarding in dynamic scale-free networks embedded in hyperbolic metric spaces. In: Proceedings of the 29th Conference on Information Communications, INFOCOM 2010, pp. 2973–2981 (2010)
13. Sarkar, R.: Low distortion delaunay embedding of trees in hyperbolic plane, http://page.inf.fu-berlin.de/sarkar/papers/HyperbolicDelaunayFull.pdf
14. Tanuma, T., Imai, H., Moriyama, S.: Revisiting hyperbolic voronoi diagrams from theoretical, applied and generalized viewpoints. In: International Symposium on Voronoi Diagrams in Science and Engineering, pp. 23–32 (2010)
15. Zeng, W., Sarkar, R., Luo, F., Gu, X.D., Gao, J.: Resilient routing for sensor networks using hyperbolic embedding of universal covering space. In: Proc. of the 29th Annual IEEE Conference on Computer Communications (INFOCOM 2010) (April 2010)

Hardness of Approximate Compaction
for Nonplanar Orthogonal Graph Drawings

Michael J. Bannister and David Eppstein

Computer Science Department, University of California, Irvine

Abstract. We show that several problems of compacting orthogonal graph drawings to use the minimum number of rows or the minimum possible area cannot be approximated to within better than a polynomial factor in polynomial time unless P = NP. However, there is a fixed-parameter-tractable algorithm for testing whether a drawing can be compacted to a given number of rows.

1   Introduction

Orthogonal graph drawing is a widely used graph drawing style for low-degree graphs, in which each vertex is represented as a point or a rectangle in an integer grid, and each edge is represented as a polyline composed out of axis-parallel line segments [4]. When used for nonplanar graphs, orthogonal drawing has several desirable properties including polynomial area, high angular resolution, and right-angled edge crossings; the last property, in particular, has been shown to aid in legibility of graph drawings [6].

Typical orthogonal graph drawing systems employ a multiphase approach [1,4] in which the input graph is *planarized* by replacing its crossings with vertices, a *topological embedding* of the graph (specifying the ordering of the edges around each vertex, but not the vertex and edge locations) is found, a flow algorithm is used to orient the edges in a way that minimizes the number of bends [10], and vertex coordinates are assigned. If vertices of degree greater than four exist, they may be expanded to rectangles as another phase of this process [1]. Finally, the drawing is improved by *compaction*, a step in which the vertices and bends of the graph are moved to new locations in order to reduce the area of the drawing while preserving its edge orientations and other features.

Some positive algorithmic results are known for the final compaction step; for instance, Bridgeman et al. [2] showed that planar orthogonal drawings in which the shapes of the faces in the drawing are restricted (so-called *turn-regular drawings*) may be compacted into optimal area in polynomial time. However, when drawing nonplanar graphs, it may not be necessary or desirable for the compaction phase to preserve a fixed planarization of the graph. If one is compacting one dimension of a drawing at a time, then for planar compaction it is only possible to map the rows of the drawing monotonically to a smaller set of rows, while for nonplanar graphs it may also be useful to permute the rows with respect to each other. This greater freedom to choose how to compact the drawing

M. van Kreveld and B. Speckmann (Eds.): GD 2011, LNCS 7034, pp. 367–378, 2012.

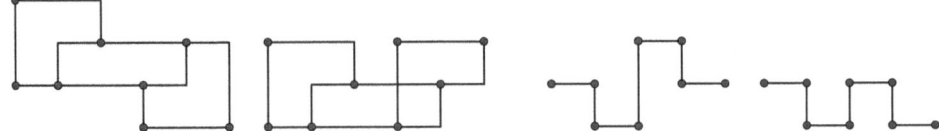

Fig. 1. Left: input and output drawings for row-by-row compaction. Right: input and output drawings for vertex-by-vertex compaction.

may lead to much greater savings in drawing area, but it also leads to greater difficulty in finding a good compaction.

As Patrignani [9] showed, even for arbitrary planar orthogonal graph drawings, compacting the drawing in a way that minimizes its area, total edge length, or maximum edge length is NP-hard. Although these results do not directly extend to the nonplanar case, NP-hardness in that case also follows from results of Eades et al. on rectilinear (bendless) drawing [3], and Maňuch et al. where certain restricted cases of rectilinear drawing are considered [8]. But since compaction is performed primarily for aesthetic reasons (a smaller area drawing allows the drawing to be viewed at a larger scale, making its features more legible), exact optimization may not be important as long as a layout with small area can be achieved. Thus, we are led to the problem of how closely it is possible to *approximate* the minimum area layout. The problem of approximate compaction for nonplanar orthogonal drawings was explicitly listed as open by Eiglsperger et al. [4], and there appears to have been little progress on it since then.

In this paper we show that nonplanar compaction is hard even to approximate: there exists a real number $c > 0$ such that, unless $\mathsf{P} = \mathsf{NP}$, no polynomial time algorithm can find a compaction of a drawing with n features that is within a factor of n^c of optimal. The main idea is to find approximation-preserving reductions from graph coloring, a problem known to be hard to approximate. We also find *fixed-parameter tractable* algorithms for finding compactions that use very small numbers of grid rows, for drawings for which such a compaction is possible.

1.1 Variations of the Compaction Problem

In the compaction problems we study, the task is to move vertices and bends while preserving the axis-parallel orientation (although not necessarily the direction) of each edge, to minimize the number of rows or area of the drawing. Our results apply either to *orthogonal drawings* (drawings in which edges may be polylines with bends, possible for any graph of maximum degree four) or *rectilinear drawings* (bendless drawings, only possible for some graphs) [3,5]: the distinction between bends and vertices is unimportant for our results.

We distinguish between three variants of the compaction problem, depending on what vertex motions are allowed. In *row-by-row compaction* (Figure 1, left), the compacted layout maps each row of the input layout to a row of the output; all vertices that belong to the same row must move in tandem. In *vertex-*

by-vertex vertical compaction (Figure 1, right), each vertex or bend may move independently, but only its y-coordinate may change; it must retain its horizontal position. In *vertex-by-vertex free compaction*, vertices or bends may move arbitrarily in both coordinate directions. In all three of these problems, edges or edge segments must stay vertical or horizontal according to their orientation in the original layout. The compaction is not allowed to cause any new intersection between a vertex and a feature it was not already incident with, nor is it allowed to cause any two edges or edge segments to overlap for nonzero length; however, it may introduce new crossings that were not previously present.

1.2   New Results

We show the following results.

- In the row-by-row compaction problem, it is difficult to compact even a drawing of a path graph (or a drawing of the two-vertex graph with many bends): if the drawing has n vertices or bends, then unless P = NP there is no polynomial time algorithm that can find a compacted drawing whose number of rows is within $O(n^{1/2-\epsilon})$ of optimal, or whose area is within $O(n^{1/2-\epsilon})$ of optimal, for any $\epsilon > 0$. Moreover, even finding drawings with a fixed number of rows is hard: it is NP-complete to determine whether there exists a compaction with only three rows.
- In vertex-by-vertex vertical compaction, there exist orthogonal graph drawings of maximum degree three such that, unless P = NP, there is no polynomial time algorithm that can find a compacted drawing whose number of rows is within $O(n^{1/4-\epsilon})$ of optimal, or whose area is within $O(n^{1/4-\epsilon})$ of optimal, where n is the number of features in the drawing, for any $\epsilon > 0$. The same result also applies in the vertex-by-vertex free compaction problem.
- For vertex-by-vertex vertical or free compaction of three-dimensional orthogonal drawings, it is not possible (unless P = NP) to approximate the minimum number of layers in any one dimension to within $O(n^{1/2-\epsilon})$ of optimal in polynomial time, for any $\epsilon > 0$, nor is it possible in polynomial time to determine whether a three-layer drawing exists.
- In row-by-row and vertex-by-vertex vertical compaction in either two or three dimensions, there is an approximation algorithm with approximation ratio $O(\sqrt{n})$, showing that some of our inapproximability bounds are tight.
- In vertex-by-vertex vertical compaction, there is an algorithm for testing whether an orthogonal graph drawing can be compacted into k rows, whose running time is $O(k!n)$. Thus, the problem is fixed-parameter tractable.

2   Preliminaries

2.1   Orthogonal Drawing

We define an *orthogonal drawing* of a graph to be a drawing in which each vertex is represented as a point in the Euclidean plane (although most of our results apply as well to drawings in which the vertices are rectangles), and each edge is represented as a polyline (a polygonal chain of line segments), with each line segment parallel to one of the coordinate axes. If each edge is itself a line segment, the drawing is *rectilinear*; otherwise, the segments of a polyline meet at *bends*. Each vertex or bend must only intersect the edges that it belongs to, and no two vertices or bends may coincide. Edges may cross each other, but only at right angles, at points that are neither vertices nor bends.

It is natural, in orthogonal drawing, to restrict the coordinates of the vertices and bends to be integers. In this case, the *width* of a two-dimensional drawing is the maximum difference between the x-coordinates of any two of its vertices or bends, the *height* is the maximum difference between y-coordinates of any two vertices or bends, and the *area* is the product of the width and height.

A *compaction* of a drawing D is another drawing D' of the same graph, in which the vertices and bends of D' correspond one-for-one with the vertices and bends of D, and in which corresponding segments of the two drawings are parallel to each other. Typically, D' will have smaller height or area than D. We distinguish between three types of compaction:

- In *row-by-row compaction*, the x-coordinate of each vertex or bend remains unchanged, and two vertices or bends that have the same y-coordinate in D must continue to have the same y-coordinate in D' (Figure 1, left).
- In *vertex-by-vertex vertical compaction*, the x-coordinate of each vertex or bend remains unchanged, but the y-coordinates may vary independently of each other subject to the condition that the result remains a valid drawing with edge segments parallel to the original drawing (Figure 1, right).
- In *vertex-by-vertex free compaction*, the x- and y- coordinates of each vertex or bend are free to vary independently of other vertices or bends.

As can be seen in Figure 1, we allow compaction to introduce new edge crossings and to reverse the directions of edge segments. These concepts generalize straightforwardly to three dimensions.

2.2   Graph Coloring and Inapproximability

In the *graph coloring problem*, we are given as input a graph and seek to color the vertices of the graph with as few colors as possible, in such a way that the endpoints of each edge are assigned different colors. Our results on the difficulty of compaction are based on known inapproximability results for graph coloring, one of the triumphs of the theory of probabilistically checkable proofs.

Lemma 1 (Zuckerman [11]). *Let $\epsilon > 0$ be any fixed constant. Then, unless* P = NP, *there is no polynomial time algorithm that can color a given n-vertex graph using a number of colors within a factor of $n^{1-\epsilon}$ of the optimal number.*

Our proofs use approximation-preserving reductions from coloring to compaction: given a graph G to be colored, we will construct a different graph G' and a drawing D of G' such that the layers in a compaction D' of D necessarily correspond to the colors in a coloring of G. With a reduction of this type, the approximation ratio for compacting D cannot be better than the approximation ratio for coloring G. However, D will in general have many more vertices and bends than the number of vertices in G: the size of D will be at least proportional to the number of edges in G, which is quadratic in its number of vertices. Therefore, although the approximation ratio will remain unchanged as a number by our reduction it will be expressed as a different function of the input size.

2.3 Notation

We write n_G, n_D, or (where unambiguous) n for the number of vertices in a graph G or drawing D and m_G, m_D, or m for its number of edges. Additionally, b_D stands for the number of bends in drawing D, $\lambda(D)$ is the number of rows in a vertex-by-vertex compaction of D, and $\bar{\lambda}(D)$ is the number of rows in a row-by-row compaction. $\chi(G)$ represents the chromatic number of graph G.

3 Hardness of Row-by-Row Compaction

As a warm-up, we start with a simplified path compaction problem in which every pair of objects on the same row of the drawing must move in tandem. Our proof constructs a drawing of a path graph such that the valid row assignments for our drawing are the same as the valid colorings of a given graph G.

Lemma 2. *Given a graph G we can construct in polynomial time a rectilinear drawing D of a path graph with $O(m_G)$ vertices, such that $\bar{\lambda}(D) = \chi(G)$.*

Proof. Find a Chinese postman walk for G; that is, a walk that starts at an arbitrary vertex and visits each edge at least once, allowing vertices to be visited multiple times. Such a walk may be found, for instance, by doubling each edge of G and constructing an Euler tour of the doubled graph. Let $u_i v_i$ be the ith edge in the walk, where $v_i = u_{i+1}$, and let $k \leq 2m_G$ be the number of edges in the walk. Additionally, choose arbitrary distinct integer numbers for the vertices of G with $\ell(v)$ being the number for the vertex v.

To construct the drawing D, for i from 0 to k, place vertices in the plane at the points $(i, \ell(u_i))$ and $(i + 1, \ell(u_i))$, connected by a unit-length horizontal edge. Additionally, for i from 0 to $k - 1$ draw a vertical edge from $(i + 1, \ell(u_i))$ to $(i + 1, \ell(v_i))$. See Figure 2 for an example of such a construction.

Two rows in the drawing conflict if and only if the corresponding vertices in G are adjacent. For every coloring of G, we may compact D by using one row for the vertices of each color, and conversely for every row-by-row compaction of D we may color G by using one color class for each row of the compaction (Figure 3). Therefore, $\bar{\lambda}(D) = \chi(G)$. Also, $n_D = 2k + 2 = O(m_G)$. □

The same drawing D can equivalently be viewed as an orthogonal drawing of the two-vertex graph K_2 with $O(m_G)$ bends. In the restricted model of compaction used in this section, horizontal compaction is disallowed, so optimizing the area of a compaction of D is the same as optimizing its number of rows.

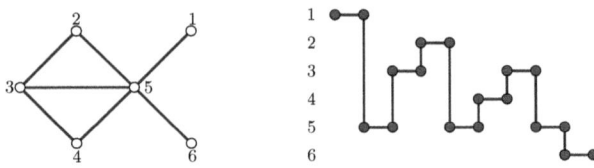

Fig. 2. Path constructed from a graph G using the walk 1, 5, 3, 2, 5, 4, 3, 5, 6

Fig. 3. The rows of a compacted drawing correspond to the colors in a coloring of G

Theorem 1. *Let $\epsilon > 0$ be any positive fixed constant, and suppose that* P $\neq$ NP. *Then there does not exist a polynomial time algorithm that approximates the number of layers or the area in an optimal row-by-row compaction of a given orthogonal or rectilinear drawing D to within a factor of $(n_D + b_D)^{1/2-\epsilon}$.*

Proof. Suppose for a contradiction that algorithm $\mathcal{A}$ can solve the row-by-row compaction problem to within a factor $\rho \leq (n_D + b_D)^{1/2-\epsilon}$ of optimal. Let $\mathcal{A}'$ be an algorithm for coloring an input graph G by performing the following steps:

1. Use Lemma 2 to construct a path drawing D from the given graph G.
2. Use algorithm $\mathcal{A}$ to compact D.
3. Color G using one color for each row of the compacted drawing.

Then the approximation ratio of algorithm $\mathcal{A}'$ for coloring is the same number ρ as the approximation ratio of algorithm $\mathcal{A}$ for compaction, whether measured by area or by number of rows. However,

$$\rho \leq (n_D + b_D)^{1/2-\epsilon} = O(m_G^{1/2-\epsilon}) = O(n_G^{1-2\epsilon}),$$

an approximation ratio that contradicts Lemma 1.   □

The same reduction, together with the NP-completeness of graph 3-colorability, shows that it is NP-complete to determine whether a given drawing D has a row-by-row compaction that uses at most three rows; we omit the details.

4   Hardness of Vertex-by-Vertex Compaction

Our hardness result for vertex-by-vertex vertical compaction follows roughly the same outline as Theorem 1: translate graph vertices into drawing features such that two features can be compacted onto the same row if and only if the corresponding graph vertices can be assigned the same color. However, direct overlaps between pairs of features would only let us represent interval graphs, which are easily colored, so instead we use an *edge gadget* depicted in Figure 4 to represent an edge between two vertices of the input graph. This gadget has six vertices and six line segments; the two vertices A and B of the gadget may be placed on two line segments representing vertices of the input graph. This connection forces the two line segments containing A and B to be placed on different rows of any compacted draw-

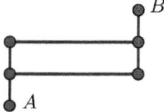

Fig. 4. Edge gadget

ing, even if these two line segments have no vertical overlap with each other: one of the two line segments must be above the central rectangle of the gadget, and the other must be below the central rectangle, although either of these two orientations is possible.

The use of these edge gadgets leads to a second difficulty in our reduction: the number of rows in the compacted drawing will depend both on the features coming from input graph vertices and the rows needed by the edge gadgets themselves. In order to make the first of these two terms dominate the total, we represent an input graph vertex by a *bundle* of θ parallel line segments, for some integer $\theta > 0$. The edge gadgets may be modified to enforce that all segments in one bundle be in different rows from all segments of a second bundle, as shown in Figure 5, while only using a constant number of rows for the gadget itself.

Figure 6 shows the complete reduction, for a graph G with five vertices and six edges, and for $\theta = 1$. Each vertex of G is represented as a horizontal black line segment (or bundle of segments, for $\theta > 1$), and each edge of G is represented by an edge gadget. The vertices of G are numbered arbitrarily from 1 to n_G, and these numbers are used to assign vertical positions to the corresponding bundles of segments in the drawing. The edge gadgets are given x-coordinates that allow them to attach to the two vertex bundles they should be attached to, and y-coordinates that place them between these two vertex bundles.

Lemma 3. *Given a graph G and a parameter θ we can construct in polynomial time an orthogonal drawing D such that the vertices of D have maximum degree 3, $n_D = O(n_G^2\theta)$, and*

$$\chi(G)\theta \leq \lambda(D) \leq \chi(G) + O(n_G)^2.$$

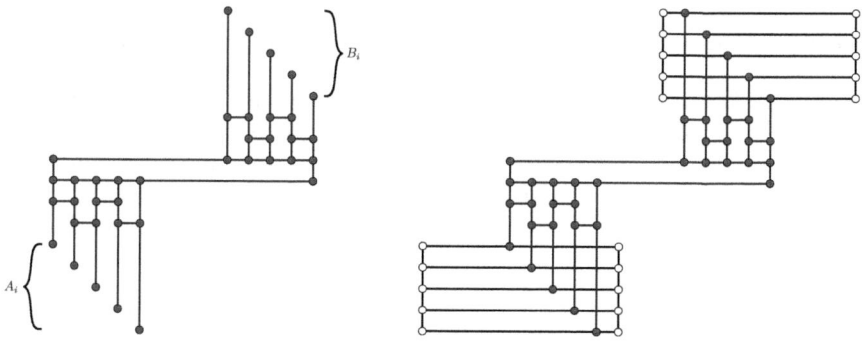

Fig. 5. The full edge gadget for $\theta = 5$.

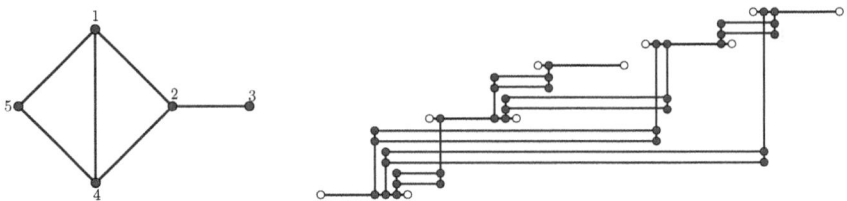

Fig. 6. Example of the complete reduction for $\theta = 1$

Proof. The construction of D is as described above. It is straightforward to verify the bounds on n_D and on the degree. If G has a coloring with χ colors, it is possible to assign the vertex bundles of D to χ sets of θ rows each, according to those colors, with an additional $O(n_G)$ rows between any two such sets to allow room for the edge gadgets to be placed without interference with each other. Therefore, $\lambda(D) \leq \chi(G) + O(n_G)^2$.

If D' is a compacted drawing of D, acyclically orient the edges of G from the vertex whose bundle is below the edge gadget to the vertex whose bundle is above the edge gadget, and assign each vertex v in G a color indexed by the length of the longest path from a source to v in this acyclic orientation. Then the number of colors needed equals the number of vertices in the longest path, and the number of rows in D' needed just for the vertices in this path is θ times the number of vertices of G in the path. Therefore, $\chi(G)\theta \leq \lambda(D)$. □

Theorem 2. *If* P $\neq$ NP, *then no polynomial time algorithm approximates the number of layers or the area in an optimal vertex-by-vertex vertical compaction of a given orthogonal graph drawing to within a factor of* $(n_D + n_B)^{1/4-\epsilon}$.

Proof. If an algorithm could achieve this approximation ratio for compaction, we could get an $O(n^{1-4\epsilon})$ ratio for coloring by applying Lemma 3 with $\theta = n_G^2$, compacting the resulting drawing, and using the coloring derived from the compaction in the proof of Lemma 3. But this would contradict Lemma 1. □

5   Hardness of Vertex-by-Vertex Free Compaction

In the reduction from the previous section, allowing the vertices to move horizontally as well as vertically does not make any difference in how much vertical compaction is possible. However, if we want to prove inapproximability for minimal-area compaction, we also need to worry about horizontal compaction. By making the width incompressible we may make the vertical compaction factor the same as the area compaction factor.

Lemma 4. *From a drawing D a drawing D' can be constructed by adding at most $O(n_D)$ vertices, such that $\lambda(D') = \lambda(D) + 1$ and D' is incompressible in the horizontal direction. If D has maximum degree three, then so does D'.*

Proof. Place a line of vertices on a new row below D; for each set of vertices with a given x-coordinate in D, add a vertex on the new row at the same x-coordinate. Connect the added vertices with horizontal edges, and add a vertical edge to connect these vertices to D at the point of D that is rightmost on its bottom row, as shown in Figure 7. This added layer conflicts with all existing horizontal layers, and forces D' to be incompressible in the horizontal direction. □

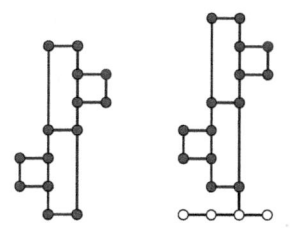

Fig. 7. Adding a row to D prevents horizontal compaction

Theorem 3. *Unless* P = NP, *it is impossible to find vertex-by-vertex free compactions with area within a factor of* $n_D^{1/4-\epsilon}$ *of optimal in polynomial time.*

6   Hardness of Three-Dimensional Compaction

Our hardness result for three-dimensional compaction follows from the construction of a drawing whose valid two-dimensional layer assignments are the same as the valid colorings of a graph G. We assign to each vertex in G a horizontal layer containing an L-shaped pair of line segments, such that when projected vertically onto a plane every two of these L shapes cross each other. For each edge in G we place a vertical edge in the drawing connecting the L shapes that correspond to the endpoints of the edge. Figure 8 shows an example.

Lemma 5. *Given a graph G we can construct in polynomial time a 3D orthogonal drawing D with maximum degree three such that* $n_D = 3n_G + 2m_G = O(n_G^2)$, *and such that the number of layers in an optimal y-compaction is* $\chi(G)$.

Theorem 4. *If* P $\neq$ NP, *then there does not exist a polynomial time algorithm that approximates the number of layers in an optimal layer compaction of a given three dimensional orthogonal drawing to within a factor of* $n_D^{1/2-\epsilon}$.

We omit the proofs, which follow the same lines as the previous results.

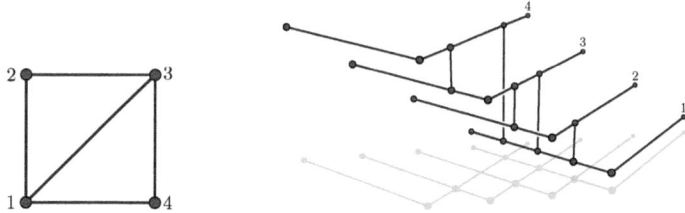

Fig. 8. Reduction from coloring to three-dimensional compaction where y is the vertical direction

7   Approximation Algorithm

In this section we show that several versions of compaction can be approximated to within a ratio of $O(\sqrt{n})$ of optimal in polynomial time. Our intent in presenting this is not so much to describe a useful compaction algorithm but rather to show that our $\Omega(n^{1/2-\epsilon})$ inapproximability bounds are nearly tight.

Our approximation method applies to both row-by-row and vertex-by-vertex vertical compaction, in two or three dimensions, with the optimization criterion being minimizing the number of rows or layers. In each case, we may form an *incompatibility graph*, where the vertices of the incompatibility graph represent sets of drawing features that must move in tandem: rows or layers, in row-by-row compaction, or connected components of the subgraph of the drawing formed by horizontal edges, in vertex-by-vertex vertical compaction. Two vertices of the incompatibility graph are connected by an edge when the drawing features they represent cannot be compacted to the same layer of the drawing, that is, when they contain parts of the drawing that are directly above one another.

Our approximation algorithm is, essentially, a standard greedy graph coloring algorithm applied to the incompatibility graph. Specifically, it performs the following steps.

1. Construct the incompatibility graph G from the given drawing D.
2. Find a *degeneracy ordering* of G by initializing an empty list L, and then repeatedly finding and adding to the end of L the vertex v minimizing the number of neighbors of v that do not already belong to L.
3. Process the vertices of G in the reverse of the ordering given by L. For each vertex, in this order, assign it the smallest positive integer that is distinct from the integers assigned to its already-processed neighbors.
4. Use the numbers assigned to the vertices in G as the coordinates of the corresponding features in a compaction of D.

To analyze this algorithm, we consider the *degeneracy* δ of G [7]. If we orient G from earlier vertices to later vertices in the degeneracy ordering described above, δ is the maximum outdegree of a vertex in the orientation. Alternatively, δ is the smallest number with the property that every set S of vertices in G includes a vertex that has at most δ neighbors in S. Let κ denote the largest number of features of D that can be pierced by a vertical line through a vertex or bend. As we now show, $\delta \leq \sqrt{2(n_D + b_D)\kappa}$. For, if $|S| \leq \sqrt{(n_D + b_D)\kappa}$, then clearly all vertices have at most $\sqrt{2(n_D + b_D)\kappa}$ neighbors in S. And, if $|S| \geq \sqrt{2(n_D + b_D)\kappa}$, then there are at most $(n_D + b_D)k$ edges in G (each vertex or bend of D contributes at most k incompatibilities) so by an averaging argument there is a vertex in S with degree at most $2(n_D + b_D)k/|S| \leq \delta$.

Theorem 5. *For 2d or 3d row-by-row or vertex-by-vertex vertical compaction, the algorithm described above computes a valid compaction whose number of rows or layers is within an $O(\sqrt{n_D + b_D})$ factor of optimal.*

Proof. No two features can overlap in the compacted drawing: for, if two features do not overlap vertically in D, they cannot overlap no matter how they are compacted, and if two features do overlap vertically then the corresponding nodes in G will be adjacent and will be assigned distinct coordinate values. Therefore, the result of the algorithm is a valid compaction.

Any valid compaction must have at least κ layers. But as we have seen, each vertex in G has $O(\sqrt{(n_D + b_D)\kappa})$ earlier neighbors in the order and each of these neighbors can only eliminate one choice from the set of possible coordinate values, so its coordinate value in the compaction is $O(\sqrt{(n_D + b_D)\kappa})$. Therefore, the approximation ratio is $O((\sqrt{(n_D + b_D)\kappa})/\kappa) = O(\sqrt{n_D + b_D})$.   □

8   Fixed-Parameter Tractability of Vertex-by-Vertex Vertical Compaction

Lemma 6. *Given an orthogonal drawing D we can compact D into k layers in $O(k!(b + n))$ time, if such a compaction is possible.*

Proof. We construct local assignments of the features into k rows via a left-to-right plane sweep. The drawing may be assumed to be in a $n \times n$ grid, so the features can be sorted in linear time. While sweeping the plane we maintain a set of those features intersecting the sweep line along with a record of valid assignments of these features into the k rows.

When a feature first intersects the sweep line we try to place it into the collection of valid assignments. If there are ℓ features intersecting the sweep line prior to the insertion, we have at most $\ell!\binom{k}{\ell}$ valid assignments to consider. In each of these valid assignments there are $k - \ell$ free rows. Altogether at most $k!$ configurations will be considered for each feature insertion. When the sweep line moves past a feature its row is freed for future use.

If at any point we cannot find any valid assignment for a new feature, we conclude that a compaction into k rows is not possible. On the other hand if the last feature can be placed into a valid assignment, then a compaction into k layers is possible. To recover the global assignment of horizontal features into rows, we may backtrack through the sets of local assignments.   □

Theorem 6. *An optimal vertex-by-vertex vertical compaction of an orthogonal drawing D can be found in $O(\lambda!(b + n))$ time where $\lambda = \lambda(D)$.*

Proof. Apply Lemma 6 for $k = 1, 2, 3, \ldots$ until finding a value of k for which a valid layering exists.   □

9   Conclusions

Our investigations have determined upper and lower bounds for several different approximation and fixed-parameter versions of the compaction problem. In some cases, our bounds are tight: we have upper and lower bounds on the approximation ratio with the same exponent. In some other cases, there remain gaps, the most important of which is in the problem with the greatest relevance for practical graph drawing: vertex-by-vertex free compaction of two-dimensional orthogonal drawings to minimize area. For this problem, we have an $\Omega(n^{1/4-\epsilon})$ lower bound on the approximation ratio, and no upper bound. Can our $O(\sqrt{n})$ approximation algorithms be extended to cover this case? Can the exponent in the lower bound be improved? We leave these questions open for future research.

Acknowledgements. This work was supported in part by NSF grant 0830403 and by the Office of Naval Research under grant N00014-08-1-1015.

References

1. Biedl, T.C., Madden, B.P., Tollis, I.G.: The three-phase method: a unified approach to orthogonal graph drawing. Int. J. Comput. Geom. Appl. 10(6), 553–580 (2000), doi:10.1142/S0218195900000310
2. Bridgeman, S.S., Di Battista, G., Didimo, W., Liotta, G., Tamassia, R., Vismara, L.: Turn-regularity and optimal area drawings of orthogonal representations. Computational Geometry: Theory and Applications 16(1), 53–93 (2000), doi:10.1016/S0925-7721(99)00054-1
3. Eades, P., Hong, S.-H., Poon, S.-H.: On Rectilinear Drawing of Graphs. In: Eppstein, D., Gansner, E.R. (eds.) GD 2009. LNCS, vol. 5849, pp. 232–243. Springer, Heidelberg (2010), doi:10.1007/978-3-642-11805-0_23
4. Eiglsperger, M., Fekete, S.P., Klau, G.W.: Orthogonal Graph Drawing. In: Kaufmann, M., Wagner, D. (eds.) Drawing Graphs. LNCS, vol. 2025, pp. 121–171. Springer, Heidelberg (2001), doi:10.1007/3-540-44969-8_6
5. Eppstein, D.: The Topology of Bendless Three-Dimensional Orthogonal Graph Drawing. In: Tollis, I.G., Patrignani, M. (eds.) GD 2008. LNCS, vol. 5417, pp. 78–89. Springer, Heidelberg (2009),doi:10.1007/978-3-642-00219-9_9
6. Huang, W., Hong, S.-H., Eades, P.: Effects of crossing angles. In: IEEE Pacific Visualization Symposium (PacificVIS 2008), pp. 41–46 (2008), doi:10.1109/PACIFICVIS.2008.4475457
7. Lick, D.R., White, A.T.: k-degenerate graphs. Canadian Journal of Mathematics 22, 1082–1096 (1970), doi:10.4153/CJM-1970-125-1
8. Maňuch, J., Patterson, M., Poon, S.-H., Thachuk, C.: Complexity of Finding Non-Planar Rectilinear Drawings of Graphs. In: Brandes, U., Cornelsen, S. (eds.) GD 2010. LNCS, vol. 6502, pp. 305–316. Springer, Heidelberg (2011), doi:10.1007/978-3-642-18469-7_28
9. Patrignani, M.: On the complexity of orthogonal compaction. Computational Geometry 19(1), 47–67 (2001), doi:10.1016/S0925-7721(01)00010-4
10. Tamassia, R.: On embedding a graph in the grid with the minimum number of bends. SIAM J. Comput. 16(3), 421–444 (1987), doi:10.1137/0216030
11. Zuckerman, D.: Linear degree extractors and the inapproximability of max clique and chromatic number. Theory of Computing 3(1), 103–128 (2007), doi:10.4086/toc.2007.v003a006

Monotone Drawings of Graphs with Fixed Embedding *

Patrizio Angelini[1], Walter Didimo[2], Stephen Kobourov[3], Tamara Mchedlidze[4],
Vincenzo Roselli[1], Antonios Symvonis[4], and Stephen Wismath[5]

[1] Università Roma Tre, Italy
[2] Università degli Studi di Perugia, Italy
[3] University of Arizona, USA
[4] National Technical University of Athens, Greece
[5] University of Lethbridge, Canada

Abstract. A drawing of a graph is a *monotone drawing* if for every pair of vertices u and v, there is a path drawn from u to v that is monotone in some direction. In this paper we investigate planar monotone drawings in the *fixed embedding setting*, i.e., a planar embedding of the graph is given as part of the input that must be preserved by the drawing algorithm. In this setting we prove that every planar graph on n vertices admits a planar monotone drawing with at most two bends per edge and with at most $4n - 10$ bends in total; such a drawing can be computed in linear time and requires polynomial area. We also show that two bends per edge are sometimes necessary on a linear number of edges of the graph. Furthermore, we investigate subclasses of planar graphs that can be realized as embedding-preserving monotone drawings with straight-line edges, and we show that biconnected embedded planar graphs and outerplane graphs always admit such drawings, which can be computed in linear time.

1   Introduction

A drawing of a graph is a *monotone drawing* if for every pair of vertices u and v, there is a path drawn from u to v that is monotone in some direction. In other words, a drawing is monotone if, for any given direction d (e.g., from left to right) and for each pair of vertices u and v, there exists a suitable rotation of the drawing for which a path from u to v becomes monotone in the direction d.

Monotone drawings have been recently introduced [1] as a new visualization paradigm, which is well motivated by human subject experiments by Huang and Eades [8] who showed that the "geodesic tendency" (paths follow a given direction) is important in comprehending the underlying graph. Monotone drawings are related to well-studied drawing conventions, such as upward drawings [5,7], greedy drawings [2,9,10], and the

* Research partially supported by the MIUR project AlgoDEEP prot. 2008TFBWL4, by the ESF project 10-EuroGIGA-OP-003 GraDR "Graph Drawings and Representations", by NSERC, and by the European Union (European Social Fund - ESF) and Greek national funds through the Operational Program "Education and Lifelong Learning" of the National Strategic Reference Framework (NSRF) - Research Funding Program: Heracleitus II. Investing in knowledge society through the European Social Fund. Work on these results began at the 6th Bertinoro Workshop on Graph drawing. Discussion with other participants is gratefully acknowledged.

M. van Kreveld and B. Speckmann (Eds.): GD 2011, LNCS 7034, pp. 379–390, 2012.
© Springer-Verlag Berlin Heidelberg 2012

geometric problem of finding monotone trajectories between two given points in the plane avoiding convex obstacles [3].

Planar monotone drawings with straight-line edges form a natural setting and it is known that biconnected planar graphs and trees always admit such drawings, for some combinatorial embedding of the graph [1]. However, the question whether a simply connected planar graph always admits a planar monotone drawing or not is still open.

On the other hand, in the *fixed embedding setting* (i.e., the planar embedding of the graph is given as part of the input and the drawing algorithm is not allowed to alter it) it is known [1] that there exist simply connected planar embedded graphs that admit no straight-line monotone drawings.

In this paper we study planar monotone drawings of graphs in the fixed embedding setting, answering the natural question whether monotone drawings with a given constant number of bends per edge can always be computed, and identifying some subclasses of planar graphs that always admit planar monotone drawings with straight-line edges. Our contributions are summarized below:

- We prove that every n-vertex planar embedded graph has an embedding-preserving monotone drawing with *curve complexity* 2, that is, the maximum number of bends along an edge is 2, and with at most $4n - 10$ bends in total. Such a drawing can be computed in linear time and has polynomial area.
- We show that our bound on the curve complexity is tight, by describing an infinite family of embedded planar graphs that require two bends on a linear number of edges in any embedding-preserving monotone drawing.
- We investigate what subfamilies of embedded planar graphs can be realized as embedding-preserving monotone drawings with straight-line edges. We prove that biconnected embedded planar graphs and outerplane graphs always admit such a drawing, which can be computed in linear time.

The paper is structured as follows. Basic definitions and results are given in Section 2. An algorithm for computing embedding-preserving monotone drawings of general embedded planar graphs with at most two bends per edge is described in Section 3. Algorithms for computing straight-line monotone drawings of meaningful subfamilies of embedded planar graphs are given in Section 4. Concluding remarks and open questions are presented in Section 5. For space reasons some proofs are sketched or omitted.

2   Preliminaries

We assume familiarity with basic concepts of graph drawing (see, e.g., [5]). Let G be a planar graph and let ϕ be a planar embedding of G. The embedding ϕ defines the set of internal faces and the outer face of G. For every vertex v of G, the embedding ϕ also defines the circular clockwise order of the edges incident to v. Graph G along with an embedding ϕ is called an *embedded planar graph*, and is denoted by G_ϕ. Any subgraph of G_ϕ obtained by removing some edges from G_ϕ is a subgraph that *preserves* the planar embedding ϕ. A *drawing of G_ϕ* is a planar drawing of G with embedding ϕ.

A *subdivision* of a graph G is obtained by replacing each edge of G with a path. A *k-subdivision* of G is such that any path replacing an edge of G has at most k internal vertices. A graph G is *connected* if every pair of vertices is connected by a path and is *biconnected* (resp. *triconnected*) if removing any vertex (resp. any two vertices) leaves G connected. In order to handle the decomposition of a biconnected graph into its triconnected components, we use the well-known *SPQR-tree* data structure [6].

A *monotone drawing* Γ of a planar graph G (of an embedded planar graph G_ϕ) is a drawing of G (of G_ϕ) such that for every pair of vertices u and v there exists a path from u to v in Γ that is monotone in some direction.

A monotone drawing of any tree T can be constructed in polynomial area by using *Algorithm DFS-based* [1], which relies on the concept of the *Stern-Brocot tree* [11,4] $\mathcal{SB}$, an infinite tree whose nodes are in bijective mapping with the irreducible positive rational numbers. Algorithm DFS-based assigns to the edges of the tree T slopes $\frac{1}{1}, \frac{2}{1}, \ldots, \frac{n-1}{1}$ (which are the first $n-1$ elements of the rightmost path of $\mathcal{SB}$) according to a DFS-visit of T. Polynomial area is ensured by the following property of $\mathcal{SB}$.

Property 1. [4,11] The sum of the numerators of the elements of the i-th level of $\mathcal{SB}$ is 3^{i-1} and the sum of the denominators of the elements of the i-th level of $\mathcal{SB}$ is 3^{i-1}.

The following property is also satisfied by any monotone drawing Γ of a tree T.

Property 2. [1] Any drawing Γ' of T such that the slopes of each edge $e \in T$ in Γ' is the same as the slope of e in Γ is monotone. Also, the slopes of any two leaf-edges e' and e'' of T in Γ are such that e' and e'' diverge, that is, the elongations of e' and e'' do not cross each other.

3   Monotone Drawings with Bends of Embedded Planar Graphs

In this section we study monotone drawings of embedded planar graphs. We remark that it is still unknown whether every planar graph admits a straight-line monotone drawing in the variable embedding setting, while it is known that straight-line monotone drawings do not always exist if the embedding of the graph is fixed [1]. We therefore investigate monotone drawings with bends along some edges, and we show that two bends per edge are always sufficient and sometimes necessary for the existence of a monotone drawing in the fixed embedding setting.

We need some preliminary definitions. An *upright spanning tree* T of an embedded planar graph G_ϕ is a rooted ordered spanning tree of G_ϕ such that: (i) T preserves the planar embedding of G_ϕ; (ii) the root of T is a vertex r of the outer face of G_ϕ; (iii) there exists a planar drawing of G_ϕ that contains an upward drawing of T such that no edge goes below r. Fig. 1(b) and (c) show two different ordered spanning trees of the embedded planar graph of Fig. 1(a): The first one is an upright spanning tree, while the second is not. Given an embedded planar graph G_ϕ, an upright spanning tree T of G_ϕ can be computed as follows. Construct any planar straight-line drawing Γ of G_ϕ. Orient the edges of G_ϕ in Γ according to the upward direction. Let r be a vertex on the outer face of G_ϕ with the smallest y-coordinate in Γ. Then, compute any spanning tree T of G_ϕ rooted at r such that the left-to-right order of the children of r in T is consistent with the left-to-right order of the neighbors of r in Γ and the left-to-right order of the children of each vertex w in T is consistent with the clockwise order of the neighbors of w in G_ϕ, computed starting from the edge connecting w to its parent in T.

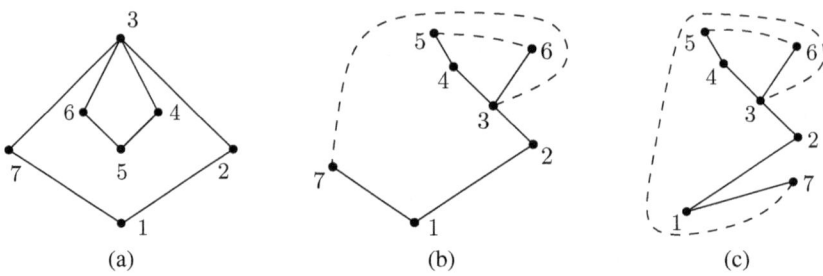

Fig. 1. (a) A drawing Γ of an embedded planar graph G_ϕ. (b) An upright spanning tree of G_ϕ. (c) A spanning tree of G_ϕ that is not upright.

Let T be an upright spanning tree of G_ϕ. The *rgbb-coloring of G_ϕ with respect to T* is a coloring of the edges of G_ϕ with four colors (red, green, blue, and black) such that: An edge is colored black if it belongs to T; an edge is colored green if it connects two leaves of T; an edge is colored red if it connects a leaf to an internal vertex of T; an edge is colored blue if it connects two internal vertices of T.

We denote by $C(G_\phi, T)$ the rgbb-coloring of G_ϕ with respect to T. We prove the following lemma.

Lemma 1. *Let G_ϕ be an embedded planar graph with n vertices, let T be an upright spanning tree of G_ϕ, and let $C(G_\phi, T)$ be the rgbb-coloring of G_ϕ with respect to T. Then we can compute a monotone drawing Γ of G_ϕ such that each black or green edge of $C(G_\phi, T)$ is drawn as a straight-line segment, each red edge has 1 bend, and each blue edge has 2 bends. The running time of the algorithm is $O(n)$ and the drawing Γ has $O(n) \times O(n^2)$ area.*

Proof. First, starting from G_ϕ and T, construct a graph G'_ϕ and an upright spanning tree T' of G'_ϕ such that: (i) G'_ϕ is a 2-subdivision of G_ϕ, (ii) T is a subtree of T', and (iii) all the edges of G'_ϕ that are not in T' connect two leaves of T'. Fig. 2(a) and (b) show a graph G_ϕ with an upright spanning tree T and the corresponding graph G'_ϕ with its upright spanning tree T' satisfying (i)–(iii). Then, the monotone drawing of G'_ϕ with curve complexity 2 is constructed by first computing a straight-line monotone drawing of G'_ϕ and then replacing each subdivision vertex with a bend; see Fig. 2(c).

Graphs G'_ϕ and T' are constructed as follows. Initialize $G'_\phi = G_\phi$ and $T' = T$. Subdivide each red edge (s, t) of G'_ϕ with a vertex k and add edge (t, k) to T', where t is the internal vertex of T'. Subdivide each blue edge (s, t) of G'_ϕ twice, with two vertices k and z, and add edges (s, k) and (t, z) to T'.

The straight-line monotone drawing of G'_ϕ is computed in two steps. First, with *Algorithm DFS-based* [1], we construct a straight-line monotone drawing of T', and then we add the remaining (non-tree) edges as straight-line segments, which results in using two segments for red edges and three segments for blue edges.

To argue the monotonicity for non-tree edges, recall that, by Property 2, it is possible to elongate the edges of T' without affecting monotonicity and planarity.

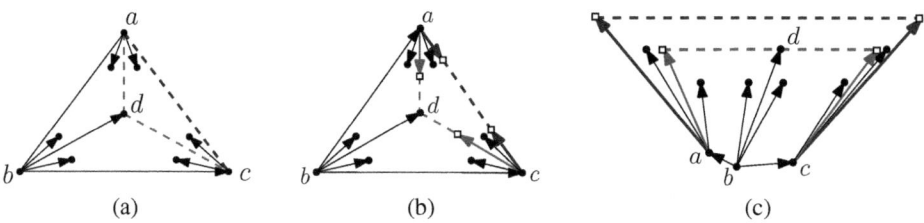

Fig. 2. (a) A graph G_ϕ with an upright spanning tree T rooted at vertex b. Solid edges belong to T, while dashed edges do not. Blue edges are thicker than red edges, which are thicker than black edges. (b) The corresponding graph G'_ϕ with its upright spanning tree T'. Solid edges belong to T', while dashed edges do not. Subdivision vertices are drawn as squares. (c) A straight-line monotone drawing of G'_ϕ that corresponds to a monotone drawing of G_ϕ with bent edges.

Further, as *Algorithm DFS-based* assigns slopes $\frac{1}{1}, \frac{2}{1}, \ldots, \frac{n-1}{1}$ to the edges of T', the elongation of each leaf-edge (u, v) intersects each vertical line $x = k$, where k is any integer value greater than the x-coordinate of u, at an integer grid point. Moreover, as by Property 2 the leaf-edge elongations diverge, such intersections appear in the same order on each vertical line $x = k'$, where k' is any integer value greater than the x-coordinate of every internal vertex of T'; see Fig. 3(a).

Another key observation is that the graph G_L induced by the leaves of T' is outer-planar and can be augmented, by adding dummy edges, to a biconnected outerplanar graph in which each internal face is a 3-cycle in such a way that the order of the vertices on the outer face is the same as the left-to-right order of the leaves of T'; see Fig. 3(b).

The vertices of G_L are assigned to levels in such a way that the end-vertices of each edge of G_L are either on the same level or on adjacent levels, as follows. The first and the last vertex in the left-to-right order of the leaves of T' have level 1. Note that, these two vertices are adjacent, as G_L is a biconnected outerplanar graph and the order of the vertices on its outer face is the same as the left-to-right order of the leaves of T'. Then, starting from this edge, consider any edge (u, v) on the outer face of the graph induced by the vertices whose level has been already assigned and consider the unique vertex w that is connected to both u and v, and whose level has not been assigned yet, if any. Note that, either u and v have the same level i or one of them has level i and the other has level $i + 1$. In both cases, assign level $i + 1$ to w, as shown in Fig. 3(b) and (c).

Let l be the number of levels of G_L. Then, place all the vertices at level i, with $i = 1, \ldots, l$, on a vertical line $x = k + l - i + 1$, where k is the x-coordinate of the rightmost internal vertex of T'. This placement, together with the fact that each such vertical line intersects the elongations of all the leaf-edges in the same order, ensures the planarity of the straight-line drawing of G_L. Further, as the order of the vertices on the outer face of G_L is the same as the left-to-right order of the leaves of T', the edges of T' do not cross any edge of G_L, hence ensuring the planarity of G'_ϕ; see Fig. 3(c).

The drawing of G'_ϕ is monotone because between any two vertices there exists a monotone path composed only of edges of T', while edges not in T' do not affect the

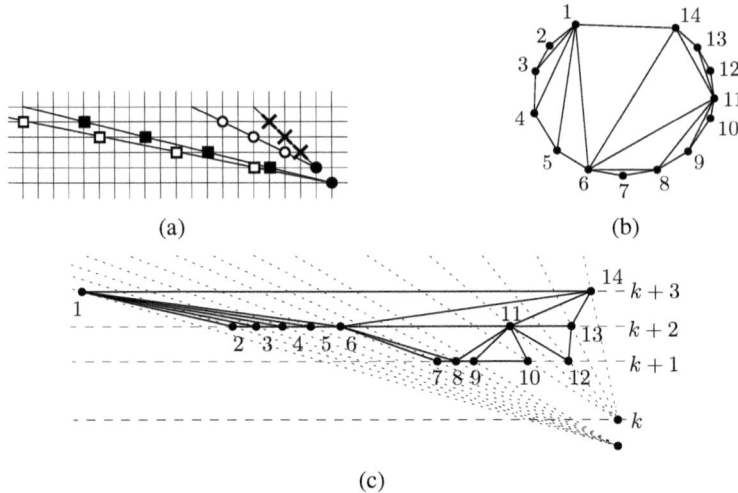

(a)

(b)

(c)

Fig. 3. For readability, the drawings in (a) and (c) are rotated to 90° and the grid unit distances in (c) are not uniform. (a) Leaf-edge elongations have integer intersections with all the vertical lines in the same order. (b) An augmented graph G_L. (c) The drawing of G_L (where $l = 3$).

monotonicity. Hence, monotonicity is maintained when dummy edges are removed. Note that, any monotone path traversing a leaf-edge of T' has the corresponding leaf as an end-vertex. If the leaf is a subdivision vertex of any non-black edge, it does not belong to G_ϕ. Hence, all the monotone paths in G_ϕ are composed only of edges of T, whose drawing is monotone since it is a subtree of T'. Therefore, the drawing of G_ϕ is monotone, each red edge has one bend, and each blue edge has two bends.

In order to compute the area of the obtained drawing, recall that *Algorithm DFS-based* [1] produces a drawing of T' in $O(n) \times O(n^2)$ area. Since the number of vertical lines added to host the drawing of G_L is equal to the number l of levels assigned to the vertices of G_L, and since l is bounded by the number of leaves, which is $O(n)$, the area of the whole drawing is still $O(n) \times O(n^2)$.

It is easy to see that the drawing can be computed in $O(n)$ time, by considering the individual steps. The computation of the three necessary graphs, T, G'_ϕ and T', can be performed in linear time. Also, the slopes of the edges of T' can be computed in linear time with *Algorithm DFS-based* [1] by constructing the Stern-Brocot tree and by performing a rightmost DFS visit of it. Further, graph G_L can be augmented in linear time. Finally, the assignment of levels to the vertices of G_L is also performed in linear time, as each vertex is considered just once and its level is assigned only based on the levels of its two neighbors. This concludes the proof of Lemma 1.   □

Note that, according to Lemma 1 there always exists a monotone drawing Γ of G_ϕ with curve complexity 2 and at most $4n - 10$ bends in total, as G_ϕ has at most $3n - 6$ edges and every spanning tree of G_ϕ has $n - 1$ edges. Using the algorithm described in Lemma 1, Γ has at most $2(3n-6-n+1) = 4n-10$ edges in total, and this upper bound

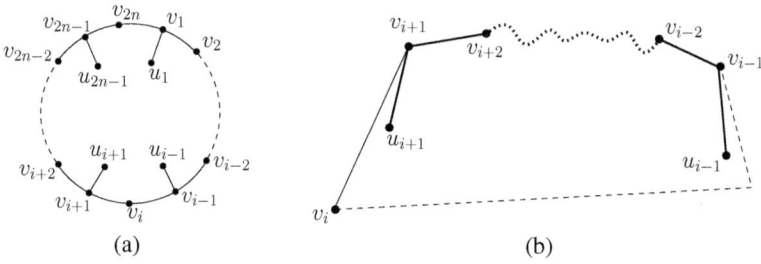

Fig. 4. (a) A graph G_ϕ with $3n$ vertices that does not admit any embedding-preserving straight-line monotone drawing. (b) Edges (v_{i-1}, v_i) and (v_i, v_{i+1}) can not be drawn as straight-line segments.

is asymptotically tight, as there exist embedded planar graphs that require a linear total number of bends in any monotone drawing. Namely, we first prove in Lemma 2 that there exist embedded planar graphs requiring at least one bend on some edges. Then, based on this lemma, we prove in Lemma 3 that there exist infinitely many embedded planar graphs whose monotone drawings require two bends on a linear number of edges.

Lemma 2. *For every $n \geq 3$ there exists an embedded planar graph G_ϕ with $3n$ vertices and $3n$ edges that does not admit any straight-line monotone drawing.*

Sketch of Proof: We describe an embedded planar graph G_ϕ that does not admit any straight-line monotone drawing (refer to Fig. 4(a)). G_ϕ consists of a simple cycle $C = v_1, \ldots, v_{2n}$ of length $2n$ and of n vertices $u_1, u_3, \ldots, u_{2n-1}$ of degree 1, called *legs*, incident to the vertices $v_1, v_3, \ldots, v_{2n-1}$ of C with odd indices, respectively. The embedding of G_ϕ is such that all the legs are inside C, that is, they are inside the unique internal face of C. As by Property 2 any two consecutive legs (v_{i-1}, u_{i-1}) and (v_{i+1}, u_{i+1}) diverge in any straight-line monotone drawing, it is not possible to connect vertices v_{i-1} and v_{i+1} by drawing edges (v_{i-1}, v_i) and (v_i, v_{i+1}) as straight-line segments. Refer to Fig. 4(b). □

The next lemma shows that there are infinitely many embedded planar graphs that require two bends per edge on a linear number of edges in any embedding-preserving monotone drawing.

Lemma 3. *For every odd $n \geq 9$ there exists an embedded planar graph G_ϕ with n vertices and $\frac{3}{2}(n-1)$ edges such that every monotone drawing of G_ϕ has at least $\frac{n-3}{6}$ edges with at least two bends and thus at least $\frac{n-3}{3}$ bends in total.*

Sketch of Proof: Refer to Fig. 5. Consider an odd integer $n \geq 9$. We construct G_ϕ iteratively. Let G_ϕ^1 be a triangle graph. Graph G_ϕ^i is constructed from G_ϕ^{i-1} as follows. Initialize $G_\phi^i = G_\phi^{i-1}$. Let (u, v, w) be a triangular internal face of G_ϕ^i. Add 6 new vertices $u_1, u_2, v_1, v_2, w_1, w_2$ and 9 new edges $(u, u_1), (u, u_2), (u_1, u_2), (v, v_1), (v, v_2), (v_1, v_2), (w, w_1), (w, w_2), (w_1, w_2)$ to G_ϕ^i in such a way that all the new vertices are inside (u, v, w). Note that the n-vertex graph G_ϕ^i is planar and has $\frac{3}{2}(n-1)$ edges. Any monotone drawing of G_ϕ has at least $\frac{n-3}{6}$ edges with at least two bends. □

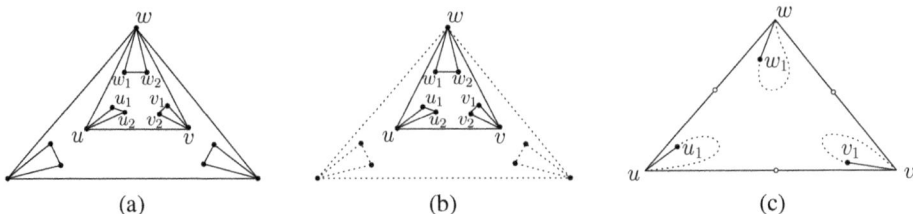

Fig. 5. (a) An example of a graph G_ϕ with $n = 15$ vertices, that coincides with a graph G_ϕ^3 constructed from G_ϕ^2 by adding vertices $u_1, u_2, v_1, v_2, w_1, w_2$ inside triangular face u, v, w. (b) A subgraph G_ϕ^t of G_ϕ induced by a triangle (u, v, w) and all the vertices inside it. (c) A subdivision (white circles) of the subgraph G_ϕ^h (solid edges) of G_ϕ^t induced by u, v, w, u_1, v_1, w_1. By Lemma 2, this subdivision does not admit any straight-line monotone drawing.

Lemma 1 and Lemma 3 together provide a tight bound on the curve complexity of monotone drawings in the fixed embedding setting. The next theorem summarizes the main contribution of this section.

Theorem 1. *Every embedded planar graph with n vertices admits a monotone drawing with curve complexity 2, at most $4n - 10$ bends in total, and $O(n) \times O(n^2)$ area; such a drawing can be computed in $O(n)$ time. Also, there exist infinitely many embedded planar graphs any monotone drawing of which requires two bends on $\Omega(n)$ edges.*

4   Monotone Drawings with Straight-Line Edges

In this section we prove that there exist meaningful subfamilies of embedded planar graphs that can be realized as straight-line monotone drawings. In particular, we prove that both the class of outerplane graphs and the class of embedded planar biconnected graphs have this property.

4.1   Outerplane Graphs

An embedded planar graph G_ϕ is an *outerplane graph* if all its vertices are on the outer face. We prove the following result.

Theorem 2. *Every outerplane graph admits a straight-line monotone drawing. Also, there exists an algorithm that computes such a drawing in $O(n)$ time and $O(n) \times O(n^2)$ area.*

Proof. Let T be an upright spanning tree of G_ϕ obtained by performing a "rightmost DFS" visit of G_ϕ; see Fig. 6(a). Consider a decomposition of G_ϕ into its maximal biconnected components. Observe that, for each maximal biconnected component B that is connected to the root of T through a cut-vertex v, T contains all the edges of B except for the internal chords (dashed edges in Fig. 6(a)) and for the leftmost edge incident to v (dotted edges in Fig. 6(a)).

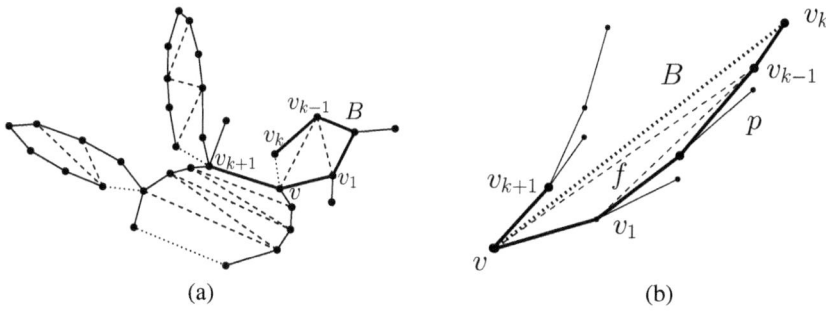

(a)   (b)

Fig. 6. (a) An outerplane graph G_ϕ and the upright spanning tree T of G_ϕ obtained by performing a "rightmost DFS" visit. Edges of T are represented as solid segments. (b) A strictly convex drawing of a maximal biconnected component B of G_ϕ.

A straight-line monotone drawing of G_ϕ is constructed by first computing a straight-line monotone drawing of T, with *Algorithm DFS-based* [1], and then reinserting the edges not in T as straight-line segments. In order to reinsert such edges, for each maximal biconnected component B, consider the path $p = (v, v_1, \ldots, v_k)$ that is composed of the edges belonging both to B and to T.

According to *Algorithm DFS-based* [1] the slopes of the edges of p are all positive and increasing with respect to the distance from v in p. Hence, path p is drawn in T as a polygonal line that is convex on the left side, that is, the straight-line segment connecting any two non-consecutive vertices of p completely lies to the left of p; see Fig. 6(b). Thus, reinserting edge (v, v_k) as the straight-line segment between v and v_k determines that (v, v_k) is the leftmost edge of B incident to v in the drawing and that the boundary of B, that is, the cycle composed of the edges of p plus (v, v_k), delimits a strictly-convex region f.

We show that f does not contain any other vertex of T. Namely, the vertex v_{k+1} such that edge (v, v_{k+1}) follows (v, v_1) in the counter-clockwise order of the edges around v in T lies outside f. This is due to the fact that, according to *Algorithm DFS-based*, the slope of (v, v_{k+1}) is greater than the slope of (v_{k-1}, v_k) which in turn is greater than the slope of (v, v_k); see Fig. 6(b).

Hence, f is an empty strictly-convex region, and the chords of B can be reinserted as straight-line segments while maintaining planarity.

The area of the drawing is the same as the area of T computed by *Algorithm DFS-based*, namely $O(n) \times O(n^2)$. The drawing can be computed in $O(n)$ time. Namely, drawing T by using *Algorithm DFS-based* takes $O(n)$ time [1], and the same holds for reinserting missing edges. $\square$

4.2 Biconnected Graphs

It is known [1] that straight-line monotone drawings of biconnected planar graphs in the variable embedding setting can always be computed. This result is obtained by means of an algorithm that exploits SPQR-trees and that preserves any given embedding, as long as the graph contains no parallel component whose poles are connected by an edge. However, this algorithm can be easily modified in order to compute monotone drawings with curve complexity 1 of every embedded biconnected planar graph, as the edges connecting the poles of a parallel component could be placed in their correct position by adding a bend, when necessary.

In this section we prove that in fact we can compute a monotone drawing of every embedded biconnected planar graph with no bends at all.

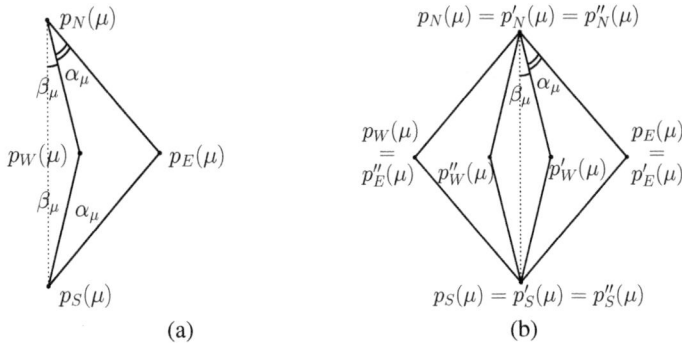

Fig. 7. (a) A boomerang. (b) A diamond

As for the variable-embedding setting case [1], our algorithm relies on a bottom-up visit of the SPQR-tree of the biconnected graph G in which at each step a drawing of the pertinent graph of the currently considered node μ is constructed inside a *boomerang* $boom(\mu)$, that is, a quadrilateral composed of points $p_N(\mu)$, $p_E(\mu)$, $p_S(\mu)$, and $p_W(\mu)$ such that $p_W(\mu)$ is inside triangle $\triangle(p_N(\mu), p_S(\mu), p_E(\mu))$ and $2\alpha_\mu + \beta_\mu < \frac{\pi}{2}$, where $\alpha_\mu = \widehat{p_W(\mu)p_S(\mu)p_E(\mu)} = \widehat{p_W(\mu)p_N(\mu)p_E(\mu)}$ and $\beta_\mu = \widehat{p_W(\mu)p_S(\mu)p_N(\mu)} = \widehat{p_W(\mu)p_N(\mu)p_S(\mu)}$; see Fig. 7(a).

In order to cope with the fixed-embedding setting, we introduce a new shape, called *diamond* and denoted by $diam(\mu)$, that is a convex quadrilateral $(p_N(\mu), p_E(\mu), p_S(\mu), p_W(\mu))$ composed of two boomerangs $boom'(\mu) = (p'_N(\mu), p'_E(\mu), p'_S(\mu), p'_W(\mu))$ and $boom''(\mu) = (p''_N(\mu), p''_E(\mu), p''_S(\mu), p''_W(\mu))$ such that $p_N(\mu) = p'_N(\mu) = p''_N(\mu)$, $p_S(\mu) = p'_S(\mu) = p''_S(\mu)$, $p_E(\mu) = p'_E(\mu)$ and $p_W(\mu) = p''_E(\mu)$; see Fig. 7(b).

A diamond is used for any P-node μ having an edge e between its poles. Namely, one of the two boomerangs composing the diamond contains the child components of μ that come before e in the ordering of the components around the poles, while the other boomerang contains the other components. Note that, since P-nodes might be contained into diamonds, the algorithm for drawing S- and R-nodes inside their own boomerangs has to be adapted to deal with this case. We have the following.

Theorem 3. *Every biconnected embedded planar graph admits a straight-line monotone drawing, which can be computed in linear time.*

5   Conclusions and Open Problems

In this paper we studied monotone drawings of graphs in the fixed embedding setting. Since not all embedded planar graphs admit an embedding-preserving monotone drawing with straight-line edges, we focused on computing embedding-preserving monotone drawings with low curve complexity. We proved that curve complexity 2 always suffices and that this bound is worst-case optimal. Furthermore, we described algorithms for computing straight-line monotone drawings for meaningful subfamilies of embedded planar graphs. All the algorithms presented in this paper can be performed in linear time and most of them produce drawings which require polynomial area.

The results in this paper naturally give rise to several interesting open problems; some of them are listed below.

Existential Questions

Problem 1. Finding meaningful subfamilies of embedded planar graphs (other than outerplane graphs and embedded biconnected graphs) that admit monotone drawings with curve complexity smaller than 2.

Problem 2. Is it possible to characterize the embedded planar graphs that admit monotone drawings with curve complexity smaller than 2?

Complexity Questions

Problem 3. Given an embedded planar graph G_ϕ and an integer $k \in \{0, 1\}$, what is the complexity of deciding whether G_ϕ admits a monotone drawing with curve complexity k?

Problem 4. Given a graph G and an integer $k \in \{0, 1\}$, what is the complexity of deciding whether there exists an embedding ϕ such that G_ϕ admits a monotone drawing with curve complexity k?

Problem 5. Given a graph G and an integer $k \in \{0, 1\}$, what is the complexity of deciding whether there exists an embedding ϕ such that G_ϕ does not admit any monotone drawing with curve complexity k?

Notice that, although Problems 3-5 are related, there is no evidence that answering one of them implies an answer for any other.

Algorithmic Questions

Problem 6. Is there any algorithm that computes monotone drawings of embedded biconnected planar graphs in polynomial area?

Problem 7. Is there any algorithm that computes monotone drawings of outerplane graphs in subcubic area?

References

1. Angelini, P., Colasante, E., Di Battista, G., Frati, F., Patrignani, M.: Monotone Drawings of Graphs. In: Brandes, U., Cornelsen, S. (eds.) GD 2010. LNCS, vol. 6502, pp. 13–24. Springer, Heidelberg (2011)
2. Angelini, P., Frati, F., Grilli, L.: An algorithm to construct greedy drawings of triangulations. J. Graph Algorithms Appl. 14(1), 19–51 (2010)
3. Arkin, E.M., Connelly, R., Mitchell, J.S.B.: On monotone paths among obstacles with applications to planning assemblies. In: Symposium on Computational Geometry, pp. 334–343 (1989)
4. Brocot, A.: Calcul des rouages par approximation, nouvelle methode. Revue Chronometrique 6, 186–194 (1860)
5. Di Battista, G., Eades, P., Tamassia, R., Tollis, I.G.: Graph Drawing. Prentice Hall, Upper Saddle River (1999)
6. Di Battista, G., Tamassia, R.: On-line planarity testing. SIAM J. Comput. 25, 956–997 (1996)
7. Garg, A., Tamassia, R.: Upward planarity testing. Order 12, 109–133 (1995)
8. Huang, W., Eades, P., Hong, S.-H.: A graph reading behavior: Geodesic-path tendency. In: PacificVis, pp. 137–144 (2009)
9. Leighton, T., Moitra, A.: Some results on greedy embeddings in metric spaces. Discrete & Computational Geometry 44(3), 686–705 (2010)
10. Papadimitriou, C.H., Ratajczak, D.: On a conjecture related to geometric routing. Theor. Comput. Sci. 344(1), 3–14 (2005)
11. Stern, M.A.: Ueber eine zahlentheoretische funktion. Journal fur die reine und angewandte Mathematik 55, 193–220 (1858)

On the Page Number
of Upward Planar Directed Acyclic Graphs[*]

Fabrizio Frati[1,2,3], Radoslav Fulek[1], and Andres J. Ruiz-Vargas[1]

[1] School of Basic Sciences - École Polytechnique Fédérale de Lausanne, Switzerland
{fabrizio.frati,radoslav.fulek,andres.ruizvargas}@epfl.ch
[2] Dipartimento di Informatica e Automazione, Università Roma Tre
[3] School of Information Technologies, University of Sydney

Abstract. In this paper we study the page number of upward planar directed acyclic graphs. We prove that: (1) the page number of any n-vertex upward planar triangulation G whose every maximal 4-connected component has page number k is at most $\min\{O(k \log n), O(2^k)\}$; (2) every upward planar triangulation G with $o(\frac{n}{\log n})$ diameter has $o(n)$ page number; and (3) every upward planar triangulation has a vertex ordering with $o(n)$ page number if and only if every upward planar triangulation whose maximum degree is $O(\sqrt{n})$ does.

1 Introduction

A *k-page book embedding* of a graph $G=(V,E)$ is a total ordering σ of V and a partition of E into subsets $E_1, E_2, \ldots, E_k$, called *pages*, such that no two edges (u,v) and (w,z) with $u <_\sigma w <_\sigma v <_\sigma z$ belong to the same set E_i. The *page number* of G is the minimum k such that G admits a k-page book embedding.

Book embeddings (first introduced by Kainen [15] and by Ollmann [19]) find applications in several contexts, such as VLSI design, fault-tolerant processing, sorting networks, and parallel matrix multiplication (see, e.g., [4,11,20,21]). Henceforth, they have been widely studied from a theoretical point of view; namely, the literature is rich of combinatorial and algorithmic contributions on the page number of various classes of graphs (see, e.g., [2,7,8,9,10,17,18]). We remark here a famous result of Yannakakis [22] stating that any planar graph has page number at most four.

Heath *et al.* [13,14] extended the notions of book embedding and page number to directed acyclic graphs (*DAGs* for short) in a very natural way: Given a DAG $G=(V,E)$, book embedding and page number of G are defined as for undirected graphs, except that the total ordering of V is now required to be a *linear extension* of the partial order of V induced by E. That is, if G contains an edge from a vertex u to a vertex v, then $u <_\sigma v$ in any feasible total ordering σ of V. The authors of [13,14] showed that DAGs with page number equal to one can be characterized and recognized efficiently; however, they proved that, in general, determining the page number of a DAG is NP-complete.

[*] Work partially supported by the Italian Ministry of Research, grant RBIP06BZW8, FIRB project "Advanced tracking system in intermodal freight transportation", by the Swiss National Science Foundation 200021-125287/1, by the ESF project 10-EuroGIGA-OP-003 "Graph Drawings and Representations", and by the MIUR of Italy, project AlgoDEEP 2008TFBWL4.

M. van Kreveld and B. Speckmann (Eds.): GD 2011, LNCS 7034, pp. 391–402, 2012.
© Springer-Verlag Berlin Heidelberg 2012

The main problem raised by Heath *et al.* and studied in, e.g., [1,6,12,13,14], is whether every *upward planar DAG* admits a book embedding in few pages. An upward planar DAG is a DAG that admits a drawing which is simultaneously *upward*, *i.e.*, each edge is represented by a curve monotonically increasing in the y-direction, and *planar*, *i.e.*, no two edges cross. Upward planar DAGs are the natural counterpart of planar graphs in the context of directed graphs. Notice that there exist DAGs which admit a planar non-upward embedding and that require $\Omega(|V|)$ pages in any book embedding [12,14]. No upper bound better than the trivial $O(|V|)$ and no lower bound better than the trivial $\Omega(1)$ are known for the page number of upward planar DAGs. It is however known that *directed trees* have page number one [14], that *unicyclic DAGs* have page number two [14], and that *series-parallel DAGs* have page number two [1,6].

In this paper we study the page number of upward planar DAGs. Before stating our results we need some background.

First, it is known that every upward planar DAG G can be augmented to an *upward planar triangulation* G' [5]. That is, edges can be added to G so that the resulting graph G' is still an upward planar DAG and every face of G' is delimited by a 3-cycle. Thus, in order to establish tight bounds on the page number of upward planar DAGs, it suffices to look at upward planar triangulations, as the page number of a subgraph G of a graph G' is at most the page number of G'. In the following, unless otherwise specified, all the considered graphs are upward planar triangulations.

Second, consider a total ordering σ of V. A *twist* is a set of pairwise crossing edges, *i.e.*, a set $\{(u_1, v_1), (u_2, v_2), \ldots, (u_k, v_k)\}$ of edges such that $u_1 <_\sigma u_2 <_\sigma \cdots <_\sigma u_k <_\sigma v_1 <_\sigma v_2 <_\sigma \cdots <_\sigma v_k$. It is straightforward that the page number of a graph G is lower bounded by the minimum over all vertex orderings σ of the maximum size of a twist in σ. Moreover, a function of the maximum size of a twist in a vertex ordering upper bounds the page number of an n-vertex graph G, as stated in the following two lemmata.

Lemma 1. *[3] Let σ be a vertex ordering of an n-vertex graph G. Suppose that the maximum twist of σ has size k. Then G admits a book embedding with vertex ordering σ and with $O(k \log n)$ pages.*

Lemma 2. *[16] Let σ be a vertex ordering of an n-vertex graph G. Suppose that the maximum twist of σ has size k. Then G admits a book embedding with vertex ordering σ and with $O(2^k)$ pages.*

Thus, in order to get upper bounds for the page number of a graph, it often suffices to construct vertex orderings with small maximum twist size.

In this paper we consider the relationship between the page number of an n-vertex upward planar triangulation G and three important graph parameters of G: The connectivity, the diameter, and the degree. We show the following results. (i) In Sect. 3, we prove that an upward planar triangulation G admits a vertex ordering with maximum twist size $O(f(n))$ if and only if every maximal 4-connected component of G does. As a corollary, upward planar 3-trees have constant page number. (ii) In Sect. 4, we prove that every upward planar triangulation G has a vertex ordering whose maximum twist size is a function of the *diameter* of G, that is, of the length of the longest directed path in G. As a corollary, every upward planar triangulation whose diameter is $o(n/\log n)$

admits a book embedding in $o(n)$ pages. (iii) In Sect. 5, we show that every upward planar triangulation has a vertex ordering with $o(n)$ page number if and only if every upward planar triangulation whose maximum degree is $O(\sqrt{n})$ does.

2 Definitions

A *directed graph* is a graph with direction on the edges. The *underlying graph* of a directed graph G is the undirected graph obtained from G by removing the directions on its edges. We denote by (u, v) an edge directed from a vertex u, which is called the *origin* of (u, v), to a vertex v, which is called the *destination* of (u, v); edge (u, v) is *incoming* v and *outgoing* u. A *source* (resp. *sink*) is a vertex with no incoming edge (resp. with no outgoing edge). A *directed cycle* is a directed graph whose underlying graph is a cycle and containing no source and no sink. A *directed acyclic graph (DAG* for short) is a directed graph containing no directed cycle. A *directed path* is a directed graph whose underlying graph is a path and containing exactly one source and one sink. The *diameter* of a directed graph is the number of vertices in its longest directed path.

A *drawing* of a directed graph is a mapping of each vertex to a point in the plane and of each edge to a Jordan curve between its end-points. A drawing is *upward* if each edge (u, v) is a curve monotonically increasing in the y-direction and it is *planar* if no two edges intersect except, possibly, at common end-points. A drawing is *upward planar* if it is both upward and planar. An *upward planar graph* is a graph that admits an upward planar drawing. A planar drawing of a graph partitions the plane into connected regions, called *faces*. The unbounded face is the *outer face*, all the other faces are *internal faces*. Two upward planar drawings of an upward planar DAG are *equivalent* if they determine the same clockwise ordering of the edges around each vertex. An *embedding* of an upward planar DAG is an equivalence class of upward planar drawings. An *embedded upward planar graph* is an upward planar DAG together with an embedding.

An *upward planar triangulation* is an upward planar graph whose underlying graph is a maximal planar graph. Consider any two upward planar drawings Γ_1 and Γ_2 of an upward planar triangulation G. Then, either Γ_1 and Γ_2 are equivalent, or the clockwise ordering of the edges around each vertex in Γ_1 is exactly the opposite of the one in Γ_2. The outer face of an upward planar drawing Γ of an upward planar triangulation G is delimited by a cycle composed of three edges (u, v), (u, z), and (v, z). Then, u, v, and z are called *bottom vertex*, *middle vertex*, and *top vertex* of Γ, respectively. Consider the two embeddings $\mathcal{E}_1$ and $\mathcal{E}_2$ of an upward planar triangulation G. Then, the bottom, middle, and top vertex of $\mathcal{E}_1$ coincide with the bottom, middle, and top vertex of $\mathcal{E}_2$, respectively. Hence such vertices are simply called the *bottom vertex of G*, the *middle vertex of G*, and the *top vertex of G*, respectively.

A *total vertex ordering* σ of a DAG G is *upward* if G has no edge (u, v) such that $v <_\sigma u$. The upward vertex orderings are all and only the vertex orderings that are feasible for a book embedding of a DAG. We say that an upward vertex ordering σ *induces* a twist of size k if G contains edges $(u_1, v_1), \ldots, (u_k, v_k)$ such that $u_1 <_\sigma \ldots <_\sigma u_k <_\sigma v_1 <_\sigma \ldots, v_k$. The *maximum twist size* of an upward vertex ordering σ is the maximum number of edges in a twist induced by σ. Two edges (u_1, v_1) and (u_2, v_2) *are nested* in σ if $u_1 <_\sigma u_2 <_\sigma v_2 <_\sigma v_1$. Two edges (u_1, v_1) and (u_2, v_2) *cross* in σ if $u_1 <_\sigma u_2 <_\sigma v_1 <_\sigma v_2$.

An undirected graph is k-*connected* if the removal of any $k - 1$ vertices leaves the graph connected. A directed graph is k-*connected* if its underlying graph is. A *maximal k-connected component* of a graph G is a subgraph G' of G such that G' is k-connected and no subgraph G'' of G with $G' \subset G''$ is k-connected. A *separating triangle C* in a graph G is a 3-cycle such that the removal of the vertices of C from G disconnects G. A separating triangle C in a graph G is *maximal* if G has no separating triangle C' such that C is internal to C'.

The *degree of a vertex* is the number of edges incident to it. The *degree of a graph* is the maximum among the degrees of its vertices. A DAG is *Hamiltonian* if it contains a directed path passing through all its vertices. An Hamiltonian DAG G has exactly one upward total vertex ordering. Moreover, if G is upward planar, then it has page number at most 2. A *plane 3-tree* is a maximal plane graph that can be constructed as follows. Let G_3 be a 3-cycle embedded in the plane. A plane 3-tree with n vertices is a plane graph that can be constructed from a plane graph G_{n-1} with $n - 1$ vertices by inserting a vertex inside an internal face of G_{n-1} and by connecting such a vertex to the three vertices incident to the face. A *planar 3-tree* is a planar graph that can be embedded as a plane 3-tree. An *upward plane 3-tree* is an upward planar DAG whose underlying graph is a plane 3-tree.

3   Page Number and Connectivity

In this section we study the relationship between the page number of an upward planar DAG and the page number of its maximal 4-connected components. We prove the following:

Theorem 1. *Let $f(n)$ be any function such that $f(n) \in \Omega(1)$ and $f(n) \in O(n)$. Consider any n-vertex upward planar triangulation G and suppose that every maximal 4-connected component of G has an upward vertex ordering with maximum twist size at most $f(n)$. Then G has an upward vertex ordering with maximum twist size $O(f(n))$.*

First, we define a rooted tree $T = (V', E')$, whose nodes correspond to subgraphs of $G=(V, E)$, which reflects the structure of separating triangles in G. Tree T is recursively defined as follows (see Fig. 1(a)). The root r of T corresponds to $G'(r) = G$. Suppose that a node a of T corresponds to a subgraph $G'(a)$ of G. If $G'(a)$ contains no separating triangle, then a is a leaf of T. Otherwise, consider every maximal separating triangle (u, v, z) of $G'(a)$; then, insert a node b in T as a child of a, such that $G'(b)$ is the subgraph of $G'(a)$ induced by the vertices internal to or on the border of cycle (u, v, z). For each node $a \in T$, denote as $V'(a)$ and $E'(a)$ the vertex set and the edge set of $G'(a)$. Further, for each node $a \in T$, let $G(a) = (V(a), E(a))$ denote the subgraph of $G'(a)$ induced by all the vertices which are not internal to any separating triangle of $G'(a)$. Note that $G(a)$ is 4-connected for every $a \in V'$.

We now define a total ordering $o(V)$ of V and we later prove that the maximum twist size of $o(V)$ is $O(f(n))$. Ordering $o(V)$ is constructed by induction on T. In the base case a is a leaf; then let $o(V'(a))$ be any total ordering of $V'(a)$ such that the maximum twist size of $o(V'(a))$ is $f(n)$. Such an ordering exists by hypothesis, since $G'(a)$ is 4-connected. In the inductive case, let $a_1, \ldots, a_m$ be the children of a in T,

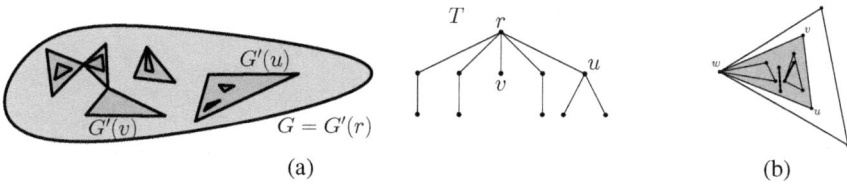

Fig. 1. (a) Tree T capturing the structure of the separating triangles in G. (b) Graph $G'(a)$; the thick edges belong to M_0.

where total orderings $o(V'(a_1)), \ldots, o(V'(a_m))$ of $V'(a_1), \ldots, V'(a_m)$, respectively, have already been computed. Compute a total ordering $o(V(a))$ of $V(a)$ such that the maximum twist size of $o(V(a))$ is $f(n)$. Again, such an ordering exists by hypothesis, since $G(a)$ is 4-connected. Next, we merge $o(V'(a_1)), \ldots, o(V'(a_m))$ with $o(V(a))$. In order to do this, we define the operation of *merging an ordering V_2 into an ordering V_1*, that takes as input two total vertex orderings $o(V_1)$ and $o(V_2)$ such that V_1 and V_2 share a single vertex v, and outputs a single total vertex ordering $o(V_1 \cup V_2)$ of $V_1 \cup V_2$ such that $o(V_1 \cup V_2)$ coincides with $o(V_i)$ when restricted to the vertices in V_i, for $i = 1, 2$, and such that every vertex of V_1 that precedes v in $o(V_1)$ (resp. follows v in $o(V_1)$) precedes all the vertices of V_2 in $o(V)$ (resp. follows all the vertices of V_2 in $o(V)$). Denote by $b(H)$, by $m(H)$, and by $t(H)$ the bottom vertex, the middle vertex, and the top vertex of an upward triangulation H, respectively. Then, ordering $o(V'(a))$ is defined as follows: Let $o_1 = o(V(a))$ and let o_{i+1} be the ordering obtained by merging $o(V'(a_i)) \setminus \{b(G'(a_i)), t(G'(a_i))\}$ into o_i, for $i = 1, \ldots, m$; then $o(V'(a)) = o_{m+1}$. Observe that $o(V'(a))$ is an upward vertex ordering because $o(V(a)), o(V'(a_1)), \ldots, o(V'(a_m))$ are and because of the definition of the merging operation.

We now prove that the size of the maximum twist induced by $o(V)$ is $O(f(n))$. Let $M = \{e_1 = (u_1, v_1), \ldots, e_k = (u_k, v_k)\}$ denote any maximal twist induced by $o(V)$. We have the following:

Claim 1. *Let a be a node of T. Let a_1 and a_2 be two distinct children of a. There is no pair of distinct edges $(u_i, v_i), (u_j, v_j)$ in M such that $(u_i, v_i) \in E'(a_1), (u_j, v_j) \in E'(a_2)$, and $\{u_i, v_i, u_j, v_j\} \cap V(a) = \emptyset$.*

Proof: Let (u^1, v^1, z^1) and (u^2, v^2, z^2) be the separating triangles of $G'(a)$ that delimit the outer faces of $G'(a_1)$ and $G'(a_2)$, where v^i is the middle vertex of $G'(a_i)$, for $i = 1, 2$. If $v^1 \neq v^2$, then, by the construction of $o(V)$, all internal vertices of $G'(a_1)$ precede all internal vertices of $G'(a_2)$ or vice versa, thus e_i and e_j do not both belong to M. Otherwise, $v^1 = v^2$. Then, again by the construction of $o(V)$, e_i and e_j are nested, thus they do not both belong to M.   □

Let r be the root of T. We assume that G is "minimal", that is, we assume that there exists no child a of r such that all the edges in M belong to $G'(a)$. Indeed, if such a child exists, graph $G = G'(r)$ can be replaced by $G'(a)$, and the bound on the size of M can be achieved by arguing on $G'(a)$ rather than on $G'(r)$. Denote by M_i, with $i = 0, 1, 2$, the subset of M that contains all the edges having i endpoints in $V(r)$. Observe that $|M| = |M_0| + |M_1| + |M_2|$, hence it suffices to prove that $|M_i| \in O(f(n))$, for

$i = 0, 1, 2$, in order to prove the theorem. By hypothesis and since $G(r)$ is 4-connected, we have $|M_2| \leq f(n)$. We now deal with the edges in M_1.

Claim 2. $|M_1| \in O(f(n))$.

Proof: First, we argue that M_1 contains at most one edge e such that an end-vertex of e is the middle vertex of an upward planar triangulation $G'(a)$, for some child a of r. Indeed, by the vertex ordering's construction, any two such edges, say e_a and e_b, are either incident to the same vertex or are such that both end-vertices of e_a come before both end-vertices of e_b in $o(V'(a))$. Thus, it is enough to bound the number of edges in M_1 whose end-vertex in $V(r)$ is the bottom vertex or the top vertex of an upward planar triangulation $G'(a)$, where a is a child of r.

Let M_1^b (resp. M_1^t) be the subset of the edges in M_1 whose end-vertex in $V(r)$ is the bottom vertex (resp. the top vertex) of an upward planar triangulation $G'(a)$, where a is a child of r. Observe, that by the above observation, $|M| \leq |M_1^b| + |M_1^t| + 1$. In the following we bound $|M_1^b|$ (the bound for $|M_1^t|$ can be obtained analogously).

Consider any edge $(u, v) \in M_1^b$, where $u \in V(r)$. We define a *corresponding edge* of (u, v) in $G(r)$ as follows. Let $a_{u,v}$ be the child of r such that $G'(a_{u,v})$ contains edge (u, v). Further, denote by $m_{u,v}$ the middle vertex of $G'(a_{u,v})$. Then, $(u, m_{u,v})$ is the corresponding edge of (u, v) in $G(r)$. Observe that edge $(u, m_{u,v})$ exists and belongs to $E(r)$. Now consider the multi-set E_1^b of the corresponding edges, that is $E_1^b = \{(u, m_{u,v}) | (u, v) \in M_1^b\}$. First, we have that, for each vertex w in $V(r)$, there exist at most two edges (z, w) in E_1^b, since each vertex in $V(r)$ is the middle vertex of at most two upward planar triangulations $G'(a_i)$, where a_i is a child of r, and since $G'(a_i)$ has at most one edge in M_1^b. If there exist two edges (z_1, w) and (z_2, w) in E_1^b, then remove one of them. Then, after such deletions, $|E_1^b| \geq |M_1^b|/2$.

Next, we prove that each vertex in $V(r)$ is an end-vertex of at most two edges in E_1^b. Namely, consider any two edges (u_1, v_1) and (u_2, v_2) in E_1^b. Then, $v_1 \neq v_2$ because of the deletions performed on E_1^b, and $u_1 \neq u_2$ as otherwise the corresponding edges in M_1^b would share a vertex, contradicting the assumption that M is a twist; thus, each vertex in $V(r)$ is the source of at most one edge in E_1^b and the sink of at most one edge in E_1^b. Since the degree of graph $(V(r), E_1^b)$ is two, there exists a subset E^* of E_1^b such that the degree of graph $(V(r), E^*)$ is one and $|E^*| \geq |E_1^b|/3$.

Finally, we have that every two edges in E^* cross. Namely, if they do not, then by the vertex ordering's construction the corresponding edges in M_1^b would not cross either, thus contradicting the assumption that M is a twist.

Since $E^* \subseteq E(r)$ and the maximum size of a twist of edges in $E(r)$ is $f(n)$, given that $G(r)$ is 4-connected, it follows that $E^* \leq f(n)$. Using $|E^*| \geq |E_1^b|/3$ and $|E_1^b| \geq |M_1^b|/2$, we get $|M_1^b| \leq 6f(n)$. Such an inequality, together with the analogous bound $|M_1^t| \leq 6f(n)$ and with $|M| \leq |M_1^b| + |M_1^t| + 1$, proves the theorem. $\square$

We now proceed by bounding the size of M_0.

Claim 3. $|M_0| \in O(f(n))$.

Proof: By Claim 1, all the edges in M_0 belong to a graph $G'(a)$, for a certain descendant a of r. Let us choose a so that the length of the path from a to r is maximized. Let w be the middle vertex of the separating triangle (u, v, w) delimiting $G'(a)$. Let a' denote the child of r which is an ancestor of a or that coincides with a. Let w' be the middle vertex of the separating triangle (u', v', w') delimiting $G'(a')$.

For any edge $(y, z) \in M_0$, we have that (y, z) "nests around w'", that is, y precedes w' and w' precedes z in $o(V)$. Indeed, if both y and z precede w' in $o(V)$ (or if they both follow w' in $o(V)$), then only the edges in $G'(a')$ can possibly cross (y, z), by the construction of $o(V)$, thus contradicting the minimality of r.

If $w \neq w'$, then $|M_0| \leq 3$, since only the edges incident to u, v and w can belong to M_0. Otherwise we have $w' = w$ (see Fig. 1(b)). Consider graph $G'(a)$; partition the edges in M_0 into two subsets, namely M_0' contains all the edges of M_0 having at least one end-vertex in $V(a)$ and M_0'' contains all the edges of M_0 having no end-vertex in $V(a)$. By definition of a and by Claim 1, $|M_0'| > 0$, as otherwise there would exist a child of a containing all the edges of M_0. However, by Claim 2 applied to $G'(a)$ and by the hypothesis of the theorem, we have $|M_0'| \in O(f(n))$. Moreover, every edge in M_0'' is in a separating triangle of $G'(a)$ having w as middle vertex; however, any such edge is nested inside any edge of M_0'; thus, since $|M_0'| > 0$, we have $|M_0''| = 0$ and hence $|M_0| \in O(f(n))$, which concludes the proof. □

Since $|M_i| \in O(f(n))$, for $i = 0, 1, 2$, it follows that $|M| \in O(f(n))$, thus proving Theorem 1. By Lemmata 1 and 2, we have the following:

Corollary 1. *If every n-vertex upward planar 4-connected triangulation has $o(\frac{n}{\log n})$ page number, then every n-vertex upward planar triangulation has $o(n)$ page number.*

Corollary 2. *Every upward planar 3-tree has $O(1)$ page number.*

4 Page Number and Diameter

In this section we study the relationship between the page number of an upward planar DAG and its diameter D. We show that upward planar DAGs with small diameter have sub-linear page number. Notice that such a result pairs the observation that graphs with diameter $n - o(n)$ have sub-linear page number as well, given that upward planar Hamiltonian DAGs have page number two. We have the following:

Theorem 2. *Every n-vertex upward planar triangulation whose diameter is at most D admits an upward vertex ordering whose maximum twist size $t(n)$ is a function satisfying $t(n) \leq aD + t(\frac{n}{2}) + b$, for some constants a and b.*

We will prove the statement for a family of upward planar DAGs that is strictly larger than the family of upward planar triangulations. Namely, we call *upward cactus* an embedded upward planar DAG G having exactly one source $s(G)$ and such that every internal face is delimited by a 3-cycle. See Fig. 2. Observe that an upward planar triangulation is an upward cactus.

Consider an upward cactus G. We call *monotone path* any directed path $P = (u_1, \ldots, u_k)$ from $s(G)$ to a sink of G. Consider an upward planar drawing Γ of G in which u_k is the vertex with highest y-coordinate. Observe that such a drawing Γ always exists because G is an upward cactus. Then, we define the *left side of P* as the

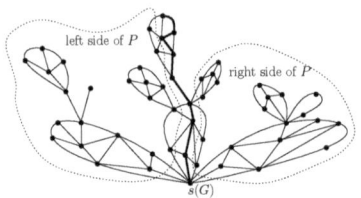

left side of P

right side of P

$s(G)$

Fig. 2. An upward cactus G. The thick edges represent a monotone path P

subgraph of G induced by all the vertices which are to the left of the Jordan curve representing P in Γ. The *right side of P* is defined analogously. Observe that the vertices of P, the vertices of the left side of P, and the vertices of the right side of P form a partition of the vertices of G. We have the following:

Claim 4. *In every n-vertex upward cactus there exists a monotone path P such that both the left side of P and the right side of P have less than $\frac{n}{2}$ vertices.*

We now prove the statement of the theorem for every n-vertex upward cactus G with diameter at most D. The proof is by induction on n. If $n \leq 3$, then in any upward vertex ordering of G the maximum twist size is 1, hence $t(3) \leq b$, for any $b \geq 1$, thus proving the base case.

Suppose that $n > 3$. By Claim 4, there exists a monotone path P in G such that both the left side of P and the right side of P have less than $\frac{n}{2}$ vertices. We now associate each vertex in the left side of P and each vertex in the right side of P to a vertex of P. Namely, we associate a vertex v in the left side of P to the vertex u_i of P such that there exists a directed path from u_i to v and such that, for every $j > i$, there exists no directed path from u_j to v. Observe that, for every vertex v in the left side of P, there exists a directed path from $s(G)$ to v, since G has a unique source, hence v is associated to exactly one vertex of P. Then, we call *left bag of u_i* the set of vertices in the left side of P which are associated to u_i, for each $i = 1, \ldots, k$. Vertices in the right side of P are associated to vertices of P analogously, thus analogously defining the *right bag of u_i*, for each $i = 1, \ldots, k$. We have the following:

Claim 5. *The subgraph G_i^L of G induced by the left bag of u_i and by u_i is an upward cactus, for every $i = 1, \ldots, k$.*

An analogous claim holds for the subgraph G_i^R of G induced by the right bag of u_i and by u_i.

Next, we construct an upward vertex ordering of G. This is done as follows. First, inductively construct an upward vertex ordering σ_i^L of G_i^L and an upward vertex ordering σ_i^R of G_i^R, for $i = 1, \ldots, k$, such that the maximum twist size of each of σ_i^R and σ_i^L is $t(\frac{n}{2})$. This is possible since G_i^L and G_i^R are upward cacti, by Claim 5, and they have less than $\frac{n}{2}$ vertices, by Claim 4. Observe that u_i is the first vertex both in σ_i^L and in σ_i^R, given that it is the only source of both G_i^L and G_i^R. Then, denote by σ_i the vertex ordering of $G_i^L \cup G_i^R$ which is obtained by concatenating σ_i^L and $\sigma_i^R \setminus \{u_i\}$. Finally a vertex ordering σ of G is obtained by concatenating $\sigma_1, \sigma_2, \ldots, \sigma_k$.

Claim 6. σ *is an upward vertex ordering.*

Next, we prove that the maximum twist size $t(n)$ of σ is at most $aD + t(\frac{n}{2}) + b$, for some constants a and b.

First, observe that the edges that have both end-vertices in P create twists of size at most two, since the graph induced by the vertices of P is upward planar Hamiltonian.

Second, we discuss the size of a twist composed of *intra-bag* edges, which are edges whose both end-vertices are associated to the same vertex of P. Consider any edge e_i^L of G_i^L and any edge e_i^R of G_i^R. Such edges do not cross. Namely, if such edges are both incident to u_i, then they do not cross by definition. If e_i^R is not incident to u_i, then both end-vertices of e_i^R come after both end-vertices of e_i^L, by construction, hence such edges do not cross. Moreover, if e_i^R is incident to u_i and e_i^L is not, then e_i^L is nested inside e_i^R, by construction, hence such edges do not cross. It follows that the maximum size of a twist of intra-bag edges is equal to the maximum twist size of σ restricted to the vertices in G_i^a for some $a \in \{L, R\}$ and some $1 \le i \le k$. By Claim 5, graph G_i^a is an upward cactus. Moreover, by Claim 4, G_i^a has at most $\frac{n}{2}$ vertices, hence the maximum size of a twist of intra-bag edges is at most $t(\frac{n}{2})$.

Third, we discuss the maximum size of a twist composed of *inter-bag* edges, which are edges whose end-vertices are associated to distinct vertices of P. We show that the maximum size of a twist composed of inter-bag edges in the left side of P is $2D$. An analogous proof shows that the maximum size of a twist composed of inter-bag edges in the right side of P is also $2D$.

Consider any two inter-bag edges (w_1, w_2) and (w_3, w_4) in the left side of P. Suppose that (w_1, w_2) and (w_3, w_4) cross in σ. Denote by $u_{j_1}, u_{j_2}, u_{j_3},$ and u_{j_4}, such that $u_{j_1} < u_{j_2}$ and $u_{j_3} < u_{j_4}$, the vertices of P vertices w_1, w_2, w_3, and w_4 have been assigned to, respectively. The following claim asserts that any two inter-bag edges (w_1, w_2) and (w_3, w_4) that cross in σ either have their sources assigned to the same vertex of P, or have their destinations assigned to the same vertex of P, or the source of one of them and the destination of the other of them are assigned to the same vertex of P.

Claim 7. *At least one of the following holds:* $j_1 = j_3 < j_2, j_4,$ *or* $j_1 < j_2 = j_3 < j_4,$ *or* $j_3 < j_4 = j_1 < j_2,$ *or* $j_1, j_3 < j_2 = j_4.$

Hence, if there are more than $2D$ inter-bag edges pairwise crossing in the left side of P, then either there are more than D inter-bag edges pairwise crossing in the left side of P such that the origins of such edges have all been assigned to the same vertex of P, or there are more than D inter-bag edges pairwise crossing in the left side of P such that the destinations of such edges have all been assigned to the same vertex of P. In the following, we discuss such two cases.

Claim 8. *Suppose that G contains inter-bag edges* $(v_1, w_1), (v_2, w_2), \ldots, (v_k, w_k)$ *in the left side of P, where* $v_1 <_\sigma v_2 <_\sigma \cdots <_\sigma v_k <_\sigma w_1 <_\sigma w_2 <_\sigma \cdots <_\sigma w_k$ *and where all the vertices w_i have been assigned to the same vertex u_l of P, for $i = 1, \ldots, k$, or all the vertices v_i have been assigned to the same vertex u_l of P, for $i = 1, \ldots, k$. Then, there exists a directed path starting at u_l and passing through $w_1, w_2, \ldots, w_k$.*

Since by hypothesis any directed path contains at most D vertices, then, by Claim 8, the maximum size of a twist of inter-bag edges sharing their destinations in the left side of P is at most D and the maximum size of a twist of inter-bag edges sharing their origins in the left side of P is at most D. Hence, by Claim 7, the maximum size of a twist of inter-bag edges in the left side of P is at most $2D$ and the maximum size of a twist of inter-bag edges is at most $4D$. Since every edge of G is either an edge having both end-vertices in P, or is an intra-bag edge, or is an inter-bag edge, it follows that the maximum size of a twist in σ is $t(n) = 2 + t(\frac{n}{2}) + 4D$, thus proving Theorem 2.

By Lemma 1, we have the following:

Corollary 3. *Every n-vertex upward planar triangulation whose diameter is $o(\frac{n}{\log n})$ has $o(n)$ page number.*

5 Page Number and Degree

In this section we discuss the relationship between the page number of a graph and its degree. We prove the following theorem.

Theorem 3. *Let $f(n)$ be any function such that $f(n) \in \Omega(\sqrt{n})$ and $f(n) \in O(n)$. Suppose that every n-vertex upward planar triangulation whose degree is $O(f(n))$ admits a book embedding with $O(g(n))$ pages, for some function $g(n) \in \Omega(1)$ and $g(n) \in O(n)$. Then, every n-vertex upward planar triangulation admits a book embedding with $O(g(n) + \frac{n}{f(n)})$ pages.*

Consider any n-vertex upward planar triangulation G. We transform G into an $O(n)$-vertex upward planar triangulation G' with degree $O(f(n))$ as follows. Fix any constant $c > 0$ and denote by $u_1, \ldots, u_k$ any ordering of the vertices of G whose degree is greater than $cf(n)$.

For $i = 1, \ldots, k$, consider vertex u_i. Suppose that u_i is an internal vertex of G, the case in which u_i is an external vertex being analogous. Since it is an upward planar triangulation, G has exactly two faces (v_1, v_2, u_i) and (v_3, v_4, u_i) incident to u_i such that edges (v_1, u_i) and (v_4, u_i) are incoming u_i and such that edges (u_i, v_2) and (u_i, v_3) are outgoing u_i. Assume, w.l.o.g., that (v_1, u_i), (u_i, v_2), (u_i, v_3), and (v_4, u_i) appear in this clockwise order around u_i. Denote by $w_1 = v_2, w_2, \ldots, w_{x-1}, w_x = v_3, w'_1 = v_4, w'_2, \ldots, w'_{y-1}, w'_y = v_1$ the clockwise order of the neighbors of u_i (see Fig. 3(a)). Remove u_i and its incident edges from G. Let $M = \lceil \frac{x}{f(n)-1} \rceil$ and $N = \lceil \frac{y}{f(n)-1} \rceil$. Insert $M + N + 2$ vertices $z_1, \ldots, z_{M+N+2}$ in G inside the cycle of the neighbors of u_i. Insert an edge from z_j to z_{j+1}, for $j = 1, \ldots, M$, insert an edge from z_{j+1} to z_j, for $j = M+1, \ldots, M+N+1$, and insert edges from z_{M+2} to $z_1, \ldots, z_M$ and from $z_{M+3}, \ldots, z_{M+N+2}$ to z_1. Insert edges from v_1 to z_1, from z_1 to v_2, from v_4 to z_{M+2}, and from z_{M+2} to v_3. Insert edges from z_j to $w_{(j-2)(f(n)-1)+1}, w_{(j-2)(f(n)-1)+2}, \ldots, w_{(j-1)(f(n)-1)}$, for $j = 2, \ldots, M+1$; insert edges from $w'_{(j-2)(f(n)-1)+1}, w'_{(j-2)(f(n)-1)+2}, \ldots, w'_{(j-1)(f(n)-1)}$ to z_{M+j}, for $j = 3, \ldots, N+2$. See Fig. 3(b).

It is easy to see that the triangulation G' obtained from G after all vertices $u_1, \ldots, u_k$ have been considered is upward planar. We have the following.

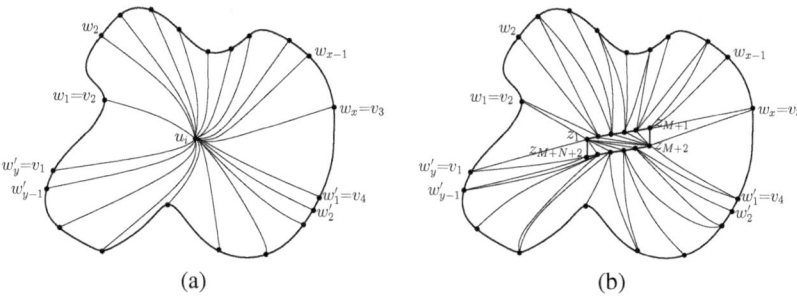

Fig. 3. (a) Neighbors of a high-degree vertex u_i. (b) Replacing u_i with lower-degree vertices, assuming $f(n) = 3$.

Claim 9. *G' has $O(n)$ vertices and $O(f(n))$ degree. Moreover, for every upward vertex ordering σ' of G', there exists an upward vertex ordering σ of G such that σ and σ' restricted to the vertices that are both in G and in G' coincide.*

We now describe how to compute a book embedding of G in $O(g(n) + \frac{n}{f(n)})$ pages. First, construct the upward planar triangulation G' as above. Second, construct a book embedding of G' into $O(g(n))$ pages. Such a book embedding exists by hypothesis, since G' has $O(n)$ vertices and $O(f(n))$ degree (by Claim 9). Denote by σ' the total ordering of the vertices of G' in the constructed book embedding. Construct any total ordering σ of the vertices of G such that σ and σ' restricted to the vertices that are both in G and in G' coincide. Such an ordering exists (and can be easily constructed) by Claim 9. The edges of G can be assigned to pages as follows: $O(g(n))$ pages suffice to accommodate all the edges that are both in G and in G'; moreover, one page can be used to accommodate all the edges incident to vertex u_i, for $i = 1, \ldots, k \in O(\frac{n}{f(n)})$. It follows that G has a book embedding in $O(g(n) + \frac{n}{f(n)})$ pages, thus proving Theorem 3.

Corollary 4. *Every n-vertex upward planar triangulation has $o(n)$ page number if and only if every n-vertex upward planar triangulation with degree $O(\sqrt{n})$ has $o(n)$ page number.*

6 Conclusions

In this paper we studied the relationship between the page number of an upward planar triangulation G and three important parameters of G: The connectivity, the diameter, and the degree. It would be interesting, in our opinion, to understand whether the statements of Theorems 1 and 2 can be referred to the page number rather than to the maximum twist size. That is: (1) Is it true that any upward planar triangulation G has page number $O(k)$ if and only if every maximal 4-connected subgraph of G has page number $O(k)$? (2) Is it true that any n-vertex upward planar triangulation G with diameter D has page number $p(n)$ satisfying $p(n) = p(\frac{n}{2}) + aD + b$, for some constants a and b?

Determining whether every n-vertex upward planar DAG has $o(n)$ page number and whether there exist upward planar DAGs with $\omega(1)$ page number remain among the most important problems in the theory of linear graph layouts.

Acknowledgments. The first author would like to thank Patrizio Angelini, Giuseppe Di Battista, and Stefano Saraulli for very useful discussions.

References

1. Alzohairi, M., Rival, I.: Series-Parallel Planar Ordered Sets Have Pagenumber Two. In: North, S.C. (ed.) GD 1996. LNCS, vol. 1190, pp. 11–24. Springer, Heidelberg (1997)
2. Buss, J.F., Shor, P.W.: On the pagenumber of planar graphs. In: Symposium on Theory of Computing (STOC 1984), pp. 98–100. ACM (1984)
3. Cerný, J.: Coloring circle graphs. Elec. Notes Discr. Math. 29, 457–461 (2007)
4. Chung, F.R.K., Leighton, F.T., Rosenberg, A.L.: Embedding graphs in books: A layout problem with applications to VLSI design. SIAM J. Alg. Discr. Meth. 8, 33–58 (1987)
5. Di Battista, G., Tamassia, R.: Algorithms for plane representations of acyclic digraphs. Theor. Comp. Sci. 61, 175–198 (1988)
6. Di Giacomo, E., Didimo, W., Liotta, G., Wismath, S.K.: Book embeddability of series-parallel digraphs. Algorithmica 45(4), 531–547 (2006)
7. Enomoto, H., Nakamigawa, T., Ota, K.: On the pagenumber of complete bipartite graphs. J. Comb. Th. Ser. B 71(1), 111–120 (1997)
8. Ganley, J.L., Heath, L.S.: The pagenumber of k-trees is $O(k)$. Discr. Appl. Math. 109(3), 215–221 (2001)
9. Heath, L.S.: Embedding planar graphs in seven pages. In: Foundations of Computer Science (FOCS 1984), pp. 74–83. IEEE (1984)
10. Heath, L.S., Istrail, S.: The pagenumber of genus g graphs is $O(g)$. J. ACM 39(3), 479–501 (1992)
11. Heath, L.S., Leighton, F.T., Rosenberg, A.L.: Comparing queues and stacks as mechanisms for laying out graphs. SIAM J. Discr. Math. 5(3), 398–412 (1992)
12. Heath, L.S., Pemmaraju, S.V.: Stack and queue layouts of posets. SIAM J. Discr. Math. 10(4), 599–625 (1997)
13. Heath, L.S., Pemmaraju, S.V.: Stack and queue layouts of directed acyclic graphs: Part II. SIAM J. Computing 28(5), 1588–1626 (1999)
14. Heath, L.S., Pemmaraju, S.V., Trenk, A.N.: Stack and queue layouts of directed acyclic graphs: Part I. SIAM J. Computing 28(4), 1510–1539 (1999)
15. Kainen, P.C.: Thickness and coarseness of graphs. Abh. Math. Sem. Univ. Hamburg 39, 88–95 (1973)
16. Kostochka, A.V., Kratochvíl, J.: Covering and coloring polygon-circle graphs. Discr. Math. 163(1-3), 299–305 (1997)
17. Malitz, S.M.: Genus g graphs have pagenumber $O(\sqrt{g})$. J. Algorithms 17(1), 85–109 (1994)
18. Malitz, S.M.: Graphs with e edges have pagenumber $O(\sqrt{e})$. J. Algorithms 17(1), 71–84 (1994)
19. Ollmann, L.T.: On the book thicknesses of various graphs. In: Hoffman, F., Levow, R.B., Thomas, R.S.D. (eds.) Southeastern Conference on Combinatorics, Graph Theory and Computing. Congressus Numerantium, vol. VIII, p. 459 (1973)
20. Rosenberg: The Diogenes approach to testable fault-tolerant arrays of processors. IEEE Trans. Comp. C-32, 902–910 (1983)
21. Tarjan, R.E.: Sorting using networks of queues and stacks. J. ACM 19(2), 341–346 (1972)
22. Yannakakis, M.: Embedding planar graphs in four pages. J. Comp. Syst. Sci. 38(1), 36–67 (1989)

Upward Point Set Embeddability for Convex Point Sets Is in $P^\star$

Michael Kaufmann[1], Tamara Mchedlidze[2], and Antonios Symvonis[2]

[1] Wilhelm-Schickard-Institut für Informatik, Universität Tübingen, Germany
`mk@informatik.uni-tuebingen.de`
[2] Dept. of Mathematics, National Technical University of Athens, Greece
`{mchet,symvonis}@math.ntua.gr`

Abstract. In this paper, we present a polynomial dynamic programming algorithm that tests whether a n-vertex directed tree T has an upward planar embedding into a convex point-set S of size n. We also note that our approach can be extended to the class of outerplanar digraphs. This nontrivial and surprising result implies that any given digraph can be efficiently tested for an upward planar embedding into a given convex point set.

1 Introduction

A *planar straight-line embedding* of a graph G into a point set S is a mapping of each vertex of G to a distinct point of S and of each edge of G to the straight-line segment between the corresponding end points so that no two edges cross each other. Planar straight-line embeddings for outerplanar graphs and trees were studied by Gritzmann *et al.* [11], Bose [4] and Bose *et al.* [5]. Cabello [6] proved that the problem to decide whether a given planar graph admits a planar straight-line embedding into a given point set is $\mathcal{NP}$-hard. Planar graph embeddings into point sets, where edges are allowed to bend, have also been studied (see, e.g., [2,7,12,14,17]).

An *upward planar directed graph* is a digraph that admits a planar drawing such that each edge is represented by a curve monotonically increasing in the y-direction. An *upward straight-line embedding* (*UPSE* for short) of an upward planar digraph G into a point set S is a mapping of each vertex of G to a distinct point of S and of each edge to the straight-line segment between its corresponding end points such that no two edges cross and for each edge (u, v) the condition $y(u) < y(v)$ holds, for the y-coordinates $y(u)$ and $y(v)$. *Upward point set embeddability* is the decision problem of whether a given digraph has an UPSE into a given point set.

$\star$ This research has been co-financed by EUROGIGA project GraDR 10-EuroGIGA-OP-003 and by the European Union (European Social Fund - ESF) and Greek national funds through the Operational Program "Education and Lifelong Learning" of the National Strategic Reference Framework (NSRF) - Research Funding Program: Heracleitus II. Investing in knowledge society through the European Social Fund.

M. van Kreveld and B. Speckmann (Eds.): GD 2011, LNCS 7034, pp. 403–414, 2012.

Upward point set embeddability was first studied by Giordano et al. [9]. The authors studied the version of the problem where bends on edges are allowed and showed that every planar st-digraph admits an upward point set embedding with at most two bends per edge. Upward point set embeddability with a given mapping, i.e., where a correspondence between the nodes and the point set is part of the input, was studied in [10,16]. Recently, straight-line drawings were studied in [1,3,8] and many interesting and partial results were presented. Among them are several results concerning upward point set embeddability of a tree into a convex point set. More specifically, several families of trees were presented, which have an UPSE into every convex point set, i.e., caterpillars, switch-trees, hourglass trees. On the other hand, it was demonstrated that the family of k-switch trees (generalization of switch-trees) does not have an UPSE into all convex point sets. An immediate question that arises from these facts is whether the existence of an UPSE of a tree into a convex point set can be efficiently tested. The contribution of this paper is an affirmative answer to this question. More specifically, we show that, given a directed tree T and a convex point set S, it can be tested in polynomial time whether T has an UPSE into S.

Recently, Geyer *et al.* [8] proved that the general *upward point-set embeddability* problem is $\mathcal{NP}$-complete even for m-convex point sets[1]. Thus one interesting open problem regarding UPSE was whether there exists a class of upward planar digraphs $\mathcal{D}$ for which the upward point set embeddability problem remains $\mathcal{NP}$-complete even for convex point sets. We answer this question in the negative by extending our UPSE algorithm for trees to the class of outerplanar graphs. Since any graph admitting a planar embedding into a convex point set is an outerplanar digraph, our result implies that the upward point-set embeddability can be efficiently solved for convex point sets and general digraphs.

For simplicity of presentation, we concentrate on the case of directed trees. In Section 2, we present the necessary notation and some basic results on UPSE, which are utilized by our tree algorithm. In Section 3, we study a restricted version of the UPSE problem which fixes the point in which the root of the tree is embedded and places restrictions on the drawing of subtrees. In Section 4, we explore the result of Section 3 and present a dynamic programming algorithm for deciding whether a directed tree has an UPSE into a convex point set. In Section 5 we state the extended result for outerplanar digraphs. Due to space constraints the proof of this result as well as some other proofs are presented in the extended version of this paper [15].

2   Notation - Preliminaries

Point Sets. Let S be a set of points on the plane. We assume that the points of S are in general position, i.e., no three of them lie on the same line. Moreover, we also assume that no two points of S share the same y-coordinate; if they do, a slight rotation of the coordinate axes can ensure that all points have distinct

---

[1] An m-convex point set can be intuitively defined as a set of m shelled, one into another, distinct convex point sets.

y-coordinates. The *convex hull* $CH(S)$ of S is the point set that is obtained as a convex combination of the points of S. A point set such that no point is in the convex hull of the others is called a *point set in convex position*, or a *convex point set*. Given a point set S, by $t(S)$ (resp., $b(S)$) we denote the top (bottom) point of S i.e., the point with the largest (resp., smallest) y-coordinate.

A *one-sided convex point set* S is a convex point set in which $b(S)$ and $t(S)$ are adjacent on the border of $CH(S)$. If $t(S)$ and $b(S)$ appear adjacent and in this order on the border of $CH(S)$ as we traverse it in the clockwise (resp., counterclockwise) direction, then the one-sided convex point set is called a *left-sided convex point set* (resp., *right-sided convex point set*). A point set consisting of at most two points is considered to be either a left-sided or a right-sided convex point set. A convex point set which is not one-sided, is called a *two-sided convex point set*.

Each given convex point set S may be considered to be the union of two specified (at the time S is given) one-sided convex point sets, one left-sided which is denoted by $L(S)$ and is referred to as the *left-side* of S, and one right-sided which is denoted by $R(S)$ and is referred to as the *right-side* of S. When there is no confusion regarding the point set S we refer to, for simplicity, we use the terms L and R instead of $L(S)$ and $R(S)$, respectively. Each of the points $b(S)$ and $t(S)$ belongs to either $L(S)$ or $R(S)$ but not both.

A subset of points of a convex point set S is called *consecutive* if its points appear consecutively as we traverse the convex hull of S in clockwise direction. Given that all points of S have distinct y-coordinates, we can refer to the first, the second, the third, etc., lowest point on the left (right) side of S. By p_i^L, $1 \leq i \leq |L(S)|$, we denote the i-th lowest point on the left side of S. Similarly, by p_i^R, $1 \leq i \leq |R(S)|$, we denote the i-th lowest point on the right side of S.

Let $S_{a..b,c..d} = \{p_i^L \mid a \leq i \leq b\} \cup \{p_i^R \mid c \leq i \leq d\}$ denote the subset of S consisting of $b-a+1$ consecutive points on the left side of S, starting from point p_a^L in the clockwise direction, and of $d-c+1$ consecutive points on the right side, starting from point p_c^R in the counterclockwise direction. For simplicity, for a one-sided point set S we use the notation $S_{a..b}$.

In this paper, we assume that queries of the form *"Find the i-th point on the left/right side of the convex point set S"* can be answered in $O(1)$ time, e.g., the points on each side of S are stored in an array in ascending order of their y-coordinates.

Trees. Consider a *directed tree* T, i.e., a directed acyclic graph whose underlying undirected structure is that of a tree. Tree T is *rooted* if one of its vertices, denoted by $r(T)$, is designated as its *root*. We then say that T *is rooted at* vertex $r(T)$. By $d^-(v)$ (resp., $d^+(v)$) we denote the in-degree (resp., the out-degree) of vertex v of T. By $d(v)$ we denote the total degree of vertex v, i.e., $d(v) = d^-(v) + d^+(v)$.

Let T be a rooted tree and let $r = r(T)$ be its root. Let $T_1^l, \ldots, T_{d^-(r)}^l, T_1^h, \ldots,$ $T_{d^+(r)}^h$ be the rooted subtrees of T obtained by removing from T its root r and r's incident arcs and having as their roots the vertices that are incident to r by either an incoming or an outgoing arc (see Figure 1.a). Trees $T_1^l, \ldots, T_{d^-(r)}^l$,

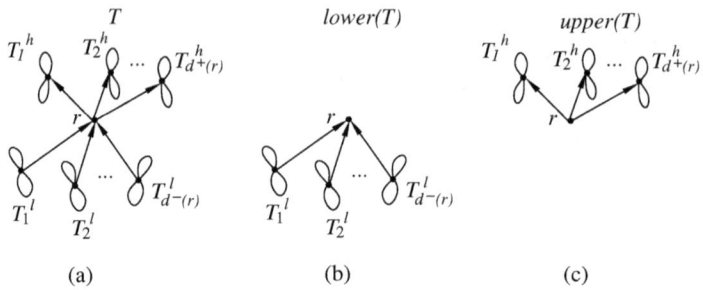

Fig. 1. (a) A rooted at vertex r tree T and its subtrees $T_1^l, \ldots, T_{d^-(r)}^l, T_1^h, \ldots, T_{d^+(r)}^h$. (b) The subtree $lower(T)$ of T. (c) The subtree $upper(T)$ of T.

$T_1^h, \ldots, T_{d^+(r)}^h$ are called the *subtrees of* T. Note that the superscripts "l" and "h" indicate whether a particular subtree of T is connected to r by an incoming to r or by an outgoing from r arc, respectively.

The rooted subtree of T consisting of T's root, r, together with $T_1^l, \ldots, T_{d^-(r)}^l$ is called the *lower subtree of* T and is also rooted at r. The lower subtree of T is denoted by $lower(T)$ (Figure 1.b). Similarly, the rooted subtree of T consisting of T's root, r, together with $T_1^h, \ldots, T_{d^+(r)}^h$ is called the *upper subtree of* T and is also rooted at r. The upper subtree of T is denoted by $upper(T)$ (Figure 1.c).

In this paper, we use the notation $\{u, v\}$ to denote arc (u, v) if $(u, v) \in T$ or arc (v, u) if $(v, u) \in T$. If u is mapped to point p and v is mapped to point q that is located below p, then we say that $\{u, v\}$ is drawn upward (downward) if $(v, u) \in T$ $((u, v) \in T)$.

2.1 Some known Results on UPSE of Rooted Directed Trees

We present some known results on UPSE of rooted directed trees that will be utilized by our algorithms. Binucci *et al.*[3] proved the following lemma concerning the placement of the subtrees of T in an UPSE of T on a convex point set.

Lemma 1 (Binucci et al. [3]). *Let T be a n-vertex directed tree rooted at r and let S be any convex point set of size n. Let $T_1, T_2, \ldots, T_{d(r)}$ be the subtrees of T. Then, in any UPSE of T into S, the vertices of subtree T_i are mapped to a set of consecutive points of S, $1 \leq i \leq d(r)$.* □

The following lemma concerns the UPSE of a rooted tree into a *one-sided* convex point set. It can be considered to be a simple restatement of a result by Heath *et al.* [13] (Theorem 2.1).

Lemma 2. *Let T be a n-vertex directed tree rooted at r and S be a one-sided convex point set of size n. Let $T_1, T_2, \ldots, T_{d(r)}$ be the subtrees of T. Then, T admits an UPSE into S so that the following are true:*

i) Each T_i, $1 \leq i \leq d(r)$, is drawn on consecutive points of S.

ii) If the root r of T is mapped to point p_r then there is no arc connecting a point of S below p_r to a point of S above p_r.

By utilizing Lemma 2, we prove the following.

Lemma 3. *Let T be a n-vertex directed tree rooted at r and S be a one-sided convex point set of size n. Then, an UPSE of T into S satisfying the properties of Lemma 2 can be obtained in $O(n)$ time. Moreover, after $O(n)$ time preprocessing, the point p_r that hosts the root r of T can be determined in $O(1)$ time (i.e., without determining the complete UPSE of T into S).*

Proof. Let $k = |lower(T)|$ be the size of subtree $lower(T)$ (rooted at r). Assuming that T was preprocessed in $O(n)$ time, k can be retrieved in constant time. It immediately follows that in an UPSE of T into S satisfying the properties of Lemma 2 there are $k - 1$ vertices of T (all belonging to $lower(T)$) that are placed below r. Thus, r is mapped to the k-th lowest point of S. This point, say p_r, can be computed in $O(1)$ time. Having decided where to place the root r, the UPSE of T can be completed in $O(n)$ time by recursively embedding the vertices of $lower(T)$ ($upper(T)$) to the points of S below (above) p_r.   □

3   A Restricted UPSE Problem for Rooted Directed Trees

In this section, we study a restricted UPSE problem that will be later on used by our main algorithm which decides whether there exists an UPSE of a given directed tree into a given convex point set.

Definition 1. *In a restricted UPSE problem for trees we are given a directed tree T rooted at r, a convex point set S, and a point $p_r \in S$. We are asked to decide whether there exists an UPSE of T into S such that (i) the root r of T is mapped to point p_r and, (ii) each subtree of T (rooted at r) is mapped to consecutive points on the same side (either L or R) of S.*

The following observation follows directly from the above definition.

Observation 1. *In a restricted UPSE of a directed tree T rooted at r into a convex point set $S = L \cup R$, where the root r of T is mapped to point $p_r \in S$, no edge enters the triangles $\triangle(t(L), t(R), p_r)$ and $\triangle(b(L), b(R), p_r)$.*

Figure 2.a shows a tree T rooted at vertex r, a convex point set S consisting of a left-sided convex point set L and a right-sided convex point set R. Tree T has a restricted UPSE only if its root r is mapped to point $p_r \in L$ (Figure 2.b). Mapping r to any other point $p \in S$ makes it impossible to map each subtree of T to consecutive points on the same side of S.

Before we proceed to describe a decision algorithm for the restricted UPSE problem, we need some more notation. Let T be a directed tree rooted at vertex r and let $\lambda = (T_1, \ldots, T_{d(r)})$ be an ordering of the subtrees of T. Let S be a convex point set and let Γ be an UPSE of T into S. We say that UPSE Γ *respects ordering* λ if for any two subtrees T_i and T_j, $1 \le i \le j \le d(r)$, that are

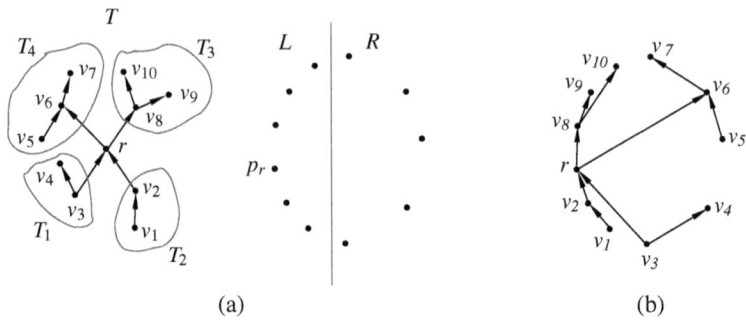

Fig. 2. (a) A tree T rooted at vertex r and a convex point set $S = L \cup R$. (b) A restricted UPSE of T into S so that r is mapped to point p_r. No restricted UPSE of T exists when r is mapped to any point other than p_r.

both mapped on the same side of S, T_i is mapped to a point set that is entirely below the point set T_j is mapped to.

Consider a tree T rooted at vertex r and let $\lambda = (T_1^l, \ldots, T_{d^-(r)}^l, T_1^h, \ldots, T_{d^+(r)}^h)$ be an ordering of the subtrees of T. Ordering λ is called a *proper ordering* of the subtrees of T if it satisfies the following properties:

(i) $|upper(T_i^l)| \le |upper(T_j^l)|$, $1 \le i \le j \le d^-(r)$, and
(ii) $|lower(T_i^h)| \ge |lower(T_j^h)|$, $1 \le i \le j \le d^+(r)$.

In Figure 2.a, ordering $\lambda_1 = (T_2, T_1, T_4, T_3)$ is a proper ordering of the subtrees of T, while ordering $\lambda_2 = (T_1, T_2, T_3, T_4)$ is not. Observe that in a proper ordering λ of T, the subtrees in the lower subtree of T appear before the subtrees in the upper subtree of T. The following lemma can be proved by reconstruction.

Lemma 4. *Let T be a n-vertex directed tree rooted at vertex r, λ be a proper ordering of the subtrees of T, and S be a convex point set of size n. Then, if there exists a restricted UPSE of T into S, there also exists a restricted UPSE of T into S that respects λ.* □

Theorem 1. *Let T be a n-vertex directed tree rooted at vertex r, L and R be left-sided and right-sided convex point sets, resp., such that $S = L \cup R$ is a convex point set of size n, and p_r a point of S. The restricted UPSE problem with input T, S and p_r can be decided in $O(d(r)n)$ time. Moreover, if a restricted UPSE for T, S and p_r exists, it can also be constructed in $O(d(r)n)$ time.*

Proof. Let $\lambda = (T_1, T_2, \ldots, T_{d(r)})$ be a proper ordering of the subtrees of T. Proper ordering λ can be computed in $O(n)$ time by a simple tree traversal that computes at the root of T the number of vertices in each subtree of $T \setminus \{v\}$ followed by a bucket sort of the sizes of the subtrees rooted at r. Since the restricted UPSE problem will be repeatedly solved on subtrees of T, we assume that T has been appropriately preprocessed in $O(n)$ time and, thus, a proper ordering of these subtrees can be then computed in $O(d(r))$ time. By Lemma 4, it is enough to test whether there exists a restricted UPSE that respects λ. Thus, we will describe a dynamic programming algorithm that tests whether there exists a restricted UPSE on input T, L, R and p_r.

Our dynamic programming algorithm uses a two-dimensional $d(r) \times |L|$ matrix M. Value $M[i, j]$ is $TRUE$ if and only if there exists a restricted UPSE of the subtree of T induced by r and $T_1, \ldots, T_i$ that uses all the j lowest points of the left-sided point set L and as many consecutive points as required in the lowest part of the right-sided convex point set R. Recall that $\{u, v\}$ denotes arc (u, v) if $(u, v) \in T$; arc (v, u) if $(v, u) \in T$; otherwise it is undefined.

For the boundary conditions of our dynamic programming we have that:

$$M[0, 0] = TRUE$$
$$M[1, j] = \begin{cases} TRUE, & \text{if } j = 0 \text{ and } p_r \notin R_{1..|T_1|} \text{ and } \{r(T_1), p_r\} \text{ is upward} \\ TRUE, & \text{if } j = |T_1| \text{ and } p_r \notin L_{1..|T_1|} \text{ and } \{r(T_1), p_r\} \text{ is upward} \\ FALSE, & \text{otherwise} \end{cases}$$

Let $\sigma = |T_1| + \ldots + |T_i|$. Value $M[i, j]$, $1 < i \leq d(r)$ and $0 \leq j \leq |L|$, is set to $TRUE$ if any of the following conditions is true; otherwise it is set to $FALSE$.

c-1: $M[i, j - 1] = TRUE$ **and** $p_r = L_{j..j}$.

This is the case where point p_r happens to be the j-th point of L. There is no need to test for upwardness of $\{r(T_i), p_r\}$ since it has been already tested when entry $M[i, j - 1]$ was filled in.

c-2: $M[i - 1, j - |T_i|] = TRUE$ **and** $p_r \notin L_{j-|T_i|+1..j}$ **and** $\{r(T_i), p_r\}$ **is upward**.

In this case, T_i is placed on L. We know that T_i fits on L since $j < |L|$, however, we must make sure that it also holds that p_r is not one of the $|T_i|$ topmost points of $L_{1..j}$.

c-3: $M[i - 1, j] = TRUE$ **and** $p_r \in R_{1..\sigma-j-|T_i|+1}$ **and** $\sigma - j + 1 \leq |R|$ **and** $\{r(T_i), p_r\}$ **is upward**.

In this case, T_i is placed to R. If p_r is one of the points in $R_{1..\sigma-j-|T_i|+1}$ then we have to make sure that at least $\sigma - j + 1$ points exist in $|R|$.

c-4: $M[i - 1, j] = TRUE$ **and** $p_r \notin R_{1..\sigma-j}$ **and** $\sigma - j \leq |R|$ **and** $\{r(T_i), p_r\}$ **is upward**.

In this case, T_i is also placed to R. However, in contrast to case **c-3**, p_r is not one of the points in $R_{1..\sigma-j}$. Thus, we only need to make sure that at least $\sigma - j$ points exist in $|R|$.

When determining the value of an entry $M[i, j]$ we need to decide whether arc $\{r(T_i), p_r\}$ is upward. In order to do that, we need to know the point to which $r(T_i)$ is mapped. By Lemma 3, this point can be computed in $O(1)$ time since T_i is mapped to $|T_i|$ consecutive points forming a one-sided convex point set.

It can be easily verified that entry $M[d(r), |L|] = TRUE$ if and only if there is a restricted UPSE of T into $L \cup R$ such that $r(T)$ is mapped to p_r.

Each entry of matrix M can be filled in $O(1)$ time. Thus, all entries of matrix M are filled in $O(d(r)|L|)$ time. The embedding, if exists, can be constructed by storing in each entry $M[i, j]$ (that was set to $TRUE$) the side ("L" or "R") in

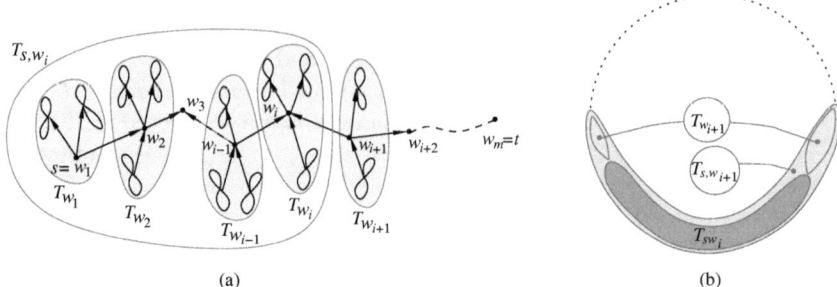

(a)   (b)

Fig. 3. (a) The decomposition of tree T based on a path between a source s and a sink t of T. (b) The structure of an UPSE of the tree T into point set S.

which T_i was placed. This information, together with the fact that the restricted UPSE respects ordering λ is sufficient to construct the embedding.   □

Denote by $\mathcal{L}(T, L, R)$ the set of points $p \in L \cup R$ such that there exists a restricted UPSE of T into $L \cup R$ where the root of T is mapped to p. The next theorem follows from Theorem 1, testing each point of $L \cup R$ as a candidate host for $r(T)$.

Theorem 2. *Let T be an n-vertex directed tree rooted at vertex r and L and R be left-sided and right-sided convex point sets, resp., such that $S = L \cup R$ is a convex point set of size n. Set $\mathcal{L}(T, L, R)$ can be computed in $O(d(r)n^2)$ time.*   □

4   The Testing Algorithm for Directed Trees

Let T be a directed tree and let S be a convex point set. In any UPSE of T into S, a source node s and a sink node t of T will be mapped to points $b(S)$ and $t(S)$, respectively. In this section, we present a dynamic programming algorithm that decides in polynomial time whether, given a n-vertex directed tree T, a source s and a sink t of T, and a convex point set S of size n, T has an UPSE into S so that s and t are mapped to $b(S)$ and $t(S)$, respectively. Applying this algorithm on all ⟨*source, sink*⟩ pairs of T, yields a polynomial time algorithm for deciding whether T has an UPSE into S.

Let s and t be a source and a sink vertex of T, respectively. Denote by $P_{s,t} = \{s = w_1, w_2, \ldots, w_m = t\}$ the (undirected) path connecting s and t in T, see Figure 3.a. By T_{s,w_i}, $1 \le i < m$, we denote the subtree of T that contains source s and is formed by the removal of edge $\{w_i, w_{i+1}\}$. By definition, $T_{s,w_m} = T$. Let $T_{w_i} = T_{s,w_i} \setminus T_{s,w_{i-1}}$, $1 < i \le m$. By definition, $T_{w_1} = T_{s,w_1}$. By Lemma 1, we know that T_{s,w_i} is drawn on consecutive points of S, call this point set S_i (see also Figure 3.b). Since s is mapped to $b(S)$, we infer that $b(S) \in S_i$. Similarly, in any UPSE of T into S, $T_{s,w_{i+1}}$ is also drawn on consecutive points of S that contain $b(S)$, call this point set S_{i+1}. Hence, $T_{w_{i+1}}$ is drawn on a set $S_{w_{i+1}} = S_{i+1} \setminus S_i$, that is, a subset of S comprised by two consecutive point sets of S, one on its left and one on its right side.

Our dynamic programming algorithm maintains a list of points $\mathcal{P}(a, b, k)$, $0 \le a \le |L|$, $0 \le b \le |R|$, $1 \le k \le m$, such that:

$$p \in \mathcal{P}(a, b, k) \iff \begin{cases} T_{s,w_k} \text{ has an UPSE into point set } S_{1..a,1..b} \text{ with} \\ \text{vertex } w_k \text{ mapped to point } p. \end{cases} \quad (1)$$

For the boundary conditions of our dynamic programming we have that $\mathcal{P}(a, b, 1) = \mathcal{L}(T_{w_1}, L_{1..a}, R_{1..b})$ where $a + b = |T_{w_1}|$. Note that since w_1 is a source, $\mathcal{P}(a, b, 1)$ is either $\{b(s)\}$ or $\emptyset$.

Our dynamic programming is based on the following recurrence relation, which allows us to add points in $\mathcal{P}(a, b, i)$. For any $1 < i \le m$ we set:

$$\begin{aligned} \mathcal{P}(a, b, i) = \{p \mid &\exists a_1, b_1 \in Z : a_1 + b_1 = |T_{w_i}| \\ &\textbf{and } p \in \mathcal{L}(T_{w_i}, L_{a-a_1+1..a}, R_{b-b_1+1..b}) \\ &\textbf{and } \exists q \in \mathcal{P}(a - a_1, b - b_1, i - 1) \\ &\textbf{and } \{p, q\} \text{ is upward} \quad \} \end{aligned} \quad (2)$$

Next we prove that the recurrence relation (2) satisfies the property described by equivalence (1). We start with the forward direction. From the boundary conditions it is true for $i = 1$. Assume that if $q \in \mathcal{P}(a-a_1, b-b_1, i-1)$ then $T_{s,w_{i-1}}$ has an UPSE into $S_{1..a-a_1,1..b-b_1}$ with vertex w_{i-1} mapped to point q. Let now $p \in \mathcal{P}(a, b, i)$. By the definition of the recurrence relation we infer that: (1) there exist $a_1, b_1 \in Z$ so that $a_1 + b_1 = |T_{w_i}|$, (2) $p \in \mathcal{L}(T_{w_i}, L_{a-a_1+1..a}, R_{b-b_1+1..b})$, which by definition of $\mathcal{L}$, means that there is a restricted UPSE of T_{w_i} into $L_{a-a_1+1..a}, R_{b-b_1+1..b}$ with w_i mapped to p, (3) $\exists q \in \mathcal{P}(a - a_1, b - b_1, i - 1)$, thus, by induction hypothesis, $T_{s,w_{i-1}}$ has an UPSE into $S_{1..a-a_1,1..b-b_1}$, and, finally, (4) edge $\{p, q\}$ is upward. Then we combine the UPSE for $T_{s,w_{i-1}}$ with the restricted UPSE for T_{w_i} in order to get an UPSE of T_{s,w_i} on point set $S_{1..a,1..b}$. By Observation 1, we have that the combined drawing is planar.

For the reversed statement we also work by induction. From the boundary conditions we know that if $T_{s,w_1} = T_{w_1}$ has an UPSE into a point set $S_{1..a,1..b}$ then $b(S) \in \mathcal{P}(a, b, 1)$, where $a + b = |T_{w_1}|$. Assume that the statement is true for $T_{s,w_{i-1}}$, i.e., if $T_{s,w_{i-1}}$ has an UPSE into a point set $S_{1..a,1..b}$ with vertex w_{i-1} mapped to q then $q \in \mathcal{P}(a, b, i - 1)$. Assume also that T_{s,w_i} has an UPSE into a point set $S_{1..a,1..b}$ with vertices s and w_i mapped to points $b(S)$ and p, respectively. By the discussion above we know that in every such embedding $T_{s,w_{i-1}}$ is mapped to consecutive points of $S_{1..a,1..b}$ that contains $b(S)$. Therefore there exist two numbers a_1 and b_1, so that $a_1 + b_1 = |T_{w_i}|$ and subtree T_{w_i} is mapped to the point set $S_{a-a_1+1..a,b-b_1+1..b}$, with vertex w_i mapped to some point p, $p \in S_{a-a_1+1..a,b-b_1+1..b}$. Moreover, by induction hypothesis, there exists $q \in \mathcal{P}(a - a_1, b - b_1, i - 1)$. So, since the edge connecting p and q is upward, by the definition of recurrence relation we infer that $p \in \mathcal{P}(a, b, i)$.

Finally we note that, an UPSE of T into S such that source s and sink t are mapped to $b(S)$ and $t(S)$, respectively, exists if and only if $\mathcal{P}(|L|, |R|, m)$ is non-empty. Note that if $\mathcal{P}(|L|, |R|, m) \ne \emptyset$, then it must hold that $\mathcal{P}(|L|, |R|, m) = \{t(S)\}$. The values $\mathcal{P}(a, b, k)$, when $0 \le a \le |L|$, $0 \le b \le |R|$, $1 \le k \le m$ are calculated by Algorithm 1.

Algorithm 1. TREE-UPSE(T, S, s, t)

input   : A directed tree T, a point set S, a source s and a sink t of T. Path
$(s = w_1, \ldots, w_m = t)$ is used to progressively build tree T from subtrees
T_{w_i}, $1 \le i \le m$.

output : "YES" if T has an UPSE into S with s mapped to $b(S)$ and t mapped
to $t(S)$, "NO" otherwise.

1. **For** $a = 0 \ldots |L|$
2.   **For** $b = 0 \ldots |R|$
3. $\mathcal{P}(a, b, 1) = \mathcal{L}(T_{w_1}, L_{1..a}, R_{1..b})$
4.   **For** $k = 2 \ldots m$ //Consider tree T_{w_k}
5. $\mathcal{P}(a, b, k) = \emptyset$
6.   **For** $i = 0 \ldots |T_{w_k}|$ //We consider the case where i vertices of T_{w_k}
                                    are placed to the left side of S
7.   **if** $(a - i \ge 0)$ **and** $(b - (|T_{w_k}| - i) \ge 0)$
8.   Let $\mathcal{L} = \mathcal{L}(T_{w_k}, L_{a-i+1..a}, R_{b-(|T_{w_k}|-i)+1..b})$
9.   //We consider all possible placements of w_{k-1}
10.   **For** each q in $\mathcal{P}(a - i, b - (|T_{w_k}| - i), k - 1)$
11.   //We consider all the possible placements of vertex w_k
12.   **For** each p in $\mathcal{L}$
13.   **if** ($\{w_{i-1}, w_i\}$ drawn on line-segment (q, p) is upward)
14.   **then** add p to $\mathcal{P}(a, b, k)$.
15. **if** $\mathcal{P}(|L|, |R|, m)$ is empty **then return**("NO");
16. **return**("YES");

Theorem 3. *Let T be a n-vertex rooted directed tree, S be a convex point set of size n, s be a source of T and t be a sink of T. It can be decided in time $O(n^5)$ whether T has an UPSE on S such that s is mapped to $b(S)$ and t is mapped to $t(S)$. If such an UPSE exists, it can be constructed within the same time bound.*

Proof. A naive analysis of Algorithm 1 yields an $O(n^7)$ time complexity. The analysis assumes that (i) the left and the right side of S have both size $O(n)$, (ii) the path from s to t has length $O(n)$, (iii) each tree T_{w_i} has size $O(n)$ and (iv) each $\mathcal{L}$-list containing the solution of a restricted UPSE problem is computed in $O(n^3)$ time. However, based on the following two observations, the total time complexity can be reduced to $O(n^5)$.

A factor of n can be saved by realizing that in our dynamic programming we can maintain a list $\mathcal{P}'(a, i)$ which uses only one parameter for the left side of the convex set (in contrast with $\mathcal{P}(a, b, i)$ which uses a parameter for each side of S). The number of points on the right side of S is implied since the size of each tree T_{s,w_i} is fixed. For simplicity, we have decided to use notation $\mathcal{P}(a, b, i)$. Another factor of n can be saved by observing that the solution of a restricted UPSE is actually $O(deg(w_i)n^2)$. Thus, summing over all i gives $O(n^3)$ in total, and not $O(n^4)$.

The UPSE of T into S can be recovered easily by modifying Algorithm 1 so that it stores for each point $p \in \mathcal{P}(a, b, k)$ the point q where vertex w_{i-1} is mapped to as well as the point set that hosts tree $T_{s,w_{i-1}}$ (i.e., its top point on the left and the right side of S).   □

By applying Algorithm 1 on all $\langle source, sink \rangle$ pairs of T we can decide whether tree T has an UPSE on a convex point set S, as the main next theorem indicates.

Theorem 4. *Let T be a n-vertex rooted directed tree and S be a convex point set of size n. It can be decided in time $O(n^6)$ whether T has an UPSE into S. If such an UPSE exists, it can also be constructed within the same time bound.*

Proof. Note that a naive application of the idea leads to the algorithm with time complexity $O(n^7)$, since there are $O(n^2)$ distinct pairs of sources and sinks. Next we explain how the overall time complexity can be reduced to $O(n^6)$. Let $P_{s,t}$ be a path from s to t, passing through m vertices, and let t' be the j-th vertex of $P_{s,t}$ that is also a sink of G. During the computation of $\mathcal{P}(a, b, m)$ corresponding to path $P_{s,t}$ we also compute $\mathcal{P}(a, b, j)$ and thus we can immediately answer whether there exists an UPSE of G into S so that s and t' is mapped to $b(S)$ and $t(S)$, respectively. Next consider a sink $\tilde{t}$ that does not belong to path $P_{s,t}$. Consider the path $P_{s,\tilde{t}}$. Assume that the last common vertex of $P_{s,t}$ and $P_{s,\tilde{t}}$ is the j-th vertex of $P_{s,t}$. In order to compute whether there is an UPSE of G into S so that s and $\tilde{t}$ are mapped to $b(S)$ and $t(S)$, respectively, we can start the computations of Algorithm 1 determined by variable k from the $j + 1$-th step (see line 4 of the algorithm). Thus, for a single source s and all possible sinks variable k changes at most n times. Since the number of different sources is $O(n)$ we conclude that the whole algorithm runs in time $O(n^6)$. □

5   Conclusions

In this paper we presented a polynomial dynamic programming algorithm that tests whether a n-vertex directed tree T has an upward planar embedding into a convex point-set S of size n. In the long version of this paper [15] we explain how our approach can be extended to the class of outerplanar digraphs, obtaining the following theorem.

Theorem 5. *Let G be a n-vertex digraph and S be a convex point set of size n. It can be decided in polynomial time whether G has an UPSE into S. Moreover, if such an UPSE exists, it can also be constructed in polynomial time.* □

Acknowledgments. We thank Markus Geyer for the useful discussions during the work on this paper.

References

1. Angelini, P., Frati, F., Geyer, M., Kaufmann, M., Mchedlidze, T., Symvonis, A.: Upward Geometric Graph Embeddings into Point Sets. In: Brandes, U., Cornelsen, S. (eds.) GD 2010. LNCS, vol. 6502, pp. 25–37. Springer, Heidelberg (2011)
2. Badent, M., Di Giacomo, E., Liotta, G.: Drawing colored graphs on colored points. Theor. Comput. Sci. 408(2-3), 129–142 (2008)

3. Binucci, C., Di Giacomo, E., Didimo, W., Estrella-Balderrama, A., Frati, F., Kobourov, S., Liotta, G.: Upward straight-line embeddings of directed graphs into point sets. Computat. Geom. Th. Appl. 43, 219–232 (2010)
4. Bose, P.: On embedding an outer-planar graph in a point set. Computat. Geom. Th. Appl. 23(3), 303–312 (2002)
5. Bose, P., McAllister, M., Snoeyink, J.: Optimal algorithms to embed trees in a point set. J. Graph Alg. Appl. 1(2), 1–15 (1997)
6. Cabello, S.: Planar embeddability of the vertices of a graph using a fixed point set is NP-hard. J. Graph Alg. Appl. 10(2), 353–366 (2006)
7. Di Giacomo, E., Didimo, W., Liotta, G., Meijer, H., Trotta, F., Wismath, S.K.: k-colored point-set embeddability of outerplanar graphs. J. Graph Alg. Appl. 12(1), 29–49 (2008)
8. Geyer, M., Kaufmann, M., Mchedlidze, T., Symvonis, A.: Upward Point-Set Embeddability. In: Černá, I., Gyimóthy, T., Hromkovič, J., Jefferey, K., Královič, R., Vukolić, M., Wolf, S. (eds.) SOFSEM 2011. LNCS, vol. 6543, pp. 272–283. Springer, Heidelberg (2011)
9. Giordano, F., Liotta, G., Mchedlidze, T., Symvonis, A.: Computing Upward Topological Book Embeddings of Upward Planar Digraphs. In: Tokuyama, T. (ed.) ISAAC 2007. LNCS, vol. 4835, pp. 172–183. Springer, Heidelberg (2007)
10. Giordano, F., Liotta, G., Whitesides, S.: Embeddability Problems for Upward Planar Digraphs. In: Tollis, I.G., Patrignani, M. (eds.) GD 2008. LNCS, vol. 5417, pp. 242–253. Springer, Heidelberg (2009)
11. Gritzmann, P., Mohar, B., Pach, J., Pollack, R.: Embedding a planar triangulation with vertices at specified positions. Amer. Math. Mont. 98, 165–166 (1991)
12. Halton, J.: On the thickness of graphs of given degree. Inf. Sci. 54, 219–238 (1991)
13. Heath, L.S., Pemmaraju, S.V., Trenk, A.N.: Stack and queue layouts of directed acyclic graphs: Part I. SIAM J. Comput. 28(4), 1510–1539 (1999)
14. Kaufmann, M., Wiese, R.: Embedding vertices at points: Few bends suffice for planar graphs. J. Graph Alg. Appl. 6(1), 115–129 (2002)
15. Kaufmann, M., Mchedlidze, T., Symvonis, A.: Upward point set embeddability for convex point sets is in P. Technical report. arXiv:1108.3092, http://arxiv.org/abs/1108.3092
16. Mchedlidze, T., Symvonis, A.: On ρ-Constrained Upward Topological Book Embeddings. In: Eppstein, D., Gansner, E.R. (eds.) GD 2009. LNCS, vol. 5849, pp. 411–412. Springer, Heidelberg (2010)
17. Pach, J., Wenger, R.: Embedding planar graphs at fixed vertex locations. Graphs and Combinatorics 17(4), 717–728 (2001)

Classification of Planar Upward Embedding[*]

Christopher Auer, Christian Bachmaier,
Franz Josef Brandenburg, and Andreas Gleißner

University of Passau, 94030 Passau, Germany
{auerc,bachmaier,brandenb,gleissner}@fim.uni-passau.de

Abstract. We consider planar upward drawings of directed graphs on arbitrary surfaces where the upward direction is defined by a vector field. This generalizes earlier approaches using surfaces with a fixed embedding in $\mathbb{R}^3$ and introduces new classes of planar upward drawable graphs, where some of them even allow cycles. Our approach leads to a classification of planar upward embeddability.

In particular, we show the coincidence of the classes of planar upward drawable graphs on the sphere and on the standing cylinder. These classes coincide with the classes of planar upward drawable graphs with a homogeneous field on a cylinder and with a radial field in the plane.

A cyclic field in the plane introduces the new class **RUP** of upward drawable graphs, which can be embedded on a rolling cylinder. We establish strict inclusions for planar upward drawability on the plane, the sphere, the rolling cylinder, and the torus, even for acyclic graphs. Finally, upward drawability remains **NP**-hard for the standing cylinder and the torus; for the cylinder this was left as an open problem by Limaye et al.

1 Introduction

Directed graphs are often used as a model for structural relations where the edges express dependencies. Such graphs are often acyclic and are drawn as hierarchies using the hierarchical approach introduced by Sugiyama et al. [22]. This drawing style transforms the edge direction into a geometric direction: all edges point upward. A graph is upward planar, for short **UP**, if it can be embedded into the plane such that the curves of the edges are monotonically increasing in y-direction with no crossing edges. **UP** is well-understood; see the comprehensive study in [5]. A graph is upward planar if and only if it is a subgraph of a planar st-graph. The graphs from **UP** admit straight-line upward drawings, which may require an area of exponential size, or upward polyline drawings on quadratic area using $O(n)$ many bends. An important result of Garg and Tamassia [10] states the **NP**-completeness of the recognition problem: Is a directed graph in **UP**? On the other hand, there are efficient polynomial time algorithms for upward planarity tests, if the graphs are given with an embedding or have a single source or are triconnected.

---

[*] Supported by the Deutsche Forschungsgemeinschaft (DFG), grant Br835/15-1.

M. van Kreveld and B. Speckmann (Eds.): GD 2011, LNCS 7034, pp. 415–426, 2012.

There were some approaches to generalize upward planarity on other surfaces using a fixed embedding of the surface in $\mathbb{R}^3$. Thomassen [23] studied graphs with a single source and a single sink on a standing cylinder. Foldes et al. [9] investigated ordered sets on the sphere and on a cylinder as a truncated sphere, and Hashemi et al. [13, 12, 7] generalized results on planarity from the plane to the sphere, including the **NP**-hardness of the recognition problem. They characterized the graphs with a spherical upward drawing as the subgraphs of the directed planar graphs with one source and one sink. Thus upward planarity and upward sphericity are distinguished by the *st*-edge connecting the single source and the single sink in the planar case. Dolati et al. [6, 8] studied upward planarity on the lying and the standing torus, and Mohar and Rosenstiehl [19] characterize toroidal maps with an upward orientation.

Planar upward drawings on the cylinder were also addressed from the viewpoint of the circuit value problem (CVP) [24, 11, 16]. In these papers the above papers were overseen, and the **NP**-hardness of upward cylindricality is stated as an open problem [16]. We solve this by using the **NP**-hardness for upward spherical and the coincidence of spherical and cylindrical upward planarity established in this paper.

In our approach we use the model of the fundamental polygon to define surfaces such as the plane, the cylinder and the torus. The plane is identified with the manifold $I \times I$, where I is the open interval from -1 to $+1$. The standing and rolling cylinder are obtained by identifying a pair of opposite sides, and the torus by a simultaneous identification of both pairs of opposite sides.

Upwardness is defined by a vector field and gives rise to the common (*strict*) increasing and the *weak* non-decreasing case. A vector field assigns a two-dimensional vector to each point (x, y) indicating the direction of the field. The basic case is the *null field* N, which assigns the null vector $(0, 0)$ everywhere. Then an upward direction becomes vacuous, and weakly upward planar coincides with planar. The *homogeneous field* H assigns the direction $(0, 1)$ and thus describes upward in y-dimension as it is commonly used. In addition, we use the *cyclic, radial* and *antiparallel fields* C, R and A, see Table 1.

Table 1. Typical fields

null	homogeneous	cyclic	radial	antiparallel
$(x, y) \mapsto (0, 0)$	$(x, y) \mapsto (0, 1)$	$(x, y) \mapsto (-y, x)$	$(x, y) \mapsto (x, y)$	$(x, y) \mapsto (0, \sin(y\pi))$

We introduce a new class of planar upward drawings on the rolling cylinder which is called **RUP**. Graphs of **RUP** may have cycles. It turns out that the rolling cylinder is stronger than the standing cylinder even for acyclic graphs. The graphs of **RUP** are related to planar recurrent hierarchies, which were introduced by Sugiyama et al. [22] as a cyclic version of their hierarchical approach and were recently studied in [4]. In recurrent hierarchies the levels are numbered from 0 to $k - 1$. The edges are upward where the difference of the levels of the vertices is computed modulo k. Hence, all cycles are unidirectional.

Another subclass of **RUP** are the graphs with a queue layout, see [1]. The input-output behavior of a queue is represented by a graph such that the behavior is legal if and only if the graph has a **RUP** embedding with all vertices placed on a horizontal line.

Our contributions are a general approach towards planar upward embeddings (Sect. 2). In Sect. 3 we unify the concepts on the sphere and establish a hierarchy for the plane, sphere, rolling cylinder and torus. Finally, the **NP**-hardness of the recognition problem is addressed.

2   Upward Embeddings with Vector Fields on Surfaces

Let $G = (V, E)$ be a simple directed graph with a finite set of vertices V and a finite set of directed edges E. A *surface* $\mathbb{S}$ is a two-dimensional differentiable manifold [20, 17]. An open interval from a to b is denoted by $]a, b[$ and a closed interval by $[a, b]$. $\langle \cdot, \cdot \rangle$ denotes the standard scalar product in $\mathbb{R}^2$. For a map $f : A \to B$ and a subset $A' \subseteq A$ denote the image of A' under f by $f[A']$. For any point $p = (p_1, p_2)$ define $x(p) = p_1$ and $y(p) = p_2$.

A *drawing* $\Gamma(G)$ on $\mathbb{S}$ is a mapping where each vertex $v \in V$ is mapped to a unique point $\Gamma(v) \in \mathbb{S}$, and each edge $(u, v) \in E$ is mapped to a piecewise continuously differentiable curve $\Gamma(u, v) : [0, 1] \to \mathbb{S}$ which starts at u and ends at v and is disjoint to the other vertex points. $\Gamma(u, v)$ does not self-intersect. When it is clear from the context, we say that $v \in V$ is *placed* at $\Gamma(v)$ and we do not distinguish between an edge $e \in E$ and its curve $\Gamma(e)$. Additionally, Γ stands for the set of points in the drawing.

Two edges $e_1 \neq e_2 \in E$ *cross* if they have a common point apart from a common endpoint. $\Gamma(G)$ is called a *plane* drawing if it is crossing-free. *Strict upward planarity* asks if a given graph admits a plane drawing where all edges are drawn monotonically increasing in a common upward direction. In the *weak* version the edges may be drawn monotonically non-decreasing. It is well-known that this makes no difference on the plane.

As outlined in Sect. 1 most prior attempts towards planar upward embeddings on the sphere, the cylinder, or the torus use a fixed embedding of the surface in $\mathbb{R}^3$ and define upward in y-direction [6, 7, 8, 11, 13, 16]. They describe the sphere and the (standing) cylinder by Cartesian coordinates $\{(x, y, z) : x^2 + y^2 + z^2 = 1\}$ and $\{(x, y, z) : x^2 + z^2 = 1, -1 \leq y \leq 1)\}$, respectively. These classes are called *spherical* and *cylindrical*. An alternative approach was used by Mohar, Rosenstiehl and Thomassen [19, 23] embedding graphs on the *flat torus* represented by its fundamental polygon. We generalize the idea by utilizing vector fields, i. e., a drawing is upward if all edge curves "go with the flow".

More formally, let $F : \mathbb{S} \to \mathbb{R}^2$ be a vector field on $\mathbb{S}$. Let $Cr(p) \subsetneq [0, 1]$ be the preimage of the bends of the curve p. $Cr(p)$ is the countable critical point set of a piecewise continuously differentiable curve $p : [0, 1] \to \mathbb{S}$. We say that p (*weakly*) *respects* F if

$$\underset{t\in[0,1]\setminus Cr(p)}{\forall} \quad \langle p'(t), F(p(t))\rangle > 0 \quad (\text{resp.} \geq 0), \tag{1}$$

where p' is the first order derivative of p. Likewise, a drawing Γ (*weakly*) *respects* F if $\Gamma(e)$ (weakly) respects F for each edge $e \in E$. Then at each point of a directed edge the angle between its tangent vector and the vector field is less (not more) than $\frac{\pi}{2}$. We call a graph *(weakly) upward embeddable on* $\mathbb{S}$ *in respect to* F if it admits a plane drawing (weakly) respecting F. We say that G *is a drawn (weakly) upward on* $(\mathbb{S}, F)$. Note that (1) holds true independently of the norm of $F(\cdot)$, i. e., only its direction is relevant.

The general definition allows for a plethora of combinations of surfaces and vector fields. From a graph-theoretic point of view many of them are equivalent in respect to upward embeddability. For reducing redundancy we consider mappings between surfaces which shall preserve the upward embeddability and so obtain equivalences.

Let $\mathbb{S}_1$ and $\mathbb{S}_2$ be smooth manifolds, i. e., locally similar to a linear space, with vector fields F_1 and F_2, respectively. Let $f : \mathbb{S}_1 \to \mathbb{S}_2$ be an injective smooth mapping between the surfaces. In the following we derive a way to express whether or not f also somehow "maps F_1 to F_2". The technique is also known as the *pushforward* of f [14]. Let z be any point in $\mathbb{S}_1$ and $p : [0, 1] \to \mathbb{S}_1$ be a smooth curve (not necessarily representing an edge) tangent to F_1 in z, i. e., $p(0) = z$ and $p'(0) = F_1(z)$. We derive how f acts on $F_1(z)$ by considering the derivative of $f(p)$ at 0,

$$(f \circ p)'(0) = (f' \circ p)(0) \cdot p'(0) = f'(p(0)) \cdot p'(0) = (f'(p(0))) \cdot F_1(z) = f'(z) \cdot F_1(z).$$

Due to the identification of the tangent space of $\mathbb{S}_1$ with $\mathbb{R}^2$ we can express $f'(z)$ by the Jacobian $J_f(z)$. From this we obtain the requirement for F_1 and F_2 of being f-*related* [14]: For each $z \in \mathbb{S}_1$, $J_f(z) \cdot F_1(z) = F_2(f(z))$, or equivalently, $F_2(z) = J_f(f^{-1}(z)) \cdot F_1(f^{-1}(z))$. As we are only interested in the direction of vectors rather than their lengths, denote by $u \simeq v$ if $u = cv$ for some positive real constant c. F_1 and F_2 are said to be f-*related up to normalization* if $F_2(z) \simeq J_f(f^{-1}(z)) \cdot F_1(f^{-1}(z))$ for each $z \in \mathbb{S}_1$. We introduce a second property to guarantee that upward embeddability is preserved.

Definition 1. *Let $\mathbb{S}_1$ and $\mathbb{S}_2$ be smooth manifolds with vector fields F_1 and F_2, respectively. We call a smooth injective homeomorphism $f : \mathbb{S}_1 \to \mathbb{S}_2$ to be* field preserving *from $(\mathbb{S}_1, F_1)$ to $(\mathbb{S}_2, F_2)$ if F_1 and F_2 are f-related up to normalization, and for any smooth curve $p : [0, 1] \to \mathbb{S}_1$,*

$$\text{sgn}\langle p'(0), (F_1 \circ p)(0)\rangle = \text{sgn}\langle (f \circ p)'(0), (F_2 \circ f \circ p)(0)))\rangle.$$

Rephrasing the above, f preserves the (non-)acuteness of the angle between a tangent vector and the vector field at any point. This gives rise to the following proposition.

Proposition 1. *Let G be a simple directed graph and let $\mathbb{S}_1$ and $\mathbb{S}_2$ be differentiable two-dimensional manifolds with vector fields F_1 and F_2, respectively. Let $\mathbb{S}_1'$ be a subset of $\mathbb{S}_1$ such that in respect to F_1, any graph upward embeddable on $\mathbb{S}_1$ is also upward embeddable on $\mathbb{S}_1'$. If G is (weakly) upward embeddable on $\mathbb{S}_1$ in respect to F_1 and there is a field-preserving map f from $(\mathbb{S}_1', F_1)$ to $(\mathbb{S}_2, F_2)$, then G is also (weakly) upward embeddable on $\mathbb{S}_2$ in respect to F_2.*

Proof. Assume G is upward embeddable on $\mathbb{S}_1$ in respect to F_1. Let Γ be a plane drawing of G on $\mathbb{S}_1'$ respecting F_1. The drawing $f[\Gamma]$ of G on $\mathbb{S}_2$ is plane as f is differentiable. It also respects F_2 as f specifically preserves the acuteness of the angles between the vector field and the tangents of the edge curves. □

Note that the well-known conformal, i. e., angle-preserving, maps are just a special case of the field-preserving maps if they relate F_1 to F_2 up to normalization. Additionally, any composition of field-preserving maps is field-preserving in respect to the corresponding manifolds and vector fields.

We define $(\mathbb{S}_1, F_1) \sim (\mathbb{S}_2, F_2)$ if and only if there are functions f and g such that f is field-preserving from $(\mathbb{S}_1, F_1)$ to $(\mathbb{S}_2, F_2)$ and g is field-preserving from $(\mathbb{S}_2, F_2)$ to $(\mathbb{S}_1, F_1)$. Proposition 1 allows us to speak of *upward embeddability of G in the equivalence class $[\mathbb{S}, F]$*. We can define the directed simple graph classes

$$[\![\mathbb{S}F]\!]_s = \{G : G \text{ is (strictly) upward embeddable on } [\mathbb{S}, F]\} \text{ and}$$
$$[\![\mathbb{S}F]\!]_w = \{G : G \text{ is weakly upward embeddable on } [\mathbb{S}, F]\},$$

where the subscripts indicate the strict or weak case. This class scheme enables us to classify and generalize prior approaches of upward planarity. We restrict ourselves to manifolds which are obtained from a square where optionally opposite sides are identified. Thus any of the considered manifolds can be represented by rectangular fundamental polygons [18]. Let $I =]-1, 1[$ and derive $I_\circ$ from I by identifying its boundaries -1 and 1. With a slight abuse of language we define the following two-dimensional manifolds as the product manifolds of I and $I_\circ$ with their natural differentiable structure: The *plane* $\mathbb{P} = I \times I$, the *standing cylinder* $\mathbb{C}_s = I_\circ \times I$, the *rolling cylinder* $\mathbb{C}_r = I \times I_\circ$, and the *torus* $\mathbb{T} = I_\circ \times I_\circ$. See Table 2 for an illustration.

A point in each of the defined manifolds can be represented by a pair (x, y). A vector field assigns a two-dimensional vector to each such pair (x, y) that defines the direction of the field at (x, y). A basic case is the null field N, which assigns the null vector $(0, 0)$ everywhere. Then any direction of the edges weakly respects the null field. Therefore, the graphs $[\![\mathbb{P}N]\!]_w$, i. e., upward embeddable in the plane and weakly respecting the null field, are exactly the planar graphs in the usual sense, denoted by **P**. Similarly, **T** $= [\![\mathbb{T}N]\!]_w$ are the toroidal graphs.

Next we consider the homogeneous field H that maps each point to $(0, 1)$. Then the upward planar graphs **UP** are exactly captured by $[\![\mathbb{P}H]\!]_s$. We additionally investigate the following graph classes: **SUP** $= [\![\mathbb{C}_sH]\!]_s$, **wSUP** $= [\![\mathbb{C}_sH]\!]_w$, **RUP** $= [\![\mathbb{C}_rH]\!]_s$, **wRUP** $= [\![\mathbb{C}_rH]\!]_w$, **UT** $= [\![\mathbb{T}H]\!]_s$, and **wUT** $= [\![\mathbb{T}H]\!]_w$, which define (weakly) upward planarity on the standing and rolling cylinder, and on the torus, respectively.

Table 2. Surfaces resulting from the cross products of I and $I_\circ$

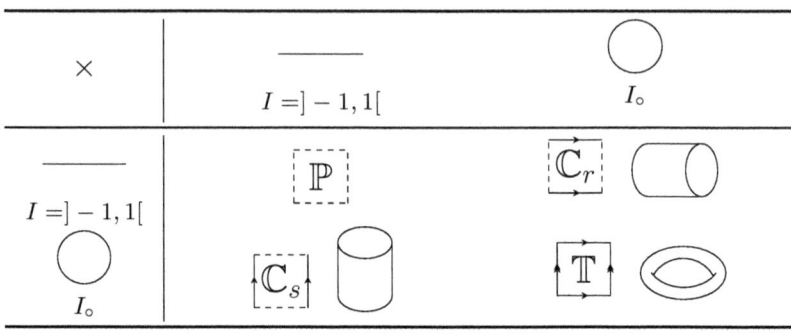

3   Classification of Upward Drawings

First we show that planar upward drawings on the sphere, the standing cylinder and the plane with the radial field coincide both in the strict and in the weak versions. Instead of proving that the spherical and cylindrical graph classes are equal according to their graph-theoretical characterizations from [12, 15], our proof makes use of the definitions from Sect. 2 by transforming the surfaces with their endowed fields into each other.

Theorem 1. *For a graph G the following statements are equivalent.*

(i) $G \in \textbf{SUP}$ $(G \in \textbf{wSUP})$
(ii) G is (weakly) spherical
(iii) G is (weakly) cylindrical
(iv) $G \in [\![\mathbb{PR}]\!]_s$ $(G \in [\![\mathbb{PR}]\!]_w)$

Proof. All of the following arguments apply to the weak and the strict case. We first show *(ii)* $\Rightarrow$ *(i)*. Consider an upward drawing Γ of G on the sphere $\mathbb{S}_1$. First assume that there is no vertex placed on the poles, i. e., with coordinates $(0, 1, 0)$ or $(0, -1, 0)$. Let $y_{\max}$ be the maximum y-coordinate of vertices of G. Note that there is no point of an edge above $y_{\max}$ as otherwise the upwardness is violated. Analogously define $y_{\min}$. Let $\mathbb{S}_1' = \{(x, y, z) : x^2 + y^2 + z^2 = 1, y_{\min} < y < y_{\max}\}$, i. e., $\mathbb{S}_1'$ is the truncated sphere [9]. We use the angle-preserving Mercator projection M [21] to map $\mathbb{S}_1'$ to the rectangle $[x'_{\min}, x'_{\max}[\times]y'_{\min}, y'_{\max}[$ in the plane. Afterwards, we scale and translate $M[\Gamma]$ to obtain a drawing in the fundamental polygon $\mathbb{C}_s$ by

$$f : (x, y) \mapsto \left(\frac{2x}{x'_{\max} - x'_{\min}}, \frac{2y}{y'_{\max} - y'_{\min}} \right) + (\Delta_x, \Delta_y), \tag{2}$$

where Δ_x and Δ_y are such that the scaled rectangle is centered at the origin.

Consider the tangent vector t at a point p on an edge curve in Γ on the surface of $\mathbb{S}_1$ and the longitudinal vector l starting at p and pointing to the north pole. As the edge curve is strictly monotonous in y-direction $\langle t, l \rangle > 0$. The same holds for the corresponding vectors $t' = (t'_x, t'_y)$ and l' in $M[\Gamma]$ since M preserves angles. Let $t'' = (t''_x, t''_y)$ and l'' be the corresponding vectors in $(f \circ M)[\Gamma]$. Note that M maps longitudinals to vertical lines. Since, up to the translation, f is a combination of scalings in x- and y-direction, we have that $l'' = (0, 1)$ after a normalization. Although f is not angle-preserving, it does not change the sign of the corresponding scalar product in $(f \circ M)[\Gamma]$ since $\langle t'', l'' \rangle = t''_x \cdot 0 + t''_y \cdot 1 = \frac{2}{y'_{\max} - y'_{\min}} t'_y = \frac{2}{y'_{\max} - y'_{\min}} \langle t', l' \rangle > 0$. Hence, the resulting edge curves respect H and we have an upward drawing of G on $(\mathbb{C}_s, H)$.

If a vertex v_N is placed at the north pole, then define $y_{\max}$ to be the maximum y-coordinate of any vertex in $V \setminus v_N$ and define $\mathbb{S}'_1$ as above. The mapping $(f \circ M)$ is applied to $\Gamma \cap \mathbb{S}'_1$ to obtain Γ'. Note that Γ' does not contain v_N. In Γ' the edges to v_N are cut at the upper side of the fundamental polygon. We additionally shrink Γ' in y-direction by $g : (x, y) \mapsto (x, \frac{1}{2} y)$. Note that in $g[\Gamma']$ all edges still respect H. In $g[\Gamma']$ we have obtained free space $B_N = [-1, 1[\times]\frac{1}{2}, 1[$ in $\mathbb{C}_s$ with no points of $g[\Gamma']$. We place v_N somewhere in B_N, e.g., at $(0, \frac{3}{4})$, and reconnect all its incident edges by straight lines, which respect the homogeneous field. A similar procedure is applied when a vertex is placed at the south pole. For the converse direction, i.e., $(i) \Rightarrow (ii)$, the proof is analogous by using the inverse of the transformation $(f \circ M)$.

For $(i) \Rightarrow (iii)$, let Γ be a drawing of $G \in [\![\mathbb{C}_s H]\!]_s$. Intuitively, we bend the fundamental polygon containing Γ such that the identified left and right sides actually mend. More formally, apply the map $f :] -1, 1[^2 \to \mathbb{R}^3 : (x, y) \mapsto (\cos x, y, \sin x)$ to Γ. As the y-coordinate is mapped onto itself and Γ respects H pointing from bottom to top, all edges in $f[\Gamma]$ increase monotonically in the y-direction of the cylinder axis. The case $(iii) \Rightarrow (i)$ follows analogously, as essentially the inverse of f can be used.

For $(i) \Rightarrow (iv)$ consider the map

$$f : \mathbb{C}_s \to \mathbb{P} : (x, y) \mapsto \frac{y + 2}{4} \cdot (\cos(\pi x), \sin(\pi x)). \tag{3}$$

Intuitively, f transforms the lateral surface of the rolling cylinder to a ring in the plane centered around the origin with inner radius $\frac{1}{4}$ and outer radius $\frac{3}{4}$. The bottom of the fundamental polygon $\mathbb{C}_s$ maps to the inner circular boundary and the top to the outer circular boundary of the ring. f is a conformal map and H is f-related to R, i.e., f preserves angles and maps H to R (see [2]). By Proposition 1 we can conclude that any graph in $[\![\mathbb{C}_s H]\!]_s$ is also in $[\![\mathbb{P} R]\!]_s$.

For $(iv) \Rightarrow (i)$, the inverse f^{-1} of f can be used. However, some care has to be taken if a vertex is placed at the origin $(0, 0)$ of $\mathbb{P}$. Then the same technique as with the sphere applies here as well.   □

Theorem 2. *A graph G is embeddable in the plane respecting the cyclic field if and only if G is embeddable on the rolling cylinder with the homogeneous field, i.e., $[\![\mathbb{P} C]\!]_s = [\![\mathbb{C}_r H]\!]_s$ and $[\![\mathbb{P} C]\!]_w = [\![\mathbb{C}_s H]\!]_w$.*

Proof. The proof is analogous to the case $(i) \Leftrightarrow (iv)$ in the proof of Theorem 1 except that for the functions f and g the coordinates x and y are swapped.   □

Hashemi et al. have shown that deciding if a graph has an upward drawing on the sphere is **NP**-complete [13]. Limaye et al. [16] stated this problem as open on the cylinder. Theorem 1 solves this problem.

Corollary 1. *Upward planarity testing on the cylinder is **NP**-hard.*

Longitudinal cycles are permitted in **RUP**, whereas **SUP** contains only acyclic graphs. Thus, **RUP** is stronger than **SUP**. Even more, this is also true if we consider only acyclic graphs.

Theorem 3. $SUP \subseteq RUP$, *even for acyclic graphs.*

Proof. Consider a graph $G \in$ **SUP** along with its drawing Γ on $\mathbb{C}_s$ with the homogeneous field. Then G is acyclic. To show that $G \in$ **RUP** we give a step-by-step transformation of Γ to a drawing on $\mathbb{C}_r$ which respects the homogeneous field H.

First we straighten Γ into a polyline drawing, which is then transformed from the standing onto the rolling cylinder while upward planarity is preserved. Cut Γ at the y-coordinates of the vertices. Each cut defines a ring of points, which are the x-coordinates of the vertices, and temporarily introduce a dummy vertex for each crossing of an edge with the cut. A *slice* consists of the region of Γ between two adjacent cuts. It has a lower and an upper ring of (dummy) vertices and a planar upward routing of segments of edges between the rings. We process slices iteratively from bottom to top. For a slice S take an edge segment connecting two (dummy) vertices, say p_1 on the lower ring and q_1 on the upper ring. Now rotate the upper ring such that p_1 and q_1 have the same x-coordinate. Replace each edge segment from a (dummy) vertex p on the lower ring to a (dummy) vertex on the upper ring by a straight line, such that the cyclic order of the incident edges of each vertex is preserved. Since two curves did not cross before, they cannot cross after the straightening, because the relative order of their endpoints on the rings with respect to (p_1, q_1) is preserved. (One can make (p_1, q_1) the boundary of the fundamental polygon.)

Now let Γ be the so obtained polyline drawing. In the remainder of the proof we need that all edges that cross the vertical line $x = -1$ leave the fundamental polygon to the right and enter it from the left, i.e., the x-value of the edge curves immediately before their crossing is positive and negative immediately afterwards. According to Lemma 5 of [3] by identifying all edges with *inner segments* a polyline drawing on $\mathbb{C}_s$ can always be transformed such that this condition holds, which we assume to hold for Γ as well.

Let $f : \mathbb{C}_s \to \mathbb{C}_r : (x, y) \mapsto \frac{1}{2}(x, y)$ be the scaling which shrinks by $\frac{1}{2}$ and consider the drawing $f[\Gamma]$ on $\mathbb{C}_r$. Since the scalar product is linear and the scaling factor $\frac{1}{2} > 0$, $f[\Gamma]$ still respects the homogeneous field H. For instance, the drawing of Fig. 1(a) is scaled to the drawing in the dotted rectangle in Fig. 1(b). It remains to show how to reconnect the formerly identical points on the left and right boundary of $f[\Gamma]$ by field-respecting edges in $\mathbb{C}_r$. Let $y_1 < y_2 < \ldots < y_k$ be the ascending y-coordinates of the points $r_i = (\frac{1}{2}, y_i)$ and $l_i = (-\frac{1}{2}, y_i)$ on the right and left boundary in $f[\Gamma]$, respectively. Define points $r'_i = (\frac{3}{4} - \frac{y_i}{4}, \frac{1}{2})$

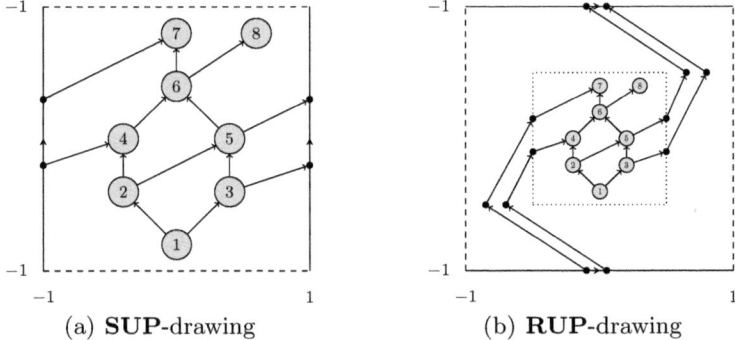

(a) **SUP**-drawing　　　　　　(b) **RUP**-drawing

Fig. 1. Transformation from the standing to the rolling cylinder

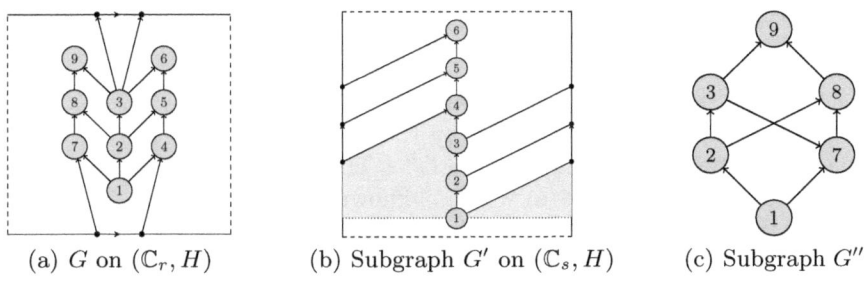

(a) G on $(\mathbb{C}_r, H)$　　　(b) Subgraph G' on $(\mathbb{C}_s, H)$　　　(c) Subgraph G''

Fig. 2. An acyclic graph $G \in \mathbf{RUP}$ but not in **SUP**

and $l'_i = (-\frac{3}{4} - \frac{y_i}{4}, -\frac{1}{2})$ with $1 \le i \le k$. Connect r_i to r'_i by a straight-line segment. Note that these segments do not intersect since $y_i < y_j \Leftrightarrow x(r'_i) > x(r'_j)$ for $i \ne j$. Analogously, connect all l'_i to l_i by non-intersecting segments. As $-\frac{1}{2} < y_i < \frac{1}{2}$, all (directed) line-segments strictly follow H. Finally, connect all r'_i to l'_i. These line-segments also strictly follow H and are non-intersecting since $x(r'_i) < x(r'_j) \Leftrightarrow x(l'_i) < x(l'_j)$. The result of the whole process applied to Fig. 1(a) is depicted in Fig. 1(b).　　　　　　　　　　　　　　　　　　　　□

Proposition 2 ([2]). *On the rolling cylinder with the homogeneous field, the class of (strictly) upward embeddable graphs coincides with the class of weakly upward embeddable graphs, i. e.,* $\llbracket \mathbb{C}_r H \rrbracket_s = \llbracket \mathbb{C}_r H \rrbracket_w$.

Forthcoming we shall establish proper inclusions among the main classes of upward drawable graphs. For the plane and the sphere this has been proved at several places and it comes from the distinction by the st-edge. The graph in Fig. 2(c) serves as a counterexample.

　　The *2-wing graph* displayed in Fig. 2(a) is an acyclic **RUP** graph which is not planar upward drawable on the sphere or the standing cylinder. It is 3-connected and due to the upward drawing its embedding is unique. Let G' be the

(a) G on $(\mathbb{P}, N)$ (b) G on $(\mathbb{C}_r, H)$

Fig. 3. A planar graph G with $G \notin \mathbf{RUP}$

subgraph of G induced by the vertices $\{1, 2, 3, 4, 5, 6\}$ which are connected by the path $P = (1, 2, 3, 4, 5, 6)$. On $(\mathbb{C}_s, H)$ the vertices along P must be placed with strictly increasing y-coordinate due to H. In Fig. 2(b) G' is drawn on $(\mathbb{C}_s, H)$ using the same embedding as in Fig. 2(a). The remaining vertices $\{7, 8, 9\}$ of G must all be placed above vertex 1, since there is path from 1 to 7, 8, and 9. Due to the uniqueness of the embedding, the vertices $\{7, 8, 9\}$ must be placed within the shaded area in Fig. 2(b). This area is homeomorphic to the plane $\mathbb{P}$. Hence, if $\{7, 8, 9\}$ could be placed within the shaded area without crossings, then the subgraph G'' of G induced by the vertices $\{1, 2, 3, 7, 8, 9\}$ would have an embedding on $\mathbb{P}$ respecting H, i.e., $G'' \in \mathbf{UP}$. However, G'' is isomorphic to the graph displayed in Fig. 2(c) which is known not to be in $\mathbf{UP}$ [5].

In $\mathbf{wSUP}$ latitudinal cycles are allowed and therefore $\mathbf{wSUP}$ properly contains $\mathbf{UP}$ and $\mathbf{SUP}$ as the latter two only allow acyclic graphs. Also $\mathbf{RUP}$ allows cycles, which implies similar proper inclusions.

The vertices of two cycles with one common vertex must have the same y-coordinate on $\mathbb{C}_s$ with H. In contrast, this graph can easily be embedded on $\mathbb{C}_r$ with H. Thus, $\mathbf{SUP} \subsetneq \mathbf{RUP}$. Further, K_5 can be embedded on the torus and, hence, $\mathbf{P} \subsetneq \mathbf{T}$. Finally, the *wheel graph* as shown in Fig. 3 shows that upward planarity on a rolling cylinder is a proper restriction over planarity. As special techniques apply, this is stated as our next lemma.

Lemma 1. $\mathbf{RUP} \subsetneq \mathbf{P}$

Proof. $\mathbf{RUP} \subseteq \mathbf{P}$ since the rolling cylinder is a surface of genus 0. For the proper inclusion consider the planar graph G depicted in Fig. 3. We show that $G \notin \mathbf{RUP}$. G has a Hamiltonian cycle $\mathcal{C} = (1, 2, 3, 4, 5, 1)$. Note that any cycle embedded on $\mathbb{C}_r$ with the homogeneous field wraps exactly once around the cylinder, i.e., its winding number is 1. Its winding number is greater 0 since otherwise its start and endpoint could not connect and it must be less than 2 since otherwise the edge curve would be self-intersecting. As all other edges in Fig. 3 follow the direction of $\mathcal{C}$ and start and end at distinct vertices of $\mathcal{C}$, their winding number on $\mathbb{C}_r$ is 0. Consider the embedding of G on $\mathbb{C}_r$ displayed in Fig. 3, where edge $(3, 1)$ is drawn dotted. $\mathcal{C}$ divides $\mathbb{C}_r$ into a left- and a right-hand region. To avoid a crossing between the edges $(1, 4)$ and $(2, 5)$, they must lie in different regions, e.g., $(1, 4)$ to the right and $(2, 5)$ to the left of $\mathcal{C}$. Now consider the region R enclosed by the edges $(1, 2), (2, 5), (4, 5), (1, 4)$, which contains vertex 3. The curve of edge $(3,1)$ must start within R and, due to the homogeneous field, must reach vertex 1 from below. Thus, the curve of edge $(3,1)$ starts within R and ends outside of R, which always causes a crossing.   □

Theorem 4. *Let **DAG** be the set of all acyclic graphs. The classes of graphs are related as follows.*

$$UP \subsetneq SUP \subsetneq RUP \cap DAG \subsetneq RUP \subsetneq UT$$

$$\begin{array}{ccccc} \| & \nparallel & & \| & \cap \\ wUP & wSUP & & wRUP & wUT \end{array} \qquad (4)$$

$$\begin{array}{cc} \nparallel & \nparallel \\ P & \subsetneq & T \end{array}$$

Finally, we classify the work of Dolati et al. [8] on upward drawings on the lying and on the standing torus, where in each case the edges respect the south-north direction. On the lying torus the south (north) pole is a ring consisting of all y-minimal (y-maximal) points of the torus. This corresponds to our notion of the antiparallel field (see Tab. 1) and the graph class $[\![\mathbb{T}A]\!]_s$. On the standing cylinder the south (north) pole is the single point with minimal (maximal) y-coordinate. In our classification this is the radial field and the graph class $[\![\mathbb{T}R]\!]_s$. The authors showed that $[\![\mathbb{T}A]\!]_s \subsetneq [\![\mathbb{T}R]\!]_s$ and state that the time complexity of deciding whether or not a graph is in (one of) the two sets is unknown.

4   Complexity

Finally we address the recognition problems for upward drawability, which are known to be **NP**-hard for the plane and sphere and, hence, the standing cylinder. It is also **NP**-hard for the torus, and still remains open for the rolling cylinder.

Theorem 5. *Deciding whether or not a graph $G \in \boldsymbol{UT}$ is **NP**-complete, even if G is connected.*

Proof. If the graph does not have to be connected, simply reduce from **UP** by adding to G a suitably directed K_7. Any embedding of the K_7 must be two-cell, so all remaining faces have genus 0. Thus $G \cup K_7 \in \mathbf{UT} \Leftrightarrow G \in \mathbf{UP}$. For connected graphs reconstruct the **NP**-completeness proof of **UP**. The constructed graph candidate for **UP** has a dedicated vertex v lying on the outside of the graph. Add an edge e from any of the K_7 vertices to v. Again, $G \cup K_7 \in \mathbf{UT} \cup \{e\} \in \mathbf{UT} \Leftrightarrow G \in \mathbf{UP}$.   □

References

1. Auer, C., Bachmaier, C., Brandenburg, F.J., Brunner, W., Gleißner, A.: Plane Drawings of Queue and Deque Graphs. In: Brandes, U., Cornelsen, S. (eds.) GD 2010. LNCS, vol. 6502, pp. 68–79. Springer, Heidelberg (2011)

2. Auer, C., Brandenburg, F.J., Bachmaier, C., Gleißner, A.: Classification of planar upward embedding. Technical Report MIP-1106, Fakultät für Informatik und Mathematik, Universität Passau (2011), http://www.fim.uni-passau.de/wissenschaftler/forschungsberichte/mip-1106.html

3. Bachmaier, C.: A radial adaption of the Sugiyama framework for visualizing hierarchical information. IEEE Trans. Vis. Comput. Graphics 13(3), 583–594 (2007)

4. Bachmaier, C., Brandenburg, F.J., Brunner, W., Fülöp, R.: Coordinate Assignment for Cyclic Level Graphs. In: Ngo, H.Q. (ed.) COCOON 2009. LNCS, vol. 5609, pp. 66–75. Springer, Heidelberg (2009)

5. Di Battista, G., Eades, P., Tamassia, R., Tollis, I.G.: Graph Drawing: Algorithms for the Visualization of Graphs. Prentice-Hall (1999)

6. Dolati, A.: Digraph embedding on t_h. In: Cologne-Twente Workshop on Graphs and Combinatorial Optimization, CTW 2008, pp. 11–14 (2008)

7. Dolati, A., Hashemi, S.M.: On the sphericity testing of single source digraphs. Discrete Math. 308(11), 2175–2181 (2008)

8. Dolati, A., Hashemi, S.M., Kosravani, M.: On the upward embedding on the torus. Rocky Mt. J. Math. 38(1), 107–121 (2008)

9. Foldes, S., Rival, I., Urrutia, J.: Light sources, obstructions and spherical orders. Discrete Math. 102(1), 13–23 (1992)

10. Garg, A., Tamassia, R.: On the computational complexity of upward and rectilinear planarity testing. SIAM Journal on Computing 31(2), 601–625 (2001)

11. Hansen, K.A.: Constant width planar computation characterizes ACC^0. Theor. Comput. Sci. 39(1), 79–92 (2006)

12. Hashemi, S.M.: Digraph embedding. Discrete Math. 233(1-3), 321–328 (2001)

13. Hashemi, S.M., Rival, I., Kisielewicz, A.: The complexity of upward drawings on spheres. Order 14, 327–363 (1998)

14. Lee, J.M.: Introduction to Smooth Manifolds. Springer, Heidelberg (2002)

15. Limaye, N., Mahajan, M., Sarma, J.M.N.: Evaluating Monotone Circuits on Cylinders, Planes and Tori. In: Durand, B., Thomas, W. (eds.) STACS 2006. LNCS, vol. 3884, pp. 660–671. Springer, Heidelberg (2006)

16. Limaye, N., Mahajan, M., Sarma, J.M.N.: Upper bounds for monotone planar circuit value and variants. Comput. Complex 18(3), 377–412 (2009)

17. Marsen, J.E., Ratiu, T., Abraham, R.: Manifolds, Tensor Analysis, and Applications, 3rd edn. Springer, Heidelberg (2001)

18. Massey, W.S.: Algebraic Topology: An Introduction. Springer, Heidelberg (1967)

19. Mohar, B., Rosenstiel, P.: Tessellation and visibility representations of maps on the torus. Discrete Comput. Geom. 19, 249–263 (1998)

20. Mohar, B., Thomassen, C.: Graphs on Surfaces. John Hopkins University Press (2001)

21. Snyder, J.P.: Map projections – a working manual. US Geological Survey, 1395 (1987)

22. Sugiyama, K., Tagawa, S., Toda, M.: Methods for visual understanding of hierarchical system structures. IEEE Trans. Syst., Man, Cybern. 11(2), 109–125 (1981)

23. Thomassen, C.: Planar acyclic oriented graphs. Order 5(1), 349–361 (1989)

24. Wegener, I.: Complexity Theory - Exploring the Limits of Efficient Algorithms. Springer, Heidelberg (2005)

Upward Planarity Testing of Embedded Mixed Graphs*

Carla Binucci and Walter Didimo

Dip. Ing. Elettronica e dell'Informazione
Università degli Studi di Perugia
{binucci,didimo}@diei.unipg.it

Abstract. A *mixed graph* has both directed and undirected edges. We study an upward planarity testing problem for embedded mixed graphs and solve it using Integer Linear Programming. Experiments show the efficiency of our technique.

1 Introduction

An *upward planar drawing* of a planar digraph G is a planar drawing of G such that all the edges are drawn as curves monotonically increasing in the vertical direction, according to their orientation. The *upward planarity testing* problem is the problem of deciding whether a planar digraph admits an upward planar drawing, and has a long tradition in graph drawing [8]. It is polynomially solvable if the planar embedding of the graph is fixed [2], while it is NP-hard in the variable embedding setting [9].

Many graphs arising from real applications have both directed and undirected edges. These types of graphs are called *mixed graphs* and have received considerable attention in the literature (see, e.g., [1]). Fig. 1(a) shows a mixed graph whose nodes represent employees of a company; the directed edges describe hierarchical relationships while the undirected edges describe collaborations. In a visual representation of a mixed graph it is still desirable that directed edges flow upward, as in Fig. 1(b). Additionally, in order to increase the readability of the layout, one may want that even the undirected edges are drawn as curves vertically monotone when possible, as in Fig. 1(c). An *upward drawing* of a mixed graph G is such that all the directed edges of G are drawn upward and all the undirected edges of G are drawn monotone in the vertical direction.

In this paper we study the following problem: *Given an embedded planar mixed graph G, decide whether G admits an upward planar drawing that preserves the planar embedding of G.* The drawing in Fig. 1(c) is an embedding-preserving upward planar drawing of the graph in Fig. 1(a), while the drawing in Fig. 1(b) is not an upward drawing, because edge (Mary, Kate) is not vertically monotone. Our problem is equivalent to decide if there exists an orientation of the undirected edges of G such that the resulting embedded digraph has an upward planar drawing. The contribution of this paper is twofold:

- We describe an ILP (Integer Linear Programming) model for the upward planarity testing problem of a planar embedded mixed graph G; the number of variables and

* Research supported in part by the MIUR project AlgoDEEP prot. 2008TFBWL4. We acknowledge Maurizio Patrignani for the useful discussion on the subject of this paper.

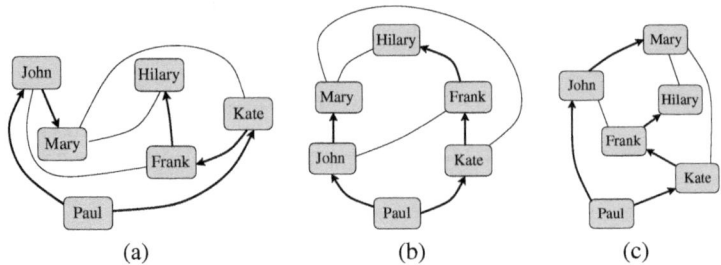

Fig. 1. (a) A planar embedded mixed graph G. (b) An embedding-preserving planar drawing of G, where the directed edges are drawn upward. (c) An embedding-preserving planar drawing of G where the directed edges are upward and the undirected edges are vertically monotone.

constraints of the model is linear in the size of G. If G has an embedding-preserving upward planar drawing, the model allows us to construct one (Section 3).
- We present an experimental study that shows how the proposed model can be solved efficiently (Section 4). Indeed, for all instances of our test suite the computation of a solution takes a few seconds, even for graphs with several hundreds of nodes.

We remark that the study of upward drawings of mixed graphs has been previously addressed by Eiglsperger *et al.* [7]. Differently from our results, they describe a heuristic that attempts to compute an upward drawing with few edge crossings. Hence, they do not start from an embedded planar graph, and the final drawing may contain crossings even if the original graph admits an upward planar drawing according to our definition.

2 Definitions and Notation

Let G be an embedded planar digraph. A *source vertex* (resp. a *sink vertex*) of G is a vertex with only outgoing edges (resp. incoming edges). A source vertex or a sink vertex of G is also called a *switch vertex* of G. A vertex v of G is *bimodal* if all its incoming edges are consecutive around v (and thus also the outgoing edges are consecutive around v). If all vertices of G are bimodal then G and its embedding are called *bimodal*. Acyclicity and bimodality are necessary but not sufficient conditions for the upward planar drawability of an embedded planar digraph [2]. Note that, if G is bimodal, the circular list of edges incident to any vertex v of G is split into two linear lists, one consisting of the incoming edges of v and the other consisting of the outgoing edges of v.

Let f be a face of G and suppose that the boundary of f is visited clockwise if f is internal, and counterclockwise if f is external. Let $a = (e_1, v, e_2)$ be a triplet such that v is a vertex of the boundary of f and e_1, e_2 are two edges incident to v that are consecutive on the boundary of f (e_1 and e_2 may coincide if G is not biconnected). Triplet a is called an *angle at v in face f*, or simply an *angle of f*. An angle $a = (e_1, v, e_2)$ of a face f is a *switch angle* of f if e_1 and e_2 are both incoming edges or both outgoing edges of v; otherwise a is a *non-switch angle*. If v is a switch vertex, all the angles at v are switch angles in the faces incident to v. We denote by $deg(v)$ the number of angles at v and by $deg(f)$ the number of angles in f.

Let Γ be an upward planar drawing of an embedded planar digraph G. Assign to each angle a of G a label S, F, or L, according to the following rules: a is labeled L if it is a switch angle that corresponds to a geometric angle larger than π in Γ; a is labeled F if it is a non-switch angle; a is labeled S otherwise. Note that, an angle is labeled S if it is a switch angle corresponding to a geometric angle smaller than π in Γ. We call this labeling the *upward labeling induced by* Γ. Given an embedded bimodal planar digraph G, an assignment $\mathcal{L}$ of labels S, F, and L to the angles of G is called an *upward planar embedding* of G if there exists an upward planar drawing Γ of G such that the upward labeling induced by Γ coincides with $\mathcal{L}$. For a given angle labeling $\mathcal{L}$ and for a given vertex v of G, we denote by $L(v)$, $S(v)$, and $F(v)$ the number of angles at v that are labeled L, S, and F, respectively; also, if f is a face of G, $L(f)$, $S(f)$, and $F(f)$ denote the number of angles of f that are labeled L, S, and F, respectively.

The next theorem characterizes the upward planar embeddings of an embedded bimodal planar digraph G. It is a consequence of the results in [2,5,6].

Theorem 1. *Let G be an embedded bimodal planar digraph and let $\mathcal{L}$ be an assignment of labels S, F, and L to the angles of G. $\mathcal{L}$ is an upward planar embedding of G if and only if the following properties hold: (a) $deg(f) - 2 = 2L(f) + F(f)$, for each internal face f of G; (b) $deg(f) + 2 = 2L(f) + F(f)$, for the external face f of G; (c) switch angles are labeled either S or L, and non-switch angles are labeled F; (d) if v is a switch vertex of G then: $L(v) = 1$, $S(v) = deg(v) - 1$, $F(v) = 0$; (e) if v is a non-switch vertex of G then: $L(v) = 0$, $S(v) = deg(v) - 2$, $F(v) = 2$.*

Given an upward planar embedding $\mathcal{L}$ of a digraph G, an upward planar drawing of G that induces $\mathcal{L}$ can be computed efficiently by using algorithms described in [2,4].

3   An ILP Model

In order to decide whether an embedded planar mixed graph $G = (V, E)$ admits an upward planar drawing, we use the characterization of Theorem 1. Namely, we want to find an orientation for the undirected edges of G and a labeling $\mathcal{L}$ for the angles of G such that the resulting digraph is bimodal and $\mathcal{L}$ is an upward planar embedding of this embedded digraph. G' will denote the digraph obtained from G by orienting its undirected edges. To decide whether G' and $\mathcal{L}$ exist we define an ILP model, whose sets, variables, and constraints are described below. We assume that the embedded digraph obtained from G by removing all the undirected edges is bimodal, otherwise we can immediately conclude that G does not have an upward planar drawing.

Sets. V is partitioned into two subsets V_{NS} and V_{PS}. Each vertex in V_{NS} has both incoming and outgoing edges, and therefore it cannot be a switch vertex of G'. Subset V_{PS} contains the remaining vertices of V; each element in V_{PS} is a potential switch vertex of G'. E is partitioned into two subsets E_D and E_U, containing the directed and the undirected edges of G, respectively. F is the set of faces of G and $adj(v)$ is the set of neighbors of v.

A denotes the set of angles of G and is partitioned into subsets A_{NS} and A_{PS}, which contain the angles of G at vertices in V_{NS} and in V_{PS}, respectively. For a vertex v and for a face f, $A(v)$ and $A(f)$ denote all angles at v and all angles in f, respectively. For a vertex $v \in V_{NS}$, $A_{NS}(v)$ is the set of angles at v. For a vertex $v \in V_{PS}$, $A_{PS}(v)$ is the set of angles at v. If $v \in V_{NS}$, we denote by e'_{out} and e''_{out} the first and the last outgoing edge of v, respectively (e'_{out} and e''_{out} may coincide). Analogously, e'_{in} and e''_{in} are the first and the last incoming edges of v. The set of angles at v formed by the edges between e'_{in} and e'_{out} in clockwise order is denoted by $A^l_{NS}(v)$. The set of angles at v formed by the edges between e''_{in} and e''_{out} in counterclockwise order is denoted by $A^r_{NS}(v)$. The set of the remaining angles at v is denoted by $A^m_{NS}(v)$.

Variables. We associate a variable ℓ_a with each angle $a = (e_1, v, e_2)$. Variable ℓ_a takes values 0, 1, or 2, which correspond to the labels S, F, or L for a, respectively.

For each edge $e = (u, v)$ we define two variables, o_{uv} and o_{vu}, that take values in the set $\{0, 1\}$ and that define the orientation of e in G'; if $o_{uv} = 1$ edge e is oriented from u to v, otherwise it is oriented from v to u.

Finally, for each angle $a = (e_1, v, e_2)$, we define a variable c_a that takes values in the set $\{-1, 0, 1\}$. This variable is used to guarantee consistency between the orientations of e_1, e_2 and the value of ℓ_a, as explained later.

Constraints. We must guarantee the properties of Theorem 1. For an internal face (resp. the external face) f of G, denote by $cap(f)$ the number of angles in f minus 2 (resp. plus 2). Properties (a) and (b) are guaranteed by the following constraints:

$$\sum_{a \in A(f)} \ell_a = cap(f), \quad \forall f \in F. \tag{1}$$

Consistency about the orientations of the edges is ensured by Constraints 2: The first constraint forces the directed edges of G to keep their orientation in G', and the second avoids that an edge can receive two distinct orientations at the same time.

$$o_{uv} = 1, \quad \forall (u, v) \in E_D, \qquad o_{uv} + o_{vu} = 1, \quad \forall (u, v) \in E. \tag{2}$$

Properties $(c) - (e)$ are guaranteed by Constraints 3 and 4.

$$\sum_{a \in A_{PS}(v)} \ell_a = 2, \quad \forall v \in V_{PS} \tag{3}$$

$$\sum_{a \in A^l_{NS}(v)} \ell_a = 1, \quad \sum_{a \in A^r_{NS}(v)} \ell_a = 1, \quad \sum_{a \in A^m_{NS}(v)} \ell_a = 0, \quad \forall v \in V_{NS} \tag{4}$$

Finally, for each angle $a = (e_1, v, e_2)$ we have to guarantee consistency between its label and the orientation of the edges e_1 and e_2. Namely, denote by v_1 the vertex of e_1 other than v, and denote by v_2 the vertex of e_2 other than v. If o_{vv_1} and o_{vv_2} have the same value (which means that e_1 and e_2 are both incoming or both outgoing v) then ℓ_a

must take a value in $\{0, 2\}$. Otherwise, ℓ_a must take value 1. This property is forced by the following constraint:

$$o_{vv_1} + o_{vv_2} = \ell_a + 2c_a, \quad \forall a \in A \tag{5}$$

We observe that Constraint 5 and the integrality constraints on variables o_{uv} and c_a, imply that variables ℓ_a always assume integer values. Hence, we can relax the integrality constraints on ℓ_a, by simply requiring that $0 \leq \ell_a \leq 2$. Also, note that the total number of variables and constraints of our model is linear in the number of angles and edges of G; therefore, since G is planar, it is linear in the number of vertices of G. The next theorem summarizes the main contribution of this section.

Theorem 2. *There exists an ILP model to decide if an embedded mixed planar graph admits an upward planar embedding, and to find one in the positive case. The number of variables and constraints of the model is linear in the number of vertices of the graph.*

4 Experimental Study

We implemented our ILP model using CPLEX and we experimented it on a large set of mixed graphs, in order to understand if it is computationally feasible in practice. We focused on two major issues: (i) What is the time required to find an upward planar embedding of an embedded mixed graph, if there exists one; (ii) what is the time required to decide whether a mixed graph admits or not an upward planar embedding. To this aim, we run the experiments on two different test suites of mixed graphs, which we refer to as MIXEDPOSITIVE and MIXEDGENERAL. MIXEDPOSITIVE contains mixed embedded planar graphs that always admit an embedding-preserving upward planar drawing. Hence, for these graphs the computation will never reject the instance, and we can measure the time required to find an upward planar embedding. MIXEDGENERAL contains mixed embedded planar graphs for which an upward planar drawing may or may not exist. From the experiments we expect that the computation is faster on those instances that do not admit a solution and that on the positive instances the time required to find an embedding increases when the number of undirected edges increases.

Each graph G in MIXEDPOSITIVE was generated by first generating an upward planar embedded digraph G' with the algorithm described in [4], and then removing the orientation on a certain percentage of edges of G'. The edges that are made undirected were selected randomly with a uniform probability distribution. Each graph G in MIXEDGENERAL was generated with the following procedure: Again, we first generated an upward planar embedded digraph G' with the algorithm in [4]. Then a planar embedded mixed graph was computed from G' by repeating the following steps until the desired percentage of undirected edges was reached: randomly choose a face f of G' and add an edge in f randomly selecting its end-vertices (multiple edges were avoided); then randomly remove from G' a directed edge, while maintaining the connectivity. Every random choice followed a uniform probability distribution. Set MIXEDPOSITIVE contains 3 graphs for each distinct triple $\langle n, d, p \rangle$, where $n \in \{100, 200, \ldots, 800\}$ is the number of vertices, $d \in \{1.4, 1.6, 1.8, 2.0\}$ is the density, and $p \in \{20, 50, 80\}$ is the percentage of undirected edges of the graph. Hence, MIXEDPOSITIVE contains

288 graphs in total. Set MIXEDGENERAL contains 10 graphs for each distinct triple $\langle n, d, p \rangle$, where n, d, and p take the same values as before. Hence, it contains 960 graphs in total.

The experiments were performed under the Windows Vista OS, on an Intel Core-Duo with 2.2 GHz and 2 GB of RAM; the computations were rather fast and confirmed our hypothesis. As expected, the CPU time for the graphs in MIXEDPOSITIVE increases when the percentage of undirected edges increases. Almost all computations required less than 4 seconds, and the maximum time of a computation was 12 seconds, for an instance with 600 vertices, 80% of undirected edges, and density 2.0.

The percentage of negative instances in MIXEDGENERAL (i.e., the percentage of graphs for which an upward planar embedding does not exist) is close to 100% for most graphs with no more than 50% of directed edges, while about half of the graphs with 80% of undirected edges admit a solution. As expected, the computation is very fast on the negative instances, while the behavior on the positive instances reflects the one for the graphs in MIXEDPOSITIVE.

5   Conclusions and Open Problems

We introduced a new upward planarity testing problem for embedded mixed graphs and we experimentally showed that this problem can be efficiently solved using Integer Linear Programming. The main open problem is to study what is the computational complexity of our testing problem in theory. Is it NP-hard? It is worth recalling that the upward planarity testing problem for embedded digraphs is polynomially solvable [2] and that polynomial-time algorithms exist for finding upward embeddings of embedded undirected graphs [6]. Another interesting problem is to design algorithmic solutions for computing the maximum upward planar subgraph for a given mixed embedded graph. We recall that the problem of computing a maximum upward planar subgraph of a planar embedded digraph is NP-hard [3].

References

1. Bang-Jensen, J., Gutin, G.: Digraphs: Theory, Algorithms and Applications, 2nd edn. Springer, Heidelberg (2009)
2. Bertolazzi, P., Di Battista, G., Liotta, G., Mannino, C.: Upward drawings of triconnected digraphs. Algorithmica 6(12), 476–497 (1994)
3. Binucci, C., Didimo, W., Giordano, F.: Maximum upward planar subgraphs of embedded planar digraphs. Comput. Geom. 41(3), 230–246 (2008)
4. Didimo, W.: Upward planar drawings and switch-regularity heuristics. Journal of Graph Algorithms and Applications 10(2), 259–285 (2006)
5. Didimo, W., Giordano, F., Liotta, G.: Upward spirality and upward planarity testing. SIAM J. Discrete Math. 23(4), 1842–1899 (2009)
6. Didimo, W., Pizzonia, M.: Upward embeddings and orientations of undirected planar graphs. Journal of Graph Algorithms and Applications 7(2), 221–241 (2003)
7. Eiglsperger, M., Eppinger, F., Kaufmann, M.: An approach for mixed upward planarization. Journal of Graph Algorithms and Applications 7(2), 203–220 (2003)
8. Garg, A., Tamassia, R.: Upward planarity testing. Order 12, 109–133 (1995)
9. Garg, A., Tamassia, R.: On the computational complexity of upward and rectilinear planarity testing. SIAM Journal on Computing 31(2), 601–625 (2001)

Combining Problems on RAC Drawings and Simultaneous Graph Drawings

Evmorfia N. Argyriou[1], Michael A. Bekos[1],
Michael Kaufmann[2], and Antonios Symvonis[1]

[1] School of Applied Mathematical & Physical Sciences,
National Technical University of Athens, Greece
{fargyriou,mikebekos,symvonis}@math.ntua.gr
[2] University of Tübingen, Institute for Informatics, Germany
mk@informatik.uni-tuebingen.de

1 Introduction and Problem Definition

We present an overview of the first combinatorial results for the so-called *geometric RAC simultaneous drawing problem* (or *GRacSim* drawing problem, for short), i.e., a combination of problems on geometric RAC drawings [3] and geometric simultaneous graph drawings [2]. According to this problem, we are given two planar graphs $G_1 = (V, E_1)$ and $G_2 = (V, E_2)$ that share a common vertex set but have disjoint edge sets, i.e., $E_1 \subseteq V \times V$, $E_2 \subseteq V \times V$ and $E_1 \cap E_2 = \emptyset$. The main task is to place the vertices on the plane so that, when the edges are drawn as straight-lines, (i) each graph is drawn planar, (ii) there are no edge overlaps, and, (iii) crossings between edges in E_1 and E_2 occur at right angles.

A closely related problem is the following: *Given a planar embedded graph G, determine a geometric drawing of G and its dual G^* (without the face-vertex corresponding to the external face) such that: (i) G and G^* are drawn planar, (ii) each vertex of the dual is drawn inside its corresponding face of G and, (iii) the primal-dual edge crossings form right-angles.* We refer to this problem as the *geometric Graph-Dual RAC simultaneous drawing problem* (or *GDual-GRacSim* for short).

2 Results

A detailed presentation of our results (including technical proofs) is available as a technical report [1]. The following theorem establishes that if two graphs always admit a geometric simultaneous drawing, it is not necessary that they also admit a GRacSim drawing.

Theorem 1. *There exists a wheel and a cycle which do not admit a GRacSim drawing.*

For the case of a path $\mathcal{P}$ and a matching $\mathcal{M}$, we can prove that a GRacSim drawing always exists. The basic idea of our algorithm is to identify in the graph induced by the union of $\mathcal{P}$ and $\mathcal{M}$ a set of cycles $\mathcal{C}_1, \ldots, \mathcal{C}_k$, $k \leq n/4$, such that:

M. van Kreveld and B. Speckmann (Eds.): GD 2011, LNCS 7034, pp. 433–434, 2012.

(i) $|E(\mathcal{C}_1)| + \ldots + |E(\mathcal{C}_k)| = n$, (ii) $\mathcal{M} \subseteq \mathcal{C}_1 \cup \ldots \cup \mathcal{C}_k$, and, (iii) the edges of cycle $\mathcal{C}_i$, $i = 1, \ldots, k$ alternate between edges of $\mathcal{P}$ and $\mathcal{M}$. The edges of the cycle collection do not cross each other, while the remaining ones introduce right-angle crossings. The following theorem summarizes our result.

Theorem 2. *A path and a matching always admit a* GRacSim *drawing on an* $(n/2 + 1) \times n/2$ *integer grid. The drawing can be computed in linear time.*

We can extend the algorithm that produces a GRacSim drawing of a path and a matching to also cover the case of a cycle $\mathcal{C}$ and a matching $\mathcal{M}$. The idea is simple. If we remove an edge from the input cycle, the remaining graph is a path $\mathcal{P}$. So, we can apply the developed algorithm and obtain a GRacSim drawing of $\mathcal{P}$ and $\mathcal{M}$, in which the insertion of the edge that closes the cycle can be done without introducing any crossings by augmenting the total area of the drawing.

Theorem 3. *A cycle and a matching always admit a* GRacSim *drawing on an* $(n + 2) \times (n + 2)$ *integer grid. The drawing can be computed in linear time.*

Corollary 1. *Let G be a simple connected graph that can be decomposed into a matching and either a hamiltonian path or a hamiltonian cycle. Then, G is a RAC graph.*

For the GDual-GRacSim drawing problem, we can show by an example that it is not always possible to compute a GDual-GRacSim drawing if the input graph is an arbitrary planar graph. This is summarized in the following theorem.

Theorem 4. *Given a planar embedded graph G, a* GDual-GRacSim *drawing of G and its dual G^* does not always exist.*

For the more restricted case of outerplanar graphs, we can state the following theorem, which is based on a recursive geometric construction that computes a GDual-GRacSim drawing of G and its dual.

Theorem 5. *Given an outerplane embedding of an outerplanar graph G, it is always feasible to determine a* GDual-GRacSim *drawing of G and its dual G^*.*

Our study raises several open problems. It would be interesting to identify other non-trivial classes of graphs, besides a matching and either a path or a cycle, that admit a GRacSim drawing. For the classes where GRacSim drawings are not possible, study drawings with bends. Study the required drawing area.

References

1. Argyriou, E.N., Bekos, M.A., Kaufmann, M., Symvonis, A.: Geometric simultaneous rac drawings of graphs. CoRR abs/1106.2694 (2011)
2. Brass, P., Cenek, E., Duncan, C.A., Efrat, A., Erten, C., Ismailescu, D., Kobourov, S.G., Lubiw, A., Mitchell, J.S.B.: On simultaneous planar graph embeddings. Computational Geometry: Theory and Applications 36(2), 117–130 (2007)
3. Didimo, W., Eades, P., Liotta, G.: Drawing Graphs with Right Angle Crossings. In: Dehne, F., Gavrilova, M., Sack, J.-R., Tóth, C.D. (eds.) WADS 2009. LNCS, vol. 5664, pp. 206–217. Springer, Heidelberg (2009)

The Open Graph Archive: A Community-Driven Effort[*]

Christian Bachmaier[1], Franz Josef Brandenburg[1], Philip Effinger[2],
Carsten Gutwenger[3], Jyrki Katajainen[4], Karsten Klein[3], Miro Spönemann[5],
Matthias Stegmaier[2], and Michael Wybrow[6]

[1] University of Passau, Germany
[2] Eberhard-Karls-Universität Tübingen, Germany
[3] Technische Universität Dortmund, Germany
[4] University of Copenhagen, Denmark
[5] Christian-Albrechts-Universität zu Kiel, Germany
[6] Monash University, Australia

1 Introduction

A *graphbase*, a term coined by Knuth [7], is a database of graphs and computer programs that generate, analyze, manipulate, and visualize graphs. The terms *graph library* and *graph archive* are often used as synonyms for this term. Our vision is to provide an infrastructure and quality standards for a public graphbase, named the Open Graph Archive, that is accessible to researchers and other interested parties around the world via the worldwide web. This paper describes the current work undertaken towards this goal; the paper is also intended to be a call for participation since this will be a community-driven effort where most of the content will be provided by users of the system.

Our motives for building this universal graphbase are similar to Knuth's motives for building the Stanford GraphBase [7]; we are just working on a larger scale. First, we want to provide standard sets of graphs to enable repeatability of experiments. We expect that the graphbase would be particularly interesting for researchers working in the areas of algorithm engineering and graph drawing. Second, we want to provide a single point of access for datasets relevant for people working with graphs. By annotating the datasets with their origin and other semantic information, we can help researchers to find publications relevant for their work. Third, a graphbase that is accessible worldwide can stimulate interesting theory development. As pointed out by Knuth [7], a graphbase can bridge the gap between theoreticians and practitioners. Fourth, the programs (and maybe also the datasets) available in a graphbase, if done well, can have a significant educational value.

Many existing collections, like the graphs available in the Stanford Graph-Base [7] and the well-known Rome graphs [3], are static and only cover a small number of data sizes, types, and properties that may be relevant for the users. In order to allow collection and exchange of interesting graphs, it is important

---

[*] This work was initiated at Schloss Dagstuhl in seminar 11191 on "Graph Drawing with Algorithm Engineering Methods".

M. van Kreveld and B. Speckmann (Eds.): GD 2011, LNCS 7034, pp. 435–440, 2012.
© Springer-Verlag Berlin Heidelberg 2012

to make the graphbase extendable. The needs of the community will certainly change over time. Expandability has been recognized as an important goal by other researchers as well (see, e.g., [1,2]), but the available data collections seem to be relevant to a limited range of users only. Our goal is to support the use in a wide variety of application areas.

2   User Needs and Requirements

In order to investigate the relevance of and the requirements for a universal graphbase we conducted a survey among 30 participants of the Dagstuhl seminar 11191, coming from the graph drawing and algorithm engineering communities. The survey solicited a variety of open-ended textual responses. In this section we summarize the most interesting and commonly recurring feedback.

Describe two most important use cases for a graph archive. The most frequent use cases were to search for graphs with specific properties, and to benchmark and compare algorithms, both mentioned by 37% of the participants. Further answers were to share datasets (27%), and to replicate experiments and compare results (23%). Since these are fundamental aspects of experimental scientific processes, we can see that a graphbase would be an important tool for researchers of graph algorithms.

What services do you expect? We proposed nine services of which the survey participants could select those they considered important. As shown in Table 1, support for tags and arbitrary comments are the most crucial features. When asked for further important services, a handful of people wanted to know which publications refer to a specific graph or collection of graphs (17%).

Which category tags and analysis properties may be useful? Participants named 20 different application domains to categorize a graph or collection of graphs, e.g., biology, social networks, geography, software engineering. Furthermore, participants named 16 graph properties, most of which can be determined automatically. The most popular properties were connectivity (60%), including the number of k-connected components, and planarity (43%), including the best known crossing number for non-planar graphs.

Table 1. Result of the multiple-choice question "Which services do you consider as critical for a graph archive?"

Add categorization tags	80%
Add comments, links, or further information	77%
Search for specific tags	77%
Automatic conversion of file formats	70%
Search for specific properties	60%
Add information on how graphs were created	60%
Add images (drawings of the graphs)	50%
Automatic analysis of graph properties	47%
Programmatic web service	23%

Name two file formats you use most. The most frequently mentioned formats were GraphML (43%) and GML (33%). Since a total of 13 different formats were named, it is evident that a universal graphbase should not rely on one specific format, but offer support for several formats, preferably even converting automatically between formats.

Existing archives and collections. Responses for existing archives showed that GraphArchive [6] from the University of Tübingen and the datasets from the DIMACS implementation challenges [5] were both known by a handful of people (20% and 13%, respectively). These numbers are quite low and might also be biased towards the archives used by the researchers that participated in the seminar. They also suggest that there is currently no commonly used and accepted graph archive service. Regarding graph collections, participants mostly worked with randomly generated graphs, as well as with the popular Rome [3] and AT&T graphs [4].

Community contributions. Several participants of the survey declared that they would be willing to provide human resources (students, testing and development time), a hardware platform, or even money. This reaffirms that there is definite interest and enthusiasm for such a system, and also that the project should take advantage of this through involvement of the community.

Technical and service requirements. The survey results and subsequent community discussions indicate that potential users agree on a core set of important features, as well as a larger list of desirable functionality. However, several questions regarding the interface, architecture, and content remain open. Below we list the most relevant issues that need to be discussed or dealt with.

Storage. Graphs must be stored persistently under a unique ID for identification and access. Should graphs be stored in their original submission format, or converted by the system or the user into a unique storage format? In file conversions it is important that as much information as possible is preserved.

Metadata. There is a variety of metadata that can be stored with a graph, e.g., creator, description of the underlying data or the generator, additional keywords, and links to corresponding experiments or publications. Some of this data should be defined as mandatory properties, whereas other parts may be added as generic text properties. Useful keywords/tags for categorization need to be defined. Some tags could be attributes for graphs or collections of graphs, and some could list their structural and semantic properties.

Searching. Based on the survey results and our own experience, we assume that a graphbase should allow the user to search using both graph properties (number of nodes, etc.) and annotations (categories, origin, etc.).

Data analysis. Automatic analysis of basic graph properties must be possible. However, we are not sure if there should be a restriction on the computational complexity of the analysis or on the size of the analyzed graphs, or if users should be allowed to upload that information, e.g., the crossing number of a graph.

Programs. In addition to datasets, it must be possible to store programs like graph generators, analyzers, or visualizers. If the graphbase contains randomly generated collections of graphs with certain attributes, it would be useful to provide access to the programs used for their generation.

Ownership and copyright. The ownership of uploaded graphs must be clear from the outset. The content should be as freely usable as possible with fair attribution to the original authors. Contributors will need to take responsibility for their submitted graphs and collections of graphs.

Existing collections. Existing popular collections should be made identifiable and accessible via the system.

Possible extensions. Further useful extensions may include the following:

- Automatic file conversion could be provided as an additional service and the programs providing these conversions could also be made available.
- A series of drawings (layouts) for submitted graphs could be provided, or even automatic layout on demand, and the programs used for drawing the graphs could be made publicly available.
- Special support for browsing collections of graphs could be provided. For this purpose a hierarchical classification system can be useful.
- Structure-based searching could be supported, e.g., find graphs containing a clique of a specific size.
- Versioning of individual graphs as well as the possibility to store a series of dynamic graphs could be supported.
- A web-service API could be provided to allow interrogation of the graphbase by computer programs, rather than via a web browser.

3   A Working Prototype: GraphArchive

In this section we give an overview of *GraphArchive*, a platform for exchanging and archiving graphs meant as a prototype for the Open Graph Archive. It is developed at the University of Tübingen and was designed as a successor to *GraphDB*, a now discontinued first attempt at creating a web-based graphbase. GraphArchive is an interactive online system built with modern web technologies. Below we list the main features of the existing prototype, followed by a short description of its software architecture. For more details, we refer to [6]. The working system can be accessed online at

http://graphdrawing.org/grapharchive/.

Main features. The features of GraphArchive, as listed below, have been chosen to support the goal of providing an open and easily accessible system.

Web-based user interface. The user interface is provided via a browser. A web portal offers all functionality that is needed to handle graphs, including uploading datasets, inspection and management of existing graphs, searching for specific graphs, and downloading datasets. Registration is performed online using a registration form, which is processed automatically. Standard techniques are used to prevent registration by spam bots.

Minimal permission management. There are no groups of users that define rights for small circles of users. Licenses for graphs limiting their usage are not encouraged in our open approach. However, if necessary, a license can be attached to a selected graph. After confirming registration by going through the opt-in e-mail process, a user has access to all graphs and can initiate queries without restrictions.

Categorization of graphs. For search queries, graphs can be assigned to the field(s) of application that they originate from. This enables researchers from different fields to use GraphArchive as a common platform.

Automatic graph analysis. After upload, graphs (with $< 100,000$ vertices) are automatically analyzed in order to provide consistent data. Consistency is very important for queries on graph properties.

Multi-criterion search. Queries can be performed on multiple parameters, specifying graph properties, categories, author, name, and upload date. Also, parameters can be added later to further narrow down the result set.

Graph visualization. An image of a graph is valuable if a user wants to visually inspect the properties of a graph. Layouts are computed automatically in the background and can also be changed after upload.

Unique links to graphs. A URI associated with each graph allows for a permanent reference to be used in publications. By giving the URI, the user can quickly jump to a particular dataset. Reference annotations can be assigned to a graph in order to highlight publications and/or websites that refer to or make use of the graph.

Visual comparison of multiple graphs. For a quick comparison of graphs, we support simultaneous presentation of multiple graphs. Properties are displayed for all graphs. Boolean properties, e.g., directed/undirected, are presented visually on a scale (properties can be shared by (a) no graph, (b) a subset of the displayed graphs, or (c) all graphs).

Several file formats. When supporting many application domains it is impossible to dictate the file format used. Therefore, we aim at supporting as many formats as possible. The system is extendable and allows for addition of further formats in the future. For downloading graphs, a user can choose the format that fits best to his or her work environment. We provide cross conversion between different formats (the users can select any supported format and the system performs the conversion automatically).

Import/export of multiple graphs. We allow upload/download of several graphs simultaneously in zip-compressed form. In an upload process, each file in a compressed archive can be optionally processed individually (for property analysis and layout computation).

Guest access for non-registered users. If a user wants to check a specific graph, he or she can access a detailed view of the graph using its URI. All properties and attributes of that graph are made visible via a guest account.

Software architecture. GraphArchive is built with common web technologies. The application is written in PHP5[1] and uses Apache2[2] for online presentation. For graph analysis and layout computation, we use the Java graph library yFiles;[3] these computations are handled in the background via PHP/Java bridge.[4] Data storage is managed by PostgreSQL database management system.[5]

More details and a descriptive walk-through showing a typical use case of the system can be found in [6]. For more news and information on the system and its current development status, please consult the system website.

4 Outlook

Our hope is to stimulate discussion on the initial system proposal and trigger community growth around the Open Graph Archive. The success of this project requires a passionate and enthusiastic community. We urge you to step up and participate by critiquing the existing system, helping the development effort, or contributing material to the graphbase.

References

1. Boisvert, R.F., Pozo, R., Remington, K., Barrett, R., Dongarra, J.J.: The matrix market: A web resource for test matrix collections. In: Quality of Numerical Software: Assessment and Enhancement. IFIP Conference Series, vol. 76, pp. 125–137. Chapman & Hall (1997), Graphs available at http://math.nist.gov/MatrixMarket
2. Davis, T.A., Hu, Y.: The University of Florida sparse matrix collection. ACM Trans. Math. Softw. 38(1) (2011), Graphs available at
   http://www.cise.ufl.edu/research/sparse/matrices/
3. Di Battista, G., Garg, A., Tamassia, R., Tassinari, E., Vargiu, F.: An experimental comparison of four graph drawing algorithms. Comput. Geom. Theory Appl. 7(5-6), 303–325 (1997), Graphs available at http://www.graphdrawing.org/
4. Di Battista, G., Garg, A., Tamassia, R., Tassinari, E., Vargiu, F.: Drawing directed acyclic graphs: An experimental study. J. Comput. Geom. Apppl. 10(6), 623–648 (2000), Graphs available at http://www.graphdrawing.org/
5. 10th DIMACS implementation challenge: Graph partitioning and graph clustering, http://www.cc.gatech.edu/dimacs10/downloads.shtml (accessed August 2011)
6. Effinger, P., Kaufmann, M., Meinert, S., Stegmaier, M.: GraphArchive: An online graph data store. Technical Report WSI-2011-03, Wilhelm-Schickard-Institut, Eberhard-Karls-Universität Tübingen (2011)
7. Knuth, D.: The Stanford GraphBase: A Platform for Combinatorial Computing. ACM Press (1994)

---

[1] See project homepage: http://www.php.net, accessed July 2011.
[2] See project homepage: http://www.apache.org, accessed July 2011.
[3] Developed by yWorks GmbH: http://www.yworks.com, accessed July 2011.
[4] See project homepage: http://php-java-bridge.sourceforge.net/pjb/index.php, accessed July 2011.
[5] See project homepage: http://www.postgresql.org/, accessed July 2011.

Drawing Graphs with Vertices at Specified Positions and Crossings at Large Angles

Martin Fink[1], Jan-Henrik Haunert[1], Tamara Mchedlidze[2],
Joachim Spoerhase[1], and Alexander Wolff[1]

[1] Lehrstuhl für Informatik I, Universität Würzburg, Germany
[2] Department of Mathematics, National Technical University of Athens, Greece

1   Introduction

In point-set-embeddability (PSE) problems one is given not just a graph that is to be drawn, but also a set of points in the plane that specify where the vertices of the graph can be placed. The problem class was introduced by Gritzmann et al. [3] twenty years ago. In their work and most other works on PSE problems, however, planarity of the output drawing was an essential requirement. Recent experiments on the readability of drawings [4] showed that polyline drawings with angles at edge crossings close to 90° and a small number of bends per edge are just as readable as planar drawings. Motivated by these findings, Didimo et al. [2] recently introduced RAC drawings where pairs of crossing edges must form a right angle and, more generally, αAC drawings (for $\alpha \in (0, 90°]$) where the crossing angle must be at least α. As usual, edges may not overlap and may not go through vertices. We investigate the intersection of PSE and RAC/αAC.

Specifically, we consider the problems $RAC\ PSE$ and $\alpha AC\ PSE$ defined as follows. Given an n-vertex graph $G = (V, E)$ and a set S of n points in the plane, determine whether there exists a bijection μ between V and S, and a polyline drawing of G so that each vertex v is mapped to $\mu(v)$ and the drawing is RAC (or αAC). If such a drawing exists and the largest number of bends per edge in the drawing is b, we say that G admits a RAC_b (or an αAC_b) embedding on S. If we insist on straight-line edges, the drawing is completely determined by a bijection between vertex and point set. If we allow bends, however, PSE is also interesting with mapping, that is, if we are given a bijection μ between vertex and point set. In order to measure the size of our drawings, we assume that the given set of n points lies on a grid Γ of size $n \times n$ and, in the output drawing, bends lie on a (potentially larger or finer) grid containing Γ. We further assume that no two points lie on the same horizontal or vertical line. We call such a point set an $n \times n$ grid point set.

2   Results

We can RAC_3 embed any graph with n vertices and m edges on any $n \times n$ grid point set using any mapping and area $O\left((n + m)^2\right)$, see Fig. 1. Here, the idea is to have crossings only between segments with slopes $+1$ and -1. By choosing the horizontal positions of the first and last bends of each edge (black boxes in

M. van Kreveld and B. Speckmann (Eds.): GD 2011, LNCS 7034, pp. 441–442, 2012.

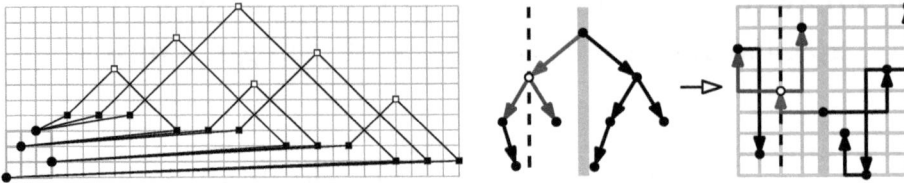

Fig. 1. RAC$_3$ embedding of K_4 **Fig. 2.** Restricted RAC$_1$ embedding

Fig. 1) so that they all have even Manhattan distance, we ensure that the middle bends (squares in Fig. 1) lie on grid points, and, thus, the drawing is valid.

In order to get 1-bend drawings (again for any mapping μ), we refine the grid by a factor of $\lambda \in O(\cot \varepsilon)$, for a small angle ε, that is, we insert, at equal distances, $\lambda - 1$ new rows and columns between each pair of consecutive old grid rows and columns, respectively. For each edge uv of the given graph G, consider the point p_{uv} in the same row as $\mu(u)$ and the same column as $\mu(v)$. Clearly, p_{uv} lies on the original grid. We place the bend of uv at one of the four new grid points that are diagonally adjacent to p_{uv}. This yields a $(\pi/2 - \varepsilon)$AC$_1$ embedding of G on the given grid point set S; the refined grid has size $O\big((\lambda n)^2\big)$.

We now turn to a restricted version of our problem where additionally every edge has to be drawn on grid lines, see Fig. 2 (right).

We show that every n-vertex binary tree admits a restricted RAC$_1$ embedding on any $n \times n$ grid point set (which is not known for the planar case). This was independently shown by Di Giacomo et al. [1]. We simply view the given tree as a search tree for the points, sorted by x-coordinate, and draw the edges (directed away from the root) such that we enter each vertex vertically and leave it horizontally to the left and to the right to its at most two children, see Fig. 2.

By an old result of Vizing (1964), every graph of maximum degree 3 can be 4-edge-colored. We exploit this to construct restricted RAC$_2$ embeddings of such graphs on any $n \times n$ grid point set even if the mapping is prescribed. We interpret each color as a direction (up, down, left, right) and, for each edge, we draw its first and third segment into this direction and the middle segment far enough.

References

1. Di Giacomo, E., Frati, F., Fulek, R., Grilli, L., Krug, M.: Orthogeodesic point-set embedding of trees. In: Speckmann, B., van Kreveld, M. (eds.) GD 2011. LNCS, vol. 7034, pp. 52–63. Springer, Heidelberg (2011)
2. Didimo, W., Eades, P., Liotta, G.: Drawing Graphs with Right Angle Crossings. In: Dehne, F., Gavrilova, M., Sack, J.-R., Tóth, C.D. (eds.) WADS 2009. LNCS, vol. 5664, pp. 206–217. Springer, Heidelberg (2009)
3. Gritzmann, P., Mohar, B., Pach, J., Pollack, R.: Embedding a planar triangulation with vertices at specified positions. Amer. Math. Mon. 98, 165–166 (1991)
4. Huang, W., Hong, S.H., Eades, P.: Effects of crossing angles. In: Proc. 7th Int. IEEE Asia-Pacific Symp. Inform. Visual (APVIS), pp. 41–46 (2008)

Viewport for Component Diagrams

Lukas Holy and Premek Brada

Department of Computer Science and Engineering, University of West Bohemia,
Univerzitni 8, Pilsen, Czech Republic

Abstract. This paper describes a viewport technique for use in the visualization of large graphs, e.g. UML component diagrams. This technique should help to work with complex diagrams (hundreds or thousands of components) by highlighting details of the important parts of the diagram and their related surroundings without losing the global perspective. To avoid visual clutter it uses clusters of interfaces and components.

1 Introduction

Although software components [2] comprise relatively large parts of systems, nowadays applications can easily consist of hundreds or thousands heavily interconnected ones. Thus their UML component diagrams become large graphs which are difficult to explore for humans. The main problem is how to show the whole diagram and provide enough detailed information at the same time. Diagrams displayed at the desired level of detail become too big to provide a sufficient overview and keep orientation; especially difficult is to trace dependencies between distant components. When displaying the whole diagram on standard screens, individual elements are hard to recognize and often there is visual clutter caused by dependency visualization.

Visualization techniques which handle the complexity, such as off-screen rendering [1], can be used instead of the traditional pan&zoom technique. This paper describes a novel approach called viewport which attempts to reconcile the above mentioned contradictory requirements and helps to explore the dependencies among components in an intuitive way.

2 Viewport for Component Diagrams

The proposed technique shows the graph (standard UML component diagram) zoomed-out to provide the appropriate overview of the complete architecture, with elements displayed without details. Besides that it shows selected components in detail inside a *viewport area* plus all their relations with other components in the diagram in an interactive border area (see Figure 1). These relations are for each component clustered into two sets: all provided interfaces (displayed as "lollipops") and all required interfaces (displayed as "sockets").

These interfaces are then connected to clustered proxy components, visually represented as rectangles with rounded corners. Each rectangle represents one

M. van Kreveld and B. Speckmann (Eds.): GD 2011, LNCS 7034, pp. 443–444, 2012.

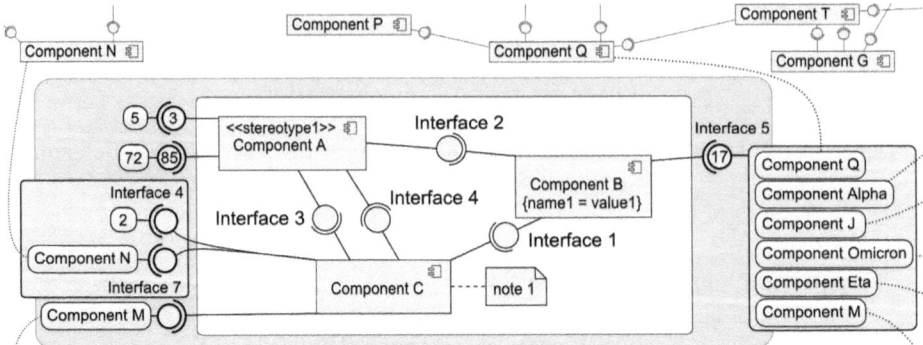

Fig. 1. Viewport for component diagrams

or more components. Numbers inside the clustered interfaces and proxy components represent a desired metric, e.g. the number of elements clustered in a given symbol. One of the key factors of our approach will be the interactivity of the border area, which should comprise user manipulation with clustering of interfaces or components, layout adjustments and selecting the components shown in the viewport.

The viewport technique should enable to explore and understand the dependencies in large diagrams by showing the context of a selected diagram subset. The clustering shall reduce the visual clutter otherwise caused by large number of relations. The proxy elements should reduce the need for the disorienting pan&zoom otherwise necessary while exploring dependencies and provide user relevant information in one place. The viewport can either be placed on a given position in the diagram (there can be more viewports in a diagram) or have a fixed position on the screen.

3   Future Work

Important part of the future research on this technique are layout algorithms for components displayed both inside and ouside of the viewport. Also, options for automatic suggestion of diagram parts suitable for displaying in viewports based on graph algorithms will be investigated.

References

1. Frisch, M., Dachselt, R.: Off-screen visualization techniques for class diagrams. In: Proceedings of the 5th International Symposium on Software Visualization, SOFT-VIS 2010, pp. 163–172. ACM, New York (2010)
2. Szyperski, C.: Component Software: Beyond Object-Oriented Programming, 3rd edn. Addison-Wesley / ACM Press (2002)

Shortest-Paths Preserving Metro Maps

Tal Milea, Okke Schrijvers, Kevin Buchin, and Herman Haverkort

Dept. Mathematics and Computer Science, TU Eindhoven, The Netherlands
{t.y.milea,o.j.schrijvers}@student.tue.nl,
k.a.buchin@tue.nl cs.herman@haverkort.net

A metro map, or subway map, is a schematic representation of a metro system of a city. The main goal of a metro map is to provide a traveler with information on which lines to take to get from station A to station B, and at which stations he needs to switch lines. It is often not beneficial to use the geographical embedding of the system, but rather a representation where the relevant information is presented as clearly as possible. There are several algorithms that aim to generate such maps [2].

One criterion that is not considered in these algorithms is whether or not the visually shortest route on the generated metro map still corresponds to the route with the shortest travel time. This could lead users to plan their travel along a route that results in a needlessly long travel time. To remedy this, we define the *theoretical planning error* (TPE) of a pair of stations to be the ratio between the travel time of the shortest route on the map and the shortest possible travel time of the metro system. This idea can be extended such that the TPE of the metro map is the maximum TPE over all pairs of stations. The theoretical planning error of a metro map can thus be defined mathematically as

$$\text{TPE} = \max_{u,v \in V} \frac{t(\operatorname{argmin}_{R \in \mathcal{R}(u,v)} \ell(R))}{\min_{R \in \mathcal{R}(u,v)} t(R)} \tag{1}$$

where V is the set of metro stations, $\mathcal{R}(u,v)$ is the set of routes from u to v, $\ell(R)$ is the perceived time it takes on the map to take route R and $t(R)$ is the actual time it takes.

Approach. We formulate the optimization of the TPE as a Mixed-Integer Program (MIP). One advantage of a MIP is that it is flexible and we can integrate our MIP with the one from Nöllenburg and Wolff [1]. The new objective function is a weighted combination of both approaches. The problem as defined in Equation 1 would result in an exponential number of constraints. The number of pairs of stations is already quadratic, but the number of possible routes between every pair can be exponential. We therefore use an iterative approach.

The process is started by generating a metro map without adding any TPE constraints. Then in each iteration the map that was generated in the previous iteration is analyzed offline, i.e. not in the MIP, to find the pair of stations that determines the TPE. For this pair, we generate a series of constraints for only two routes, the route that *appears* to have the shortest travel time, and the one that really has the shortest travel time. We add these constraints to the MIP

M. van Kreveld and B. Speckmann (Eds.): GD 2011, LNCS 7034, pp. 445–446, 2012.
© Springer-Verlag Berlin Heidelberg 2012

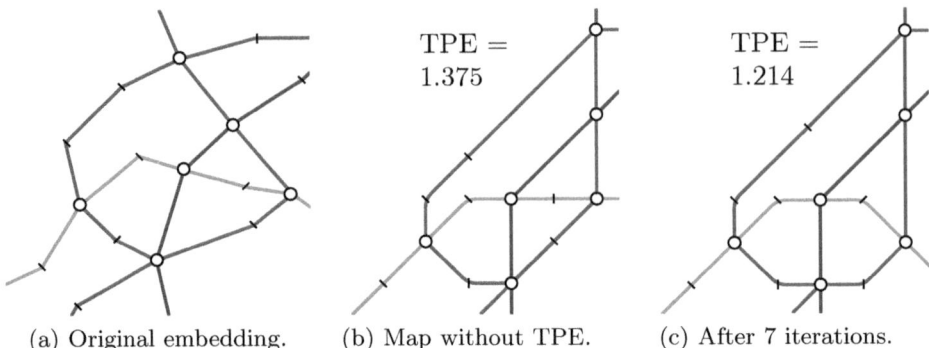

(a) Original embedding.   (b) Map without TPE.   (c) After 7 iterations.

Fig. 1. Results for part of the Vienna metro system

and solve it again. Since the TPE adds to the objective function, it is up to the linear program solver to see whether it can make the second route shorter (or more commonly whether it can make the first one larger) on the map while complying with the aesthetic requirements. The process terminates when the TPE of the metro map is 1 and therefore all shortest routes on the metro map correspond to the routes with the shortest travel time.

Results and Future Work. In Figure 1, a part of the results for the Vienna metro system can be seen. From Figures 1b and 1c it follows that small changes can in fact decrease the theoretical planning error (from 1.375 to 1.214). However, even though we have tried to limit the number of constraints in the program, the MIP solver CPLEX runs out of memory after 7 iterations. This cannot be solely attributed to the number of constraints as the original program has 2379 and iteration 7 contains 5816 constraints. We are currently looking into this issue.

We have verified that our approach indeed decreases the TPE. However, user studies are required to see whether or not this truly influences the planning capabilities of users. Additionally we consider allowing bends between adjacent stations to see if this increases the quality of the map.

Acknowledgments. We would like to thank Martin Nöllenburg for making the MIP generation code from [1] available to us.

References

1. Nöllenburg, M., Wolff, A.: Drawing and labeling high-quality metro maps by mixed-integer programming. IEEE Transactions on Visualization and Computer Graphics 17, 626–641 (2011)
2. Wolff, A.: Drawing subway maps: A survey. Informatik - Forschung und Entwicklung 22, 23–44 (2007)

Challenger, a New Way to Visualize Data

Remus Zelina[1], Sebastian Bota[1], Siebren Houtman[2],
Jaap Jan van Assen[2], and Bas Hattink[2]

[1] Meurs HRM, Baia Mare, Romania
{rzelina,seby}@meurs.ro
[2] Meurs HRM, Woerden, The Netherlands
{s.houtman,j.assen,b.hattink}@meurshrm.nl

Abstract. Challenger is a software product that provides fast and online data visualization. This is done by visualizing data (graphs) as a network. Both force based and modularization algorithms are used and experimented with. Challenger facilitates fast and easy understanding of complex data. This is not only a matter of showing one 'perfect' visualization, but rather of letting users browse, analyze and 'play' visually with (subsets of) data interactively.

Keywords: Challenger, Force-Directed, Clustered Graph.

1   Introduction

This abstract[1] presents Challenger, a software product, originally designed to visualize organizations. Organizations were considered as a network of employees, customers and projects (nodes) and links (i.e. 'working on project x') between them (edges). Visualizing makes it easier to understand and interpret the (relations in) organizations. However, as networks grow bigger, it turned out to be complex to draw a meaningful visualization within a reasonable amount of time. Challenger as it is now, is the result of our research on data visualization and our attempt to put this knowledge in a useful and fast application. The solution is useful in general to visualize complex data.

2   Implementation

We use different versions of force based algorithms [1], [2], [3] and modularization algorithms [4], [5], [6]. We split the data (graph nodes) into "almost cliques" by maximizing a global criterion. An almost clique is a subset of nodes of the graph that has many internal links and less external links. The global criterion is a value referring to a certain collection of almost cliques that covers the entire graph and it's

---

[1] This abstract accompanies a poster that can be obtained by contacting us at http://www.meurs.ro/challenger. Here one can also find additional information and demonstrations of Challenger.

M. van Kreveld and B. Speckmann (Eds.): GD 2011, LNCS 7034, pp. 447–448, 2012.

formula puts in evidence the above phenomenon. The best known formula for this criterion is the Modularity (Q) [4]. It is defined to be the fraction of edges that fall within the given cliques minus the fraction of edges that are expected to fall within the clique if the edges were distributed at random. Splitting a graph into almost cliques leads to information loss (due to external links). To reduce this we consider influences of all almost cliques over a given node, not only those of the almost clique that contains the given node. To optimize speed, we experimented with and optimized different algorithms in different combinations and architectures. The application is web based, and at the moment we achieve performances about 250.000 nodes and 1.000.000 edges in less than 60 seconds including data transfer.

More important than showing a single visualization, Challenger offers possibilities to visually browse, analyze and redraw the graph. This improves the usability. Challenger offers possibilities to:

- mix a number of entity types (nodes) and link types (edges) and combine them in a single graph;
- involve characteristics of elements in the graph (i.e. gender when persons are the elements) by entering the values (i.e. male; female) as 'virtual' nodes and put edges - when relevant - between them;
- filter, search and redraw based on characteristics of nodes as well as weights of edges;
- further analyze specific modules;
- create sub graphs and visualizations based on modules, selections of nodes or 1st and 2nd degree relations of selected nodes;
- zoom in and zoom out and keep track in a mini map of the total graph;
- toggle visualization modes.

References

1. Fruchterman, T., Reingold, E.: Graph Drawing by Force-directed Placement. Software-Practice and Experience 21(11), 1129–1164 (1991)
2. Kamada, T., Kawai, S.: An algorithm for drawing general undirected graphs. Information Processing Letters 31(1), 7–15 (1989)
3. Noack, A.: Energy Models for Graph Clustering. Journal of Graph Algorithms and Applications 11(2), 453–480 (2007)
4. Clauset, A., Newman, M.E.J., Moore, C.: Finding community structure in very large networks. Physical Review E 70, 066111 (2004)
5. Wakita, K., Tsurumi, T.: Finding community structure in mega-scale social networks. In: Proceedings of the 16th International Conference on World Wide Web, Banff, Alberta, Canada, pp. 1275–1276. ACM, New York (2007)
6. Blondel, V.D., Guillaume, J.L., Lambiotte, R., Lefebvre, E.: Fast unfolding of community hierarchies in large networks. J. Stat. Mech. (10), P10008 (2008)

Graph Drawing Contest Report

Christian A. Duncan[1], Carsten Gutwenger[2], Lev Nachmanson[3], and Georg Sander[4]

[1] Louisiana Tech University, Ruston, LA 71272, USA
duncan@latech.edu
[2] Technische Universität Dortmund, Germany
carsten.gutwenger@tu-dortmund.de
[3] Microsoft, USA
levnach@microsoft.com
[4] IBM, Germany
georg.sander@de.ibm.com

Abstract. This report describes the 18th Annual Graph Drawing Contest, held in conjunction with the 2011 Graph Drawing Symposium in Eindhoven, the Netherlands. The purpose of the contest is to monitor and challenge the current state of graph-drawing technology.

1 Introduction

As in recent years, this year's Graph Drawing Contest was divided into the offline contest and the online challenge. The offline contest had three categories: two dealt with angular resolution and one was a composers graph, kindly provided by Tom Sawyer Software. The data sets for the offline contest were published months in advance, and contestants could solve and submit their results before the conference started. For the two angular resolution categories, the submitted drawings were judged using visual comparison with emphasis foremost on angular resolution, particularly the worst-case deviation of the angular resolution from the perfect angular resolution value. The composers graph data set represented a very large graph, and the task was to combine graph drawing algorithms with appropriate techniques for complexity reduction (such as filtering and varying the graphical attributes) to create an illuminating visualization (one or more images, possibly with commentaries, or a movie). It was not a requirement to present the entire data set.

The online challenge took place during the conference in a format similar to a typical programming contest. Teams were presented with a collection of challenge graphs and had approximately one hour to submit their highest scoring drawings. This year's topic was the same as in the previous year, namely to minimize the length of the longest edge in a planar orthogonal grid drawing.

Overall, we received 30 submissions: 12 submissions for the offline contest and 18 submissions in the online challenge.

2 Angular Resolution

For the two categories in this topic, our primary concern was angular resolution. The *angular resolution* of any vertex in a drawing is the smallest angle formed by its

M. van Kreveld and B. Speckmann (Eds.): GD 2011, LNCS 7034, pp. 449–455, 2012.

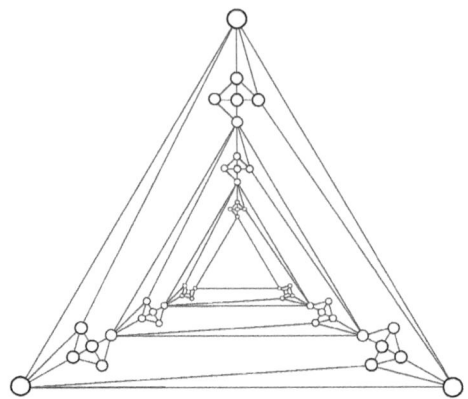

Fig. 1. First place, Angular Resolution, Category A

adjacent edges. When the edges are drawn as curved arcs it is measured with regards to the tangent at that vertex. In addition, we required that the graph should use a reasonably small grid area. The vertices did not have to be on an integer grid but the *vertex resolution*, the ratio between the distance of the closest two vertices and the farthest two vertices, should still remain relatively low. Both contest graphs were highly symmetric so any exploitation of that feature was also taken into consideration.

2.1   Category A: Straight-Line Planar

The first data set for the angular resolution topic was a planar graph with 48 nodes and 102 edges, and contestants had to create a drawing of the graph in the plane, without crossings, and using only straight-line edges. Whereas many bad examples of angular resolution use sequences of nested triangles, this graph contained only two nested triangles whose removal created a collection of outerplanar graphs.

We received only one valid submission in this category, and hence the winner was Hanley Weng from the University of Sydney; see Fig. 1. The layout was created by starting with a drawing produced by a variation of the spring-embedder algorithm, which was then manually modified such that the recursive symmetric structure of the graph was emphasized.

2.2   Category B: Curved Drawings

The second graph had 15 nodes and 45 edges and was not planar. The drawing, again in the plane, could have as many crossings as necessary but the angles of the crossings were also taken into consideration. In addition, the drawings could use curved arcs. There could be as many bends as needed but during judging both the number of bends and smoothness of the bends were taken into consideration and weighed against the gain in overall angular resolution.

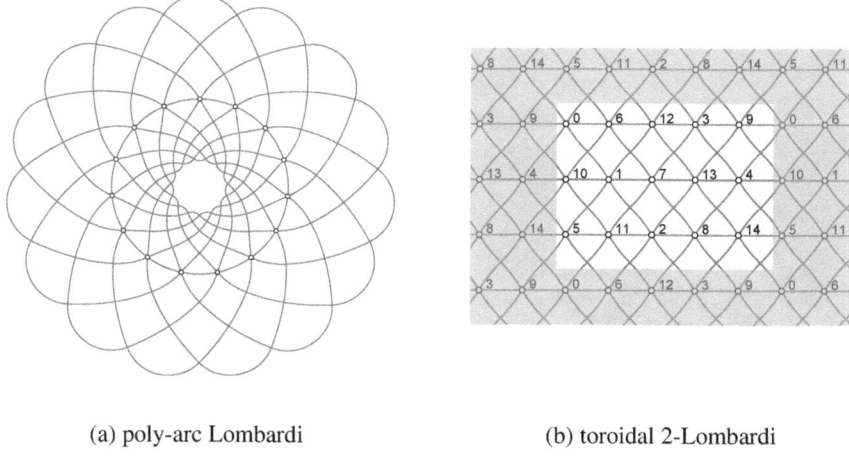

(a) poly-arc Lombardi   (b) toroidal 2-Lombardi

Fig. 2. First place, Angular Resolution, Category B

We received 5 submissions (8 drawings), from which two submissions with poly-arc Lombardi drawings [1,2] and perfect angular resolution looked quite the same (see Fig. 2(a)), but one of these submissions also contained a different visualization on the torus shown in Fig. 2(b). Though this is not really a drawing in the plane, it perfectly reflects the structure of the graph. Therefore, this pair of drawings by Maarten Löffler, David Eppstein, Michael Goodrich (UC Irvine) and Stephen Kobourov (University of Arizona) was judged to be the winning submission in this category.

3 Composers Graph

The composers graph was a large directed graph, where the nodes represented Wikipedia articles about composers, and the edges represented links between these articles. The graph had 3405 nodes and 13832 edges.

We received 5 submissions for the composers graph, including several high-quality submissions with movies, graph analysis, and specialized tools. The winning submission came from Remus Zelina, Sebastian Bota, Siebren Houtman, and Robert Ban (Meurs, Romania). The submission was comprised of an A0 poster of the graph's largest connected component (see Fig. 3) with 2743 nodes and 13769 edges, a dynamic web page that allowed one to browse and analyze the graph, as well as a movie showing the dynamic web page in action. For many of the nodes, photos of the corresponding composers were added. The underlying graph layout was mainly computed using force-directed techniques. The layout revealed several modules (shown in different colors[1]) corresponding to specific eras like Renaissance and Baroque, or genres like Russian composers and troubadours.

---

[1] This is only easily visible on the electronic version.

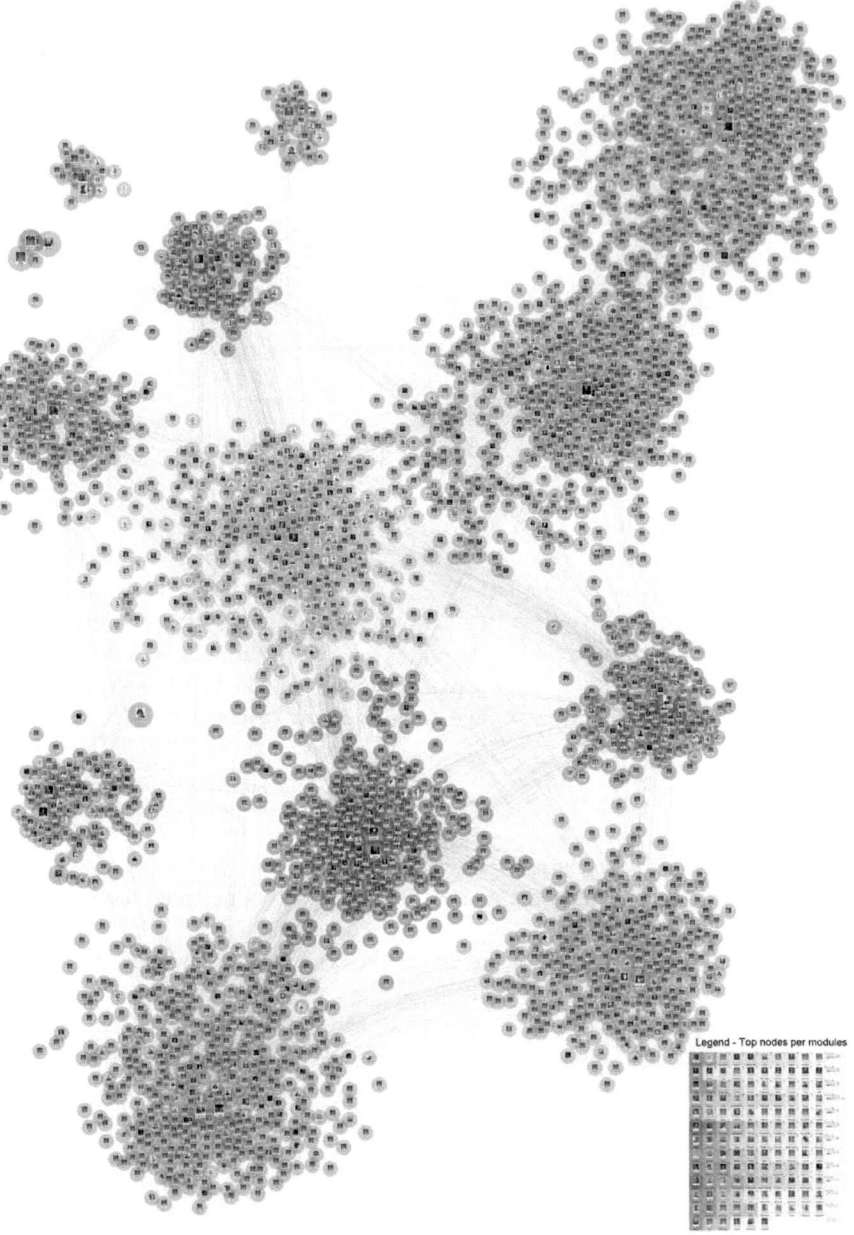

Fig. 3. First place, Composers Graph

4   Graph Drawing Challenge

The online challenge, which took place during the conference, dealt with minimizing the longest edge in a planar orthogonal grid drawing. The longest edge can be a bottleneck for many applications; hence, minimizing its length is important. The challenge graphs were planar and had at most four incident edges per node. The task was to place nodes and edge bends on integer coordinates so that the edge routing is orthogonal and the layout contains no crossings or overlaps. At the start of the one-hour on-site competition, the contestants were given six graphs with an initial legal planar layout with very long edges. The goal was to rearrange the layout to reduce the length of the longest edge. Only the length of the longest edge was judged; other aesthetic criteria, such as the number of edge bends or the area, were ignored.

The contestants could choose to participate in one of two categories: *automatic* and *manual*. To determine the winner in each category, the scores of each graph, determined by dividing the longest edge length of the best submission in this category by the longest edge length of the current submission, were summed up. If no legal drawing of a graph was submitted (or a drawing worse than the initial solution), the score of the initial solution was used.

In the automatic category, contestants received graphs ranging in size from 59 nodes / 85 edges to 1532 nodes / 2296 edges and were allowed to use their own sophisticated software tools with specialized algorithms. Manually fine-tuning the automatically obtained solutions was allowed. Six teams were rated in this category (2 manual teams accidentally solved the automatic graphs and were rated in both categories). The two top-scoring teams used the OGDF [3] graph drawing library for obtaining an initial solution using flow-based bend minimization and compaction techniques combined with their own heuristics to optimize the solution. With a score of 5.05, the winner in the automatic category was Sergey Pupyrev from Ural State University.

The 14 manual teams solved the problems by hand using IBM's *Simple Graph Editing Tool* provided by the committee. They received graphs ranging in size from 9 nodes / 17 edges to 150 nodes / 186 edges. Three of the larger input graphs were also in the automatic category, and the best manual teams scored similar (for two graphs) and better (for one graph) than the automatic teams. With a score of 3.82, the winner in the manual category was the team of Maarten Löffler from UC Irvine and Martin Nöllenburg from Karlsruhe Institute of Technology who found the best results for three of the six contest graphs.

Fig. 4 shows the initial layout and the best automatically obtained result of one challenge graph with 120 nodes and 146 edges. Fig. 5 shows the challenge graph used in both categories for which the manual teams found a better solution than the automatic teams (longest edge length one compared to two obtained by the automatic teams). Finally, Fig. 6 shows the only graph for which the judges know a better solution than the best solution found during the contest (6 compared to 11 found by the team of Till Bruckdorfer and Philip Effinger from Tübingen University). This graph was only used in the manual category.

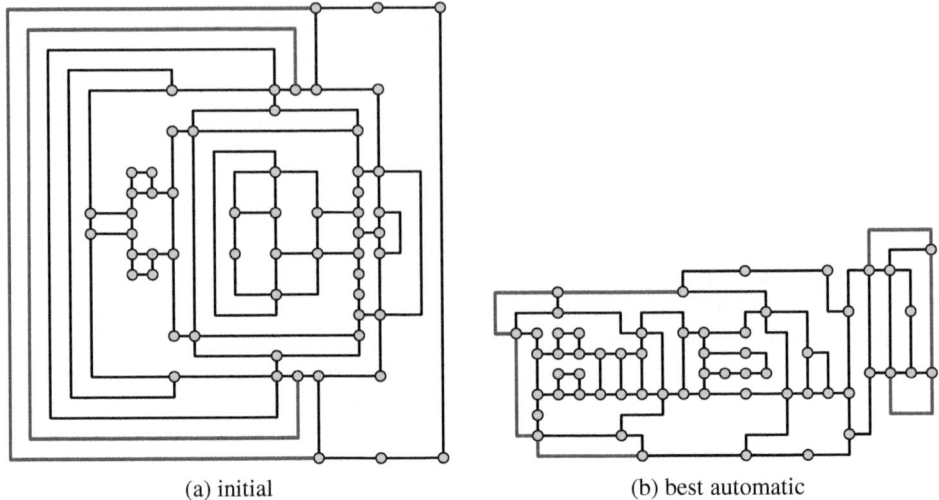

(a) initial   (b) best automatic

Fig. 4. Challenge graph with 59 nodes and 85 edges: (a) initial layout (longest edge length: 52) and (b) best automatic result obtained by Sergey Pupyrev (longest edge length: 6)

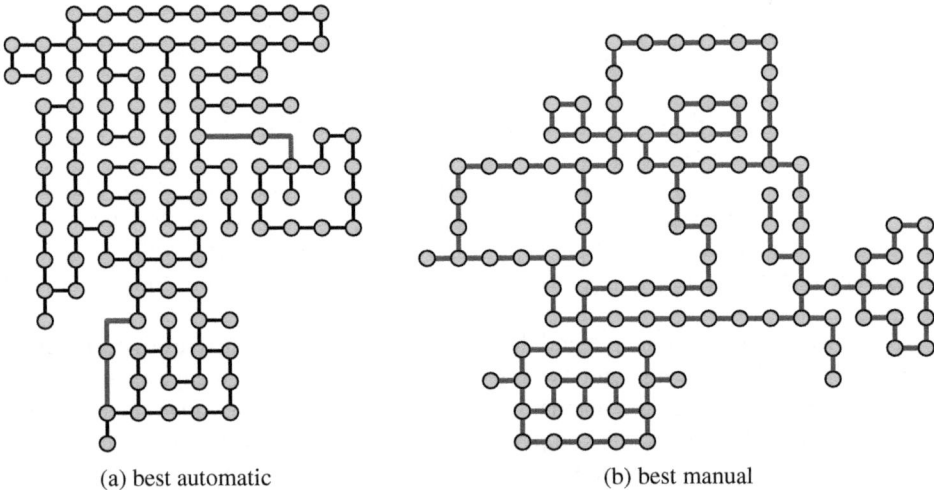

(a) best automatic   (b) best manual

Fig. 5. Challenge graph with 110 nodes and 118 edges: (a) best automatic result by team Gronemann, Mallach, and Schmidt (longest edge length: 2) and (b) best manual result by team Löffler and Nöllenburg (longest edge length: 1)

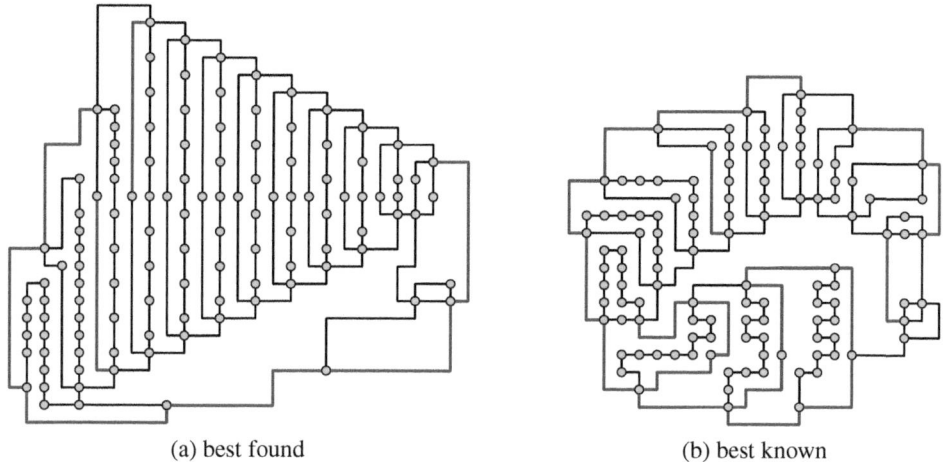

(a) best found   (b) best known

Fig. 6. Challenge graph with 118 nodes and 144 edges: (a) best solution found by team Bruck-dorfer and Effinger with longest edge length 11 and (b) best known result with longest edge length 6

Acknowledgments. The contest committee would like to thank the generous sponsors of the symposium and all the contestants for their participation. Further details including submitted videos and winning images can be found at the contest website, http://www.graphdrawing.de/contest2011/results.html.

References

1. Duncan, C.A., Eppstein, D., Goodrich, M.T., Kobourov, S.G., Löffler, M.: Planar and poly-arc Lombardi drawings. In: van Kreveld, M., Speckmann, B. (eds.) GD 2011. LNCS, vol. 7034, pp. 308–319. Springer, Heidelberg (2011)
2. Duncan, C.A., Eppstein, D., Goodrich, M.T., Kobourov, S.G., Nöllenburg, M.: Lombardi Drawings of Graphs. In: Brandes, U., Cornelsen, S. (eds.) GD 2010. LNCS, vol. 6502, pp. 195–207. Springer, Heidelberg (2011)
3. OGDF. The open graph drawing framework, http://www.ogdf.net

Author Index